D0464327

925

Estimating for the Building Trades

The late JOSEPH STEINBERG was Chairman of the Construction Technology Department of the New York Community College. His broad experience in the building trades included architectural and structural draftsman, chief estimator, construction superintendent, expediter, and consultant. He was a lecturer and co-author of *Practices and Methods of Construction* and *Construction Estimating*.

MARTIN STEMPEL has been teaching construction technology since 1947. He has broad experience as an estimator and consultant in all phases of construction. Previous to this he was a general construction superintendent, project manager, and in U. S. Coast Guard construction. He is co-author of *Practices and Methods of Construction*.

Estimating for the Building Trades

JOSEPH STEINBERG

Late Professor of Construction, Construction Technology Department, New York City Community College

MARTIN STEMPEL

Professor of Construction, Construction Technology Department, New York City Community College

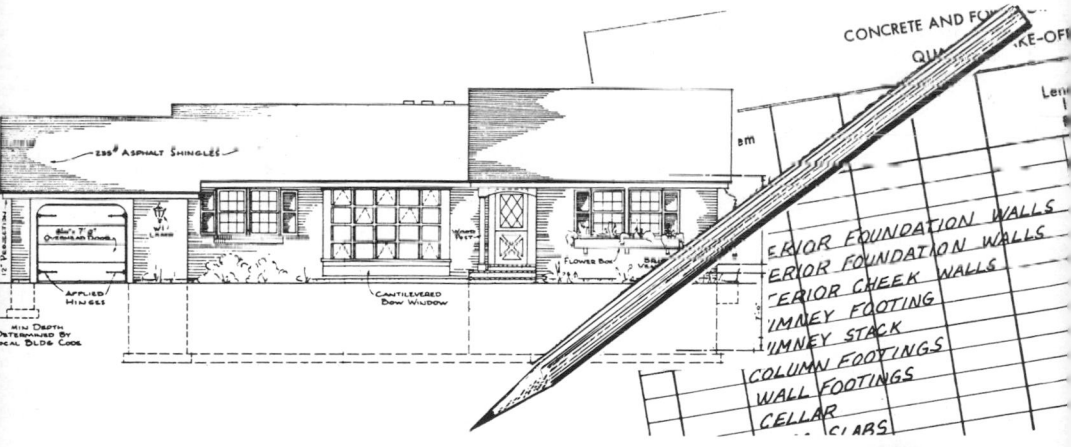

published by

AMERICAN TECHNICAL SOCIETY
Chicago, Illinois 60637

Copyright © 1965, 1973 by American Technical Society

1st Edition
   1st Printing 1965
   2nd Printing 1967
   3rd Printing 1969
   4th Printing 1970
   5th Printing 1972
2nd Edition
   6th Printing 1973
   7th Printing 1974

All Rights Reserved

No portion of this book may be reproduced by *any* process
such as photocopying, recording, storage in a retrieval system or
transmitted by any means without permission of the publisher.

Library of Congress Card Catalog No.: 72-80554

ISBN: 0-8269-0537-4

Printed in the United States of America

PREFACE
to the Second Edition

Estimating is defined in the dictionary as a means used to evaluate, to fix a cost, and to calculate a worth—all in an approximate manner. Contractors and estimators avoid "approximating" as much as possible; that is, they attempt to arrive at a total cost or a unit cost more by accurate methods than by guessing. A contractor needs accurate construction cost and overhead expense estimates in order to determine how low he can bid for a job and still be able to make a fair profit.

This second edition of *Estimating for the Building Trades* has been thoroughly revised with the hope that the basic information on the estimating process will aid the student, contractor, and builder in making quick, accurate quantity take-offs and labor-time estimates. The authors' broad experience in practical estimating and teaching were used to develop the step-by-step methods used in this book. Illustrations were chosen to show details which do not always appear on plans and details which could cause trouble for the inexperienced estimator.

The emphasis is on accurate analysis of the working drawings and specifications. Contracts have been lost because the submitted bids were too high or too low. Why? In many cases, later examinations of work sheets have shown that the quantity take-offs were wrong, thus leading to errors in pricing and rejected bids. Careless and hasty study of specifications and working drawings have led to inaccurate quantity take-offs.

Although material and labor costs vary among different regions, and for this reason no actual costs are given, the detailed estimating *methods* contained herein *will* apply everywhere.

The chapters on "Plan Reading," "Specifications," and "Review of Square and Cubic Measure" will be refresher material for some readers, but some of it will be new information for other readers. This material must be understood before estimates are attempted because it is fundamental to all estimating.

A set of house plans is included and used for many of the sample estimates throughout the book. Estimating larger structures involves the same basic methods used for these plans. The student who can estimate materials and labor for this house with little or no difficulty will have little trouble estimating larger structures once he has had some practical estimating experience.

The early chapters are detailed and emphasize estimating *methods*. The remaining chapters emphasize the *materials* peculiar to each of the building trades.

This edition has been revised and expanded to include contemporary materials and methods of estimating. The chapters on "Framing," "Electrical Wiring," and "Plumbing" have been completely rewritten in accordance with up-to-date standards and methods. Other chapters include new materials, such as acrylic window panes and modular plumbing and new labor methods such as those used in installing heating and air conditioning units.

A chapter on "Recapitulation" has been included at the end of the book to synthesize and simplify the rather involved process of estimating. It will also serve as a summary of the previous chapters and a check list before submitting a final bid.

A special appendix, "Hourly Estimates of Labor" has been added and correlated to the "Estimating Labor" section of the chapters. Other new appendices include "Grading of Lumber and Softwood Lumber Standards" and "Metric-to-English and English-to-Metric Conversions."

We express our appreciation to:

> Elmer W. Sundberg for his technical advice in the preparation of the chapters on framing;
>
> Kenneth L. Gebert for his valuable assistance in the rewriting of the chapter on electrical wiring;
>
> Morris Babbin for his aid in the plumbing and recapitulation chapters and for his advice throughout the manuscript.

In addition, we acknowledge the outstanding contributions of Patricia L. Reband, Revising Editor, for her careful analysis and preparation of the manuscript for publication in cooperation with the author, Martin Stempel.

The Publishers

CONTENTS

ESTIMATING
AND THE ESTIMATOR

The successful builder depends on accurate estimates of construction costs. Estimating these costs is an exacting process based on a thorough knowledge of the various trades involved in the construction industry.

WHAT IS ESTIMATING?

Estimating for the building trades is divided into two major phases: *quantity take-offs* and *labor pricing*. An estimate includes the costs of raw materials, mechanical equipment, and labor necessary for a construction job. When bidding for a contract, the contractor's overhead and profit are added to the estimate. It is almost impossible for one individual to do the quantity take-offs and labor estimates for all the various trades which are involved in the construction of a building. There are men in the construction industry who specialize in estimating one or two similar trades (e.g. electrical estimators and plumbing estimators). These specialists combine their estimates in order to determine the total construction cost of a building.

General contractors usually employ these various specialists to do all their estimating and pricing. Sometimes a general contractor works in close cooperation with a group of subcontractors who estimate their own costs and then submit the costs to the general contractor. Subcontractors usually include the cost of *material, labor, machinery, overhead,* and *profit.* A general contractor compares these estimates with his own rough estimates which are based on his experience and judgment — and determines a total price for the construction (after including his own costs and allowance for profit). He is then ready to bid for the building contract.

KINDS OF ESTIMATING

1. Area or Volume Method. This is usually based on the type of structure. It is a very vague and rough estimate—

one that is not advised. It may be used as a guide as a comparison to the actual cost of the building.

2. Guesstimate. This method is one that is not advised since it is based on personal opinion, personal experience, and, in most cases, only for one particular area of the country. As stated previously, the times call for hard, cold, accurate facts and figures.

3. Detailed Quantity Take-off. This method is based on a detailed, item by item breakdown of material, labor and time of operation which can be done only with the accurate use of the plans and the detailed reading of the specifications. Here accuracy, experience, previous similar work and previous job records play an important role.

4. Unit Cost Estimate. This method evolves over a period of time in a particular trade, in a particular area where the estimator has accurately taken quantities of many jobs and is able to arrive at a figure based on a square foot of finish floor or a lineal foot of base molding or a cubic yard of plain concrete.

Of the four types of estimates listed above, the Detailed Quantity Take-off is the only one recommended.

THE ESTIMATOR

The estimator is a person who must assume a great amount of responsibility. He is highly trained or experienced and is familiar with all phases of building construction and general contracting. He is persistent, progressive, observant, and patient. The good estimator will acquire these traits and find that he will also acquire the respect of those with whom he deals in the building industry.

Experience. There is no substitute for experience in the building industry. The estimator should have many years of practical experience in the field which he is estimating, in order to arrive at a competitive and fair bid.

The estimator should be aware that insurance and sales tax increase the cost of the job. The amount of coverage for insurance and the amount for sales tax vary from location to location and should be checked by the estimator. Intensive practical experience gives the builder, contractor, or estimator an insight into the construction plans and permits him to calculate the *hidden* items which are not clearly evident on the plans but are necessary to carry the plans to completion. For example, a plastering estimator not only must estimate the area to be plastered, but must be aware of the heights at which the plasterers will be working. An 8-foot high plaster wall does not require special scaffold rigging but a 20-foot-high wall (such as a mezzanine room) will require many special set-ups of the same scaffold. Rigging the scaffold for the higher wall will require much more time—perhaps many more hours of labor. Both rooms may have the same total wall and ceiling area, but you can see that different prices will be bid for the rooms. The difference will be in the labor estimate. The estimator calculates the time it will take to readjust the scaffold and bring materials up to the men at the different heights.

Mathematical Ability. It is important that the estimator possess some mathematical ability, especially in computational and tabulating work.

The use of desk calculators and slide rules will speed the work. (Note: The slide rule is not recommended unless you are sure where to place the *decimal points* in the answers. Serious errors can result if you are inexperienced in slide rule operation.) In many offices, the estimator does not do his own tabulating. This is usually done by an apprentice or an individual experienced in the operation of a desk calculator.

Neatness. The estimator should be neat and clean both in his work and personal habits. Experience has shown that a neat, clean individual usually does work of a similar quality. This is a very necessary and desirable trait in estimating because other people will need to read the estimate. There have been cases where the estimator's work was shown to him months later and he was unable to explain the figures and results shown on his tally sheets. This situation can be avoided when the estimator is neat and clean.

Need for Good Estimating. Today more than ever before, the competition that is found in construction has decreased the profit to a point where accurate estimating is a most vital need for survival of the builder or contractor. The estimator who uses guesswork and hearsay to estimate a job is a thing of the past.

Accurate estimating is the foundation upon which a successful business is built. It supplies the builder or contractor with an accurate list of materials and labor on which to make a bid. In addition to the bidding of the job the estimate should give the following information:

1. Accurate take-off of material should be provided so that the material can be ordered as needed for the job.

2. Some kind of time limitation should be placed on the job so that it can be scheduled.

3. The number of men and the skills

Item	Man Hours	Unit Cost Labor	Total Cost of Labor	Unit Cost of Material	Total Cost of Material	Total Cost
LABOR & MATERIAL COST SHEET						

Fig. 1-1. A sample estimating form.

required for the job should be included.

Standardization. The estimator must follow a pattern which will be reflected in his completed work. This habit has been the incentive for organizations to compile standard forms for estimating and pricing, and to train all their new personnel in their own particular system. The estimating forms are often used as actual forms for ordering material and equipment. See Fig. 1-1.

Knowledge. The estimator must know the various construction materials and their uses. He must also be acquainted with equivalent materials and acceptable substitutes which are available in his area. Obviously the estimator must know the prevailing prices of materials and their equivalents — or how to quickly obtain the prices. On many sets of plans, the architect specifies the materials to be used. If the estimator has a thorough knowledge of the materials, he can save his company much money. Not only can he save money by knowing the available materials (in his area), but by knowing of equivalent materials. This is especially important if a particular brand should change in quantity available (or price) between the time of the estimate and the actual construction. *It should be fully understood that if the contractor cannot obtain the specified material, he must ask for permission to substitute another material.* This should be granted (in writing) by the builder, owner, or architect or, in some cases, all the concerned parties.

Keeping Records. A good estimator is a man who knows how to keep records and has the patience to file them over a long period of time. Since estimating consists of quantity take-offs and labor pricing, the estimator should maintain a continuous record of the costs of the various operations during the construction of a particular project. He compiles this record from the daily and weekly progress reports which are sent into the office by the field men. If he has records of similar jobs within a given time span, he is able to determine the average costs for labor, material, and equipment for almost any situation which will arise on a similar job in the future.

Average costs are sometimes used when an estimate is needed quickly. For a particular kind of building, only the *square feet* of floor area or *cubic yards* of volume are necessary in order to determine the average costs. This method seems simple, but remember that it is based upon the average costs of previous projects. Thus accurate records are a necessity for this kind of estimating.

The estimator must be exceedingly thorough. He must know exactly what the working drawings illustrate, and he must be able to understand the specifications. Understanding the specifications is half the battle in the pricing and estimating of a job. Applying the specifications to the job is equally important. If there are items which are not clear in the specifications (or are not to be priced), the estimator lists these on an *exception sheet* or an *omission sheet*.

The estimator knows that the specifications take precedence over the working drawings. For example, the basement or foundation drawing of a small home might show the basement slab poured on the ground. The specifications, however, state that the slab is to be placed on a bed of gravel. The specifications would be followed. If the contractor wanted to pour the slab on the ground, he must obtain permission from the architect and/or owner.

HOW TO PREPARE AN ESTIMATE

The estimator is given a complete set of plans and specifications. He starts to set up the job after briefly browsing through the plans and specifications. If it is possible he should visit the site to place the job conditions clearly in his mind. Such items as existing structures, access to and from the job, trees, shrubs, and the neighborhood may play an important part in the cost of the job. It is also important that he be aware of labor and material availability.

In setting up the job, the estimator reads the specifications carefully and on a bid sheet or master bid sheet, Fig. 1-2, he lists the section of the specs and all items to be estimated and priced. He then sets up individual manila folders for each section of the specs. He then sends cards or makes phone calls inviting subcontractors to bid on their phase of the work and, at the same time, letting them know when and where the plans may be seen for their individual take-off.

When all this is complete, the estimator and his staff may begin to take-off (estimate) and price each one of the sections. These items are then entered in the columns of the master bid sheet as the budget bid (sometimes called the house bid). When the subcontractors' bids for the sections come in, they are placed in the quoted bid column. The house bid and the quoted bid column should compare favorably. If they are far apart, it is wise to call in the subcontractor in question and try to compare quantities and, if possible, the pricing to see if any mathematical errors can be detected and so try to resolve the spread of the bid and bring them closer into line. This accomplished, all the items priced and checked, and all profit, overhead, and taxes included, the job is now ready for final review and a definite bid.

It should be noted that the estimator does not always use the lowest bid. He should use the lowest *bonafide bid* or the bid that he feels can complete the job successfully. It is not a good or ethical practice to take a low bid which might have been gotten through an error and then try to get the job at that low bid. This could result in problems and, in some cases, financial disaster to the subcontractor and other interested parties.

The estimator should have taken into consideration:

1. Plans and Specifications
2. A. Construction Check List
   B. Material Take-off
   C. Labor Take-off
3. Summary Sheet
4. Direct Cost of Construction.

PROJECT _____

MASTER BID SHEET

SHEET #   OF #

SPEC. NO.	SUBCONTRACT DESCRIPTION	BUDGET BIDS	QUOTED BIDS	BIDS USED
Sec	General Conditions			
1	Demolition			
2	Clearing of Site			
3	Excavation			
4	Concrete and Concrete Superstructure			
5	Brickwork			
6	Carpentry			
7	Millwork			
8	Lath and Plaster			
9	Ceramic Tile			
10	Wood Flooring			
11	Resilient Tile			
12	Hardware			
13	Painting and Decorating			
14	Site Work			
15	Plumbing			
	Heating			
	Air Conditioning			
16	Electrical			
17	Elevators and Lifts			
	NOTE: This is a sample sheet. The actual sections and titles are taken from specifications.			

Fig. 1-2. Sample Master Bid Sheet.

SUMMARY

You can see that accurate estimates are necessary if you wish to make a profit. Estimates are based on quantity take-offs and labor prices; to these are added equipment costs, overhead costs, and an allowance for profit.

The builder who wishes to be an estimator must possess certain traits if he is to be successful. He must be systematic and able to think in an orderly and logical manner. Mathematical ability, neatness, standardization, and complete and accurate records are also necessary.

All these personal traits will have little meaning unless the estimator is able to read and understand specifications and plans. Remember that there is no substitute for experience. One must be familiar with all phases of the construction for which estimating is done.

2

PLAN READING

~~~~~~

A set of plans is a group of drawings and specifications for a structure. The plans and specifications indicate dimensions, kinds of materials, number and location of rooms, number and location of windows and doors, and any other information pertinent to creating the structure.

WORKING DRAWINGS

The working drawings and specifications are equally important because one cannot estimate without having both. A set of working drawings may have many sheets and consist of:

Site Plan and Building Land. These sheets are usually titled: Site, Sheet No. 1; Site, Sheet No. 2; and so forth for as many sheets as may be necessary.

Architectural. These sheets are usually titled: A Sheet No. 1, A Sheet No. 2, and so forth for as many sheets as may be necessary.

Structural. This type of drawing includes the structural steel and the concrete. These are usually titled: S Sheet No. 1, S Sheet No. 2, and so forth for as many sheets as may be necessary.

Details. These drawings show the details and the blow-ups of units of work that may be on each of the preceding sheets of the drawings. These sheets are titled: Detail Sheet No. 1, Detail Sheet No. 2, and so forth for as many sheets as may be necessary.

Electrical. These are similarly titled: E-Sheet No. 1, E-Sheet No. 2, and so forth.

Mechanical. M Sheet No. 1, M Sheet No. 2, and so forth.

Plumbing and Heating. Pl & H Sheet No. 1, Pl & H Sheet No. 2, and so forth.

Air Conditioning. A Cond Sheet No. 1, A Cond Sheet No. 2, and so forth.

There may be many more sections to a complete set of drawings. The number depends upon the size and type of building and its location.

The working drawings are usually drawn on tracing paper so that reproductions can be made (in the form of blueprint, blue line, brown line, or black line copies). The original plans and specifications are duplicated and distributed among bidders, estimators, contractors, and other interested parties. Each of the interested parties will have identical information about the proposed structure.

The terms *prints, blueprints,* and *drawings* are all used here to refer to the *working drawings*. These terms are used as equivalent in the building trades. Other trade terms will be used in the balance of this book; the estimator must become familiar with these terms.

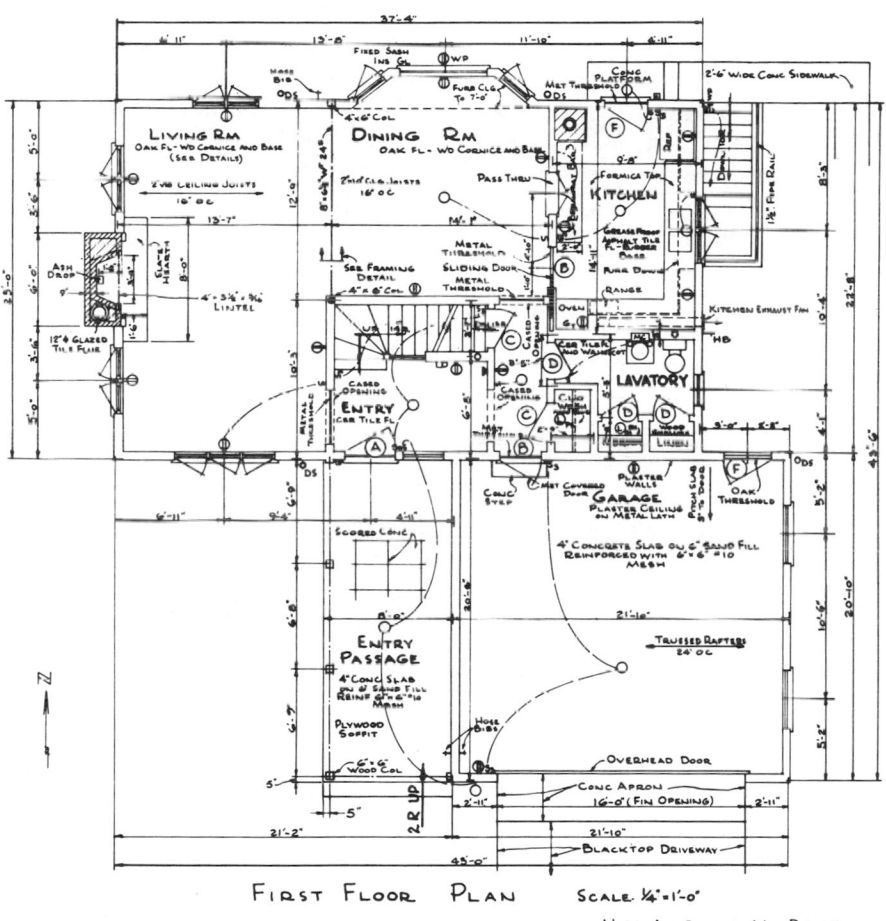

FIRST FLOOR PLAN   SCALE ¼"=1'-0"

PUBLISHER'S NOTE: THIS DRAWING WAS ORIGINALLY DRAWN TO THE SCALE SHOWN. THE DRAWING WAS REDUCED TO FIT THE PAGE AND CAN NO LONGER BE SCALED.

NOTE: ALL EXTERIOR WALL DIMENSIONS ARE TO OUTSIDE FACE OF STUDS AND CENTERLINES OF WINDOWS AND DOORS. INTERIOR DIMENSIONS ARE TO CENTERLINES OF PARTITIONS.

Fig. 2-1. A typical floor plan.

HOUSE PLAN A (5 plates at the end of this chapter) is a set of drawings for a single unit residence. These drawings will be referred to throughout this chapter in order to illustrate various points in the reading of working drawings.

Plan Views. Every set of plans includes a plan view of each floor of the house. (Plan views are also known as *floor plans;* the meaning is the same.) Fig. 2-1 shows a typical plan view.

The plan view looks directly down on any particular floor or foundation. (See Plates 1 and 2 of HOUSE PLAN A also.) It shows the room arrangement, chimneys, fireplaces, stairs and closet. It shows the location of various devices, as well: plumbing fixtures, lighting outlets, heating apparatus, and mechanical appliances. Each floor has its own plan.

Elevations. An elevation view (Fig. 2-2) of a building is what you see, for example, if you stand directly in front of it and look at the front side. This view is similar to a photograph, but it lacks perspective or depth. Four outside elevation views are required for most buildings (see Plates 3 and 4 of HOUSE PLAN A). When the term *elevation* is used alone, it usually refers to an *outside elevation*. There are, however, *interior elevations*. An interior elevation is shown in Plate 5, "Kitchen Cabinet Elevations."

Sections. Plan views, to a large extent, represent simplified sectional views. An example of sectioning might involve a watermelon. The outer sur-

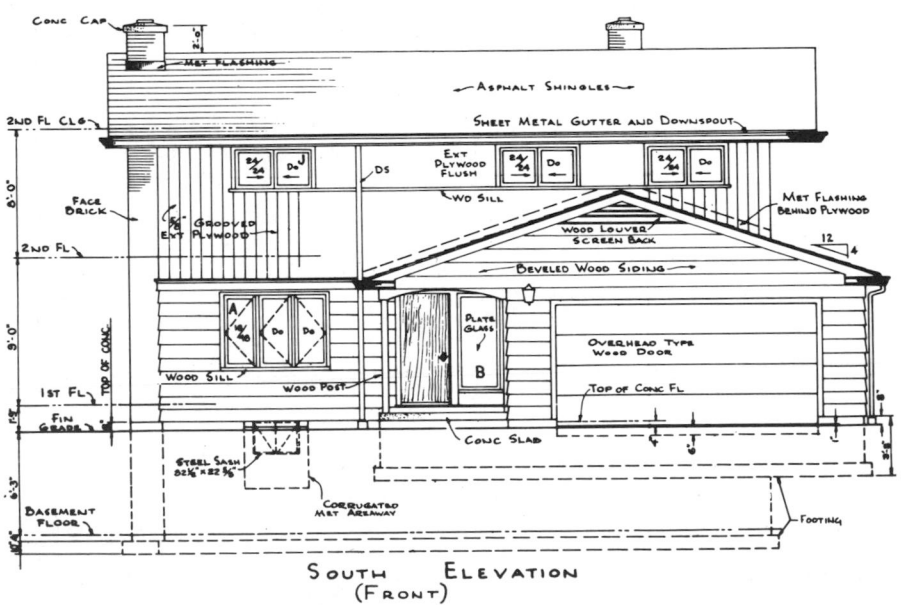

Fig. 2-2. A typical elevation view.

face (rind) gives no indication of the interior. By cutting the melon in half you can see the interior structure—red meat and black seeds. The cut surface of the melon can be called a section because it shows the makeup or construction of the melon, including the skin thickness, distribution of the seeds, and the apportionment of red meat.

Sections can be taken from either a plan view or from an elevation view. Plan view sections are called *cross sections* or simply *sections*. Sections taken from elevation views are called *longitudinal sections* or *sectional elevations*.

Section drawings (Fig. 2-3) are made to illustrate structural details not shown in the plan or elevation drawings. Look at Plate 5 of HOUSE PLAN A and note the parts of the drawing entitled "Section . . ." and "Cross Section. . . ." These sectional drawings were made to expose the interior and to show proportions, arrangements, and composition.

The wall section (like Fig. 2-3), is the most important part of the working drawings for the estimator. It shows most of the materials that go into the building. It eliminates a lot of guesswork and acts as a check list so the estimator has less chance of making errors and omitting materials that will have to be furnished to complete the building.

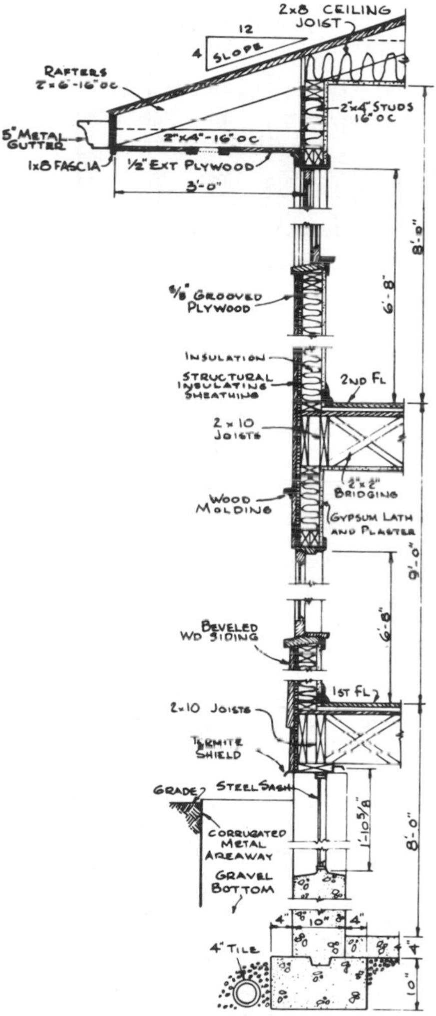

Details. Often parts of the drawings need to be amplified in size for purposes of clarification. These parts are then drawn to a larger scale and are called detail drawings. They may be made of parts of floor plans, sectional drawings, or elevations, and they may be distributed throughout the various pages of drawings or grouped together as detail sheets. Figs. 2-4 through 2-6 and the Plumbing Diagram on Plate 5 of HOUSE PLAN A are all detail drawings.

SECTION A-A

Fig. 2-3. Typical wall section.

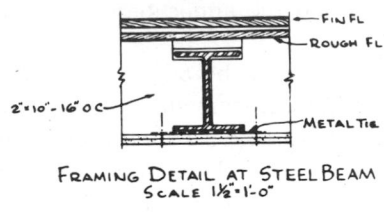

Fig. 2-4. Beam framing detail.

Schedules. Separate schedules for doors and windows are shown with the first-floor plans. References to window openings are sometimes indicated by numbers and references to doors by letters. This practice helps to keep the drawings from becoming cluttered with too many details which often

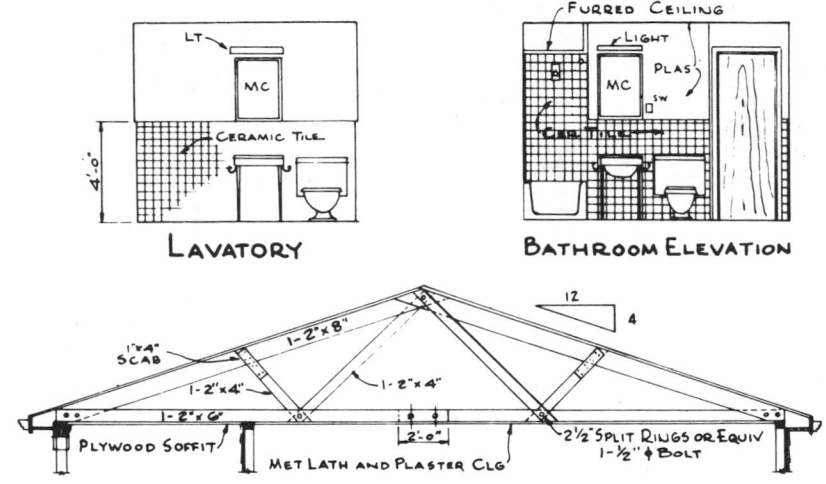

Fig. 2-5. Lavatory, bathroom and trussed rafter details.

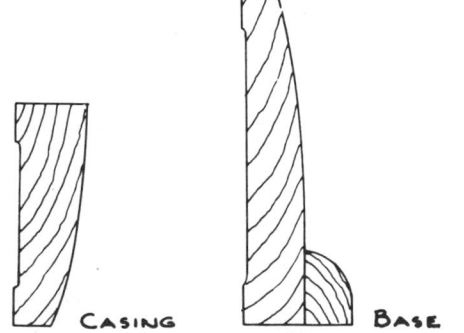

Fig. 2-6. Detail of typical trim (full size).

Door Schedule for Entire House			
Mark	Size	Amt Req'd	Remarks
A	5'-0"x 6'-8"x1¾"	1	Exterior Flush Door
B	2'-8"x 6'-8"x1¾"	7	Flush Doors 1-Sliding 1-Metal Covered
C	2'-6"x 6'-8"x1⅜"	4	Flush Door
C₁	2'-6"x 6'-8"x1⅜"	2	Louvered
D	2'-4"x 6'-8"x 1⅜"	4	Flush Door
D₁	2'-4"x 6'-8"x1⅜"	1	Louvered
E	1'-5"x 6'-8"x1⅜"	1	Bi-Fold Louvered
F	2'-10"x6'-8"x 1¾"	2	Exterior 2 Lights
G	2'-8"- 6'-8"x1¾"	1	Exterior 2 Lights

Fig. 2-7. Typical door schedule.

make the instructions difficult to read. Fig. 2-7 illustrates a typical door schedule.

SCALE

Drawings must be made smaller than the actual size of the building they represent. Working drawings are made to a predetermined ratio of the actual dimensions. This practice is known as drawing to scale. In order to scale, it is necessary to adopt some unit of measurement so that the ratio of drawing dimensions is constant. (The ratio should be stated on the plans.)

A common twelve-inch ruler is a scale that provides a unit of measure. The twelve inches are marked and divided into halves, quarters, eighths, and sixteenths. Architects, however, use an *Architect's Scale* which may be either flat or three-sided. Fig. 2-8 shows a three-sided scale. The three-sided scale has one edge marked in inches, as does the common twelve-inch rule. The other edges contain 10 scales. These are the ⅛ and ¼, the 1 and ½, the ¾ and ⅜, the 3/16 and 3/32,

and the 1½ and 3-inch scale. To explain these scales, the ¼-inch scale will be used.

Note the edge marked "¼". This edge is read from right to left. It is divided into spaces each ¼ inch apart. At one end, one of the ¼ inch spaces has been divided into halves, quarters, and twelfths. One of the ¼-inch spaces represents 1 foot, or 12 inches. One of the halves, in that space divided into 12 equal parts, represents 6 inches, and so forth.

Suppose in drawing a plan view you are confronted with the problem of drawing a line 8 feet long according to the ¼-inch scale. To do this, you would just count off 8 spaces on the scale. As another example, suppose you must draw a line representing a dimension of 6'-6", using the same ¼-inch scale. First count off 6 of the ¼-inch spaces; then add 6 spaces from that space which is divided into 12 equal parts. (This is the distance from A to B in Fig. 2-9.) In drawing to the ¼-inch scale, therefore, you are simply substituting quarter inches for feet.

The other scales operate in the same manner as the ¼-inch scale. Every part of a drawing must be drawn accu-

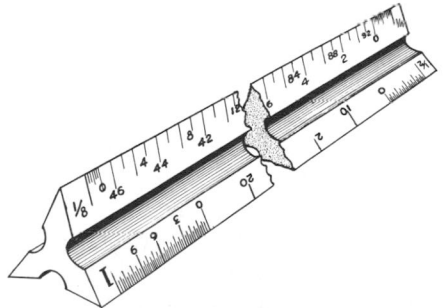

Fig. 2-8. A three-sided architect's scale.

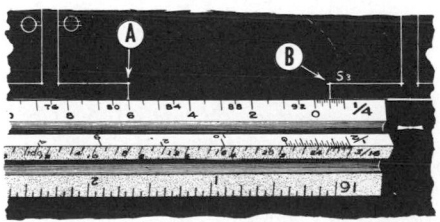

Fig. 2-9. Scaling a working drawing.

rately according to whatever scale is being used, unless that portion of the drawing is labeled "not to scale." This procedure enables the designer to make his drawings in exact ratio to the proposed building.

Generally, drawings are made using the ¼-inch scale. However, in designing large buildings, a smaller scale is sometimes used. Conversely, small details are often drawn to a large scale.

DIMENSIONS

Dimensions are the all-important guide for the estimator and builder. The prints should show dimensions for all rooms, windows and door locations, heights, and so forth. Fig. 2-10 illustrates typical dimension lines. This can be seen in Plates 1 through 5.

Look at Plate 1 and locate the 16'-6" dimension in the lower right corner. If this dimension had not been shown, you could find it by using simple arithmetic. Looking at the other dimensions, you see that the total length of the house front is 64'-0". Two partial lengths are given as 16'-0" and 31'-6". Subtracting the sum of these two partial lengths from the total length, you can find the missing dimension:

$$
\begin{array}{r}
64'\text{-}0'' \\
-47'\text{-}6'' \quad (16'\text{-}0'' + 31'\text{-}6'') \\
\hline
16'\text{-}6''
\end{array}
$$

Note that 64'-0" = 63'-12". This makes it easy to subtract 47'-6". Converting to decimal fractions makes the calculations even easier. (This is discussed in Chapter 4.)

Missing dimensions can be found by scaling, that is, by using a set of dividers to measure the desired dimension on the prints and then placing the dividers along a printed scale. Another method of finding the missing dimensions is by using a ruler (preferably an *Architect's Scale*) and measuring the dimensions directly from the prints. Direct measuring should be avoided, however, because of inevitable inaccuracies in drawing and because the prints can shrink.

STRUCTURAL CODE

Architects, builders, and estimators make use of what might be called a "structural code." This code includes symbols and abbreviations—the symbols by and large have the same meaning in most parts of the world. For example, if a group of American, Swedish, French, and Australian architects looked at Plates 1 through 5 they would all recognize these as working drawings and they would be able to translate the symbols into their own languages.

Symbols. Prints must indicate a great amount of information on a relatively small piece of paper. This means that the data given on the print must in some way be reduced to a fraction of its actual size. A brick wall, for example, may actually be a foot thick, but it will appear only a fourth (¼) of an inch thick on the print. In addition to indicating thickness (which is done by scale drawing—explained in a later section) the print must also indicate the material in the wall. In this case, the material is brick. No drafts-

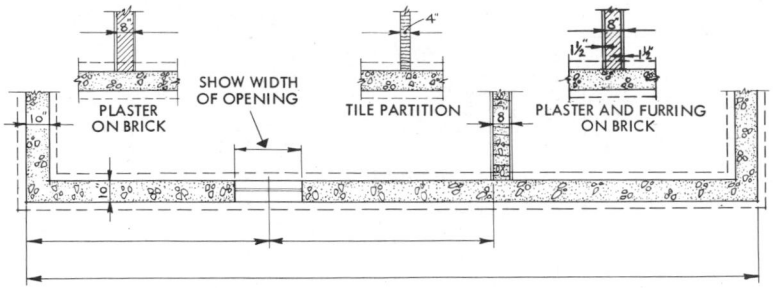

CONCRETE FOUNDATION

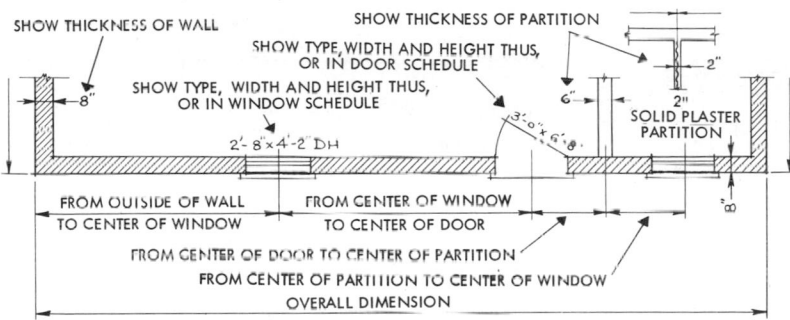

BRICK WALL

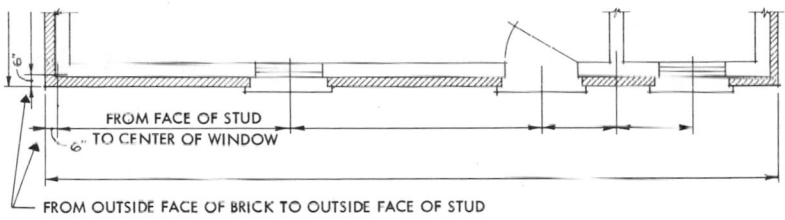

BRICK VENEER

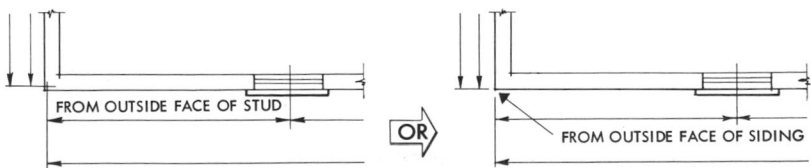

FRAME WALL

Fig. 2-10. Typical dimension lines used on drawings.

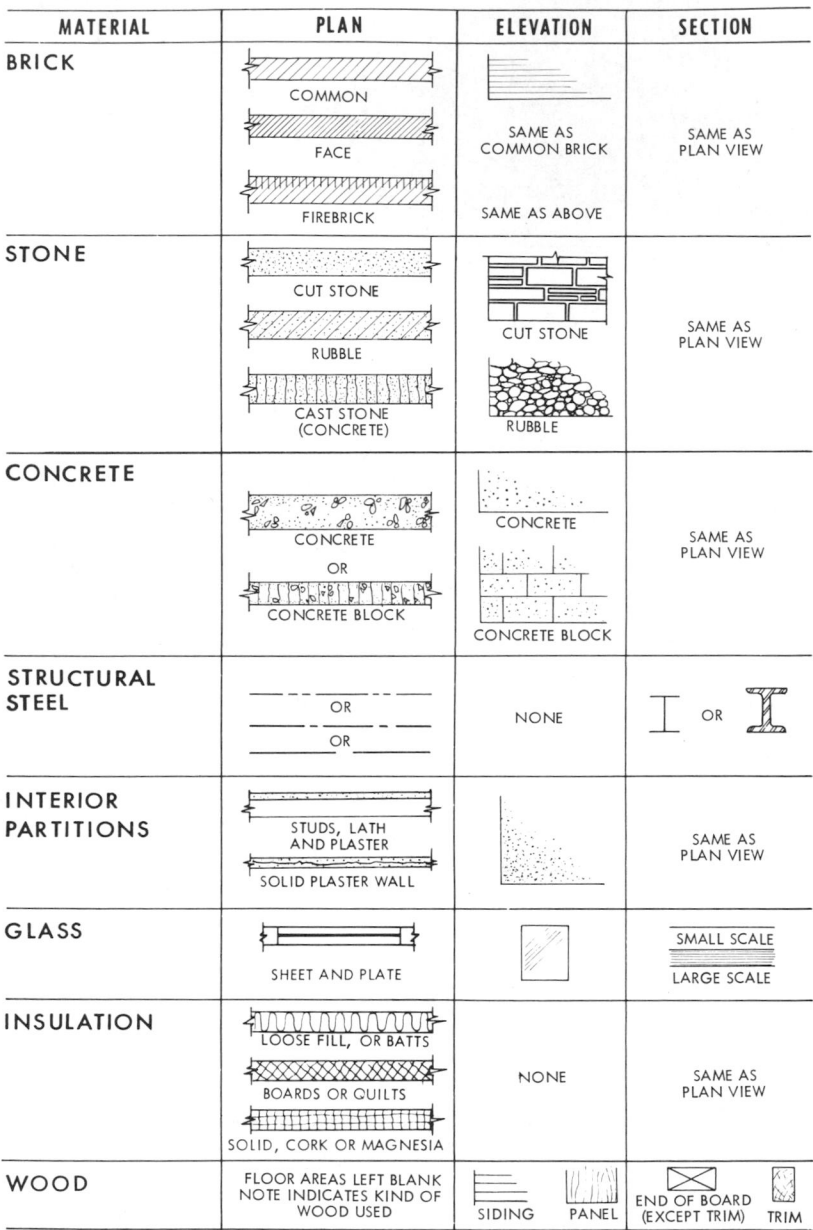

Fig. 2-11. Material Symbols.

MATERIAL	PLAN	ELEVATION	SECTION
SHEET METAL FLASHING	INDICATE BY NOTE		HEAVY LINE SHAPED TO CONFORM
EARTH	NONE	NONE	
ROCK	NONE	NONE	
SAND	NONE	NONE	
GRAVEL OR CINDERS	NONE	NONE	
FLOOR AND WALL TILE			
SOUNDPROOF WALL		NONE	NONE
PLASTERED ARCH	p. a.	DESIGN VARIES	SAME AS ELEVATION VIEW
GLASS BLOCK IN BRICK WALL			SAME AS ELEVATION VIEW
BRICK VENEER	ON FRAME	SAME AS BRICK	SAME AS PLAN VIEW
CUT STONE VENEER	OR / ON BRICK / ON CONCRETE BLOCK	SAME AS CUT STONE	SAME AS PLAN VIEW
RUBBLE STONE VENEER	OR / ON FRAME / ON BRICK / ON CONCRETE BLOCK	SAME AS RUBBLE	SAME AS PLAN VIEW

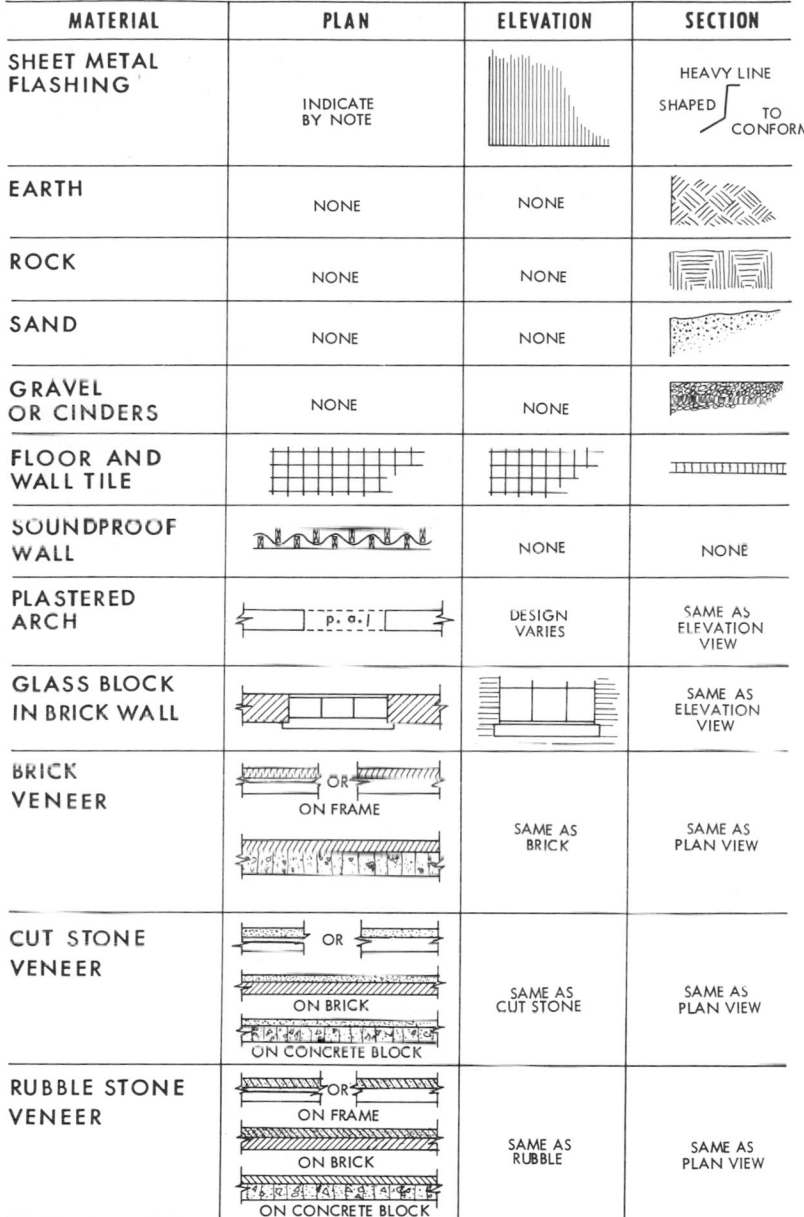

Fig. 2-11B. Material Symbols.

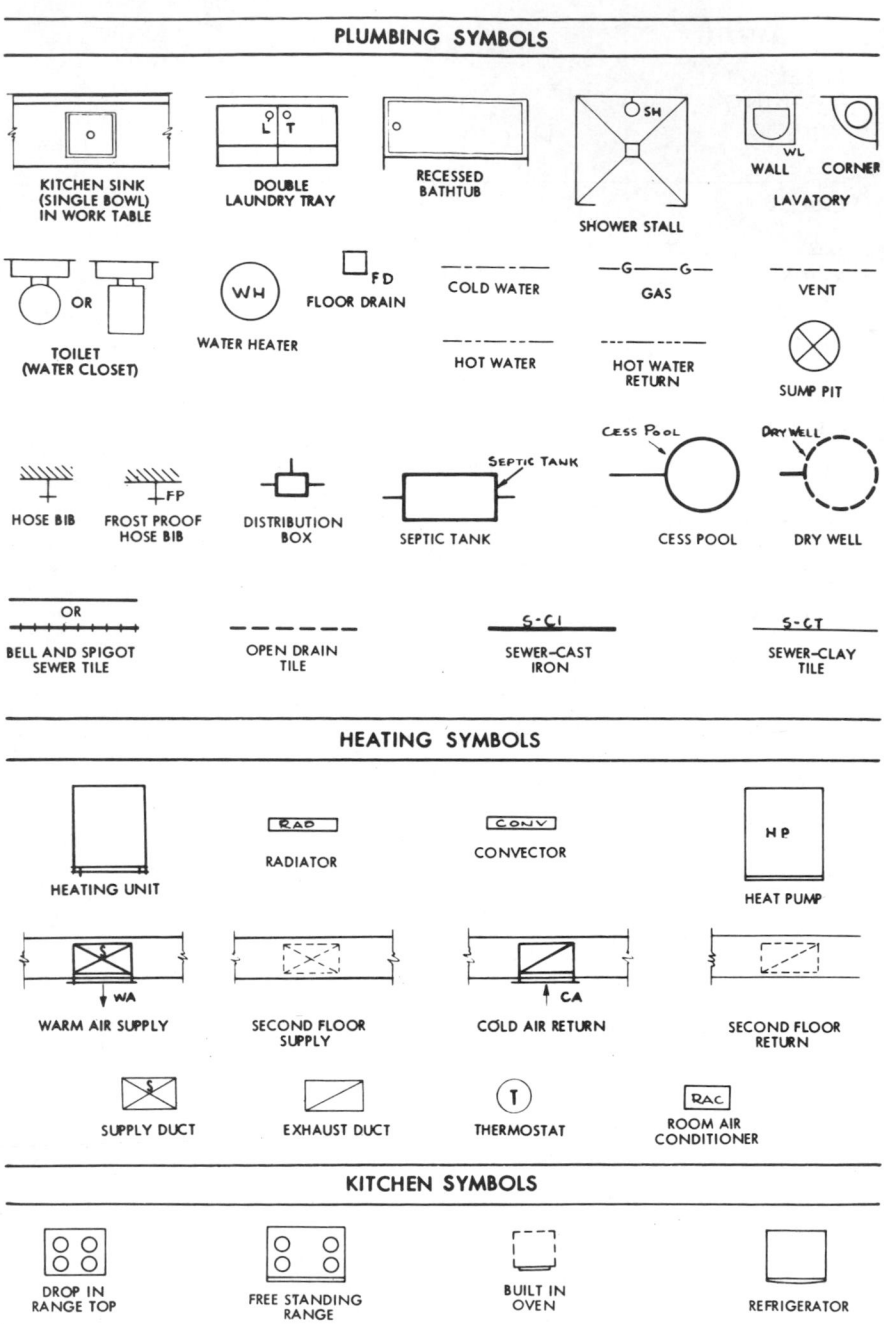

Fig. 2-12. Common plumbing, heating and kitchen symbols.

ELECTRICAL SYMBOLS

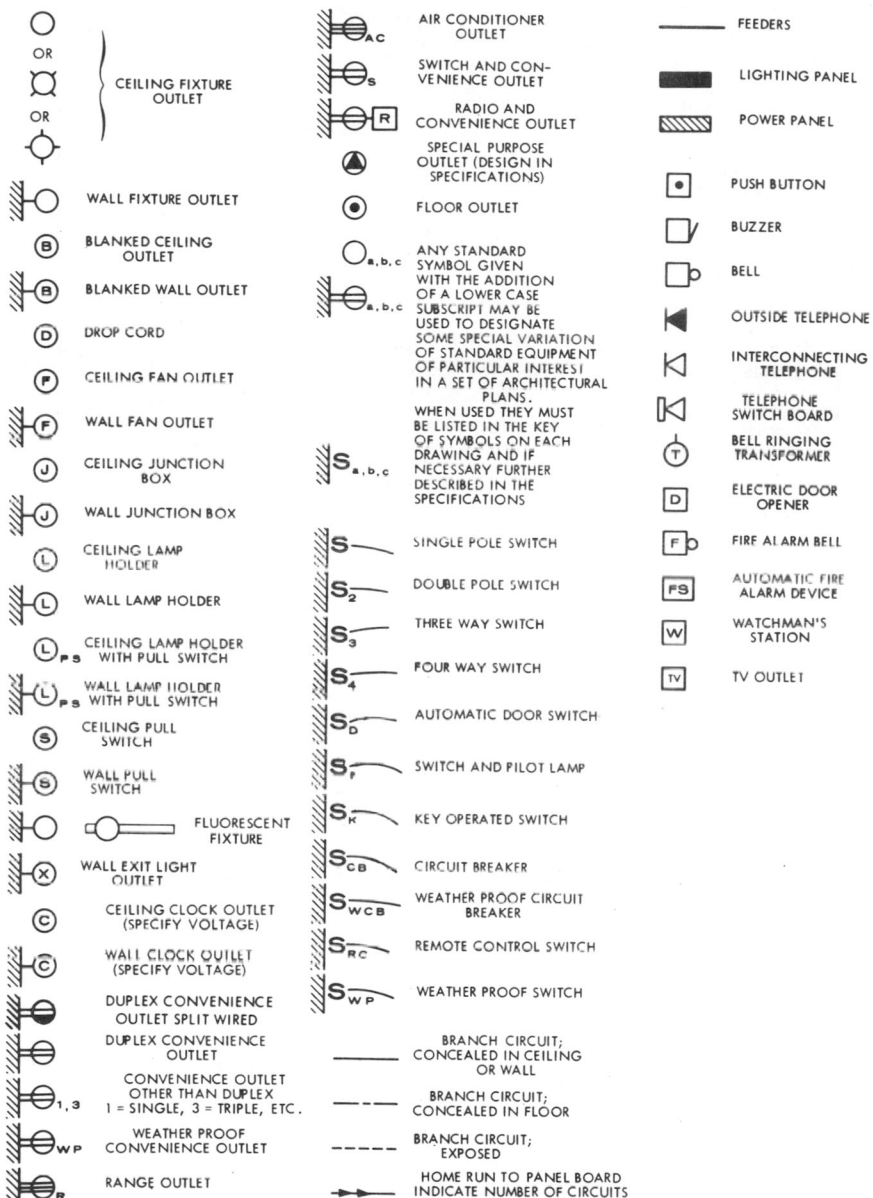

Fig. 2-13. Common electrical symbols.

ABBREVIATIONS

Access Door	AD	Dining Room	DR
Access Panel	AP	Direct Current	DC
Acoustic	ACST	Dishwasher	DW
Aggregate	AGGR	Ditto	DO, "
Alternating Current	AC	Door	DR
Aluminum	AL	Double-Acting Door	DAD
Anchor Bolt	AB	Double-Hung Window	DHW, DH
Angle	$\angle$	Double Strength Glass	DSG
Apartment	APT	Down	DN, D
Asbestos	ASB	Downspout	DS
Asbestos Board	AB	Drain	D, DR
At	@	Drain Board	DB
Barrel	BBL	Drawing	DWG
Basement	BSMT	Dressed & Matched	D & M
Bathroom	B	Drip Cap	DC
Bath Tub	BT	Dryer	D
Beam	BM	Electric Panel	EP
Bedroom	BR	Electric	ELEC
Blueprint	BP	Elevation	ELEV, EL
Bolts	BT	End-to-End	E to E
Boundary	BDY	Entrance	ENT
Brass	BRS	Excavate	EXC
Brick	BRK	Expansion Joint	EXP JT
Broom Closet	BC	Exterior	EXT
Building	BLDG	Finish	FIN
Building Line	BL	Finished Floor	FIN FL
Cabinet	CAB	Firebrick	FBRK
Casing	CSG	Fireplace	FP
Cast Iron	CI	Fireproof	FPRF
Catch Basin	CB	Fireproof Solid Core	FPSC
Ceiling	CLG	Flashing	FL
Cellar	CEL	Floor	FL
Cement	CEM	Flooring	FLG
Center	CTR	Fluorescent	FLUOR
Center Line	CL, $\cent$	Flush	FL
Center Matched	CM	Foot or Feet	FT, '
Center-to-Center	C to C	Footing	FTG
Ceramic	CER	Foundation	FDN
Cesspool	CP	Frame	FR
Channel	CHAN	Fresh Air Intake	FAI
Cinder Block	CIN BL	Full Size	FS
Circuit Breaker	CIR BKR	Furring	FURR
Clean Out	CO	Games Room	GR
Clear	CLR	Galvanized	GALV
Clear Glass	CL GL	Galvanized Iron	GI
Closet	CLO, CL, C	Garage	GAR
Cold Air	CA	Gas	G
Cold Water	CW	Gage	GA
Collar Beam	COL B	Glass	GL
Column	COL	Glass Block	GL BL
Concrete	CONC	Grade	GR
Concrete Block	CONC B	Grade Line	GL
Conduit	CND	Gypsum	GYP
Copper	COP	Hall	H
Counter	CTR	Hardware	HDW
Cubic Feet	CU FT	Height	HGT, H, HT
Cubic Yards	CU YDS, $\cent$	High Point	H PT
Cut Out	CO	Hot Air	HA
Dampproofing	DP	Hot Water Tank	HWT
Detail	DET	Inches	IN., "
Diagram	DIAG	Inside Diameter	ID
Diameter	DIA, ϕ	Insulation	INS
Dimension	DIM	Interior	INT
Dinette	DT	Iron	I
Dining Alcove	DA	Jamb	J

ABBREVIATIONS

Joist Space	JS	Rubber	RUB
Kick Plate	KP	Rubber Tile	R TILE
Kitchen	K	S–Beam	S
Kitchen Cabinet	KC	Scale	SC
Kilowatt	KW	Screen	SCR
Knocked Down	KD	Section	SECT
Landing	LDG	Sewer	SEW
Lath	LTH	Sheathing	SHTHG
Laundry	LAU	Sheet	SH
Lavatory	LAV, L	Shelving	SHELV
Leader	L	Shiplap	SHLP
Length	L, LG, lgth	Shower	SH
Length Overall	LOA	Siding	SDG
Level	LEV	Single Strength Glass	SSG
Light	LT	Sink	SK, S
Limestone	LS	Sliding	SLID
Line	L	Soil Pipe	SP
Linen Closet	LIN, L CL	Specifications	SPEC
Lining	LN, Lng	Square	SQ, sq.
Linoleum	Lino	Square Inch	SQ IN , sq. in.
Living Room	LR	Square Foot	SQ FT , sq. ft.
Louver	LV	Stairs	ST
Low Point	LP	Standard	STD
Maximum	MAX	Steel	STL
Medicine Cabinet	MC	Stone	STN
Metal	MET, M	Steel Sash	SS
Minimum	MIN	Storage	STG
Miscellaneous	MISC	Switch	SW, S
Mixture	MIX	Telephone	TEL
Molding	MLDG	Terra Cotta	TC
Mortar	MOR	Thermostat	THERMO
Movable Partition	M PART	Thick or Thickness	THK, T
Mullion	MULL	Threshold	TH
Number	NO., #	Thousand	M
On Center	OC	Through	THRU
Opening	OPNG	Tongue & Groove	T & G
Outlet	OUT	Top Hinged	TH
Outside Diameter	OD	Tread	TR, T
Panel	PNL	Unexcavated	UNEXC
Partition	PTN	Utility Room	UR
Perpendicular	PERP	Vent	V
Plaster	PLAS, PL	Ventilating or Ventilation	VENT
Plate	PL	Vertical	VERT
Plumbing	PLMB	Vinyl Tile	V TILE
Porch	P	Volt	V, v
Pound	LB, #	Volume	VOL, V
Pounds per Square Foot	PSF	W–Beam	W
Powder Room	PR	Washing Machine	WM
Precast	PRCST	Washroom	WR
Prefabricated	PREFAB	Water Closet	WC
Pull Chain	PC, P	Water Heater	WH
Pull Switch	PS	Waterproofing	WP
Radiator	RAD	Watts	W, w
Radius	R, r	Weather Stripping	WS
Recessed	REC	White Pine	WP
Refrigerator	REF	Width	W, WTH
Reinforced	REINF	Window	WDW
Revision	REV	Window Radiator	WR
Riser	R	Wire Glass	W GL
Rivet	RIV	Wood	WD
Roof	RF	Wrought Iron	WI
Roof Drain	RD	Yard	YD, yd.
Roofing	RFG	Yellow Pine	YP
Room	RM, R		

man would want to draw a likeness of each individual brick and no firm of architects would find it economical to do so. To facilitate drawing and make drawing less expensive to produce— a standard symbol is used. Figs. 2-11 through 2-13 show some of the symbols used in working drawings. The estimator should learn to recognize these symbols quickly and accurately. (*Warning:* Some localities and trades use other symbols. Be sure to check the symbols given here against local custom. The symbols presented here, however, are widely used and recognized.)

Abbreviations. Like symbols, abbreviations are used to save time and to conserve space. Again as with symbols, there are non-standard abbreviations in use, but their meanings usually become clear as you study the prints on which they appear.

STRUCTURAL CONVENTIONS

Windows and Doors. Fig. 2-14 shows some of the commonly used types of windows. Fig. 2-15 illustrates how these windows would be shown on drawings. Fig. 2-16 shows how doors are indicated on drawings. These illustrations are self-explanatory and should be studied carefully. Now look at Plates 1 through 5 of HOUSE PLAN A to identify the window and door types used in this building.

The dimensions of windows and doors should be given on the plan views, as in Plate 1, or contained in a "Window and Door Schedule" on one of the sheets of drawings.

Chimneys. Chimneys are drawn accurately to scale at the various floor levels. The flues can be shown at each level, together with flue dimensions and a heavy line symbol indicating a flue lining. The elevation view should give the height of chimneys above the roof, and any decoration features. If the chimney is partially outside, the elevation should give its shape and dimensions. Figs. 2-17 and 2-18 show typical conventions for a fireplace and chimney. In such conventions the fireplace, flues, and so forth should be dimensioned fully. The number of flues depends on fireplaces, furnaces, and so forth, which must be accommodated.

Areaways. Fig. 2-19 is an illustration of an areaway. An areaway provides light and ventilation for basement areas. An areaway wall is a masonry or corrugated metal retaining wall which is built around one or more basement windows. The dimensions and specifications should give all necessary information concerning location, width, thickness, depth, and so forth.

Stairs. Fig. 2-20 shows elevation and plan conventions used for stairs. Fig. 2-21 is the basement stairs for Plate 1. These illustrations are what is usually found on a print. The detail drawings should indicate the treads, risers, and other details for the complete stairway. Dimensions on the plan views and detail drawings should give the estimator complete information about the stairs. The written specifications will still take precedence over the working drawings.

Fixtures. A fixture is an attachment

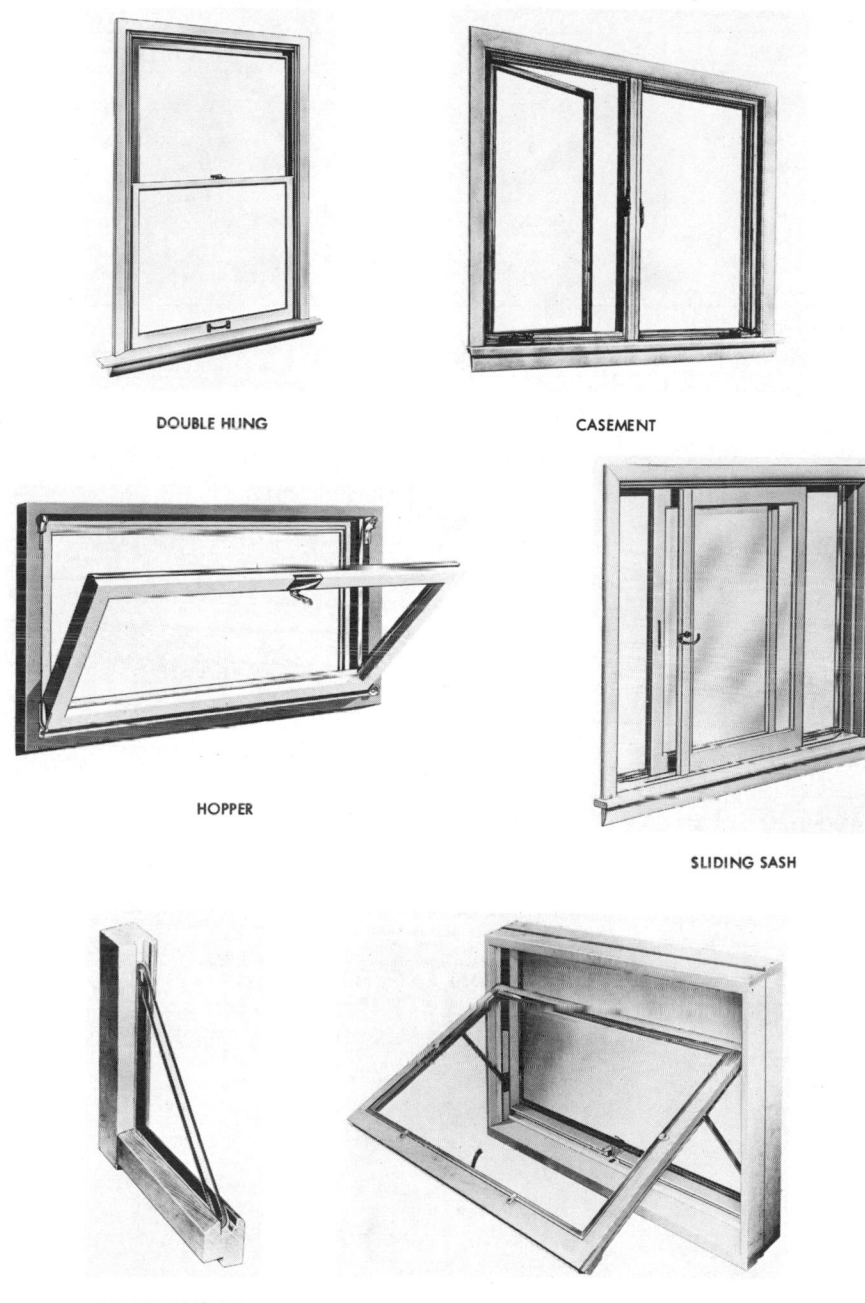

Fig. 2-14. Windows commonly used for residential construction.

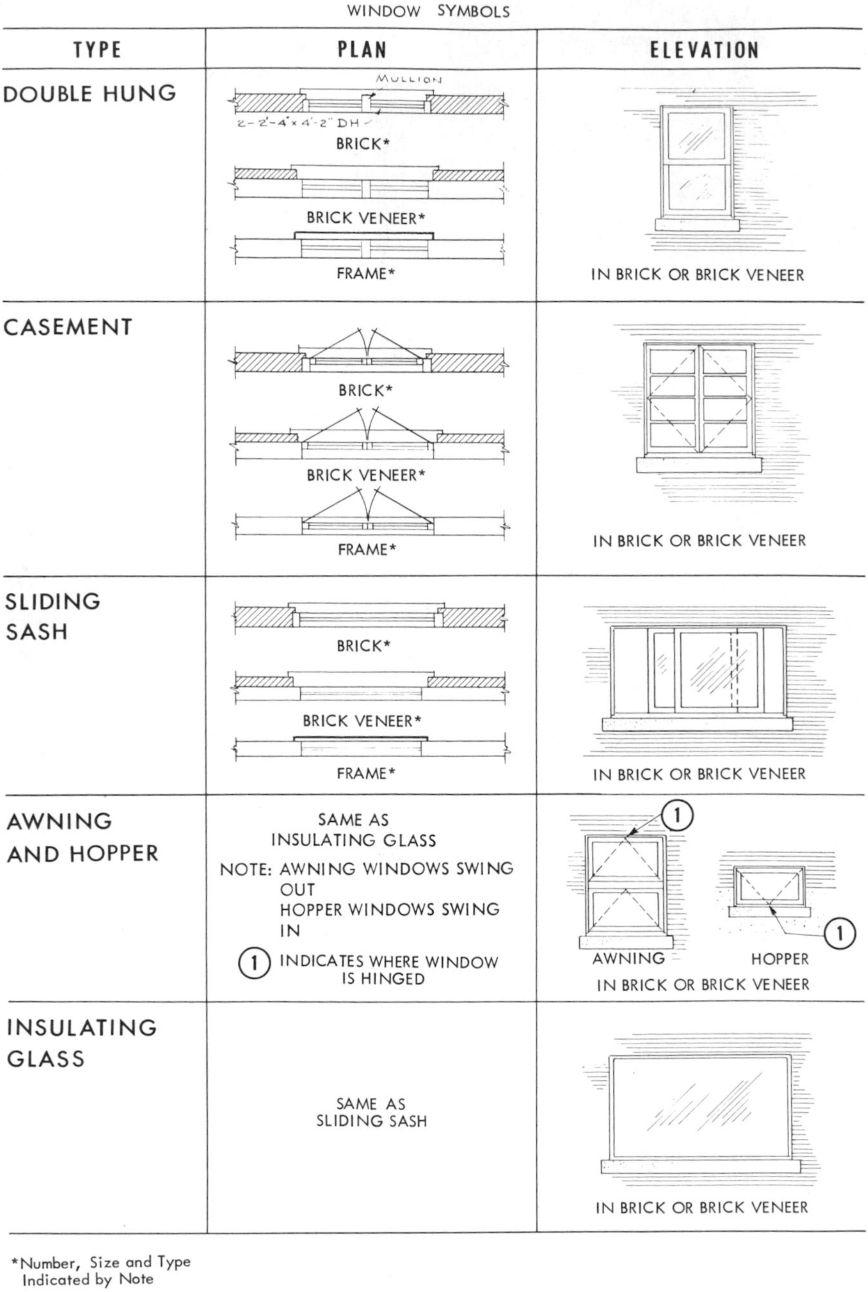

WINDOW   SYMBOLS

TYPE	PLAN	ELEVATION
DOUBLE HUNG	MULLION 2-2'-4"x 4'-2" DH BRICK* BRICK VENEER* FRAME*	IN BRICK OR BRICK VENEER
CASEMENT	BRICK* BRICK VENEER* FRAME*	IN BRICK OR BRICK VENEER
SLIDING SASH	BRICK* BRICK VENEER* FRAME*	IN BRICK OR BRICK VENEER
AWNING AND HOPPER	SAME AS INSULATING GLASS NOTE: AWNING WINDOWS SWING OUT HOPPER WINDOWS SWING IN ① INDICATES WHERE WINDOW IS HINGED	AWNING HOPPER IN BRICK OR BRICK VENEER
INSULATING GLASS	SAME AS SLIDING SASH	IN BRICK OR BRICK VENEER

*Number, Size and Type
 Indicated by Note

Fig. 2-15. Window symbols.

DOOR SYMBOLS

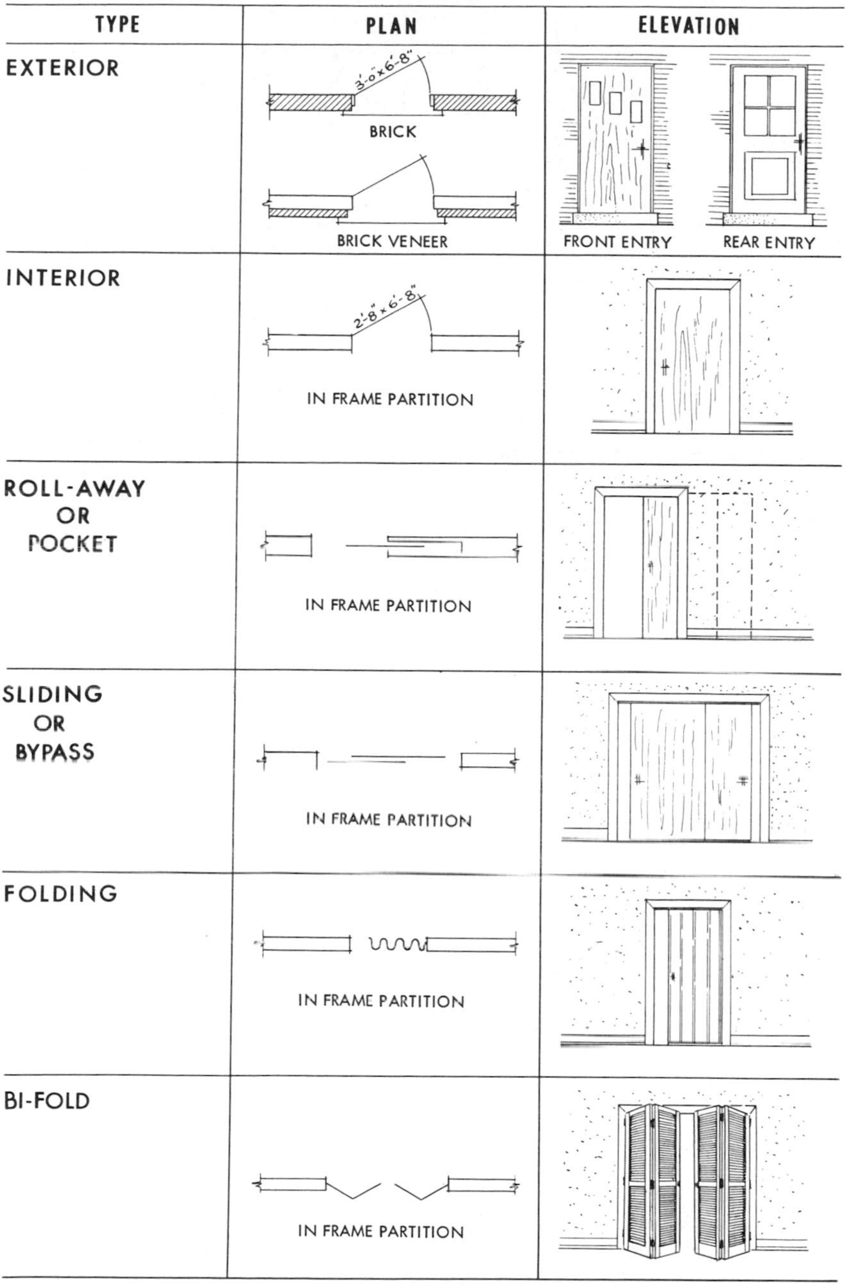

TYPE	PLAN	ELEVATION
EXTERIOR	3'-0" x 6'-8" BRICK BRICK VENEER	FRONT ENTRY REAR ENTRY
INTERIOR	2'-8" x 6'-8" IN FRAME PARTITION	
ROLL-AWAY OR POCKET	IN FRAME PARTITION	
SLIDING OR BYPASS	IN FRAME PARTITION	
FOLDING	IN FRAME PARTITION	
BI-FOLD	IN FRAME PARTITION	

Fig. 2-16. Door symbols.

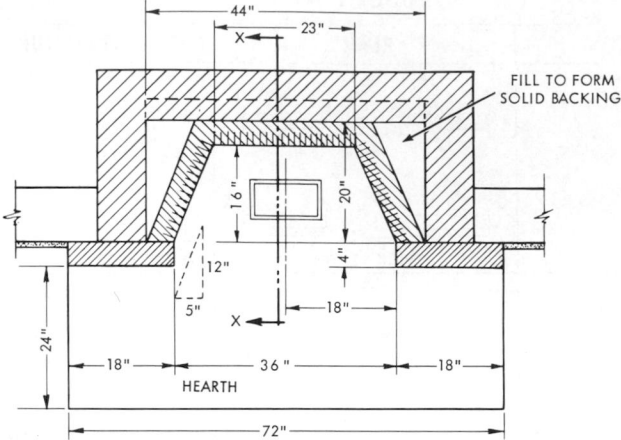

Fig. 2-17. Plan view of fireplace.

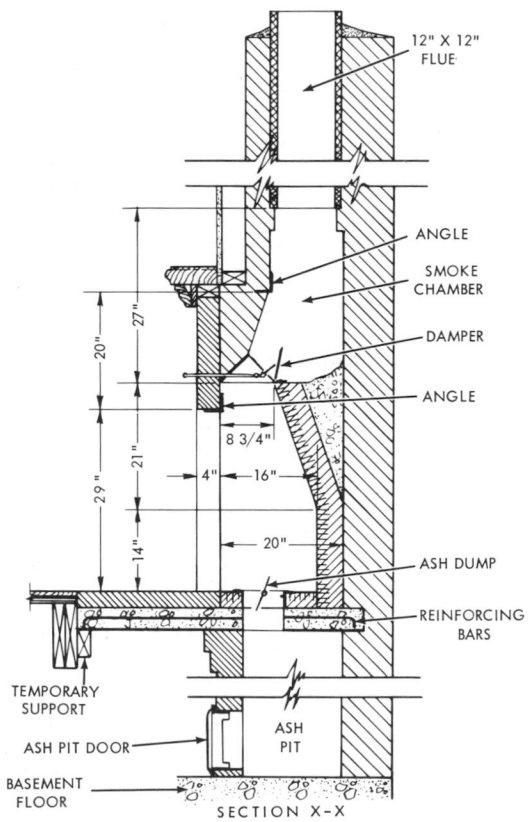

Fig. 2-18. Elevation section view of fireplace and chimney.

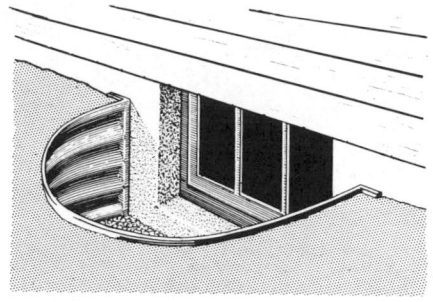

Fig. 2-19. Areaway with corrugated metal retaining wall.

to a building; it can be an electrical device such as a ceiling lamp holder, or it can be any of the various parts of the plumbing system.

Bathroom fixtures include lavatories, toilets (also called water closets), bathtubs, showers, and medical cabinets and tile floors. These conventions must be drawn accurately to scale. The written specifications tell the style, quantity, and color.

Kitchen fixtures such as an electric

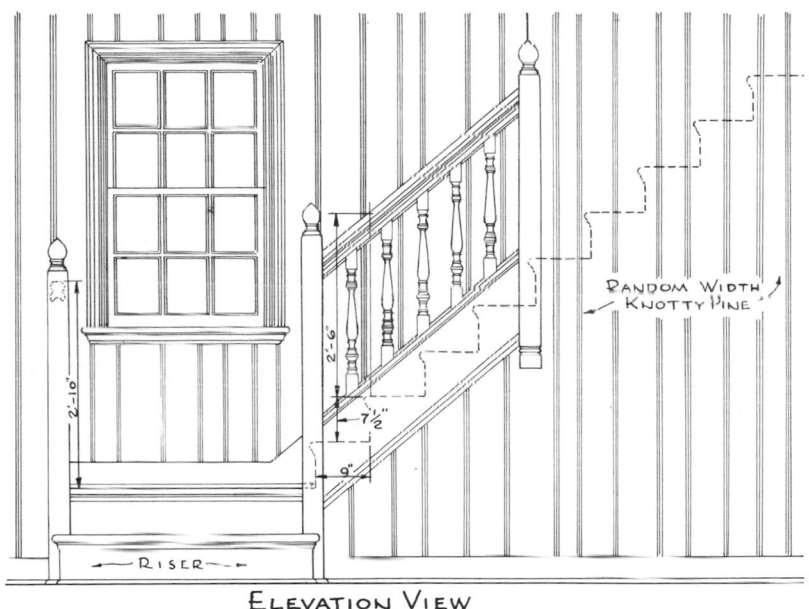

ELEVATION VIEW

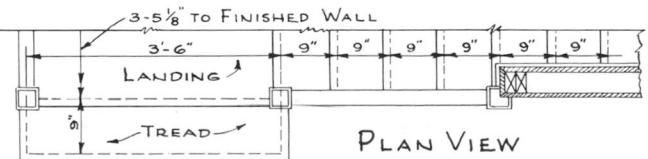

PLAN VIEW

Fig. 2-20. Elevation and plan views of closed stringer stairway.

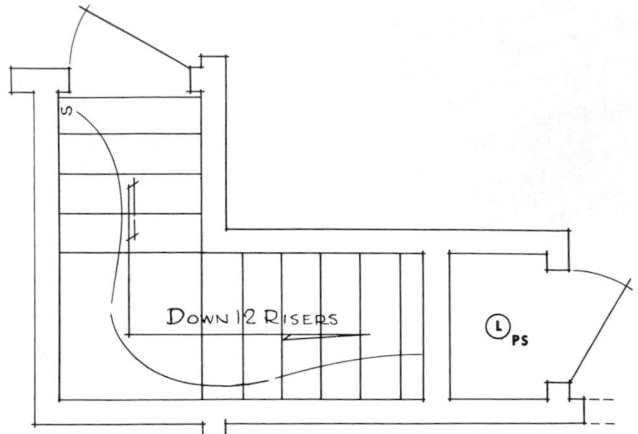

Fig. 2-21. Basement stairs of House Plan A.

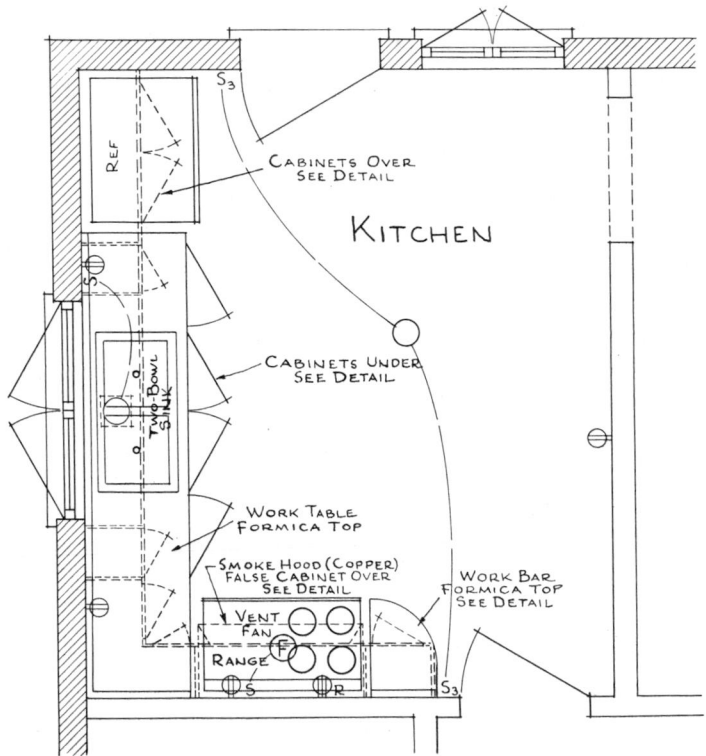

Fig. 2-22. Floor plan of a kitchen, showing locations of fixtures.

range or gas stove, refrigerator, and table and cabinets can all be indicated by rectangles drawn to scale. These are placed in their proper positions and sometimes named on the drawings as in Fig. 2-22. The written specifications should give the style, quantity, color, and so forth.

Laundry fixtures can be indicated by rectangles or circles representing washtubs and washing machines. Ironing boards, washing machines, and dryers can also be represented by rectangles.

Sewers and Drains. Sewer pipes and water pipes are usually represented by small circles or by dot-and-dash lines. Catch basins (cisterns) are indicated by large circles. Downspouts (also called "leaders") are represented by small circles on plan views and by parallel lines on elevations.

Remember that the plans and specifications must be used *together*.

A sketch of the finished building constructed from House Plan A.

IIOUSE PLAN A

The set of plans on the following pages are for a single-unit residence. The plans were originally drawn to 1/48 actual size but were further reduced to fit this book. These sample plans are referred to in the following chapters.

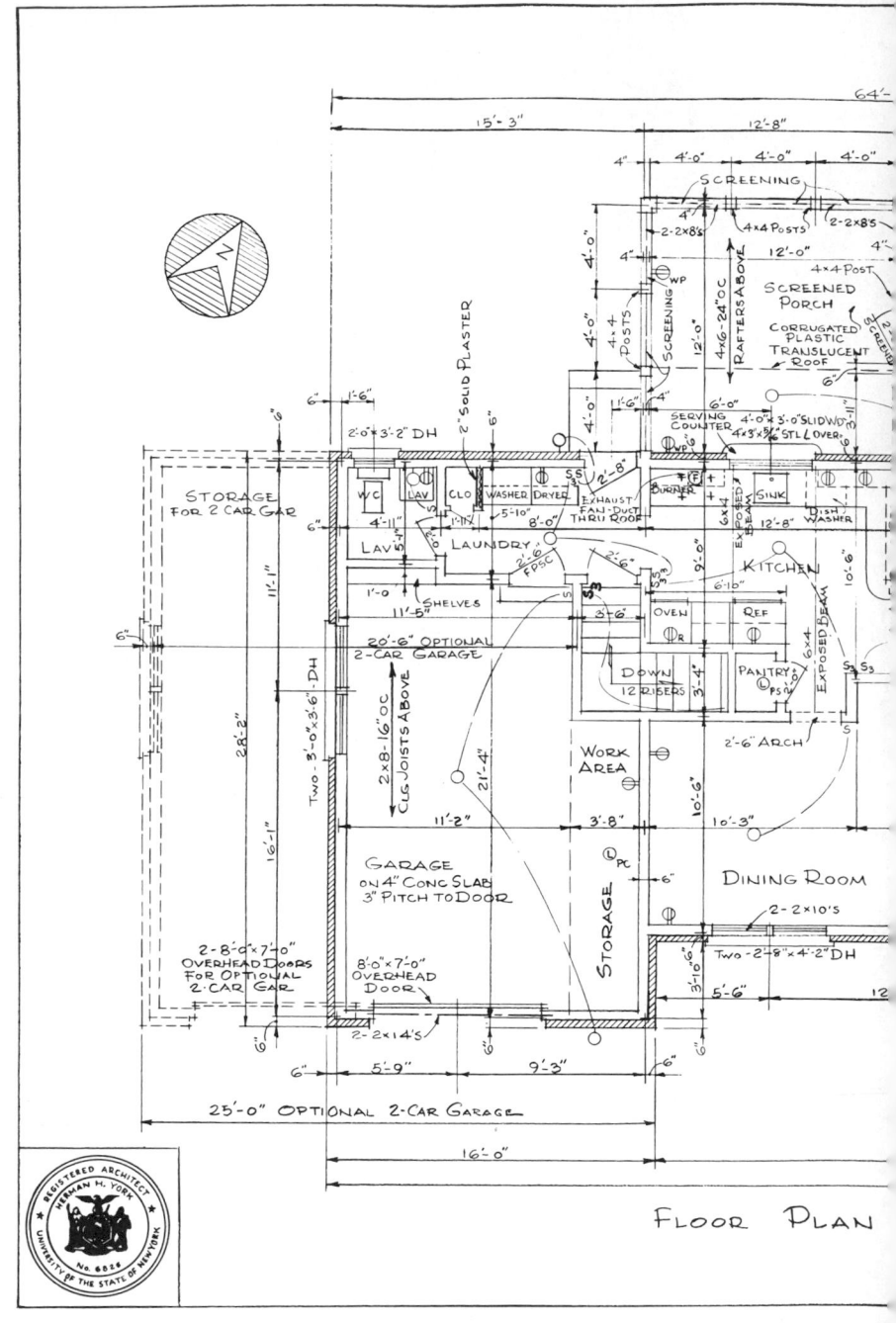

FLOOR PLAN

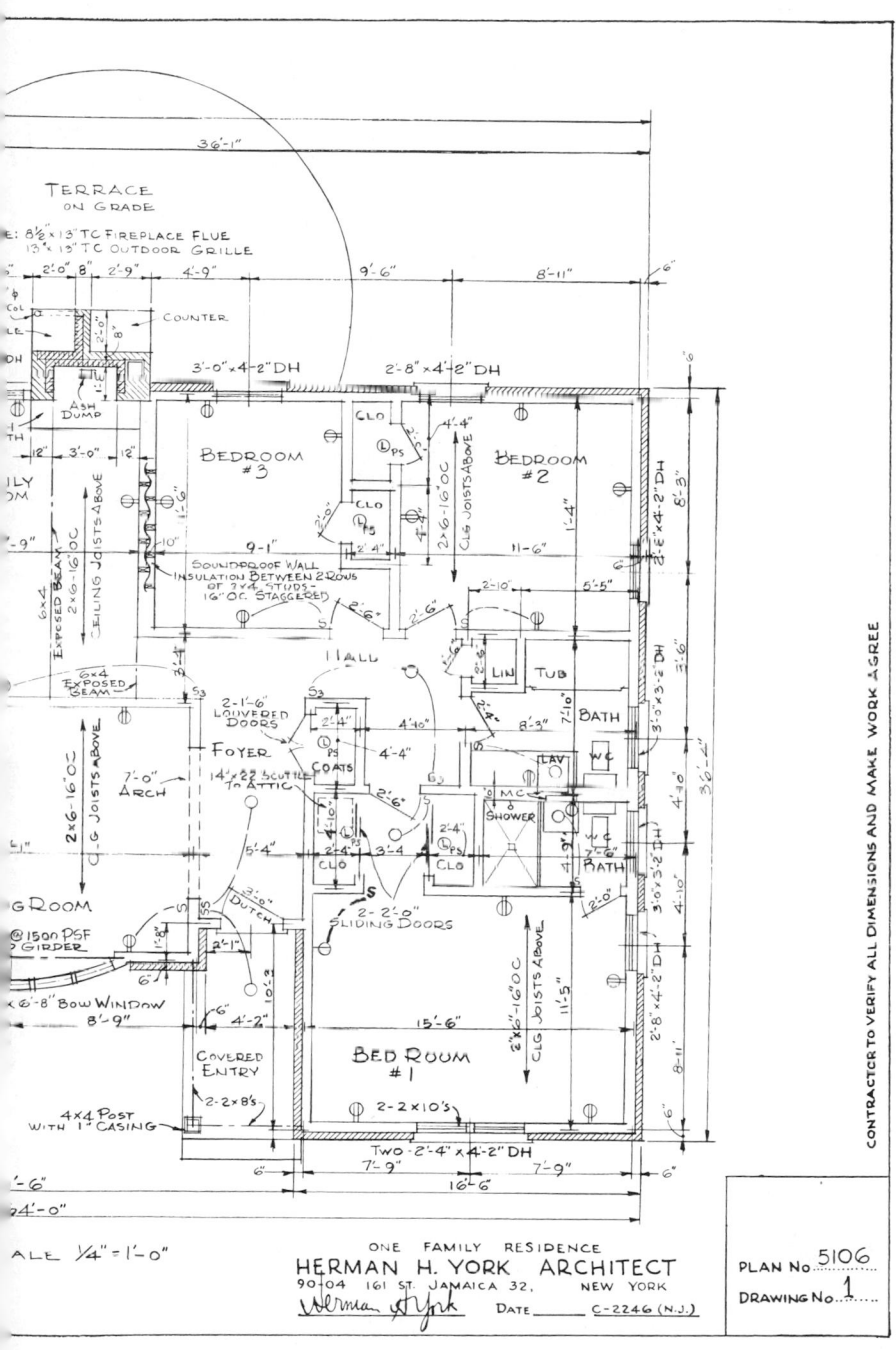

ONE FAMILY RESIDENCE
HERMAN H. YORK ARCHITECT
90-04 161 ST. JAMAICA 32, NEW YORK
DATE _____ C-2246 (N.J.)

PLAN No. 5106
DRAWING No. 1

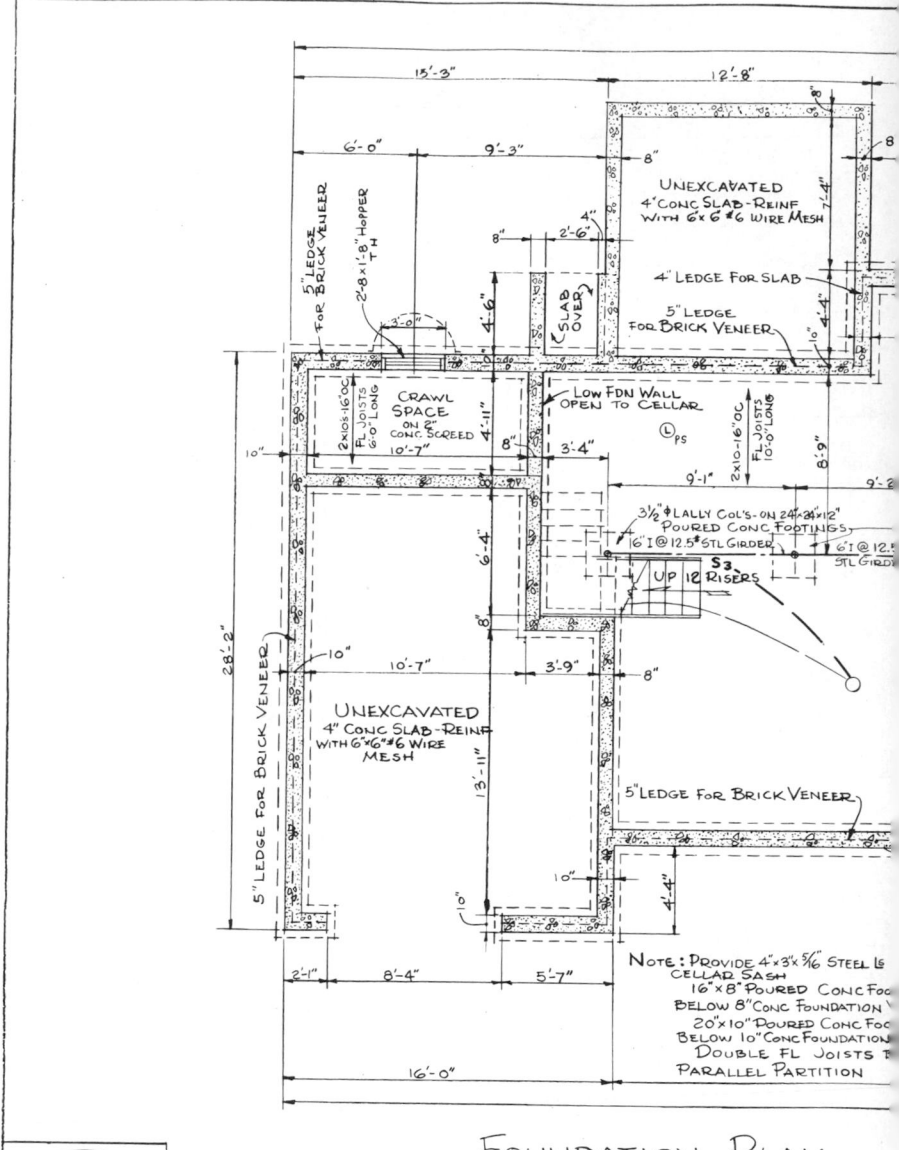

FOUNDATION PLAN

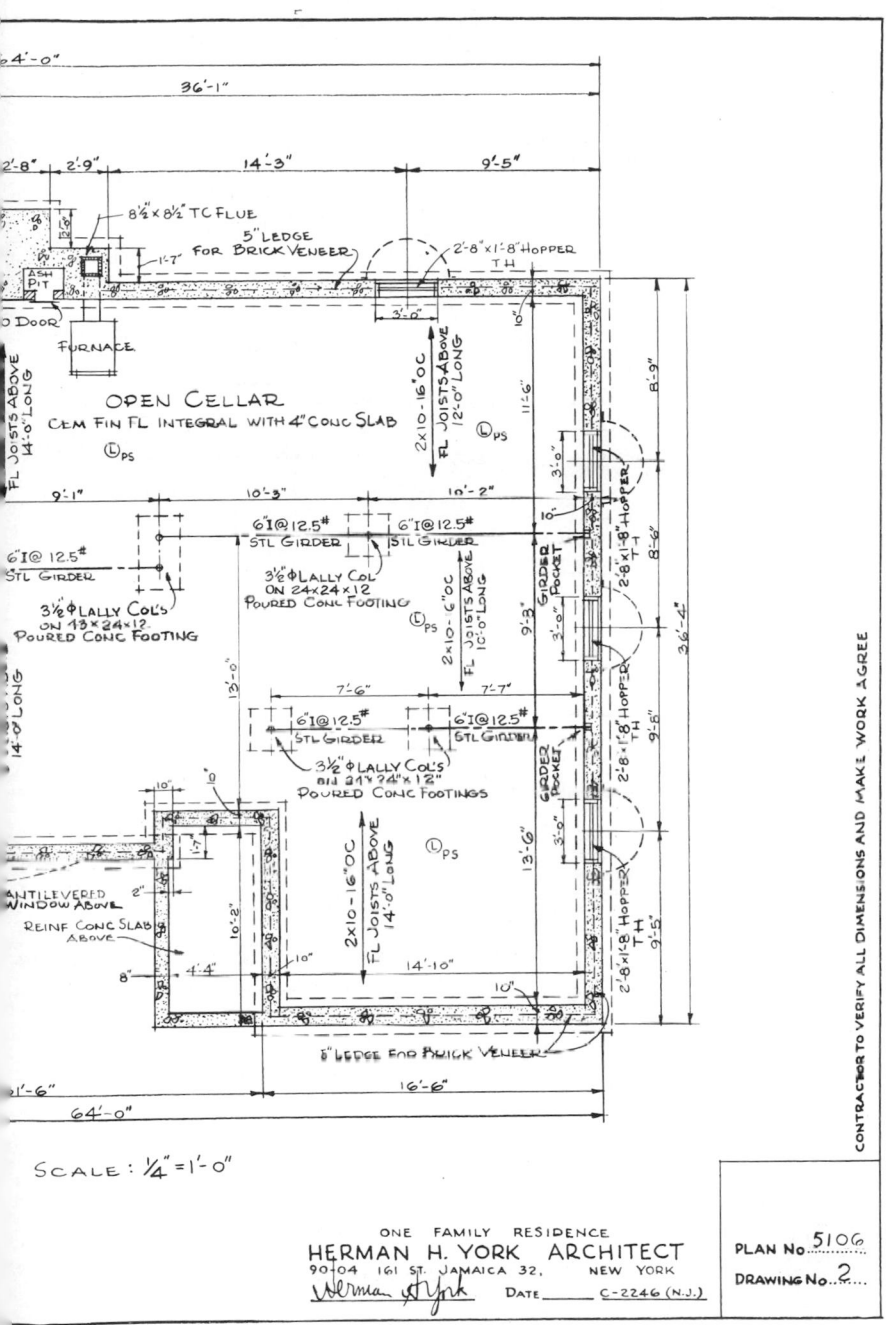

SCALE: ¼"=1'-0"

CONTRACTOR TO VERIFY ALL DIMENSIONS AND MAKE WORK AGREE

ONE FAMILY RESIDENCE
HERMAN H. YORK ARCHITECT
90-04 161 ST. JAMAICA 32, NEW YORK
Herman H. York   DATE _____ C-2246 (N.J.)

PLAN No. 5106
DRAWING No. 2

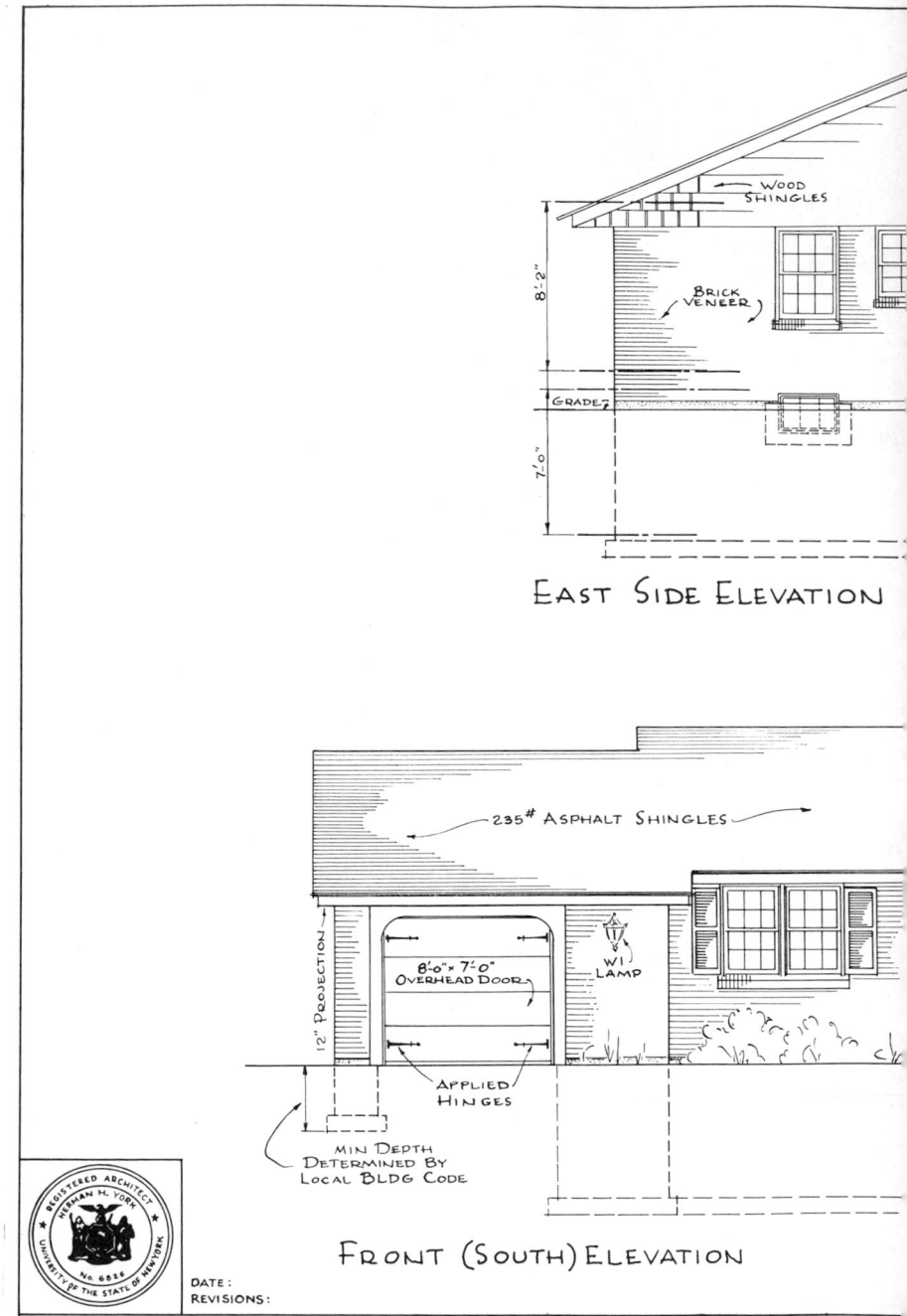

EAST SIDE ELEVATION

FRONT (SOUTH) ELEVATION

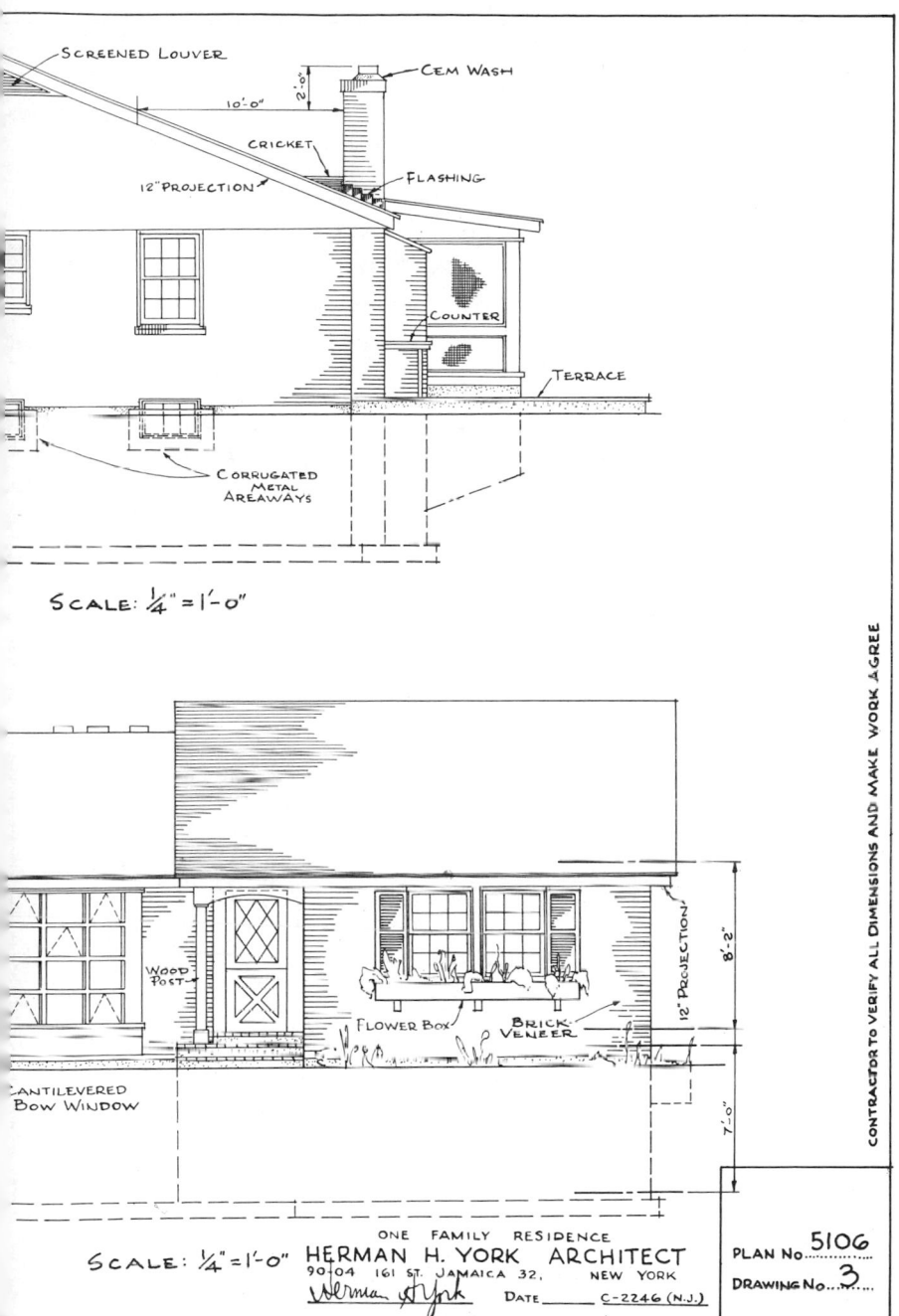

SCREENED LOUVER

CEM WASH

10'-0"   2'-0"

CRICKET

FLASHING

12" PROJECTION

COUNTER

TERRACE

CORRUGATED
METAL
AREAWAYS

SCALE: ¼" = 1'-0"

WOOD
POST

FLOWER BOX

BRICK
VENEER

12" PROJECTION

8'-2"

7'-0"

CANTILEVERED
BOW WINDOW

SCALE: ¼" = 1'-0"

ONE FAMILY RESIDENCE
HERMAN H. YORK ARCHITECT
90-04 161 ST. JAMAICA 32, NEW YORK
Herman H York   DATE_____ C-2246 (N.J.)

PLAN No. 5106
DRAWING No. 3

CONTRACTOR TO VERIFY ALL DIMENSIONS AND MAKE WORK AGREE

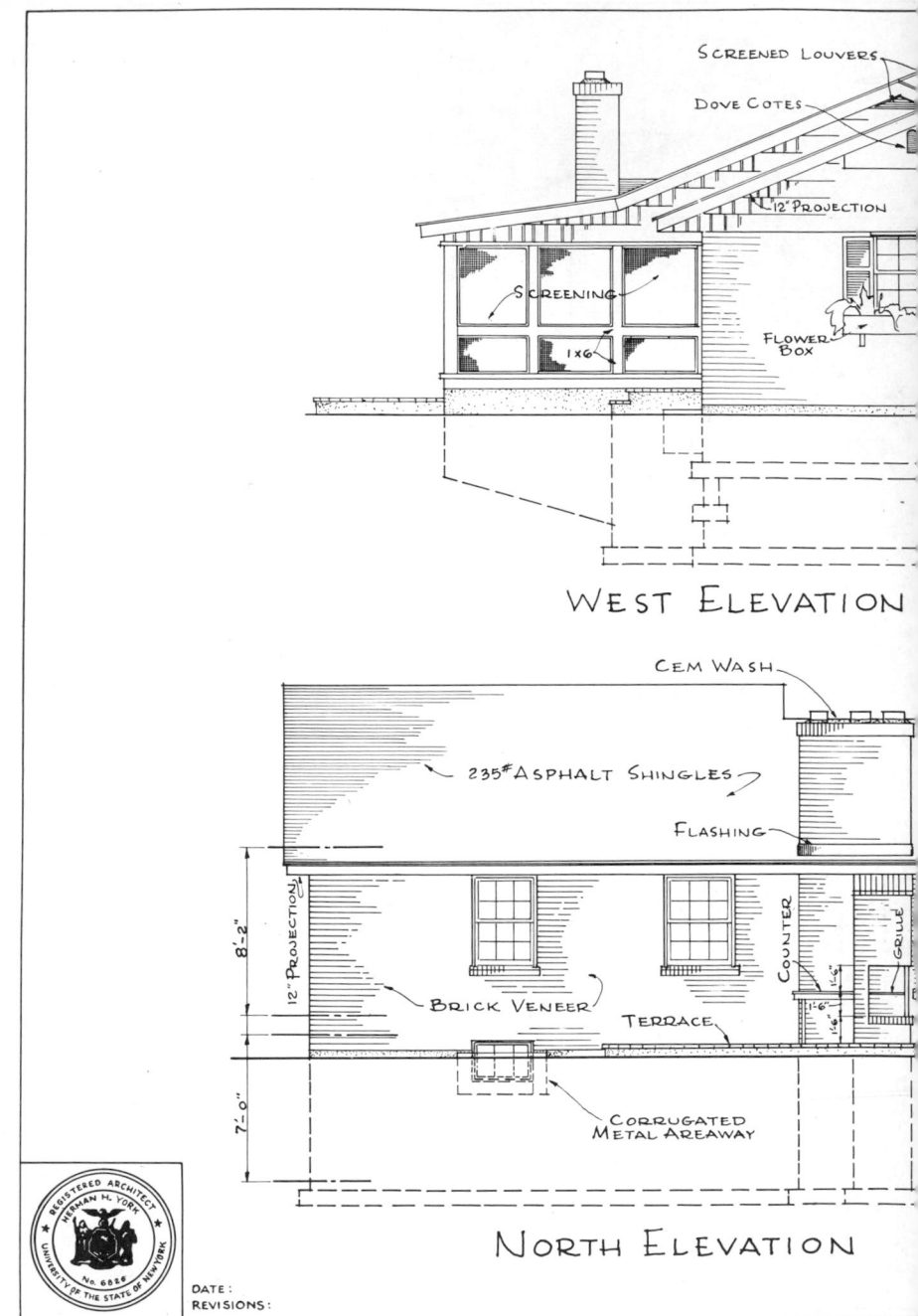

WEST ELEVATION

NORTH ELEVATION

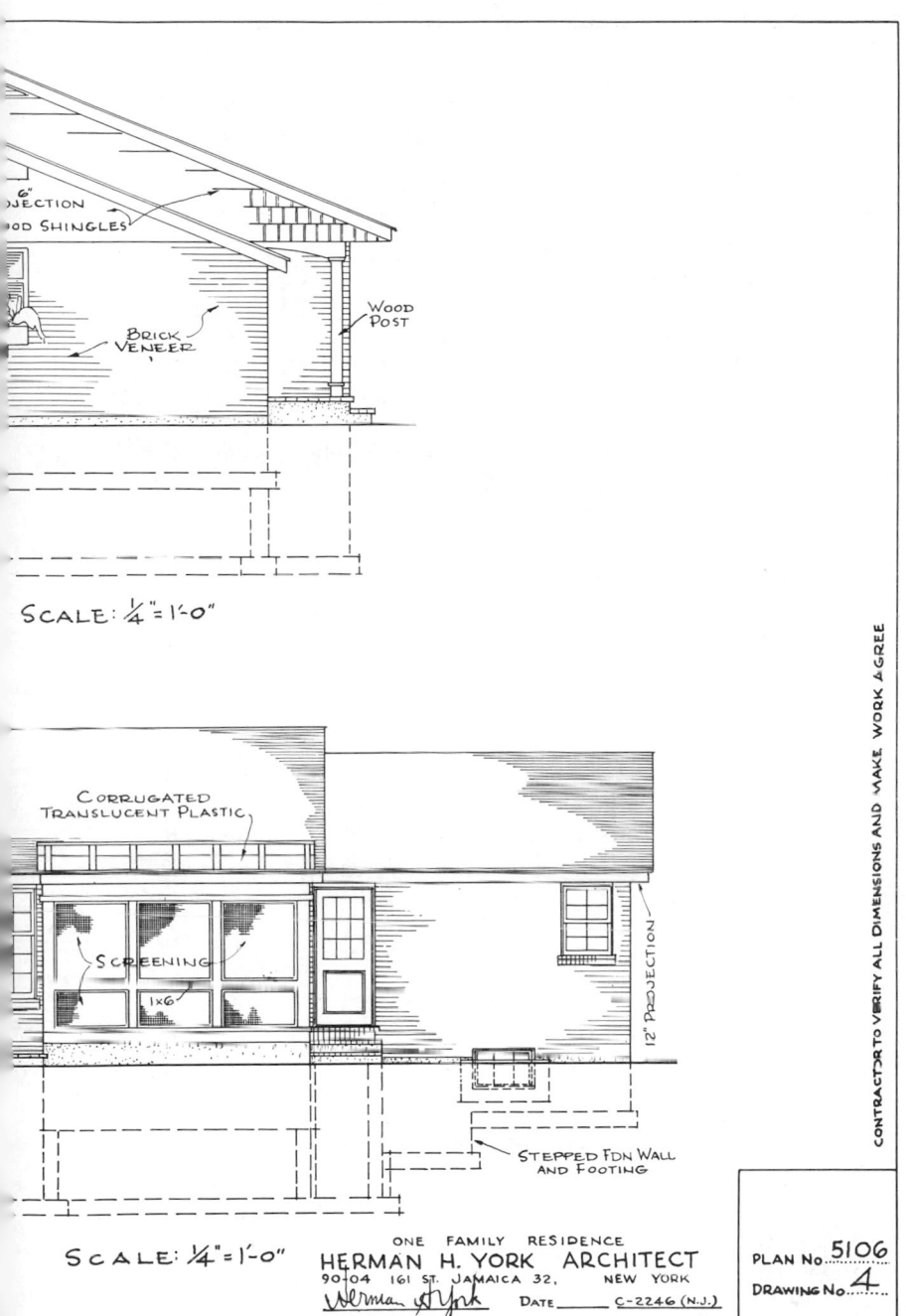

6"
PROJECTION
WOOD SHINGLES

WOOD
POST

BRICK
VENEER

SCALE: ¼"=1'-0"

CORRUGATED
TRANSLUCENT PLASTIC

SCREENING

1×6

12" PROJECTION

STEPPED FDN WALL
AND FOOTING

SCALE: ¼"=1'-0"

ONE FAMILY RESIDENCE
HERMAN H. YORK ARCHITECT
90-04 161 ST. JAMAICA 32, NEW YORK
Herman H York   DATE_____   C-2246 (N.J.)

PLAN No. 5106
DRAWING No. 4

CONTRACTOR TO VERIFY ALL DIMENSIONS AND MAKE WORK AGREE

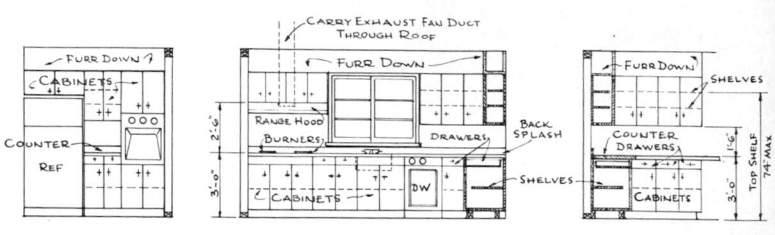

KITCHEN ELEVATIONS   SCALE: ¼"=1'-0"

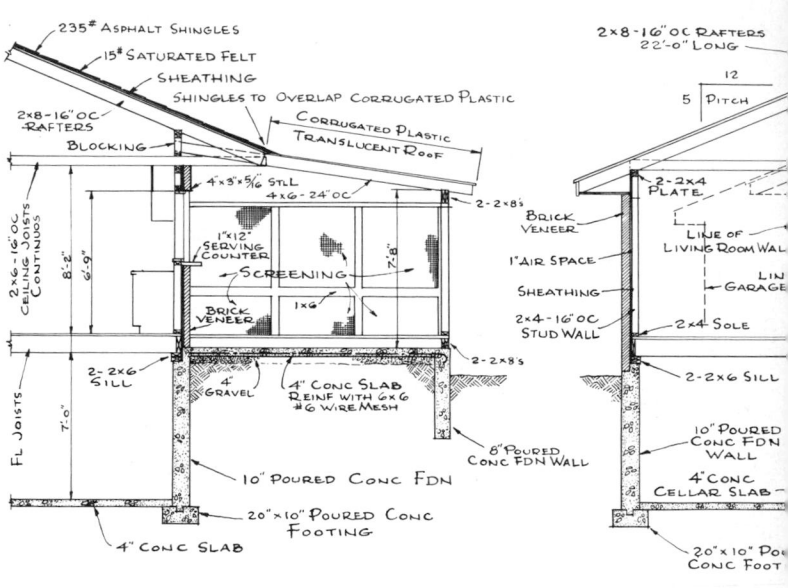

SECTION AT SCREENED PORCH   SCALE: ¼"=1'-0"

DATE:
REVISIONS:

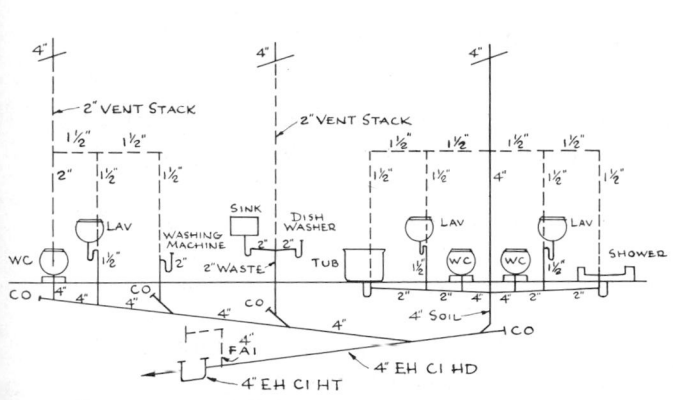

PLUMBING DIAGRAM — NOT TO SCALE

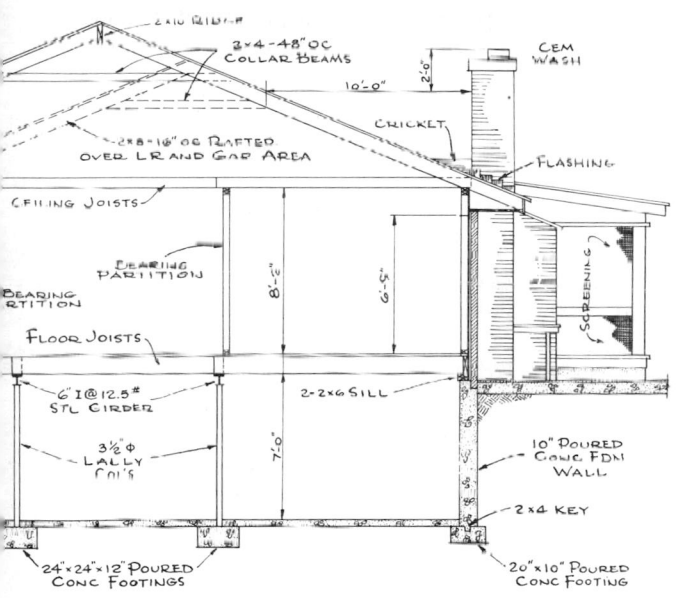

SECTION AT BEDROOMS   SCALE: ¼"=1'-0"

ONE FAMILY RESIDENCE
HERMAN H. YORK ARCHITECT
90-04 161 ST. JAMAICA 32, NEW YORK
DATE_____ C-2246 (N.J.)

PLAN No. 5106
DRAWING No. 5

CONTRACTOR TO VERIFY ALL DIMENSIONS AND MAKE WORK AGREE

Questions and Problems

The following questions refer to plates 1 through 5 of House Plan A.

1. What are the interior dimensions of the foundation for the screened porch?
2. What size is the door leading to the basement?
3. Is any steel specified?
4. Is all of the basement floor at the same level?
5. What are the main dimensions of the foundations?
6. How much excavation is necessary?
7. How are footings indicated?
8. What is the depth of the basement floor?
9. How deep are the foundations under those areas not excavated?
10. Of what material are the areaway walls to be constructed?
11. What size and kind of basement windows are indicated?
12. How many fireplaces are to be constructed?
13. How many steps are necessary between the basement and the first floor?
14. How wide are the basement stairs?
15. Are lintels necessary over the basement windows?
16. Of what material are the garage walls to be constructed?
17. What are some of the fireplace details?
18. How many electrical baseboard outlets are there in the Family Room?
19. How many ceiling lights are there in the basement?
20. What are the details of the front entrance?
21. What are the details of the screened porch?
22. What size and type of windows are used on the first floor?
23. What type and size of opening leads from the foyer to the living room?
24. How many flues are visible in the chimney at the basement level?
25. What fixtures are located in the bathrooms?
26. What size doors are used for the bedrooms?
27. What size closet doors are used?
28. What type of louvers are shown?
29. What are the details of a typical wall section?
30. What type of outside wall construction is used?
31. What type of door is used for the garage?
32. How is the top of the chimney finished?
33. What is the roof pitch?
34. What type of shingles are to be used?
35. Are shutters indicated?

3

SPECIFICATIONS

~~~~~~

A set of plans consists of specifications *and* working drawings. *Written specifications supply the contractor with a word description of units of work, types and methods of installation, and all other items that could not, or do not, appear on the drawings. The specifications also explain units that appear which are not fully described by the working drawings.*

SPECIFICATIONS

To avoid misunderstanding and confusion, all the various methods, requirements and materials for the construction of a building are listed on one or more sheets of the specifications. These "specs" are part of the set of plans, and a set of working drawings without specs is not complete.

The specs always accompany the working drawings. At no time should they be separated from the drawings, for constant reference is made from the specs to the drawings and from the drawings to the specs.

The statement, "Work shall be done in a workmanlike manner and in accordance with the working drawings," is usually found in the builder's contract. This means that the contractor must follow both the specs and the drawings in their entirety, as given to

him by the architect. He must not make changes for any reason. Lack of materials, lack of "know-how," misinterpretation of drawings or specs does not excuse him from the requirements of this contract. Before he can deviate from the plans, he must obtain a change order or revision to the drawing or specs. *One unauthorized change could upset the design or the strength of a unit of construction.*

The architect makes every effort to cover all items concerned with the construction, but sometimes he unintentionally leaves out some phase or item that on the surface does not appear to be important. Nevertheless, the contractor, once he starts his job, is held responsible for providing the omitted phase or item under the statement, "Work shall be done in a workmanlike manner. . . ."

When preparing his material for bid

purposes the estimator should examine both the working drawings and the specs very carefully. Not only the units of construction but also the *General Conditions of the Specifications* (often shortened to "General Conditions") should be examined carefully; some of the very important dollar items are outlined in this section.

This chapter will examine some phases of the specs and interpret them in layman's language.

SPECIFICATION ANALYSIS

The detailed design of a building is the architect's job. There are meetings between the architect and the owner or developer to determine the needs of the owner or tenant. The architect, after deliberation and conferences, prepares preliminary sketches of the building and submits them to the owner for approval. The owner suggests such changes as he thinks are required and returns the sketches to the architect for the final work on the plans.

In order for a building to be erected from a set of plans, the needs of the structure, the materials that go into the building of the structure, and the methods by which these materials are placed and erected must be clearly specified. This is so that all parties concerned with the building will understand the architect's concept of the structure.

The architect or developer submits both the specifications and working drawings to several contractors for bid purposes. The specifications are considered the law for the builder. If there is a difference between the working drawings and the specifications, the specifications will take precedence over the drawings and govern the materials and methods which will be used for the job.

The specifications are divided into sections which cover the requirements of each trade involved in building the structure. It is understandable that some building specifications will have more sections than others. There is, however, a basic set of specifications which is usually required to carry the average building to completion. This chapter will discuss the most pertinent parts of the specifications to give the reader a good idea of the need for careful analysis of every section, every paragraph, and every line. The misreading of one word, one comma, or one paragraph may distort the entire meaning and requirements of a particular section of the specifications and can lead to a faulty bid or a dangerous deviation from the plans.

General Conditions. The General Conditions section includes *overhead* expenses which do not come under any specific trade. The General Conditions section is included in the specifications because the construction of a building involves many people and many trades. Construction also involves city and county agencies, banks and insurance companies. It is essential that the wishes of the parties involved be stated in writing so that every contractor who is interested in bidding for the job is aware of the requirements. It gives each contractor a fair and equal chance to arrive at the lowest possible bid.

GENERAL CONDITIONS SUMMARY SHEET

JOB _____ DATE _____

Superintendence	$ .00
Engineer & Rod Men	.00
Clerks & Job Stationery	.00
Waterboys	.00
Permits	.00
Security Requirements	.00
Job Site Inspection	.00
Mason Superintendents	.00
Sheds & Office	.00
Toilet Facilities	.00
Heating	.00
Lighting	.00
Telephone	.00
Traveling Expenses	.00
Enclosures	.00
Ladders & Chutes	.00
Cutting & Patching	.00
Signs	.00
Photographs	.00
Surveys	.00
Protection to Adj. Property	.00
Protection to Adj. Utilities	.00
Protection to Existing New Work	.00
Samples	.00
Shop Inspections	.00
Cold Weather Protection	.00
Barricades	.00
Small Tools	.00
Material Hoists	.00
Rental of Equipment	.00
Gasoline & Oil	.00
Trucking	.00
Freight	.00
Scaffolding	.00
Storage Facilities	.00
Fire Extinguishers	.00
Periodic Cleaning & Final	.00
Special Hazard Insurance	.00
Special Overtime Work	.00
Glass Breakage	.00
Bonds	.00
Progress Charts	.00
Glass Cleaning	.00
Special Allowances	.00
Conc. Cylinder Tests	.00
Aluminum Protection	.00
Taxes	.00
Pumping & Baling	.00
Temporary Water	.00
	.00
TOTAL	$ .00

Fig. 3-1. Sample General Conditions summary sheet. (Chicago Estimating Service)

If the needs and wishes of the various organizations are not clearly stated in the General Conditions, some contractors might bid too low by avoiding requirements that are not clearly stated. This could cause the contract to be awarded to a contractor who would not receive enough money to carry out the architect's intentions; he may not break even or make reasonable profit.

Fig. 3-1 is a typical "General Conditions Summary Sheet" used by an estimator. The most important items in this section of the specifications are discussed in the following sections.

Insurance, Contributions and Bonds. The contractor should not begin his work until he has obtained and paid in full for all insurance required under the contract and until the insurance policies have been approved by the architect, the owner, or both. The policies should be issued by a financially responsible firm and the amount of coverage should be sufficient and adequate to cover any occurrences during the construction of the building. Receipts for complete payment of premiums must be shown to the architect, owner, or both, to indicate that the policy is in force *for the particular work* of the contractor submitting the policy.

There are many types, forms and limits of insurance coverage. It is necessary that the estimator, when preparing a bid, be aware of the fact that the insurance coverage of the various policies will affect the final cost of construction. The estimator must keep in mind that the costs of these policies vary with the size, location and extent of the job. This being the case, the unit which he would apply to the estimate would necessarily have to vary in form and in percentage. The following types of insurance should be considered by the contractor prior to preparing his bid.

Workmen's Compensation. Each of the United States and territories and the Canadian provinces require this type of insurance protection. Federal laws cover federal employees, private employees in the District of Columbia, and longshoremen and harbor workers. Workmen's Compensation provides employees with certain benefits in the event of injury or death and (in some states) certain diseases.

A number of states have severe penalties for failure to carry this type of insurance; fines, civil damage awards, and even imprisonment of the employer are possible.

Contractors' Public Liability Insurance. The contractor or subcontractor must also carry public liability insurance covering bodily injury or death suffered as a result of any accident occurring from or by reason of, or in the course of operation of, the construction project. This includes accidents occurring by reason of omission or act of the contractor or any of his subcontractors, or by people employed by the contractor or subcontractor. The extent of the coverage is usually determined by the size of the job and/or by the architect or the owner.

Fire and Extended Coverage. The contractor must have and maintain insurance against loss by fire, windstorm, tornado, and so forth, to cover any damages done while the work is in progress. This insurance covers a wide range of items and equipment. Special clauses are usually written into the policy by an experienced insurance agent who has a thorough background in construction coverage liability.

Fire rates vary among areas according to the rating organization having jurisdiction.

As an example of fire rates, the General Rating Rule for Fire Resistive Buildings in the course of construction as stated by the City Division of the New York Insurance Rating Organization is as follows:

1. Base Rate: .17 ($0.17 per hundred dollars)
2. Height: Add .005 for each story of planned height in excess of 25 stories; maximum height charge to be .15.
3. Protection: Above rate may be reduced 60 percent if the following four items are complied with:
   (a) Approved installation of dry line standpipe equipment with one or more properly located siamese connections [also called "Y" connectors] with due regard to accessibility and complete compliance with regulations of the Fire Department.
   (b) Elevator ready for use with competent operator at all times, in buildings exceeding 150 feet in height.
   (c) Approved watchman on premises and so warranted.

(d) Acceptable and adequate first aid supplies provided.

For illustrative purposes let us assume a ten story, fire resistive building with a 100 percent completed value of $1,000,000.

Co-insurance is the dividing of the contractor's insurance among different companies, resulting in a lower insurance rate.

Completed Value Builders Risk rate would be 55 percent of the 100 percent co-insurance Builders Risk rate:

1. Base Rate: .170
2. Height: no charge
3. Protection: assuming full compliance with the four items listed; less 60 percent   −.102
                                              .068

Less 10 percent for 100 percent co-insurance   −.006
100 percent co-insurance Builders Risk Fire rate   .062
Completed Value Builders Risk rate (55 percent)   .034
Annual Premium =
$1,000,000 $\times$ .034/100 = $340

Failure to comply with the four items listed under "Protection" would change the rate as follows:

1. Base Rate: .170
2. Height: no charge
3. Protection: no credit
   Less 10 percent for 100 percent co-insurance   −.017
100 percent co-insurance Builders Risk Fire rate   .153
Completed Value Builders Risk rate (55 percent)   .084

Annual Premium =
$1,000,000 $\times$ .084/100 = $840

The actual premium savings for compliance with the four items listed under Protection is $500, hardly enough to pay for these items, but the added protection they afford against accident is worth much more.

Social Security Payments. The contractor must pay the taxes and assessments for Social Security Insurance and Unemployment and Old Age Benefits. The amount of money to be paid is determined by the wages that the employees of the general contractor receive and also the employees of his subcontractors. The general contractor must usually accept responsibility for these assessments; he usually will release the owner from any responsibility for assessments which might be defaulted by himself or his subcontractors. Some other contributions which must be paid and yet are not required by law would be welfare dues to local unions as well as other local and miscellaneous fees.

To summarize, the general contractor must include in his bid some percentage to cover these and other taxes and assessments.

Performance Bond. When estimating a construction job, the owner and the architect must be assured that the lowest, competent bidder will accept the contract as bid. In order to insure this, the architect or the owner, or both, may require that the subcontractor or contractor post a performance bond. This performance bond can be for the entire cost of the job or a percentage

of the job, depending upon the requirements of the architect.

Completion Bond. Many contingencies such as weather, labor, supply of material and so forth make it necessary for the owner to protect himself to the extent that the contractor, once he has started, must guarantee completion of the job. This is usually taken care of by the posting of a completion bond. A completion bond is usually for the cost of the total estimate of the particular job.

Insurance Summary. As an estimator, you are concerned only with the amount of money that it costs your company to secure insurance policies, make contributions and to furnish the various bonds which are required by contract or law. Whether the percentage of insurance is 5 percent, 8 percent, or 3 percent of the total cost of the job, it is necessary that you consider and enter the insurance costs on your estimate.

Job Meetings. It is necessary for efficient operation that the contractor, the architect and the owner meet periodically and discuss items such as payment progress, changes that might be necessary, discrepancies in plans and specifications and other items. It is standard procedure for most large construction companies to hold job meetings to discuss the problems. A stenographer usually takes complete minutes of the meetings, duplicates them, and makes them available to all parties directly concerned with the construction. Meetings are another cost of construction which does not show up anywhere in the building or on the plans, but must be taken into consideration by the estimator since they are a part of the General Conditions of many contracts.

Permits and Inspection Fees. This section requires a contractor or subcontractor to pay fees for a license to practice his trade. He must obtain all necessary permits required in the locality where the construction takes place.

Line and Grade. The section of the specifications known as "Line and Grade" requires engineering services under which the contractors, at their own expense, establish lines and grades required for the installation of their work.

The contractor must lay out his own work and be responsible for all lines, elevations and measurements for the work. He must verify the figures shown on the drawings before laying out the work, and he is responsible for any error resulting from his failure to exercise this precaution. He also must verify lines and grades furnished him by the subcontractor for general construction and is held responsible for any errors resulting from failure to exercise such precaution.

Rubbish Removal. The following statement is quoted from an actual specification: "Each contractor shall clean and gather daily all debris from the construction site and place the debris in containers or appropriate cans which are provided by the contractor at the entrance of the building unit or units, and have same removed by a bonafide rubbish removal contractor."

This item may seem insignificant, but it isn't if you are estimating a one hundred unit apartment building. When you consider the refrigerator cartons, the packings around the kitchen cabinets, and the cardboard containers and the plastic coverings for the bathtub and showers, this rubbish situation becomes a large problem. Although this cost compared to the total construction cost does not seem large, we wish to emphasize that no contractor should willfully overlook or give away money.

Rubbish removal in most areas is governed by city and state codes. A careful study of codes, plans and building site can prevent rubbish removal from becoming a very costly problem.

As an example, it may be necessary to construct a rubbish chute. The cost of erecting a suitable rubbish chute for a multiple unit dwelling increases in relation to the height of the building. It can be related to the cost of building the forms for a very long, large concrete column. A rubbish chute is as costly to construct and in many cases more costly because it must be anchored securely to the building at intervals and at floors and must be self-supporting. The average column in a building only runs from floor to floor or from floor to mezzanine or balcony. The breakout or rupturing factor for the column is much less than for a rubbish chute. If a rubbish chute collapses it could mean the loss of life to many construction men on the site. Since it must be made safe, anchored securely, and built firmly, it will be expensive. The removal of dirt and rubbish from a building site and structures must be entered into the General Conditions of the contract to make sure that your costs are complete.

Window Cleaning. This item, sometimes neglected by the specifications, is a very important and costly item to the contractor. If the estimator omits this unit it will cost the contractor money he did not plan to spend. Windows are marked with chalk, whitewash, or other methods so that all trades will know there is a pane of glass in the sash and thereby avoid breakage. When the building is complete, and before the turnover to the owner, it must be presentable. The contractor must clean each and every window in the building. An estimator should include this cost even though it may be omitted from the specifications.

Winter Protection. The contractor must protect his work from the elements. This includes snow, sleet, rain, windstorms, cyclones, tornadoes, hail and flood. The elements must be taken into consideration for the entire time the building will be under construction. The most costly item is winter protection. Winter protection includes the rigging of canvas or plastic around the structure and heating the inside by some temporary means, such as salamanders (cans with open fires). The heating and the canvas or plastic are necessary if the various trades are to work during inclement weather. Fig. 3-2 illustrates this practice. Heating is also necessary for the proper curing of concrete. The duration of the job must be considered for winter protection. If the job is to be protected for one or more winters, the cost on the

General Conditions must be entered and taken into serious consideration.

Watchmen. On most large jobs it is necessary for the contractor to provide adequate watchmen for general patrolling of the site, both day and night, including Saturdays, Sundays and holidays. The watchman is necessary to prevent the theft of materials. He is also necessary to prevent trespassing on the construction site which can lead to bodily injury, for which the general contractor could be held responsible. A watchman is another costly item which must be entered into the General Condition costs to give you a complete picture of the job.

Inspection and Testing. This item is an important section of the specifications. Under it the contractor must furnish facilities and assistance for inspection, examination and tests that the building inspectors may require. He must also secure for the building inspectors free access to factories and plants in which any materials are being manufactured or prepared, and to all parts of the construction. The contractors may be required to give the inspectors advance notice of preparation, manufacture, or shipment of any materials. This section becomes very important when concrete is used; innumerable slump tests and samples may be required.

Field Office. This item is no longer the well-constructed shanty that took several carpenters possibly a week to construct. Today, the field office is usually a beautiful mobile home which contains not only office furniture, but also has toilet facilities, water, elec-

Fig. 3-2. A protective enclosure used to prevent rapid evaporation and to provide proper curing of exterior portland cement. (Portland Cement Association)

tricity, calculators, typewriters and all the necessities of a modern office. This cost may seem trivial, for this mobile home can be used from job to job, but what one fails to take into consideration is the two, three or four months wait between jobs. The cost must then be prorated (divided proportionately) and can become large. Therefore, in figuring your estimate, always take care of this item.

Project Sign. This item consists of a sign naming the general contractor, the architect and the project. Not only must the sign be erected, the words lettered with paint and protected by a surface that will withstand the elements, but it must also be kept in a good state of repair throughout the life of the project. We recommend that this item be taken into consideration when estimating.

Site Fence. A fence is a costly thing to erect. Not only must you use a surveyor to get the proper line, but you must bring a fence to security. Security means vertical, horizontal and sometimes diagonal support so that a storm or childish horseplay cannot do any damage. As an example, we quote from a specification which states:

"The entire area must be enclosed by a fence 7'-0" high above the respective grades as indicated on drawings which are attached to this specification. Should there be any trees, the contractor shall also include the cost of maintaining and providing fences for such trees. These are covered by additional drawings. If there is any existing

fence at the site, such fencing may be used to an extent. If the fence is built along the curb, free access from the street to hydrants, fire and police alarm box and light standards must be maintained."

The above paragraph should give the reader an idea of what he may run into when constructing a fence around a proposed construction site. The structure might be a block square and the fence, therefore, four blocks long. Provisions must also be made for entry of trucks, workmen, and so forth. This fence must be constructed and possibly painted to withstand the elements.

Temporary Light and Power. The contractor must provide, on the construction site, light and power for the various trades and subcontractors under the jurisdiction of his contract. The light and power must meet with approval of the county, township or city in which the building is to be erected. The installation of proper size meters and proper height electrical poles can become a costly item. This cost is generally determined by the length of live wire which must be brought in from existing lines to the central point of the construction site. Temporary light and power is just another one of the many items which the estimator must price in relation to the size and duration of his project.

Temporary Plumbing. The contractor must provide temporary heating and toilet facilities.

Temporary heat is a very costly item. It includes such things as heating and ventilating, radiators, piping and other units required to heat the structure while under construction. This is the contractor's responsibility and therefore becomes a building cost. The large dollar item consists of labor and fuel. The estimator should give strict attention to the specifications of this unit.

Temporary toilet facilities range from the simple "outhouse" to elaborate, multiple unit water closet and lavatory installations, depending on the size, location, and type of construction.

Local and state laws, as well as union regulations, must be considered in estimating this unit of cost. In "highrise" construction it may be necessary to provide at least one water closet for every 30 persons, located no more than four stories above or below the place of work. These toilet facilities must be sheltered from view, from weather, and from falling objects.

The estimator must check laws and union regulations carefully; even if this item is omitted from the specifications the contractor might still be required to pay the cost of temporary heat and toilet facilities.

Progress Schedules and Records. The General Conditions covered so far will normally be found in the specifications for large construction projects. This is by no means a complete list. For example, some jobs may call for progress photographs. This item entails the hiring of a competent commercial photographer who must periodically submit photographs to the owner and the architect to illustrate the good construction practices being used.

Progress Schedules. These are called for on most large jobs. The making of such a schedule requires the services of a competent man who knows the times required for each phase of the work and each trade.

Progress Records. This is simply a record of the work which has been completed. It is usually based on daily work reports.

Temporary plumbing, discussed in the previous section, often requires daily work reports which must be submitted to inspectors. The inspectors may go over these reports and not be satisfied with the fuel consumption or attendance of engineers and require the contractor to put on additional men or other items. Additional costs such as this can be discovered while the job is under construction.

There are, of course, many more items which the estimator must consider. Even if these items do not appear in the specifications, the contractor and estimator will have to include their costs in the estimate if it is to be accurate and complete, and result in a profit for the contractor.

THE CONSTRUCTION SPECIFICATIONS INSTITUTE FORMAT

Advances in technology and methods of processing data have often made the writing of specifications cumbersome, confusing, or even obsolete. The Construction Specifications Institute in Washington, D.C., has made an attempt to remedy this situation with the publication of the *CSI Format*. The *Format's* value lies in its potential for unifying and universalizing the writing of specifications. It offers a logical framework for the specifier to work with as well as a standard for the many persons who must read the specifications.

A sort of "Dewey Decimal System" of construction specification, the *Format* is comprised of four major groupings:

Bidding Requirements
Contract Forms
General Conditions
Specifications (Technical)

Within this last grouping of *Specifications,* 16 permanent *divisions* are found. These divisions are constant in sequence and short in name. "Divisions" do not name units of work but rather relationships of units of work. The units of work are the "sections" within the divisions.

These divisions were established by considering construction relationships: materials, trades, functions and locations of specified work. The 16 divisions of Specifications in the *Format* are:

Division 1—General Requirements
Division 2—Site Work
Division 3—Concrete
Division 4—Masonry
Division 5—Metals
Division 6—Wood and Plastics
Division 7—Thermal and Moisture Protection
Division 8—Doors and Windows
Division 9—Finishes
Division 10—Specialties
Division 11—Equipment
Division 12—Furnishings
Division 13—Special Construction
Division 14—Conveying Systems
Division 15—Mechanical
Division 16—Electrical.

THE UNIFORM SYSTEM

The *Uniform System,* also published by CSI, is essentially the same as the *Format*. The 16-division arrangement is almost identical. In addition, the *Uniform System* offers a system for filing data and a cost accounting guide that correspond to the organization provided in the *Format*. The *Uniform System* is properly labeled "Buildings," however, and excludes projects of a strictly engineering nature.

The *Uniform System* is rapidly gaining acceptance and has recently been adopted for use by *Sweet's Catalogue of Construction Information,* as well as by the AIA's *Architectural Graphic Standards*.

Sample Specifications

The specification shown in the following pages is referred to a "fill-in-form."[1] *This is a basic form which may be used for all residences. The architect adds the exact specifications for the particular residence. Specifications are arranged, as far as possible, in the order in which the various trades will work on the structure.*

The specifications should spell out the responsibilities of both parties. For example, in the specification form given below the contractor is required to provide liability and workmen's compensation insurance. The owner is to provide fire and windstorm insurance during construction. What the owner will do is commonly introduced by "will."

The working drawings and the specifications function together as a whole. What is mentioned in either is considered to be in both. All items which are necessary for the completion of the structure, even though not mentioned, are considered to be included. If the specifications and the working drawings are in conflict, normally the specifications take precedence. It is the responsibility of the architect, of course to prevent such conflict. Exactness is especially necessary since the specifications and drawings are used for making the estimate.

The specifications are binding on all parties, including the sub-contractors. Normally, modifications can be made only by mutual agreement.

---

1. This specification was designed by the Miller Planning Service, Kalamazoo, Michigan.

SPECIFICATIONS

The contractor shall provide all necessary labor and materials and perform all work of every nature whatsoever to be done in the erection of a residence for _____ as owner in accordance with the specifications and drawings.

The location of the residence will be as follows:_____.

GENERAL

All blank spaces in these specifications that apply to this building are to be filled in. All blank spaces that do not apply to be crossed out. The general conditions herein set forth shall apply to any contract given under these specifications and shall be binding upon every sub-contractor as well as general contractor.

The plans, elevations, sections, and detail drawings, together with these specifications, are to form the basis of the contract and are to be of equal force. Should anything be mentioned in these specifications and not shown in the drawings, or vice versa, the same shall be followed as if set forth in both, as it is the intent of these specifications and accompanying drawings to correspond and to embody every item and part necessary for the completion of the structure. In the event that items which are normally part of a complete house are omitted from both the plans and specifications, it is expected that they will still be supplied as part of the general contract. Example: area walls, bathroom towel bars, soap dishes, etc., and similar items. The contractor shall comply with all health and building ordinances that are applicable.

EXCAVATION AND GRADING

The contractor shall do all necessary excavating and rough grading. The excavation shall be large enough to permit inspection of footings after the foundation has been completed. All excess dirt shall be hauled away by the contractor. Black surface loam to be piled where directed by the owner for use in grading and will be bulldozed into place by the general contractor. All subsoil and top soil required for fill and/or rough grading shall be paid for by the owner and bulldozed into place by the general contractor. The finish grading shall be done by the owner. Grade level shall be established by the owner, who will also furnish a survey of the lot showing the location of the building. The finish grading, seeding, sodding, and landscaping shall be done by the owner unless specified as follows: _____.

CONCRETE FOOTINGS

Footings shall be of concrete mixed in the proportion of one part Portland cement, three parts clean, coarse, sharp sand and five parts of gravel or crushed rock. Concrete shall be machine mixed with clean water to the proper consistency and shall be placed immediately after mixing. Footings shall be thoroughly protected with hay or straw in freezing weather. All footings shall be set below the frost line and rest on firm soil and shall be flat and level on the underside. Footings shall be of sizes shown on plan.

BASEMENT WALLS

Basement walls shall be of poured concrete _____ inches thick. Poured walls shall be straight, level, and plumb,

OR

basement walls will be constructed of _____ inch concrete blocks per plan, of approved quality. Blocks shall be laid in a full bed of mortar, composed of one part of cement to three parts of sand. Mortar joints shall be filled thoroughly with cement mortar, neatly pointed on both sides.

All walls shall have uniform bearing for framing, being straight, plumb, and level. Beam fill to be placed as shown on the plans. Waterproof basement walls with two coats of waterproofing applied according to manufacturer's specifications. Mortar drippings shall be cleaned from footings, and a cement cove trowelled into place.

BASEMENT FLOOR

Basement floor shall be of _____ inch concrete, poured monolithically, with a trowelled finish. Thoroughly tamp the base concrete into place and carefully pitch to floor drains; .004" Polyethylene shall be placed under basement floors under recreation rooms, bedroom, and bathrooms, etc.

CRAWL SPACES

Crawl spaces shall have a skim coat of concrete over .004" Polyethylene. Minimum clearance between joists and slab shall be two feet.

CEMENT WALKS AND STEPS

All cement walks shall be four inches thick, of widths and in locations shown on plans and shall be poured monolithically with a trowelled finish. The steps at the front and rear entrances shall be of wood, cement, or brick construction as indicated on the plan. If rain leaders are not connected to sewer, provide concrete splash blocks for each rain leader.

CHIMNEYS

Chimneys shall be constructed of common brick with face brick top. Provide tile flue lining of size and extent shown on plans for all flues. Thimbles and cleanouts shall be built in as required. If fireplace is required, furnish and install ash dump, damper and cleanouts, fire brick for lining and hearth, facing and outer hearth of material selected by owner. Hearth to be supported on concrete slab of size shown on plan.

All flues shall be cleaned of mortar drippings, and the chimney shall be capped with a concrete cap as shown on plans -- minimum thickness at thinnest point two inches.

Concrete hearth support to be fireproof; hearth floor to be _____.

Chimneys shall be flashed and counter-flashed where they pass through the roof.

Face of fireplace opening, if masonry, shall be _____.

Mantel shelf, if masonry, shall be _____.

Incinerator, if any, shall be _____.

Damper to be _____.

BRICK WORK

All brick work, if any, shall be laid in cement and lime mortar, with all bricks well bedded and shoved into place, with both vertical and horizontal joints on straight lines. Joints to be of color selected by owner.

The price allowed for face brick is $_____ per 1,000. Any cost above that amount will be borne by the owner. Lintels to be properly placed above all openings where masonry is shown above.

TILE WORK

The contractor shall furnish and set all tile in a neat and workmanlike manner. Recessed towel bars, paper holder, and soap dish shall be furnished by tile contractor.

The following areas shall be covered with ceramic tile: (Mark "yes" or "no".)

Bathroom floors _____

Bathroom walls to a height 4 ft. above floor _____

Bathroom tub areas to a height 5 ft. above tub _____

Showers to a height 6'-6" above floor _____

Front entrance hall floor _____

Vanity cabinet tops _____

Other _____

CARPENTER WORK

The contractor shall and will provide all necessary labor and perform all carpenter work of every nature whatsoever to be done. He shall lay out all work and be responsible for all measurements and keep a competent foreman in charge. All work shall be done in a workmanlike manner, level, straight, plumb, and true and strictly in accordance with the plans and specifications.

GIRDERS AND COLUMNS

Girders or supporting beams and columns shall be as required by the size of the building and shall be of the size and location shown in the plans.

JOISTS

First floor joists 2" x ___   O. C. Grade #1 Btr. Fir - 15 to 25% #2 permitted.

Second floor joists 2" x ___   "   "   "   "

Ceiling joists   2" x ___   "   "   "   "

Rafters   2" x ___   "   "   "   "

Collar ties   2" x ___   "   "   "   "

Valley rafters   2" x ___   "   "   "   "

Double joists under partitions and around all openings.

All dimension material covered in the above specification is Douglas Fir, as graded by the West Coast Lumber Inspection Bureau.

STUDDINGS AND PARTITIONS

Studdings shall be sized 2 x 4's, spaced 16" on centers, single plate on bottom and double plate on top of each wall or partition. 2 x 4's shall be doubled around all openings and shall be _____.

BRIDGING

First and second floor joists shall have one row of 1 x 3 wood or approved metal bridging for all spans of 8 to 14 feet. All spans over 14 feet shall have two rows, all fastened securely to joists at each end.

ROUGH FLOORING

Sub-flooring shall be _____ C-D plywood securely nailed. All joints shall be made on joists.

SHEATHING

Outside walls shall be covered with securely nailed _____.

Roof sheathing shall be _____.

SIDING

Siding, if any, to be _____.

ROOFING

Shingles for roof to be _____ laid _____ inches to weather, using galvanized nails.

BUILT-UP ROOFS

Built-up roofs shall be installed to a 15 year specification and covered with gravel, slag, or white aggregate.

INSULATION AND PAPER

Sidewall insulation to be _____.

Top floor ceiling to be insulated with _____.

Building paper under shingles to be _____.

Building paper over sheathing to be _____.

All exterior walls shall be covered on the inside edge of the studs from floor to plate with .004" Polyethylene. Application shall be continuous, with cutouts for windows, elec. openings, etc., being made after lath is applied.

If hot water heat is used, ceilings shall be covered as above -- application to be made to the bottom edge of the ceiling joists.

OUTSIDE FINISH

All lumber required for outside finish shall be _____.

All exterior siding, cornice, and miscellaneous trim shall be woodlife dipped and nailed with non-rusting nails.

WINDOW AND DOOR FRAMES

All window and outside door frames as shown on plans shall be of sound, clear pine, free from objectionable defects. Door sills shall be oak.

Assembled window units, if any, shall be per plan.

Assembled door units, if any, shall be _____.

WINDOWS, STORM SASH, AND SCREENS

All windows and sash shall be as shown on plans. Storm sash and screens shall be as required for patented units. Double-hung windows and _____ units shall have _____.

FINISHED FLOORS

Finished floors in living room and dining room to be _____.

Finished floors in family room to be _____.

Finished floors in bedrooms to be _____.

Finished floors in front entry to be _____.

Finished floors in kitchen to be _____.

Finished floors in bathrooms to be _____.

Finished floors in toilet to be _____.

Finished floors in _____ to be _____.

Finished floors in rear entry to be _____.

Finished floors in attic to be _____.

All hardwood floors shall be properly nailed and machine sanded to a smooth, even surface. Place 30 lb. red rosin paper between sub-floor and finish floor. Floors under linoleum shall be securely nailed with screw type or annular threaded nails.

SLATE FLOORS

Allow _____ per sq. ft. for slate floors in areas shown on plan.

NON-RIGID FLOOR COVERINGS
(not including carpeting)

Allow   per sq. yd. for linoleum-type floor covering where required.
Allow   per sq. ft. for homogenous vinyl floor covering where required.
Allow   per sq. ft. for vinyl asbestos floor covering where required.

Other _____.

INSIDE FINISH

Trim in the living room, dining room, and front entry shall be _____.

Trim in the kitchen and rear entry shall be _____.

Trim in _____ shall be _____.

Trim in the bedrooms, bathrooms, and hall shall be _____.

DOORS

All of the inside doors shall be 1 3/8" thick as follows: _____.

The front door shall be 1 3/4" thick of _____.

The remaining outside doors shall be 1 3/4" thick of _____.

Provide combination storm and screen doors for all outside doors.

Provide scuttle door to attic and plumbing access doors.

JAMBS AND CASINGS

All inside jambs shall be 3/4" thick and of kinds specified above. Casings shall be 3/4" thick of stock design.

STAIRS

Stairs leading from first to second floor shall be as shown in plans with _____

risers and _____ treads. Basement stairs shall have _____ risers and _____

treads. Three stair horses shall be provided for each stair and shall be _____

x _____. Wall stringers shall or shall not be housed. All stairs to be equipped

with hand rails.

CABINET WORK

Built-in medicine cabinet in bathroom -- allow $ _____.

Size of mirrors _____.

Kitchen cabinets shall be _____ and be placed as shown on plan. Mantel and mantel

shelf, if any, shall be _____. Kitchen counter tops to be _____. Splash back to

be _____. Edging material to be _____. Linen cabinet to be _____. Other

cabinet work, if any, shall be as follows: _____.

CLOSETS

All closets shall have one shelf and one clothes rod.

HARDWARE

The contractor shall furnish all rough hardware, such as nails. The amount to be allowed for finish hardware is $_____. Any cost in excess of that amount will be paid by the owner.

ELECTRICAL WORK

Contractor shall provide all necessary labor and material and perform all electrical work of every nature whatsoever to be done. All work to comply with local ordinances. Provide ____ openings, plus special equipment outlets as listed below, plus door bells on _____.

Provide 100 amp. service entrance facilities. All lights and switches to be placed as indicated on plans or as directed by owner. Provide special equipment outlets for

stove _____, water heater _____, washer _____, dryer _____, water pump

_____, bathroom exhaust fan _____, kitchen exhaust fan _____. Heating plant

by heating contractor. Other _____.

ELECTRIC FIXTURES

Electric light fixtures to the value of $_____ shall be furnished by contractor. Any cost in excess of this amount will be paid by the owner, and any cost lower than this amount listed is to be credited to the owner.

Allow $_____ for built-in range and oven.

LATHING, PLASTERING, AND GROUNDS

Lath all walls, ceilings, etc., with gypsum or rocklath applied strictly in accordance with manufacturer's directions. Lath all interior corners throughout with 6 inch strips of angle-shaped metal lath. Provide galvanized corner beads on all exterior corners. All plaster to be two-coat work of a standard brand of hard wall plaster mixed in accordance with manufacturer's directions and shall be straight and true. Finish for each of the several rooms shall be as called for in room finish schedule. Plastering contractor shall repair all defects and do all patching necessary to leave the work in good condition.

Finish to consist of the following: _____.

	Ceilings	Walls
Living room	_____	_____
Kitchen	_____	_____
Bedrooms	_____	_____
Halls	_____	_____
Bathrooms	_____	_____
Dining rooms	_____	_____
Front entry	_____	_____
Rear entry	_____	_____
Basement	_____	_____
Garage	_____	_____

STUCCO

If stucco is required, it shall be three-coat work applied over self-furring, galvanized, expanded metal lath weighing 3.2 lbs. per sq. yard. Final coat to be of a color selected by owner. Provide a waterproof paper on sheathing under metal lath.

PLUMBING

Contractor shall provide all labor and material and perform all plumbing work of every nature whatsoever to be done. The fixtures shall be as follows:

_____ Bath tub _____

_____ Toilet combination _____

_____ Size _____ Lavatory _____

_____ Kitchen sink _____ Size and specification _____

_____ Gallon hot water heater (gas, electric, or oil)

Two-compartment laundry tray, swing faucet _____

Dishwasher _____   Garbage disposer _____

Floor drain _____   Sill cocks _____

Towel bars _____   Soap dish _____

Paper holder _____   Shower bath _____

All of the above shall be properly installed and all connections thoroughly tested, and shall be installed according to local ordinance. Hot and cold water connections shall be made with bath tub, shower, lavatory, kitchen sink, and laundry tray. Water connections shall be made with water main in the street, sewer connection shall be made with sewer in the street, gas connections shall be made with gas main in the street, all to be paid for by the contractor. All meters will be paid for by the owner. Private sewage disposal systems, if required, shall be provided and installed in accordance with local codes.

Private water supply systems, if required, shall be provided and installed according to the owner's instructions. Allow $ _____ .

HEATING

Contractor shall and will provide all necessary labor and material and perform all heating work of every nature whatsoever to be done, including the installation of _____ heating system of sufficient size to properly heat all parts of the house in coldest weather. If hot air system is to be used, it is to be installed according to the code of the National Society of Heating and Ventilation Engineers.

If hot water, steam, or any other heating system is to be used, such installation shall consist of the following: _____ .

SHEET METAL WORK AND FLASHING

Contractor shall and will provide all necessary labor and materials and perform all sheet metal work of every nature whatsoever to be done, including gutters under all eaves with suitable conductors. All joints to be well soldered and securely fastened, and all work to be done in a neat and workmanlike manner. Gutters to be 26 gauge galvanized iron _____ type.

Down spouts to be _____ . Proper _____ flashing shall be provided wherever necessary. Clothes chute, if shown on plan, shall be lined with _____ .

INTERIOR PAINTING

All woodwork to be carefully cleaned of finger marks, stains, and other defects before any oil, filling, paint, or varnish is applied, and all rough spots to be sandpapered smooth before being filled with colored putty to match color desired. Finish to consist of the following: _____ .

	Walls and ceilings	Trim
Living room	_____	_____
Dining room	_____	_____
Dinette	_____	_____
Kitchen	_____	_____
Rear entry	_____	_____
Bedroom	_____	_____
Bedroom	_____	_____
Hall	_____	_____
Front entry	_____	_____
Family room	_____	_____

All hardwood floors shall be sanded, filled, and varnished two coats, excepting _____ .

EXTERIOR PAINTING

All exterior woodwork shall have _____ coats of prepared paint of colors to be selected by owner. All sash and trim to be neatly traced. All knots and other defective work to be shellacked and all nail holes to be puttied before applying last coat. All exposed sheet metal shall have one coat of red lead and two coats of finished coat paint. All paint products used on the house, both exterior and interior, shall be manufactured by a reputable firm, suitable for the surface to which they are to be applied, and shall be applied according to the manufacturer's specifications.

EXTRAS OR CREDITS

Any deviation from these specifications or plans involving an extra charge or a credit must be agreed upon in writing between the contracting parties before the change is made. The contractor shall not take advantage of any discrepancies in the drawings and specifications. If any discrepancies are found, they shall be referred to the owner or the architect and be corrected before any contract is entered into.

INSURANCE

The contractor shall provide liability insurance and workmen's compensation insurance in full until completion of the building. Fire and windstorm insurance during construction will be provided by the owner.

CLEANING UP

The contractor shall not remove all debris from the premises when the job is completed. The contractor shall not clean all window glass when job is completed.

Driveway _____ .

Sidewalks _____ .

The Federal Housing Administration Standard Forms

The Federal Housing Administration was established through the National Housing Act to encourage "improvements in housing standards and conditions, to provide a system of mutual mortgage insurance, and for other purposes." The FHA does not provide money for loans but insures the loan to protect the loaning agency in case of foreclosure. Consequently, the FHA may set up minimum standards for well-planned, safe and soundly constructed homes.

The FHA issues several publications in accord with its standards and provides standard fill-in forms such as the following.

FHA Form 2005
VA Form 26-1852
Rev. 3/68

U. S. DEPARTMENT OF HOUSING AND URBAN DEVELOPMENT
FEDERAL HOUSING ADMINISTRATION

For accurate register of carbon copies, form
may be separated along above fold.   Staple
completed sheets together in original order.

Form Approved
Budget Bureau No. 63-R0055

Plan 329 (includes 2 car garage)

☐ Proposed Construction

DESCRIPTION OF MATERIALS

No. _____

(To be inserted by FHA or VA)

☐ Under Construction

Property address _____ City _____ State _____

Mortgagor or Sponsor _____ _____
                             (Name)   (Address)

Contractor or Builder _____ _____
                            (Name)   (Address)

INSTRUCTIONS

1. For additional information on how this form is to be submitted, number of copies, etc., see the instructions applicable to the FHA Application for Mortgage Insurance or VA Request for Determination of Reasonable Value, as the case may be.

2. Describe all materials and equipment to be used, whether or not shown on the drawings, by marking an X in each appropriate check-box and entering the information called for in each space.   If space is inadequate, enter "See misc." and describe under item 27 or on an attached sheet.

3. Work not specifically described or shown will not be considered unless required, then the minimum acceptable will be assumed.   Work exceeding minimum requirements cannot be considered unless specifically described.

4. Include no alternates, "or equal" phrases, or contradictory items.   (Consideration of a request for acceptance of substitute materials or equipment is not thereby precluded.)

5. Include signatures required at the end of this form.

6. The construction shall be completed in compliance with the related drawings and specifications, as amended during processing.   The specifications include this Description of Materials and the applicable Minimum Property Standards.

1. **EXCAVATION:**

Bearing soil, type ___Sand and Gravel___

2. **FOUNDATIONS:**

Footings: concrete mix ___2500 #___ ; strength psi _____ Reinforcing _____

Foundation wall: material ___Poured Conc. 8"___ Reinforcing _____

Interior foundation wall: material _____ Party foundation wall _____

Columns: material and sizes ___3" adj. Post___ Piers: material and reinforcing _____

Girders: material and sizes ___1 Beam 7" at 15.3 #___ Sills: material _____

Basement entrance areaway _____ Window areaways _____

Waterproofing ___Tar___ Footing drains _____

Termite protection ___Shield at Brick___

Basementless space: ground cover _____ ; insulation _____ ; foundation vents _____

Special foundations _____

Additional information: _____

3. **CHIMNEYS:**

Material ___Face Brick___ Prefabricated (make and size) _____

Flue lining: material __Vitrified Clay__ Heater flue size __8 x 12__ Fireplace flue size __12 x 12__
Vents (material and size): gas or oil heater __5" G. I.__ ; water heater __3" G. I.__
Additional information: _____

4. FIREPLACES: OPTIONAL
Type: ☒ solid fuel; ☐ gas-burning; ☐ circulator (make and size) _____ Ash dump and clean-out __10__
Fireplace: facing __Brick__ ; lining __Fire Brick__ ; hearth __Ceramic__ ; mantel __None__
Additional information: _____

5. EXTERIOR WALLS:
Wood frame: wood grade, and species __Cedar #2__ ☒ Corner bracing. Building paper or felt __15# Felt__
Sheathing __Asphalt Impregnated Fiberboard__ ; thickness __½"__ ; width __4 x 8__ ; ☒ solid; ☐ spaced ___ o. c.; ☐ diagonal; _____
Siding __Aluminum__ ; grade ___ ; type ___ ; size ___ ; exposure __8"__ ; fastening __Nailed__
Shingles ___ ; grade ___ ; type ___ ; size ___ ; exposure ___ ; fastening __Per Mfg.__
Stucco ___ ; thickness ___ "; Lath ___ ; weight ___ lb.
Masonry veneer __Face Brick $60/M__ Sills __Lime Stone__ Lintels ___ Base flashing ___
Masonry: ☐ solid ☐ faced ☐ stuccoed: total wall thickness ___ "; facing thickness ___ "; facing material ___
Backup material ___ ; thickness ___ "; bonding ___
Door sills ___ Window sill ___ Lintels ___ Base flashing ___
Interior surfaces: dampproofing, ___ coats of ___ ; furring ___
Additional information: ___
Exterior painting: material __Exterior lead and oil__ ; number of coats __3__
Gable wall construction: ☐ same as main walls; ☐ other construction __Aluminum siding in gable__

6. FLOOR FRAMING: __2 x 10 - 16" o.c.__
Joists: wood, grade, and species __#2 Fir__ ; other ___ ; bridging __1 x 3__ ; anchors ___
Concrete slab: ☒ basement floor; ☐ first floor; ☐ ground supported; ☐ self-supporting; mix __5 sk.__ ; thickness __3__ ";
reinforcing ___ ; insulation ___ ; membrane ___
Fill under slab: material __Sand__ ; thickness __4__ ". Additional information: ___

7. SUBFLOORING: (Describe underflooring for special floors under item 21.)
Material: grade and species __1/2" plyscore__ ; size __4 x 8__ ; type __Plyscore__
Laid: ☒ first floor; ☐ second floor; ☐ attic ___ sq. ft.; ☐ diagonal; ☐ right angles. Additional information: __Solid__

8. FINISH FLOORING: (Wood only. Describe other finish flooring under item 21.)

LOCATION	ROOMS	GRADE	SPECIES	THICKNESS	WIDTH	BLDG. PAPER	FINISH	
First floor	Liv. Rm. Bedrooms Hall	#1	Oak	25/32	2½	Slaters	Felt	Bruce
Second floor		Common	Shorts					
Attic floor	Family Rm. sq. ft.		Oak	Ranch	Plank		Prefinished	

Additional information: ___

DESCRIPTION OF MATERIALS

9. PARTITION FRAMING:
Studs: wood, grade, and species ___Cedar #2___ size and spacing ___2 x 4   16" o.c.___ Other _____
Additional information: _____

10. CEILING FRAMING:
Joists: wood, grade, and species ___Trusses 24" o.c.___ Other _____ Bridging _____
Additional information: _____

11. ROOF FRAMING:
Rafters: wood, grade, and species ___Fir 24" o.c. Trusses 24" o.c.___ Roof trusses (see detail): grade and species _____
Additional information: _____

12. ROOFING:
Sheathing: wood, grade, and species ___Fir Plyscore 3/8"___ ; size ___4 x 8___ ; type ___C.D.___ ; ☒ solid; ☐ spaced _____ " o.c.
Roofing ___Asphalt shingles___ ; grade ___C___ ; weight or thickness ___235___ ; size ___12x36___ ; fastening ___Mfg. specs.___
Underlay _____
Built-up roofing _____ ; number of plies _____ ; surfacing material _____
Flashing: material ___26 G. I.___ ; gage or weight _____ ; ☐ gravel stops; ☐ snow guards.
Additional information: _____

13. GUTTERS AND DOWNSPOUTS:
Gutters: material ___G. I.___ ; gage or weight ___24___ ; size ___4"___ ; shape ___O. G.___
Downspouts: material ___G. I.___ ; gage or weight ___24___ ; size ___3"___ ; shape ___Rect.___ ; number _____
Downspouts connected to: ☐ Storm sewer; ☐ sanitary sewer; ☐ dry-well. ☒ Splash blocks: material and size ___Concrete 12 x 24___
Additional information: _____

14. LATH AND PLASTER:
Lath ☐ walls, ☐ ceilings: material _____ ; weight or thickness _____ Plaster: coats _____ ; finish _____
Dry-wall ☒ walls, ☒ ceilings: material ___Gypsum___ ; thickness ___1/2___ ; finish ___Smooth___ ;
Joint treatment ___Rust proof metal corners, back block ceiling joints; tape cement and sand___

15. DECORATING: *(Paint, wallpaper, etc.)*

ROOMS	WALL FINISH MATERIAL AND APPLICATION	CEILING FINISH MATERIAL AND APPLICATION
Kitchen		
Bath		
Other		
All Rooms	Alkyd Resin stipple - 2 coats	Same - 2 coats

Additional information: _____

16. INTERIOR DOORS AND TRIM:
Doors: type ___Flush and folding___ ; material ___Birch, Pine___ ; thickness ___1-3/4 & 1-3/8___
Door trim: type ___Casing 1-3/4___ ; material ___W. P.___ Base: type ___Wood___ ; material ___W. P.___ ; size ___2-1/4___
Finish: doors ___Flush - Seal and Varnish___ ; trim ___W. P.___
Other trim *(item, type and location)* _____
Additional information: _____

17. WINDOWS:

Windows: type __Double hung__ ; make __Grand Rapids__ ; material __W. Pine__ ; sash thickness __1-3/8__

Glass: grade __SS__ ; □ sash weights; □ balances, type __Spring__ ; head flashing ___

Trim: type __Casing__ ; material __W. P.__ Paint __Enamel__ ; number coats __2__

Weatherstripping: type __Friction__ ; material __SS__ Storm sash, number ___

Screens: □ full; ☒ half; type __Alum.__ ; number __9__ ; screen cloth material __alum.__

Basement windows: type __2-lite__ ; material __steel__ ; screens, number __0__ ; Storm sash, number __0__

Special windows __Wood picture - per elev.__

Additional information: ___

18. ENTRANCES AND EXTERIOR DETAIL:

Main entrance door: material __W. P.__ ; width __3'__ ; thickness __1-3/4__. Frame: material __Wood__ ; thickness __5/4__"

Other entrance doors: material __W. P.__ ; width __2'8"__ ; thickness __1-3/4__. Frame: material __W. P.__ ; thickness __5/4__"

Head flashing __Galv.__ Weatherstripping: type __Friction__ ; saddles __Aluminum__

Screen doors: thickness ___"; number ___ ; screen cloth material ___ ; Storm doors: thickness ___"; number ___

Combination storm and screen doors: thickness __1__"; number __2__ ; screen cloth material __Aluminum__

Shutters: □ hinged; ☒ fixed. Railings __Per Elev.__ , Attic louvers __2 - 14 x 20 metal__

Exterior millwork: grade and species __1 - W. P.__ Paint __Exterior Grade__ ; number coats __3__

Additional information: ___

19. CABINETS AND INTERIOR DETAIL:

Kitchen cabinets, wall units: material __Birch W. P.__ Sp. feet of shelves __25__ ; shelf width __12"__

Base units: material __Birch W. P.__ counter top __16 sp. ft.__ ; edging __Formica__

Back and end splash __Ceramic full Back Splash__ Finish of cabinets __Stain and Varnish__ ; number coats __3__

Medicine cabinets: make __Miami Carey 621__ ; model __Ideal 501__

Other cabinets and built-in furniture __48" Formica Vanity__

Additional information: ___

20. STAIRS:

STAIR	TREADS		RISERS		STRINGS		HANDRAIL		BALUSTERS	
	Material	Thickness	Material	Thickness	Material	Size	Material	Size	Material	Size
Basement	Fir	2 x 10			Fir	2 x 10	W. P.	1-3/4	Round	
Main										
Attic										

Disappearing: make and model number ___

Additional information: ___

2

21. SPECIAL FLOORS AND WAINSCOT:

	LOCATION	MATERIAL, COLOR, BORDER, SIZES, GAGE, ETC.	THRESHOLD MATERIAL	WALL BASE MATERIAL	UNDERFLOOR MATERIAL
FLOORS	Kitchen and Dining area - Vinyl Tile			W.P. 2¼	5/8 Ply
	Bath	Ceramic	Marble	Ceramic	Cement
	½ Bath	Vinyl Tile			5/8 Ply

	LOCATION	MATERIAL, COLOR, BORDER, CAP. SIZES, GAGE, ETC.	HEIGHT	HEIGHT OVER TUB	HEIGHT IN SHOWERS (FROM FLOOR)
WAINSCOT	Bath	Ceramic Tile in Mastic with bull nose			
		Tub recess only		6'	6'

Bathroom accessories: ☐ Recessed; material _____ ; number ____ ; ☒ Attached; material _Ceramic_ ; number _9_
Additional information: _____

22. PLUMBING:

FIXTURE	NUMBER	LOCATION	MAKE	MFR'S FIXTURE IDENTIFICATION NO.	SIZE	COLOR
Sink	1	Kitchen	Townsend	Steel enamel double compartment	32 x 21	Colored
Lavatory	2	Bath	Rheem Richmond	Roundell & Richeleu	18" & 19x17	Colored
Water closet	2	Bath	Rheem Richmond	G2210		Colored
Bathtub	1	Bath	Rheem Richmond		5'	
Shower over tub △	1		Chrome pltd.Brass			
Stall shower △						
Laundry trays	1	Basement	Mustee-single	Compartment fiberglass	24 x 24	

△☒ Curtain rod   △☐ Door   ☐ Shower pan: material _____
Water supply: ☒ public; ☐ community system; ☐ individual (private) system.★
Sewage disposal: ☐ public; ☐ community system; ☒ individual (private) system.★
★Show and describe individual system in complete detail in separate drawings and specifications according to requirements.
House drain (inside): ☐ cast iron; ☐ tile; ☒ other _copper_   House sewer (outside): ☐ cast iron; ☒ tile; ☐ other _____
Water piping: ☐ galvanized steel; ☒ copper tubing; ☐ other _____   Sill cocks, number _2_
Domestic water heater: type _Automatic_ ; make and model _Republic_ ; heating capacity
_25.2_ gph. 100° rise. Storage tank: material _____ ; capacity _40_ gallons.
Gas service: ☒ utility company; ☐ liq. pet. gas; ☐ other _____   Gas piping: ☒ cooking; ☒ house heating.

Footing drains connected to: ☐ storm sewer; ☐ sanitary sewer; ☐ dry well. Sump pump; make and model _____
_____; capacity _____; discharges into _____

23. HEATING:

☐ Hot water. ☐ Steam. ☐ Vapor. ☐ One-pipe system. ☐ Two-pipe system.
  ☐ Radiators. ☐ Convectors. ☐ Baseboard radiation. Make and model _____
  Radiant panel: ☐ floor; ☐ wall; ☐ ceiling. Panel coil: material _____
  ☐ Circulator. ☐ Return pump. Make and model _____; capacity _____ gpm.
  Boiler: make and model _____ Output _____ Btuh.; net rating _____ Btuh.
Additional information: _____
Warm air: ☐ Gravity. ☒ Forced. Type of system __Perimeter__
  Duct material: supply __Sheet Metal__; return __Sheet Metal__ Insulation _____, thickness _____ ☐ Outside air intake.
  Furnace: make and model __Lennox__ Input __120,000__ Btuh.; output __96,000__ Btuh.
  Additional information: _____
☐ Space heater; ☐ floor furnace; ☐ wall heater. Input _____ Btuh.; output _____ Btuh.; number units _____
  Make, model _____ Additional information: _____
Controls: make and types __Minneapolis-Honeywell Automatic__
Additional information: _____
Fuel: ☐ Coal; ☐ oil; ☒ gas; ☐ liq. pet. gas; ☐ electric; ☐ other _____; storage capacity _____
  Additional information: _____
Firing equipment furnished separately: ☐ Gas burner, conversion type. ☐ Stoker: hopper feed ☐; bin feed ☐
  Oil burner: ☐ pressure atomizing; ☐ vaporizing _____
  Make and model _____ Control _____
  Additional information: _____
Electric heating system: type _____ Input _____ watts; @ _____ volts; output _____ Btuh.
  Additional information: _____
Ventilating equipment: attic fan, make and model _____; capacity _____ cfm.
  kitchen exhaust fan, make and model _____
Other heating, ventilating, or cooling equipment __Rangpire Model 523 Ductless__

24. ELECTRIC WIRING:

Service: ☒ overhead; ☐ underground. Panel: ☒ fuse box; ☐ circuit-breaker; make _____ AMP's __100__ No. circuits __7 x 1__
Wiring: ☐ conduit; ☐ armored cable; ☒ nonmetallic cable; ☐ knob and tube; ☐ other _____
Special outlets: ☐ range; ☐ water heater; ☐ other __Furnace-Laundry circuit__
☒ Doorbell. ☐ Chimes. Push-button locations __Front and Rear Doors__ Additional information: _____

25. LIGHTING FIXTURES:

Total number of fixtures __11 up / 5 down__ Total allowance for fixtures, typical installation, $__100.00__
Nontypical installation _____
Additional information: _____

3

DESCRIPTION OF MATERIALS

DESCRIPTION OF MATERIALS

26. INSULATION:

Location	Thickness	Material, Type, and Method of Installation	Vapor Barrier
Roof			
Ceiling	3"	Blown in fiberglass	
Wall	1/2	Asphalt impregnated fiberboard & 1-1/2 fiberglass with vapor barrier	
Floor			

HARDWARE: *(make, material, and finish.)* Front and Rear - Quikset 400 Std.
Passage - Reliant 700
Bath Passage - Chrome 710

SPECIAL EQUIPMENT: *(State material or make, model and quantity. Include only equipment and appliances which are acceptable by local law, custom and applicable FHA standards. Do not include items which, by established custom, are supplied by occupant and removed when he vacates premises or chattles prohibited by law from becoming realty.)*

27. MISCELLANEOUS: *(Describe any main dwelling materials, equipment, or construction items not shown elsewhere; or use to provide additional information where the space provided was inadequate. Always reference by item number to correspond to numbering used on this form.)*

Birch Cabinets — Flush Birch Closet Doors
Brush Coat Basement — 100 Ampere Service
Double Compartment Sink — Screens
Built-in Range and Oven — Family Room paneled with Abitibi
48" Formica Vanity — Gun Stock
Formica Edged Countertops — Glass Doorwall
Miami-Carey Sliding Mirrors
Hood and Fan
Aluminum Siding

PORCHES: Floating slabs, post hole footings on large porch elev.

TERRACES:

42 x 42 on ground slab

GARAGES:

19 x 20 Garage, 16' Taylor Door, 42" footing-
Brick front one side, over half, drywall sidewall & Ceiling trusses 24 o.c.,
aluminum siding on end and rear; 12 x 19 Family Room rear of Garage

WALKS AND DRIVEWAYS: 16' at Garage, 16' at street

Driveway: width _____ ; base material ____Sand____ ; thickness ___4__"; surfacing material ___Concrete___ ; thickness __4__"

Front walk: width __3'__ ; material __Concrete__ ; thickness __4__". Service walk: width __2'__ ; material __Conc.__ ; thickness __4__"

Steps: material _____ ; treads _____"; risers _____". Cheek walls _____

OTHER ONSITE IMPROVEMENTS:

(Specify all exterior onsite improvements not described elsewhere, including items such as unusual grading, drainage structures, retaining walls, fence, railings, and accessory structures.)

Finish grade entire lot

LANDSCAPING, PLANTING, AND FINISH GRADING:

Topsoil __4__" thick: ☐ front yard; ☒ side yards; ☒ rear yard to __rear of lot__ feet behind main building.

Lawns *(seeded, sodded, or sprigged)*: ☒ front yard _____ ; ☒ side yards _____ ; ☐ rear yard _____

Planting: ☐ as specified and shown on drawings; ☐ as follows:

__1__ Shade trees, deciduous, __1-1/2__ " caliper.   _____ Evergreen trees. _____ ' to _____ ', B & B.

_____ Low flowering trees, deciduous, _____ ' to _____ '   _____ Evergreen shrubs. _____ ' to _____ ', B & B.

_____ High-growing shrubs, deciduous, _____ ' to _____ '   _____ Vines, 2-year _____

_____ Medium-growing shrubs, deciduous, _____ ' to _____ '

_____ Low-growing shrubs, deciduous, _____ ' to _____ '

IDENTIFICATION.—This exhibit shall be identified by the signature of the builder, or sponsor, and/or the proposed mortgagor if the latter is known at the time of application.

Date_____   Signature_____

    Signature_____

FHA Form 2005
VA Form 26-1852

4

4

REVIEW OF SQUARE AND CUBIC MEASURE

~~~~~~

As you begin to study this chapter you might ask yourself, "Is it necessary to know all these figures and shapes and their areas and volumes"? For accurate quantity take-offs and estimates, the answer is "Yes," as you will realize when you work the problems. You will be reviewing mensuration, *which is the branch of mathematics dealing with* length, area, *and* volume. *Problems at the end of each section will illustrate the relevant types of calculations.*

Remember that all units must be the same when you do the calculations. For example, you cannot multiply feet by inches or yards by feet. When the dimensions are given in feet, the area is in square feet *and the volume is in* cubic feet.

Decimal fractions are used whenever possible. (See Table I for the decimal equivalents.) For example, instead of using 2'6" or 2½', it is easier to use 2.5'. By converting to decimals, the calculations are made easier and you can also use slide rules or adding and calculating machines. Using the slide rule or these machines helps to reduce errors and can save much time and effort.

SQUARE MEASURE

A *plane* is a flat surface bounded by straight or curved lines and having no thickness. All building sites may be considered plane surfaces. The figure is a *polygon* if bounded by straight lines. The sum of the sides of a polygon, that is, the distance around, is the *perimeter*. The perimeter of a circle is called the *circumference*. Typical plane figures are illustrated in Fig. 4-1.

The main point to remember about plane figures is that they are considered as having area only. (That is, they have no depth.) The surface of a roof, floor, or sidewalk is a plane figure when regarded solely in terms of its boundary lines. Area can therefore be thought of as describing the extent of

TABLE I. DECIMAL EQUIVALENTS OF INCHES IN FEET

1 INCH	= 0.08 FEET	7 INCHES	= 0.58 FEET
2 INCHES	= 0.17 "	8 "	= 0.67 "
3 "	= 0.25 "	9 "	= 0.75 "
4 "	= 0.33 "	10 "	= 0.83 "
5 "	= 0.42 "	11 "	= 0.92 "
6 "	= 0.5 "	12 "	= 1.0 "

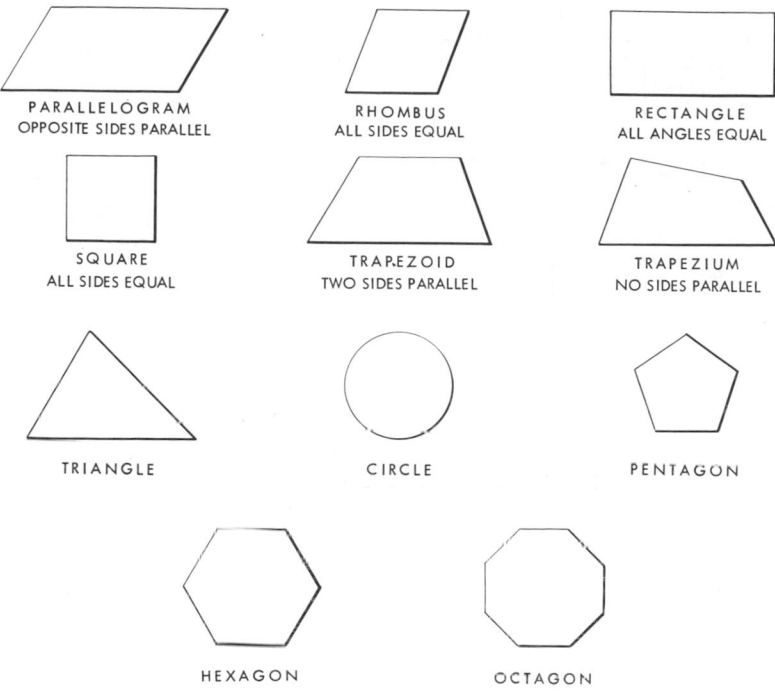

Fig. 4-1. Typical plane figures.

a plane figure. When you say that a farm contains 160 acres you are thinking only in terms of the surface area occupied by the farm. You are not considering depth of the soil. The same idea applies to all plane figures.

Areas are calculated in terms of square feet, square yards, acres, square miles, and so forth. See Table II. Generally, the estimator is concerned with square feet and square yards.

TABLE II. SQUARE OR SURFACE MEASURE

144 SQUARE INCHES	= 1 SQUARE FOOT
9 SQUARE FEET	= 1 SQUARE YARD
30 1/4 SQUARE YARDS	= 1 SQUARE ROD
160 SQUARE RODS	= 1 ACRE
43,560 SQUARE FEET	= 1 ACRE
4,840 SQUARE YARDS	= 1 ACRE
640 ACRES	= 1 SQUARE MILE = 1 SECTION
36 SECTIONS	= 1 TOWNSHIP

Fig. 4-2 illustrates square measure. The part bounded by A, B, C, and D represents one square foot (*sq. ft.*). It is a *square* foot because each of its four sides is one foot long. You know that 1 foot equals 12 inches, so in the square $ABCD$ the sides are each divided into 12 equal parts, or "inches." If lines are drawn horizontally and vertically, you can see that the square $ABCD$ is divided into 144 smaller squares, each equal to 1 square inch (*sq. in.*). The area is the length times the width, or 12 inches $\times$ 12 inches equals 144 square inches; thus 144 square inches equals 1 square foot.

The larger square in Fig. 4-2, $AEFG$, represents one square yard

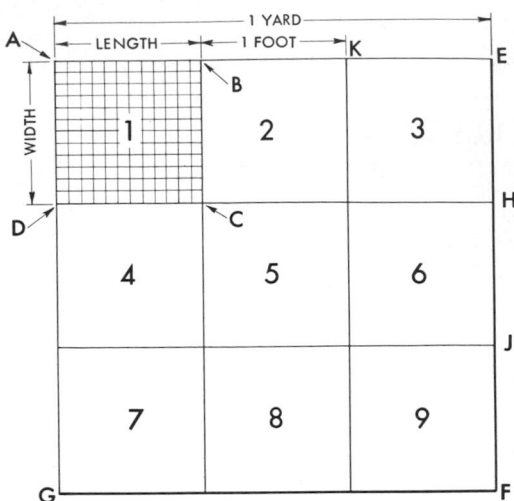

Fig. 4-2. A square yard contains 9 square feet. A square foot contains 144 square inches.

(*sq. yd.*). It is a square yard because all sides are one yard, or three feet long. The distances *AB, BK, KE, EH, HJ,* and *JF* are each one foot long. Since the square marked 1 equals one square foot, a square yard contains nine square feet.

Rectangles and Squares. *The area of a rectangle or square is equal to the product of its length and width.*

$$A = l \times w$$

where A = area
l = length
w = width

NOTE: Since the sides of a *square* are equal, the formula becomes

$$A = s \times s = s^2*$$

Example 1. Calculate the area of a rectangular floor whose length is 13'-6" and whose width is 10'-8".

---

*The small figure "2" means "squared"; e.g. $3^2 = 3 \times 3 = 9$.

Solution: First convert to decimals (use Table I).

$$13'\text{-}6'' = 13.5' \qquad 10'\text{-}8'' = 10.667'$$

(Round off to two places in the computation.)

$$\text{Area, } A = l \times w$$
$$= 13.5' \times 10.67'$$
$$= 144 \text{ sq. ft.}$$

Example 2. How many *square yards* are there in the floor of Example 1?
Solution: 9 sq. ft. = 1 sq. yd.

$$\text{Area, } A = \frac{144}{9} = 16 \text{ sq. yd.}$$

Triangles. A *triangle* is a polygon enclosed by three straight lines called sides, Fig. 4-3. Many building elements have a triangular shape. Two examples are the gabled roof and most shed roofs (elevation views).

Triangles are named by reading the letters at each angle just like a rec-

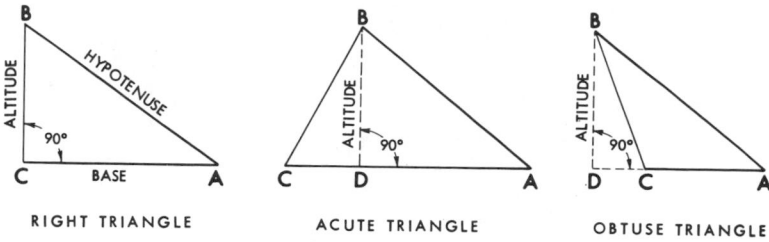

Fig. 4-3. Types of triangles.

tangle or other polygon. The order in which the triangles are read makes no difference; hence the triangles in Fig. 4-3 could each be named *ABC, ACB, BAC, BCA, CAB,* or *CBA.*

The *base* of a triangle is the side upon which the triangle is supposed to stand. Any side may be taken as the base.

The *altitude* of a triangle is the perpendicular drawn from the vertex to the base. It may be necessary to extend the base of an "obtuse" triangle (Fig. 4-3) so that the perpendicular will meet it.

A *right-angled* triangle, often called a *right triangle,* is one that has a right angle. The longest side (opposite the right angle) is called the *hypotenuse,* and the other sides are often referred to as *legs.* In a right triangle if two angles are equal then two legs will be equal. One leg may be considered as the base and the other leg as the altitude of the triangle.

Right Triangle Law. Pythagoras, a Greek mathematician and philosopher (about 580 B.C.), formulated the *Pythagorean Theorem,* also known as the *Right Triangle Law:*

The square of the hypotenuse of a right triangle is equal to the sum of the squares of the other two sides.

$$c^2 = a^2 + b^2$$

where c = hypotenuse
a = altitude
b = base

It is always possible to find the length of one side of a right triangle if the other two sides are known.

$$a^2 = c^2 - b^2$$
$$b^2 = c^2 - a^2$$

Example 3. A tool shed has a roof with a rise of 6 feet and a run of 10 feet. What is the length of the rafter? (That is, in arithmetical terms, what is the hypotenuse in a triangle with one leg of 6 feet and the other 10 feet?)

Solution: Substitute in the formula,

$$c^2 = a^2 + b^2$$
$$c^2 = 6^2 + 10^2$$
$$= 36 + 100$$

$$c = \sqrt{136}$$
$$= 11.66 \text{ ft., or } 11'\text{-}8''$$

NOTE: A table of square roots is included in Appendix C.

Area of Triangles. *The area of a triangle is equal to one-half the prod-*

uct of the base and the altitude. (See Fig. 4-3.)

$$Area = \frac{1}{2} \times b \times a$$
$$\text{where } b = \text{base}$$
$$a = \text{altitude}$$

Example 4. Given that the base of a triangle is 10 feet and the altitude is 8 feet. Find the area.

Solution: Substitute in the formula,

$$Area = \frac{1}{2} \times b \times a$$
$$Area = \frac{1}{2} \times 10 \times 8$$
$$= 40 \text{ sq. ft.}$$

Example 5. Refer to Fig. 4-4. The base of the triangle is 18 feet and each side is 15 feet. Find the area.

Solution: First find the altitude, *BD*. The line *BD* is a perpendicular, and in an *isoceles* triangle (two sides equal) it divides the triangle into equal right triangles; thus triangle *ABD* equals triangle *BCD*, and *AD* equals *DC*.

$$AD = DC = 9 \text{ ft.}$$
$$a^2 = c^2 - b^2$$
$$(BD)^2 = (AB)^2 - (AD)^2$$
$$= 15^2 - 9^2$$

$$= 225 - 81$$
$$= 144 \text{ sq. ft.}$$
$$BD = \sqrt{144} = 12 \text{ ft.}$$

The area is now calculated:

$$Area = \frac{1}{2} \times b \times a$$
$$= \frac{1}{2} \times 18 \times 12$$
$$= 108 \text{ sq. ft.}$$

Circles. The circle, Fig. 4-5, is a common shape in the building trades. Structural columns, pipes, and fixtures involve the circle.

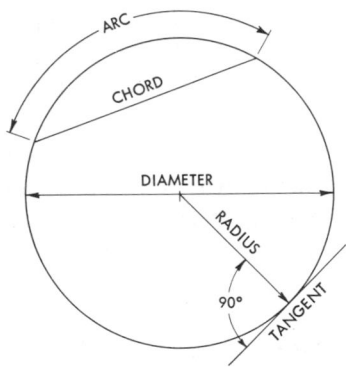

Fig. 4-5. The various parts of a circle.

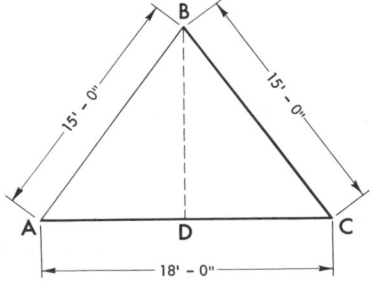

Fig. 4-4. An isoceles triangle has two sides of equal length.

A diameter of a circle is a straight line drawn through the center of the circle, ending at two points on the *circumference (perimeter)* of the circle.

A radius of a circle is a straight line joining the center and the circumference. All radii of the same circle are equal and their length is always one-half that of the diameter.

An arc is any part of the circumference of a circle.

INSCRIBED SQUARE   INSCRIBED HEXAGON   INSCRIBED TRIANGLE

Fig. 4-6. Inscribed polygons.

A tangent to a circle is a straight line touching the circumference only at one point. A radius drawn to this point forms a right angle with the tangent.

A chord is a straight line joining the extremities of an arc. When a number of chords form the sides of a polygon, the polygon is said to be inscribed in the circle, Fig. 4-6.

The circumference of a circle is equal to π times the diameter, or π times twice the radius.

$$C = \pi \times d$$
$$C = \pi \times 2r$$

where C = circumference
d = diameter
r = radius
π = 3.1416 or 22/7

The area of a circle is equal to the circumference multiplied by one-half the radius, or π times the radius squared.

Equation 1 — $A = C \times r/2$
Equation 2 — $A = \pi \times r^2$

Example 6. Find the area of the base of a circular column whose radius is 44 inches.

Solution 1: Find the circumference and multiply by one-half of the radius.

$$C = \pi \times 2 \times 44$$
$$= 276.46 \text{ in.}$$
$$A = C \times r/2$$
$$= 276.46 \times 44/2$$
$$= 6{,}082.1 \text{ sq. in.}$$

Solution 2: This solution uses the second formula for circle area.

$$A = 3.1416 \times 44^2$$
$$= 3.1416 \times 1{,}936$$
$$= 6{,}082.1 \text{ sq. in.}$$

Area in *square feet:*
$$\frac{6{,}082.1}{144} = 42.2 \text{ sq. ft.}$$

Trapezoids. *The area of a trapezoid is equal to the product of the altitude and one-half the sum of the bases.* (The trapezoid is shown in Fig. 4-1.)

$$A = a \times \frac{b_1 + b_2}{2}$$

where a = altitude
b_1 = lower base
b_2 = upper base

Example 7. Find the area of a lot

shaped like a trapezoid whose bases are 80 feet and 60 feet and whose altitude is 30 feet.

Solution: Substitute in the formula.

$$A = 30 \times \frac{80 + 60}{2}$$
$$= 30 \times 70$$
$$= 2,100 \text{ sq. ft.}$$

Hexagons. The estimator will occasionally find it necessary to calculate the area of a pentagon, hexagon, or more complex polygon when doing quantity take-offs. There are formulas for rapidly determining these areas, but they apply only to *regular* polygons, that is, polygons whose sides are *equilateral* (equal length) and whose angles are *equiangular* (equal angles). It is possible, however, to find the area of a polygon by dividing it into triangles and adding together the areas of the triangles.

Triangles are used to find the area of the regular hexagon in Fig. 4-7. Diagonals *BE, AD,* and *CF* divide the

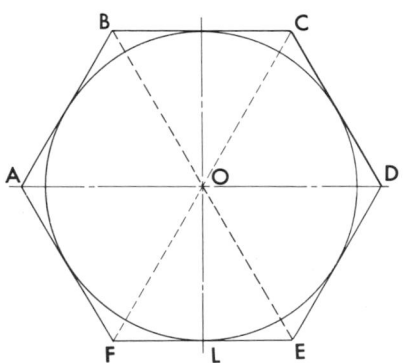

Fig. 4-7. A regular hexagon with inscribed circle.

hexagon into six equal triangles. Line *OL* is the *apothem.* It is perpendicular to *FE* and is the altitude of triangle *FOE.* In a regular hexagon, the sides of the triangles are equal: $OF = FE = OE,$ and the apothem is the radius (r) of the inscribed circle.

The area of triangle *FOE* is one-half *FE* times *OL,* or

Area of $FOE = \frac{1}{2} \times FE \times OL.$

Since there are six triangles in the hexagon, the area of the hexagon is six times the area of triangle *FOE,* or

$$A = 6\,(\frac{1}{2} \times FE \times OL)$$
$$= 3 \times FE \times OL.$$

The area of a regular hexagon is equal to three times the product of the apothem and one side.

$$A = 3 \times s \times r$$
where $s =$ one side of hexagon
$r =$ apothem

Example 8. The perimeter of a regular hexagon is 36 feet. What is the length of the apothem? (Refer to Fig. 4-7.)

Solution: Each side of a regular hexagon is one-sixth of the perimeter, or 6 feet. *OF* and *OE* are therefore 6 feet. The apothem, *OL,* divides the base of triangle *FOL* into two equal parts. $FL = LE = 3$ feet. The apothem is one leg of the right triangle *FOL* and is found by using the Right Triangle Law.

$$OL^2 = OF^2 - FL^2$$
$$= 6^2 - 3^2$$
$$= 27$$
$$OL = \sqrt{27} = 5.19 \text{ ft.}$$

Example 9. What is the area of a regular hexagon whose sides are 2 yards in length and whose apothem is 1.7 yards?

Solution: Substitute directly into the formula for area.

$$A = 3 \times s \times r$$
$$= 3 \times 2 \times 1.7$$
$$= 10.2 \text{ sq. yd.}$$

Irregular Shapes. To find the area of an irregular shape, the usual procedure is to divide the area into smaller areas of common shapes, calculate the areas and then add them together. This was done in the previous section; the area of a hexagon was found by dividing it into triangles.

Fig. 4-8 is the outline of the house shown in the plans at the end of Chapter 2. It is an irregular figure which was divided (as indicated) into common shapes: rectangles, a triangle, and a circle. The area of each is calculated separately and then added together to obtain the total area.

Fig. 4-9 is an irregularly-shaped figure which is bounded on three sides by straight lines. The widths of the figure at five points are measured and indi-

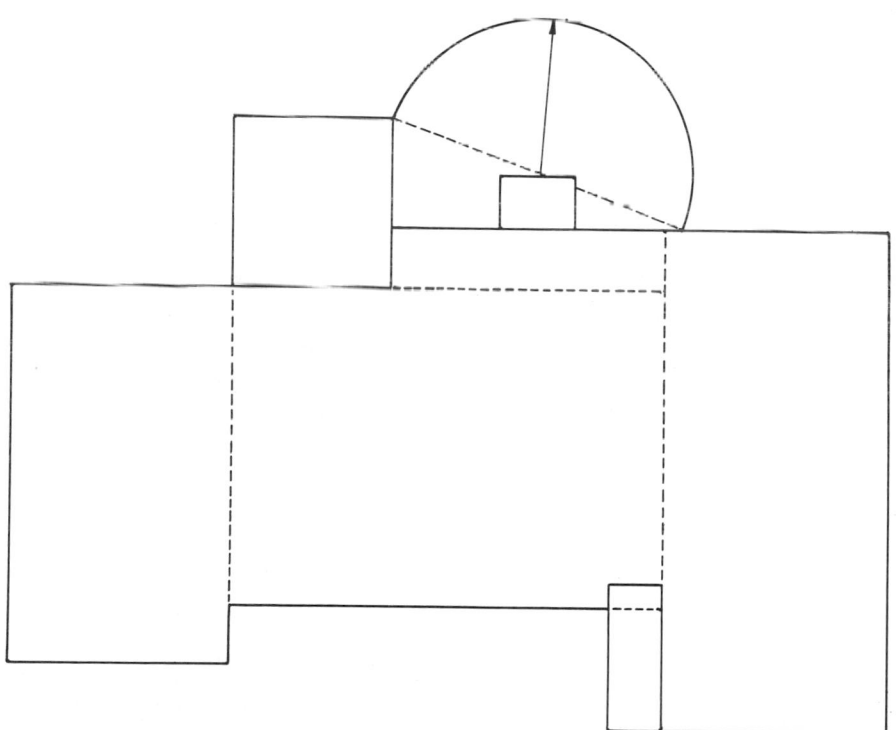

Fig. 4-8. The area of House Plan A is found by dividing the area into convenient shapes.

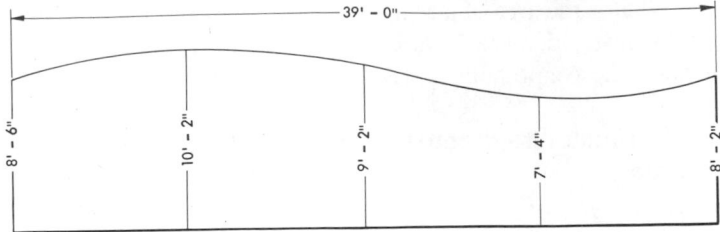

Fig. 4-9. The area of an irregular shape can be closely approximated.

cated as shown. A typical method of calculating the area is as follows.

The sum of the widths is 43.33 feet. (Use Table I to convert inches to decimal fractions.) The mean or *average*

width is $\dfrac{43.33}{5}$ or 8.67 feet. The area

is the length times the average width.

$$A = 39 \times 8.67$$
$$= 338 \text{ sq. ft.}$$

NOTE: Greater accuracy is obtained by using more widths when initially determining the average width.

Table III summarizes the formulas for circumference, length, and area.

CUBIC MEASURE

A solid is a body, such as a sack of cement, a building, or a cabinet. Fig. 4-10 illustrates other, less specific solids. All calculations relating to solids are based on those shown in Fig. 4-10.

Plane figures are considered in terms of the area they cover. Another consideration exists for solids, namely, *depth* (or *height* or *thickness*). To visualize the difference between a plane figure and a solid, consider a sheet of paper as a plane surface and a book as a solid. The book is made of many sheets of paper in layers. For the paper alone, only length and width are considered; but the book has thickness and that must also be considered. The comparison is immediately apparent if you imagine a sheet of paper so thin

TABLE III. AREA FORMULAS

PARALLELOGRAMS

A = l x w

A = b x a

(for squares)
A = s^2

TRIANGLES

A = (b x a)/2

Right Triangle Law:

$c^2 = a^2 + b^2$

CIRCLES

C = π x d

C = π x 2r

A = C x r/2

A = π x r^2

TRAPEZOIDS

A = a x (b_1 + b_2)/2

HEXAGONS

A = 3 x s x r

A = area
a = one side or leg or altitude
b = second side or base
c = third side or hypotenuse
b_1 = lower base
b_2 = upper base
s = length of one side
r = radius or apothem
C = circumference
d = diameter
π = 3.14 or 22/7

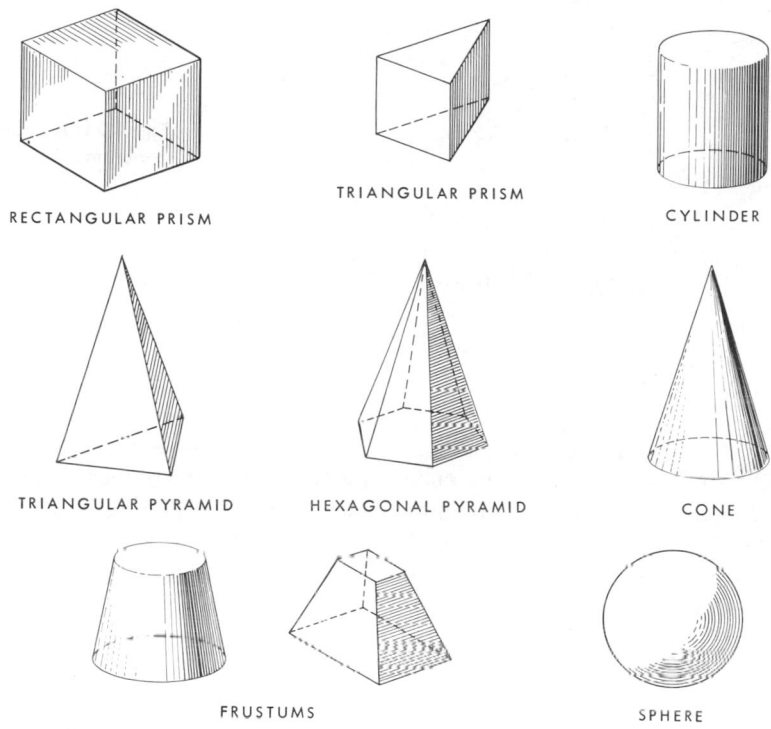

RECTANGULAR PRISM

TRIANGULAR PRISM

CYLINDER

TRIANGULAR PYRAMID

HEXAGONAL PYRAMID

CONE

FRUSTUMS

SPHERE

Fig. 4-10. Typical solids.

as to have no discernible thickness. A plane surface lacks *volume*.

Cubic, or cubage, means volume and gives the size of a body in terms of its bulk. The volume of the sand or earth used to cover a lot to a depth of two feet is specified in terms of cubic measure because it has depth or thickness as well as length and width. The same is true if the lot is *excavated* to a depth of two feet. Thus a foundation has length, width, and thickness and contains so many *cubic* yards of concrete. A room has length, width, and height and therefore has a certain cubage or volume. The same is true

for a sand or gravel truck. Its capacity is specified in terms of cubic yards or other units of volume.

Fig. 4-11 shows a large *cube*. A cube is a solid whose six sides are all equal in length and width. The cube represents one cubic foot *(cu. ft.)* in volume because each side (such as *AB, AN,* and *AD*) is equal to one foot. If you keep in mind that the volume of a cube is length times width times thickness you will easily see that 1 ft. × 1 ft. × 1 ft. = 1 *cubic* ft. *(Thickness* can be called *height* or *depth.)*

Lay off horizontal and vertical lines one inch apart on all six faces of

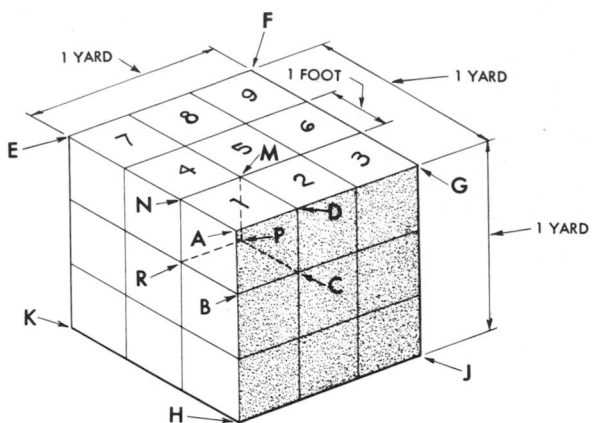

Fig. 4-11. A cubic yard contains 27 cubic feet.

ABCDNMPR. Cut or saw this cube into 12 equal layers or slices as illustrated in Fig. 4-12. Note slice *CDMP* which is numbered 1. If the small squares in this slice were counted, their sum would be 12 × 12, or 144 square inches. Since each of these slices is one inch deep, the volume of slice number 1 is 144 × 1, or 144 cubic inches. There are 12 equal slices, so the total volume is 12 × 144, or 1,728 cubic inches *(cu. in.)*. This illustrates the fact that 1 cubic foot contains 1,728 cubic inches.

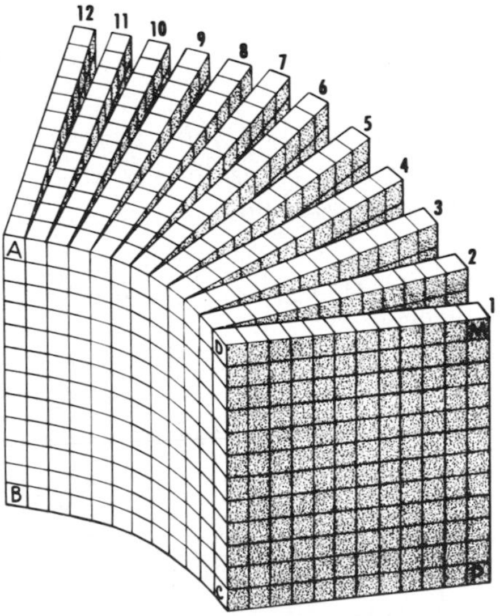

Refer again to Fig. 4-11. Note that the large square *AEFG* contains nine smaller squares. The large square is also three layers high. Each layer is one square yard and therefore contains nine square feet. Since each layer is one foot deep, the three layers are three feet deep, and each layer contains nine cubic feet. The three layers contain 3 × 9, or 27 cubic feet, which is the number of cubic feet in a cubic yard *(cu. yd.)*. Each side of the cube is one yard long.) Table IV lists some of the units of cubic measure.

Fig. 4-12. Illustrating that 1,728 cubic inches are contained in 1 cubic foot.

TABLE IV. CUBIC MEASURE

1,728 CUBIC INCHES	= 1 CUBIC FOOT
27 CUBIC FEET	= 1 CUBIC YARD
128 CUBIC FEET	= 1 CORD

Prisms. A *prism* is a solid whose bases or ends *(top* and *bottom)* are equal in area, similar in shape, and parallel in plane surfaces. The *lateral faces* or sides of a prism are parallelograms. Fig. 4-13 illustrates some typical prisms.

Solution: The volume is found by direct substitution in the formula.

$$V = 12 \times 12 \times 10$$
$$= 1,440 \text{ cu. ft.}$$
$$\frac{1,440}{27} = 53.33 \text{ cu. yds.}$$

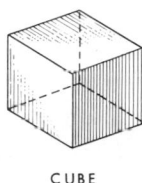
CUBE

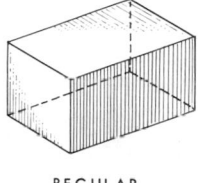
REGULAR
PARALLELEPIPED

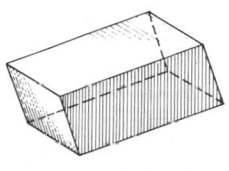
OBLIQUE
PARALLELEPIPED

Fig. 4-13. Typical prisms.

The lateral area of a prism is the combined area of all its lateral faces.

The total area of a prism is the combined area of the lateral faces and the bases.

Total Area =
lateral area + area of the bases

NOTE: The total area of a cube is six times the area of one side, or
$A = 6s^2$ where s = length of side
The volume, V, of any prism is found by multiplying the area of a base by the altitude.

$$V = \text{base area} \times a$$

Example 10. Calculate the volume of a room whose dimensions are 12 feet long by 12 feet wide and whose ceiling height is 10 feet. Convert your answer to cubic yards.

Example 11. What is the volume of an A-frame house, which has the shape of a triangular prism, whose height is 20 feet? Each end of the prism is a triangle whose base is 8 feet and whose altitude is 5 feet.

Solution: First find the area of the base by using the rule for calculating the area of a triangle,

$$A = \tfrac{1}{2} \times b \times a$$
$$A = \tfrac{1}{2} \times 8 \times 5$$
$$= 20 \text{ sq. ft.}$$

The volume is the area of the base times the altitude.

$$V = 20 \times 20$$
$$= 400 \text{ cu. ft.}$$

Cylinders. A *cylinder* is illustrated in Fig. 4-14. A cylinder, such as a pipe, post, or column, may have any altitude, *a,* and any radius, *OA.*

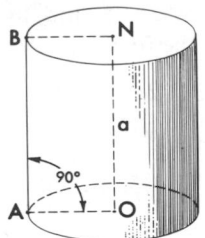

Fig. 4-14. A right circular cylinder.

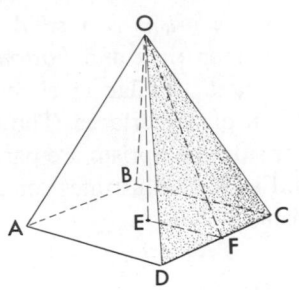

Fig. 4-15. A pyramid.

The lateral area of a cylinder is found by multiplying the circumference of the base by the altitude.

$$Lateral\ Area = C \times a$$

The volume of a cylinder is found by multiplying the area of its base by its altitude.

$$V = base\ area \times a$$

Example 12. A cylindrical form is 6 feet in diameter and 8 feet long. How many cubic yards of concrete are required to fill it?

Solution: The base area is found by using the formula $A = \pi \times r^2$

$$Base\ Area = 3.1416 \times 3^2$$
$$= 28.27\ sq.\ ft.$$

The volume is $\quad V = 28.27 \times 8.0$
$$= 226.16\ cu.\ ft.$$

$$\frac{226.16}{27} = 8.38\ cu.\ yds.$$

Pyramids. A *pyramid,* Fig. 4-15, is a solid whose base is a polygon and whose sides are triangles. The triangles meet at a common point to form the vertex, O. The altitude is OE and the slant height is OF.

The *slant height* of a regular pyramid (one whose base is a regular poly-

gon) is a line drawn on a side from the vertex to the center of any one side of the base, OF in Fig. 4-15. In other words, it is the altitude of one of the triangles which form the sides.

The *lateral edges* of a pyramid are the intersections of the triangular sides.

The lateral area of a pyramid is equal to the perimeter of the base multiplied by one-half the slant height.

Lateral Area =

$$perimeter\ of\ base \times \frac{slant\ height}{2}$$

The volume of a pyramid is one-third its base area times its altitude.

$$V = \tfrac{1}{3} \times base\ area \times a$$

Example 13. Find the volume of a regular hexagonal pyramid whose altitude is 12 feet. One side of the base is 6 feet. Refer to Figs. 4-7 and 4-10.

Solution: First find the area of the base. Refer to the section on "Hexagons." The apothem, $r,$ of the hexagon bisects one side of the hexagon; half of one side is 3 feet. The apothem is now found by the Right Triangle Law:

$$r^2 = 6^2 - 3^2$$
$$= 27$$
$$r = \sqrt{27} = 5.196 \text{ ft.}$$

Substituting in the formula for area of a hexagon,

$$A = 3 \times s \times r$$
$$= 3 \times 6 \times 5.196$$
$$= 93.528 \text{ sq. ft.}$$

The volume of the pyramid is found by substituting in the volume formula:

$$V = \tfrac{1}{3} \times \text{base area} \times a$$
$$= \tfrac{1}{3} \times 93.528 \times 12$$
$$= 374.112 \text{ cu. ft.}$$

Cones. A *cone* is a solid whose base is a circle and whose lateral surface tapers to a point—the vertex or top, Figs. 4-10 and 4-16. A cone may be considered a pyramid with numerous sides or faces, each side so small that the surface or lateral area has no edges and appears smooth.

The lateral area of a cone is the area of the tapering side.

The rules for finding areas and volumes of cones are the same as those for areas and volumes of pyramids.

The lateral area of a cone is the circumference of the base multiplied by one-half the slant height.

$$Lateral\ Area = C \times \frac{\text{Slant height}}{2}$$

The volume of a cone is one-third the product of its base area and altitude.

$$V = \tfrac{1}{3} \times \text{base area} \times a$$

Example 14. Both the circumference of the base and the slant height of a cone are 25 inches. Find the volume.

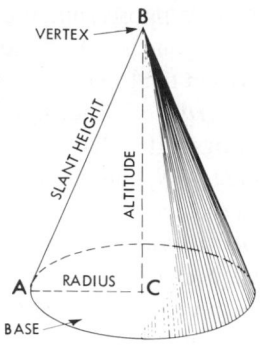

Fig. 4-16. A right circular cone.

Solution: Finding the volume requires both altitude and base area. Neither are given, so they must be calculated. (See Fig. 4-16.) First find the radius, then calculate the base area.

$$C = \pi \times d$$
$$d = C/\pi$$
$$= 25/3.1416$$
$$= 7.95 \text{ in.}$$
$$r = 3.98 \text{ in.}$$
$$Base\ Area = \pi r^2$$
$$= 3.14 \times 3.98^2$$
$$= 49.76 \text{ sq. in.}$$

The altitude of the cone is found by using the Right Triangle Law. The slant height, AB, is 25 inches. This is the hypotenuse of right triangle ABC in Fig. 4-16. The radius, AC, is 3.98 inches. Altitude BC is the second leg of the triangle.

$$AB^2 = 25^2 - 3.98^2$$
$$= 625 - 15.84$$
$$= 609.16$$
$$AB = 24.7 \text{ inches}$$

Substituting in the formula for volume,

$$V = \tfrac{1}{3} \times 49.76 \times 24.7$$
$$= 409.8 \text{ cu. in.}$$

Frustums. Suppose that the top of a pyramid or cone is cut off parallel to its base. When the top part is removed it leaves a *frustum,* Fig. 4-17. Concrete footings often have this shape. The *altitude, MN,* and the slant height, *OL,* are indicated.

The lateral area of a frustum of a right regular pyramid is one-half the sum of the perimeters of the two bases times the slant height.

The total area of a frustum is the sum of the lateral area and the two bases.

Total Area =
    lateral area + base area + A_B

The volume of a frustum is the sum of the areas of the two bases added to the square root of the product of the two bases and the result multiplied by one-third of the altitude.

$$V =$$
$$(A_1 + A_2 \times \sqrt{A_1 \times A_2}) \times \frac{a}{3}$$

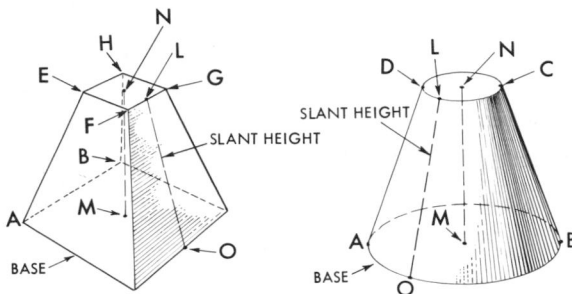

Fig. 4-17. Frustums of a pyramid and cone.

Lateral Area =
    ½ × $(P_1 + P_2)$ × slant height
where P_1 = Perimeter of one base
        P_2 = perimeter of other base

Since a cone may be considered as a pyramid with numerous sides so small that the surface appears smooth, a similar rule is used for finding the lateral area of a frustum of a cone.

The lateral area of a frustum of a right regular cone is one-half the sum of the circumferences of the bases multiplied by the slant height.

Lateral Area =
    ½ × $(C_1 + C_2)$ × slant height
      where C_1 = circumference
           of one base
       C_2 = circumference
           of other base

Example 15. What is the volume of the frustum of a square pyramid whose altitude is 15 feet? The sides of the lower and upper bases are 8 feet and 2 feet long respectively.

Solution: First find the areas of the bases.

$$A_1 = 8^2 = 64 \text{ sq. ft.}$$

$$A_2 = 2^2 = 4 \text{ sq. ft.}$$

$$A_1 \times A_2 = 64 \times 4 = 256$$

$$\sqrt{A_1 \times A_2} = \sqrt{256} = 16 \text{ sq. ft.}$$

The volume is now found by substituting in the formula:

$$V = (64 + 4 + 16) \times \frac{15}{3}$$

$$= 84 \times 5$$

$$= 420 \text{ cu. ft.}$$

Spheres. A *sphere* is a solid bounded by a curved surface every point of which is equally distant from the center. In other words, it is a perfectly round ball, Fig. 4-18.

A plane is *tangent* to a sphere when it touches the sphere at only one point, as plane *PNQ* (seen edgewise) in Fig. 4-18, which touches the sphere at *N*.

When a plane cuts through a sphere, the section is a circle such as plane *ACBD*. If the plane cuts through the center of the sphere, the resulting section is a *great circle* such as *ACBD*, *ANBM*, and *NCMD*. A *small circle*, *LFE*, is formed when the plane does not pass through the center.

The *circumference* of a sphere is the circumference of a great circle.

The surface area of a sphere is π times the diameter squared.

$$S = \pi \times d^2$$

where S = surface area

Since $d = 2r$,

$$S = \pi \times 4r^2$$

The volume of a sphere equals the surface area multiplied by one-third of the radius.

$$V = S \times \frac{r}{3}$$

Since $S = 4r^2 \times \pi$, you can substitute for S in the formula:

$$V = 4 \times r^2 \times \pi \times \frac{r}{3}$$

$$= \frac{4}{3} \times \pi \times r^3$$

Example 16. Find the volume of a sphere whose radius is 5 feet.

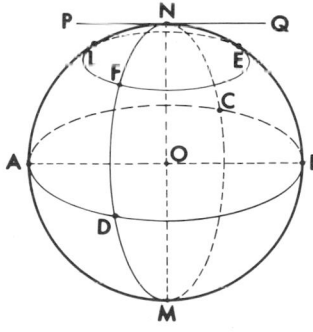

Fig. 4-18. A sphere.

TABLE V. VOLUME FORMULAS

PRISMS

Total Area = lateral area + base area

Total Area = 6 x s² (for squares)

V = base area x a

CYLINDERS

Lateral area = C x a

V = base area x a

s = length of one side
a = altitude
C = circumference
S = total surface area
V = volume

PYRAMIDS

Lateral Area = base perimeter x slant height/2

V = base area x a/3

FRUSTUMS

Lateral Area = (base perimeter + top perimeter) x slant height/2   (right pyramid)

Lateral Area = (base C + top C) x slant height/2   (right circular cone)

Total Area = lateral area + base area

V = (Base area + top area + √(base area x top area)) x a/3

SPHERES

$$S = \pi \times d^2$$

$$S = \pi \times 4r^2$$

$$V = \pi \times 4r^3 / 3$$

Solution: Substitute directly into the formula for volume.

$$V = \frac{4}{3} \times \pi \times 5^3$$

$$= \frac{4}{3} \times 3.1416 \times 125$$

$$= 523.6 \text{ cu. ft.}$$

Table V contains formulas for the areas and volumes of the most common solids.

BOARD MEASURE

Lumber is usually computed by *Board Measure,* B.M., the unit being a square foot one inch thick. Any number less than one inch thick is usually computed as one inch thick. (One exception to this is plywood; it is measured in square feet because it is sold in the form of panels.)

Framing Square Method. The back of a blade of a typical *framing square* is shown in Fig. 4-19. On the back of this blade is the Board Measure, where eight parallel lines along the length of the blade are shown and divided at every inch by cross lines. Under 12, on the outer edge of the blade, are found the various lengths of the boards, as 8, 9, 10, 11, 13, and so forth. For example, take a board 14 feet long and 9 inches wide. To find the number of board feet, look under 12, and find 14; then follow this space to the cross-line under 9, the width of the board; here is found 10 feet 6 inches, indicating the number of board feet of the board.

Calculation Method. The usual method of calculating the B.M. of lumber is to multiply the length in feet

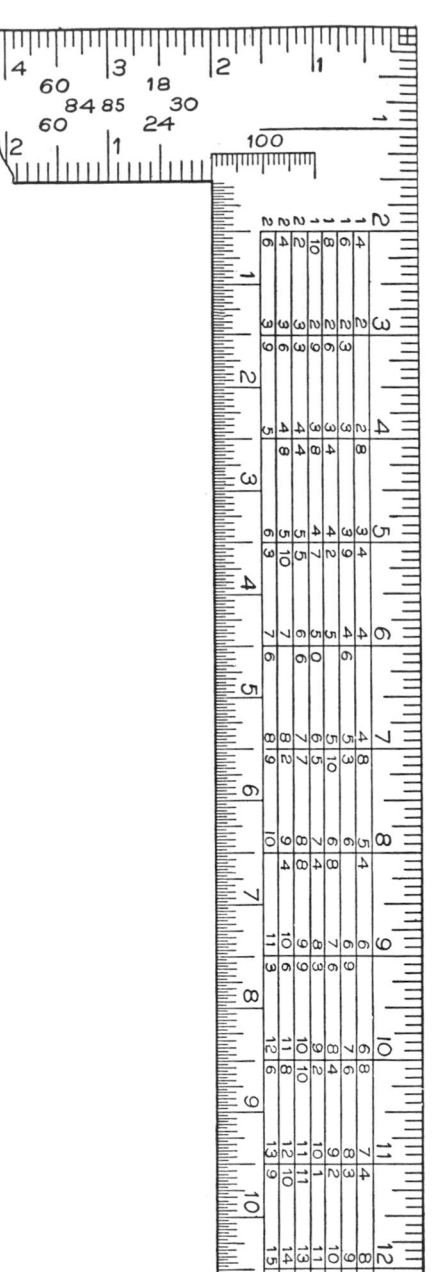

Fig. 4-19. A framing square. The back of the blade is used for board measure calculations.

by the width and thickness in inches, and divide the product by 12. This is a slow process and is a violation of our rule about multiplying feet by inches (discussed at the beginning of this chapter). The following system, as shown by examples, is recommended as a quicker method of calculation.

Example	No. of Pieces	Size	Length	Board Feet
A.	1	$2'' \times 8''$	30'	40
B.	1	$4'' \times 10''$	18'	60
C.	1	$10'' \times 10''$	36'	300
D.	1	$20'' \times 20''$	60'	2000

A.

2×8 equals 16; 16 divided by 12 equals 16/12 or 4/3. When this is multiplied by the length (30') the answer is 40, the Board-Measure.

$$\frac{2 \times 8}{12} \times 30 = 40 \text{ board feet}$$

B.

4×10 equals 40; 40 divided by 12 equals 10/3; multiplying 18 by 10/3 is 60.

$$\frac{4 \times 10}{12} \times 18 = 60 \text{ board feet}$$

C.

$$\frac{10 \times 10}{12} \times 36 = 300 \text{ board feet}$$

D.

$$\frac{20 \times 20}{12} \times 60 = 2,000 \text{ board feet}$$

The following is a list of *standard sizes* of boards and their *conversion factors*. The conversion factors are constants which were found by multiplying the width of the board by the thickness of the board and then dividing by 12. The *linear footage* (length)

Size	Constant
1×3	1/4
1×4	1/3
1×6	1/2
1×8	2/3
1×10	5/6
1×12	1
2×3	1/2
2×4	2/3
2×6	1
2×8	1-1/3 or 4/3
2×10	1-2/3 or 5/3
2×12	2
3×3	3/4
3×4	1
3×6	1-1/2 or 3/2
3×8	2
3×10	2-1/2 or 5/2
3×12	3
4×4	1-1/3 or 4/3
4×6	2
4×8	2-2/3 or 8/3
4×10	3-1/3 or 10/3
4×12	4
8×8	5-1/3 or 16/3
10×10	8-1/3 or 25/3
12×12	12
14×14	16-1/3 or 49/3
16×16	21-1/3 or 64/3
18×18	27
20×20	33-1/3 or 100/3
22×22	40-1/3 or 121/3
24×24	48

of the board multiplied by the conversion factor is the B.M.

A Variation in Method. A convenient method for computing B.M. is as follows:

For all 12 ft. lengths multiply width by thickness.

For all 14 ft. lengths multiply width

by thickness and add 1/6 of the resulting total.

For all 16 ft. lengths multiply width by thickness and add 1/3 of the resulting total.

For all 18 ft. lengths multiply width by thickness and add 1/2 of the resulting total.

For all 20 ft. lengths multiply width by thickness and add 2/3 of the resulting total.

For all 22 ft. lengths multiply width by thickness and add 5/6 of the resulting total.

For all 24 ft. lengths multiply width by thickness and double the resulting total.

Some objection may be taken to the use of 2/3 and 5/6, but often you may be able to substitute 1/6, 1/3, or 1/2 as in the following examples:

1) You need 10 pieces of 1 × 18 boards, each 22 feet long. In order to avoid the awkward fraction 5/6, you reverse the 18 and the 22. In this problem consider that you need 10 pieces of 1 × 22's, each 18 feet long. In this way you may use the formula for 18′ lengths, with its more convenient fraction of 1/2, without affecting the solution.

2) Similarly, 16 pieces of 1 × 22's each 20 feet long may be considered as 20 pieces of 1 × 22's, each 16 feet long.

The above system is very convenient when calculating lumber from 12 to 24 feet long, particularly where odd widths and thicknesses often occur.

Converting Board-Measure to Lineal Feet. Simply reverse the multiplier used to bring lineal feet to Board-Measure; in other words, multiply board feet by 12 and divide by thickness and width.

Example 17. How many lineal feet are there in 1,000 board feet of 2 × 8?

Solution:

$$\frac{12 \times 1,000}{2 \times 8} = 750 \text{ lineal feet}$$

Example 18. Car orders frequently call for a specified amount of sizes containing special lengths. Before proceeding to load, it is necessary to find the number of pieces required. Find the number of pieces in:

1,000 ft. B.M. 2 × 4—14
1,000 ft. B.M. 2 × 4—16
1,000 ft. B.M. 2 × 4—20

Solution: Change the Board-Measure to lineal feet as shown in Example 1; then divide the length into lineal feet. The result is the number of pieces.

$$\frac{12 \times 1,000}{2 \times 4} = 1,500 \text{ lineal feet}$$

$$\frac{1,500}{14} = 107 \text{ pcs.} \quad \frac{1,500}{16} = 94 \text{ pcs.}$$

$$\frac{1,500}{20} = 75 \text{ pcs.}$$

107 pcs. 2 × 4—14 containing   998 ft. 8 in. B.M.
94 pcs. 2 × 4—16 containing 1,002 ft. 8 in. B.M.
75 pcs. 2 × 4—20 containing 1,000 ft.   B.M.
—————————————————————
276   3,001 ft. 4 in. B.M.

Questions and Problems

1. One leg of a right triangle is 1' long and the other leg is 3' long. What is the length of the third side? (Answer: 3.16')
2. A right triangle has a hypotenuse 7" long and a leg which is 2" long. Find the length of the third side. (Answer: 6.71")
3. A stairway 35' long reaches the top of a wall 28' high. How far is the foot of the stairs from the base of the wall?
(Answer: 21.0')
4. Refer to the obtuse triangle in Fig. 4-3. Let the base AC equal 9' and the side AB equal 15'. A perpendicular, BD, intersects line AC at D. The distance from C to D is 3'. Find the area of triangle ABC.
(Answer: 40.5 sq. ft.)
5. Find the circumference of a drain pipe which is 3.5" in diameter. (Answer: 11.0")
6. Find the area (in sq. yds.) of the base of a silo with a radius of 10'. (Answer: 34.91 sq. yd.)
7. How many square feet are there in a patio which is in the shape of a half-circle, and whose radius is 75"?
(Answer: 61.36 sq. ft.)
8. Find the area of a trapezoid whose bases are 12.25' and 16.25' and whose altitude is 12.0'. (Answer: 171.0 sq. ft.)
9. What is the volume of a rectangular prism whose height is 9'?

The dimensions of the base are 3'-6" × 2'-0".
(Answer: 63.0 cu. ft.)
10. A circular excavation is to be 20 yards in diameter and 4 yards deep. What is the volume of the excavation?
(Answer: 1256.64 cu. yds.)
11. What is the volume of a regular triangular pyramid whose altitude is 20'? The sides of the base are each 15' long.
(Answer: 649.5 cu. ft.)
12. Find the volume of a cone whose slant height is 20' and whose base is 22' in circumference.
(Answer: 253 cu. ft.)
13. What is the volume of the frustum of a cone, the radiuses of the bases being 4' and 2', and the altitude 10'?
(Answer: 293.22 cu. ft.)
14. Find the surface area and volume of a spherical storage tank whose diameter is 24'.
(Answer: S = 1,809.56 sq. ft., V = 7,238.25 cu. ft.)
15. Find the number of board feet contained in a board 10' long and 7" wide. Use the framing square in Fig. 4-19.
(Answer: 5'-10")
16. Find the B.M. of the following pieces of lumber. Each is a single piece:
(a) width = 4", thickness = 2", length = 24"
(Answer: 16)

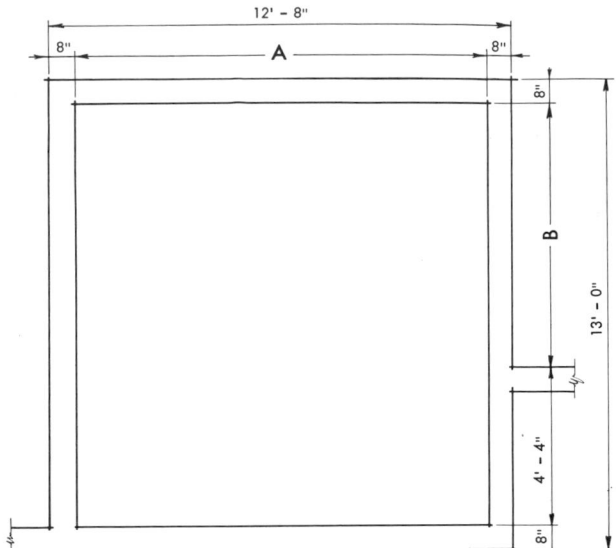

Fig. 4-20. Finding missing dimensions.

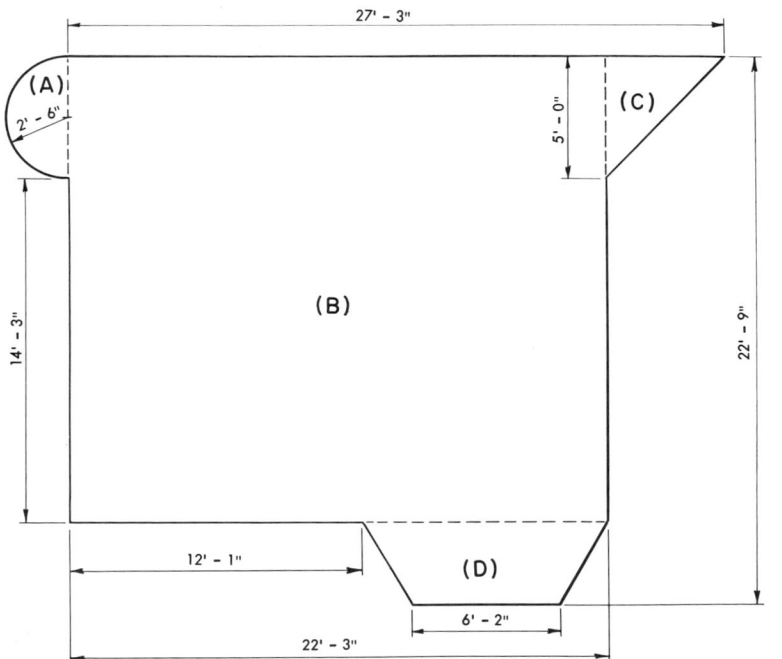

Fig. 4-21. Finding the area of an irregular figure.

(b) width = 8″, length = 30′-0″, thickness = 5″

(Answer: 100)

(c) thickness = 8″, width = 8″, length = 60′-0″

(Answer: 320)

17. Find the missing dimensions in Fig. 4-20. (Answers: A = 11′-4″, B = 7′-4″)

18. Find the area of the irregular figure shown in Fig. 4-21. (Break the area into simple shapes.)

(Ans.: A = 9.82 sq. ft.

(Ans.: B = 428.31 sq. ft.

(Ans.: C = 12.5 sq. ft.

(Ans.: D = 28.6 sq. ft.

Total area = 479.23 sq. ft.)

5

SURVEYING
AND EXCAVATION

A set of plans for which an estimate is being made usually includes a plot or location plan on which a surveyor has previously noted the exact lot position in addition to the elevations. ("Elevation" as it is used in this chapter refers to the distance above sea level or above an arbitrary elevation of 0 or 100.) The estimator must have a general understanding of surveying principles in order to compute excavation yardage accurately in all cases.

If a lot is nearly level the calculation of excavation yardage is quite simple. Only the elevation under the basement floor and the surface elevation need be known in order to compute the volume of the excavation. Or, if the lot is level and it is known that the elevation must be 6'-4" deep the computation is equally simple, as will be shown later in this chapter. If a lot is sloping or irregular in surface the calculation of excavation yardage is much more complicated, and it becomes necessary to have more detailed elevation data.

This chapter will acquaint you with general principles relating to excavation calculations and with surveying operations and principles required for the excavation calculations.

SURVEYING

Topographic Maps. Frequently, topographic maps are made as a preliminary step to construction work and are called plot, survey, or location plans. These maps (as they are technically called) should show all details which will be of interest to the architect, the estimator, the contractor, and the owner. Such details include trees, elevations, terraces, slopes, building location, lot lines, street and sidewalk lines, sewer lines, water lines, and utility lines. The estimator is primarily interested in the location of the house, the elevations, and the character of the surface of the ground. In fact, if a set of plans does not contain such a map and if the lot is irregular in elevation it will be necessary for the estimator to have a map made or to visit the site and make such a map himself.

Contours. A contour is a line joining points which have the same elevation. An elevation is a known distance above or below a given level, such as sea level or an arbitrary elevation of 0 + 00 or 100 + 00, the latter being used to avoid negative numbers. Every town and city has one or more permanent points whose elevation has been established and from which elevations on a given lot may be determined. In the larger cities these points of known elevation are situated at convenient locations in different parts of the city. Sometimes they are marked on buildings and sometimes by iron rods fixed in concrete bases. The exact number of feet is determined and marked on a *bench mark* or master marker. At various points in the city, other markers give the elevation at their location. Elevations vary because of slopes or other irregularities of the surface. In Chicago, which is practically level, the variation is slight, while at Davenport, Iowa, located among the rolling hills of the Mississippi River Valley, elevation readings differ sharply.

Some cities have what is known as a *city datum* from which all elevations or levels can be calculated.

The simplest way to understand the nature and variability of contours is to picture them as elevation marks on a cone. This is illustrated in Fig. 5-1. Suppose a right circular cone, Fig. 5-1-*A,* is 6 feet high and is placed in a

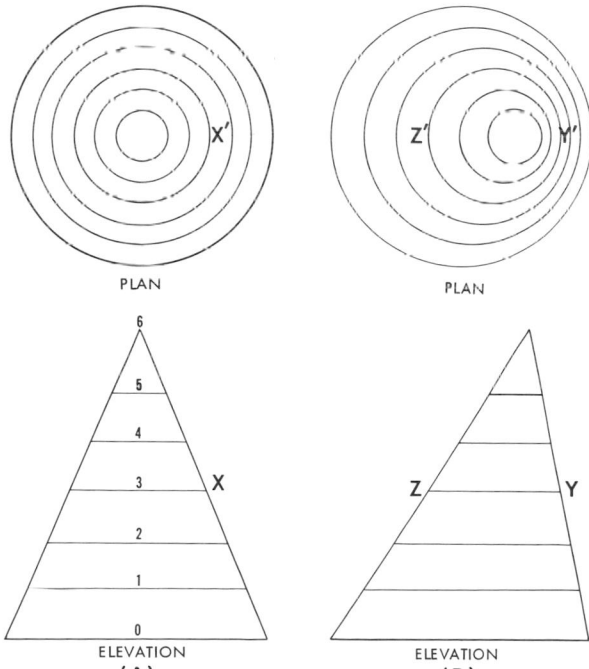

Fig. 5-1. A method of representing contours.

tank of water. When the water just covers the bottom of the tank, the contour is the line marking the outer circle in the plan view above Fig. 5-1A. As the water is raised by 1-foot increments the contours will be indicated by concentric circles. If the cone is inclined, the contours will be as in the plan view above Fig. 5-1B. The evenly spaced contours at X' indicate a uniform slope; where farther apart, as at Z', they indicate a more gentle slope; and where closer together, as at Y', they indicate a steep slope.

Fig. 5-2 shows a typical contour sketch in which the lower part describes typical contours and the upper part indicates the point of elevation.

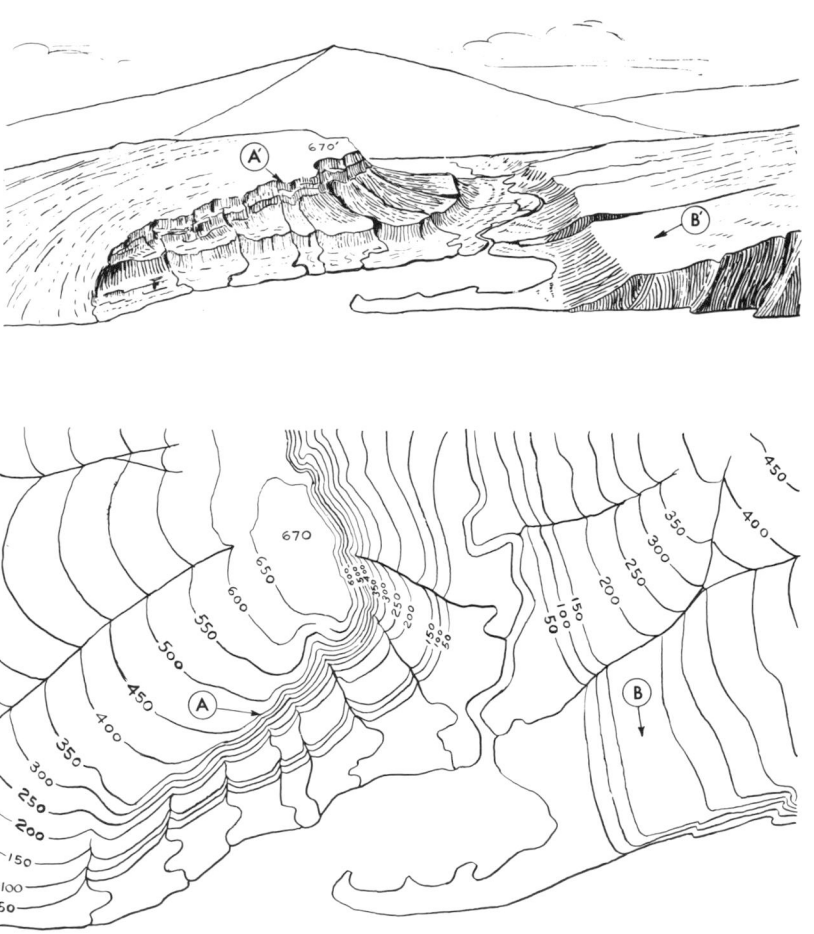

Fig. 5-2. Contour sketch.

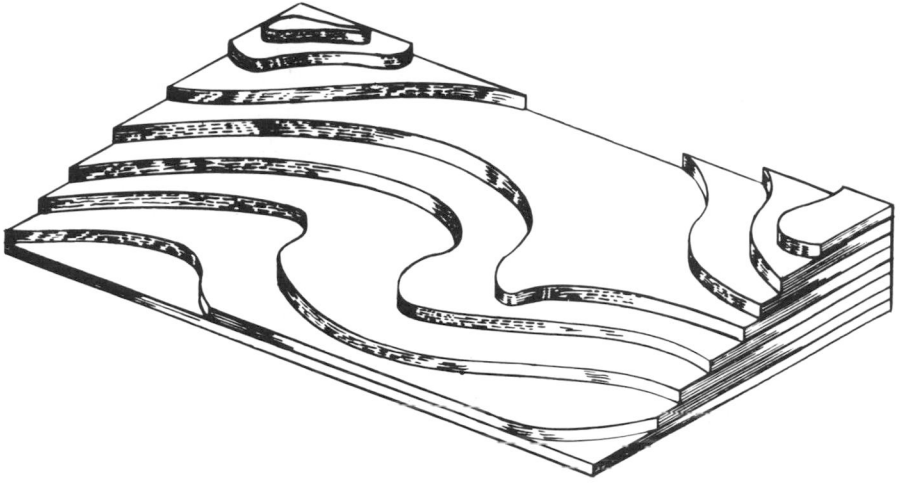

Fig. 5-3. A contour model.

For example, the contours at *A* are close together and indicate the terrace shown at *A'*. The wider contour interval at *B* indicates the flat surface at *B'*.

Fig. 5-3 will further enable you to understand contours. If parts that are intended to show the various elevations (levels) of a given area are cut from cardboard and assembled successively, a contour model such as that shown in Fig. 5-3 is obtained. Each part is scaled to the level of the contour interval it represents. Plaster or clay in natural colors may be applied to the cardboard to give the model the finished appearance of a relief map.

Contour Interval. A certain uniform vertical increment, or contour interval, is always used in making a map, the interval varying with the scale of the map and the character of the country. For the closest of detail work on a larger scale, such as might be used for landscape gardening, a con-

tour interval of 1 or 2 feet might be necessary. But for ordinary purposes 5 feet or more is used as the contour interval. The topography for preliminary railroad surveys usually is done with a 5-foot contour interval. The topographic maps made by the United States Geological Survey, which are plotted at a scale of approximately 1 or 2 miles per inch, are made with a contour interval varying from 20 to 100 feet, the 100-foot interval being necessary where steep mountain ranges are to be represented.

Principles of Contours. There are certain principles pertaining to contours which must be kept in mind by the estimator in order to avoid making errors in his work. Barring a few exceptional cases which are mentioned later, the following principles are observed: (1) a contour is always a continuous line enclosing an area; (2) since that area may and frequently does run off the map, the contour may run from one edge of the map to any other point on the edge; (3) a contour line never stops abruptly; and (4) contour lines do not cross each other or merge into each other.

The exceptions to the above rules occur only where the contours run into an artificial vertical wall or a precipice which actually is vertical. In the case of an overhanging cliff the contours might actually cross each other very slightly, but with rare exceptions (which are readily recognized where they occur) the above principles are used for most contour maps.

Contours are shown in the same manner on maps of large or small scale; the interval alone varies. The contour elevations are always given with reference to some datum plane, such as sea level, and the elevations are always a multiple of the contour interval. For example, if the contour interval is 20 feet and the datum plane is sea level, then each contour elevation will be a multiple of 20 even though the whole tract is far above sea level. The contours might be at elevations of 660, 680, 700, 740, etc. Then the 600-, 700-, or 800-foot contours would be made extra heavy and could be distinguished more readily, even in steep places where the contours would be very close together.

The contours should be numbered at frequent intervals so that the elevation of a contour at any point may be determined by following it for a very short distance.

Location Plan. Fig. 5-4 illustrates a typical *location plan* or *plot diagram*

for House Plan A, the single-unit residence shown on the working drawings in Chapter 2. The plot diagram indicates such things as property lines, easements, walks and driveways, and grade levels. It can be considered a "bird's-eye view" of the land surrounding the house.

In Fig. 5-4, the elevations at different points on the property are indicated. The northeast corner is highest (115.7′) and the southwest corner is lowest (110.7′), indicating a slight slope. The house itself is to be built at a level of 112.6′. The first floor level is raised almost two feet to 114.5′.

If Fig. 5-4 had been made in Chicago, using actual sea level elevations, the elevations shown would be increased by approximately 500 feet each.

With this type of map the estimator can easily compute the yardage of the lot (unless the surface is very irregular). Methods of calculation will be explained later in this chapter.

Determining Contours. One method of determining contours which is commonly used is called the *square method*. This method consists in establishing a grid of squares or rectangles over the desired lot or area and determining the elevation of each corner by means of a *level-transit,* a *wye-level,* or a *dumpy level.*

The points where the contours cross the side of the squares are next determined, usually with a hand level and a metal tape. Suppose one corner, Fig. 5-5, is at elevation 803.2 feet and the next corner, 812.5 feet. If a 3-foot contour interval is being used, three

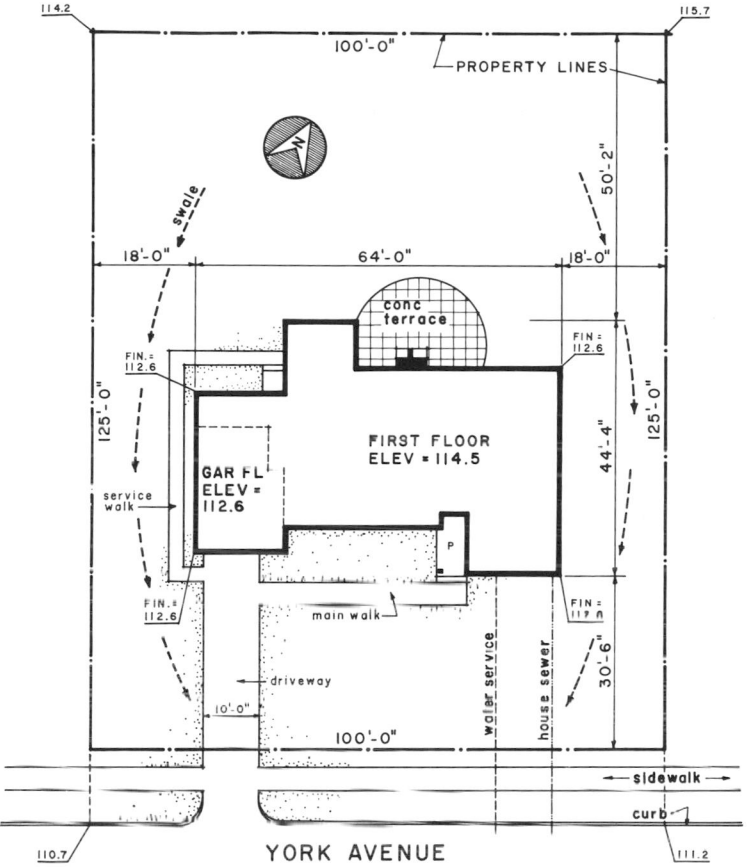

Fig. 5-4. Plot Plan of House Plan A.

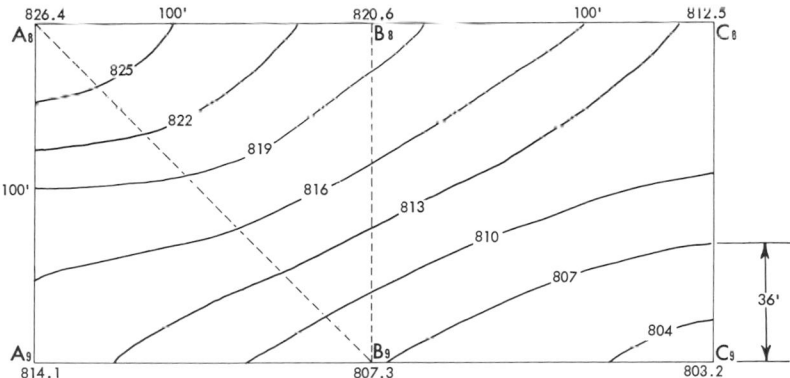

Fig. 5-5. The square method of determining contours.

contours will cross the line between these corners—804, 807, and 810. The hand level is used to take a plus (+) sight of, say 5.1 feet on the lower corner. The elevation of the line of sight is then 803.2 + 5.1, or 808.3, and a minus (−) sight of 808.3 −807.0, or 1.3 feet, is then required to locate a point on the 807 contour. The rod is therefore moved along the line between the corners until it gives a reading of 1.3 feet; and the contour is located by taking the distance, 36 feet, between the rod and the corner.

In many cases it is not necessary to locate the points where the contours cross the sides of each square, as they may run practically straight for several squares. In other cases, such as that shown in the left-hand square, Fig. 5-

5, bends occur at points inside the squares, and careful work requires that the contours be located on the diagonal (dotted) lines. Careful sketches made approximately to scale in the field are essential.

The square method is probably the most accurate method of contouring and is used where a small interval, 2 or 3 feet, is required. Such a small interval is used in maps for landscape gardening and similar work and requires close accuracy of location.

Another method of determining contours is called the *cross-section* method. A traverse consisting of straight lines between instrument stations is run with a transit or other angle measurer and a tape. At certain intervals, stakes are placed and the

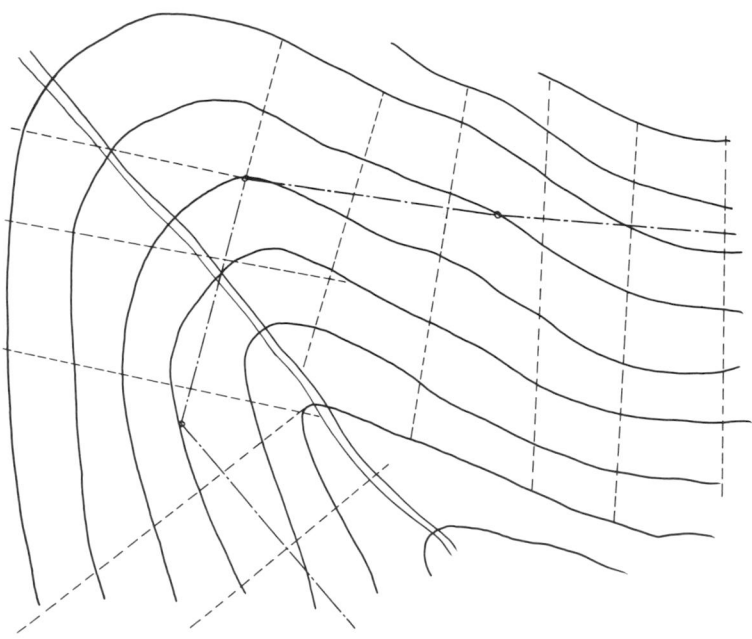

Fig. 5-6. The cross-section method of determining contours.

elevations found with a level-transit. Cross lines are then run at right angles to the traverse lines as shown in Fig. 5-6. The elevation of the stations being known from the traverse lines, the contour points on the right-angle lines are determined in exactly the same way as in the square method.

EXCAVATION

Briefly, excavation is the digging of a hole in the ground to provide room for engineering improvements. These improvements include footings, foundations, basements, sub-stories, ramps, exterior stairways, areaways, pipe trenches, pits for septic tanks, and traps.

The estimator, while considering his drawings and specifications, prior to his actual quantity take-off, should carefully study all the General Conditions. Once he has these clearly in mind he can start to do the quantity take-off. The quantity take-off for excavation is usually calculated in cubic yards. Should the excavation require the removal of a few additional inches of soil it would be calculated in square feet.

The estimator must not only do a very careful quantity take-off, but should study the job for such money-saving items as *topsoil*. This valuable substance is found on many building sites. He can then sell any unneeded soil to other contractors or landscapers. This is only one example of how a careful estimator can save or earn money for his contractor. Careful study of the site, topographic maps, plans,

and specifications will reveal other opportunities for saving money.

Further savings can be realized with respect to *fill*. If the ground is lower in certain spots than the finished grade shown on the drawings, the contractor may have to buy fill to bring the level to the specified grade. This is not a usual occurrence but represents a possible cost to the contractor. The required fill will often take many more yards of earth than has been excavated. The estimator must consider this possibility in his calculations; failure to do so will lower the contractor's profit.

Other items which the estimator must take into account are the costs involved in *backfilling, utility line excavation,* and *cradling.* (These are explained in the following section, "Special Terms.") This means that the estimator must carefully study *all* parts of the specifications and drawings and still realize that there might be other *unspecified* factors which must be taken into consideration.

Special Terms. The following terms should be understood by the estimator before he attempts to do a quantity take-off for excavation:

General Excavation. All excavation, other than rock and water removal, which can be done by available mechanical equipment or any general piece of equipment such as bulldozer, clam shell cranes, backhoes, scrapers, power shovels, and loaders, is considered as general excavation. The use of trucks to remove the excavated material is also included under this heading.

Special Excavation. All excavation done by hand, by special machine, by blasting, or by the use of special methods is considered as special excavation. One type of special excavation is a pipe trench for a water line. This would be done by a trencher, commonly called a "ferris wheel." Another type of special excavation is required for the installation of telephone, electrical, and utility poles. This type of excavation might require a vertical boring rig mounted on a bulldozer.

Sheet Piling. Vertical retaining walls of wood or steel are used when excavating adjacent to an existing structure or when excavating near a heavily traveled roadway where there is vibration from passing vehicles. These vertical retaining walls are called sheet piling; they retain any horizontal thrust of earth which might cause the excavation walls to slide, Fig. 5-7.

Wood planking placed closely together and steel sheets (often of an interlocking construction) are the most common forms of sheet piling. In some cases, where the sheet piling is to become a permanent part of the structure, precast concrete tongue-and-groove planking is used.

There are several methods of installing sheet piling:

1. If the earth is soft enough to permit driving, a *pile driver* with a drop hammer rig or single stroke hammer is used to drive the piles to a "point of refusal" which will prevent the sheet piles from being pushed over by any horizontal thrust of earth.

2. Another method of placing sheet piling is by *pneumatic hammer*. This is done by hand on short-driven piles.

3. Another method is to place the

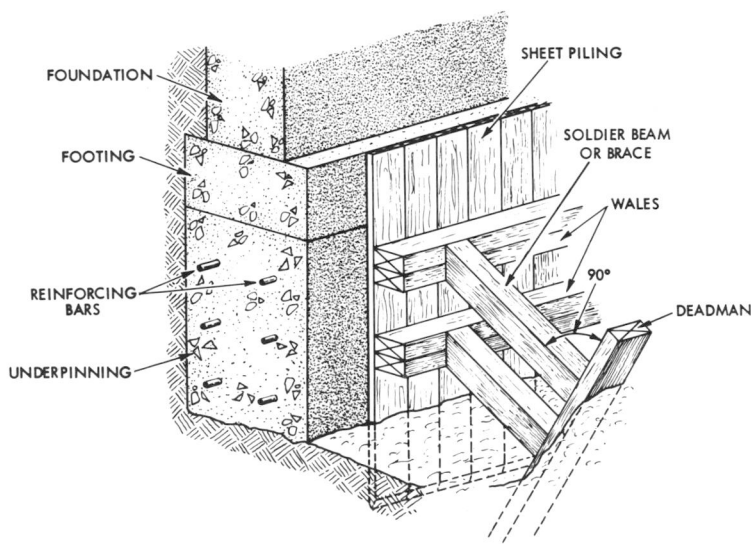

Fig. 5-7. Sheet piling and underpinning.

sheet pile into a hole which was hand dug. The piles are then *braced, shored,* or *backfilled.*

Underpinning. The laws of some states require underpinning when a building is constructed adjacent to an existing structure and when the foundation of the proposed building goes below the foundation line of the existing structure. It is necessary to bring the foundation footing of the existing structure down to the level of the proposed building. (Underpinning is illustrated in Fig. 5-7.) This means that when excavating for a building 10 feet lower than an existing structure, you must pour a concrete foundation footing, 10 feet deep, the entire length of the existing structure which faces the proposed building. This is a costly and time-consuming job.

The estimator should also know the local laws and the soil conditions at the site. *Remember that the contractor is held responsible for any damage to existing structures.*

Soil and Rock. For estimating purposes, soil and rock may be classified as follows:

Light soil is a granular substance which can be readily shoveled by hand without the aid of machines. This includes gravel, sand, coarse sand, and fine sand.

Medium soil includes all soils which can be loosened by picks, shovels, and drag line scraper. This includes the cohesive soils such as clay and adobe.

When clay is encountered, the excavation is usually made about 4 inches deeper than for sand, sandy soil, or gravel. This allows for fill material such as crushed gravel.

Heavy soil includes all soils which can be loosened by picks, but are hard to loosen with shovels. Machinery such as backhoes and dippers must be used. Compacted gravel, small stones, and boulders are in this classification.

Hard pan includes boulders, clay, cemented mixtures of sand and gravel, and other substances which are difficult to loosen with a pick. It is advisable to use light blasting to loosen any type of cemented soil, thereby making it easy for a power shovel to pick it up and load it on a truck.

Rock includes soft rock (*shale*), medium hard rock (*slate*), and hard sound rock (*schist*). Rocks require heavy blasting which is expensive. Sub-surface drawings should be studied to determine if rock is present under a site; its presence may mean additional excavation costs.

Angle of Repose. The slope which a given material will maintain without

TABLE 1. ANGLES OF REPOSE

SOIL	DEGREES FROM HORIZONTAL		
	DRY	MOIST	WET
SAND	20–35	30–45	20–40
EARTH	20–45	25–45	25–30
GRAVEL	30–48	.....	.....
GRAVEL, SAND, AND CLAY	20–37	.....	.....

sliding is called the angle of repose. Soils vary greatly in their ability to hold in place at the edge of an excavation. A clay soil, unless subjected to the action of running water, will stand practically vertical at the edge of an excavation; a soil with a high sand content slides so that the bank of an excavation becomes a long slope. Table I lists angles of repose for some common soils.

Cradling. Cradling is the temporary supporting of any existing utility lines in or around the excavated area. On many construction jobs it is not possible to remove, cut, or demolish utility lines; the lines must be properly protected so that the utilities of the surrounding area are not interrupted. Cradling is accomplished by placing timbers under the utility lines (usually pipes) and then jacking up the timbers as the excavation goes deeper; more timbers are placed under the pipe until there is a tower of timbers from the bottom of the excavation to the underside of the utility line. The utility lines are thus kept at their original level.

Well Points. This system is used to *dewater* an excavation by means of horizontal header pipes which are connected by a swing joint to vertical pipes which are driven into the lower ground water table. These vertical or riser pipes will then drain the ground

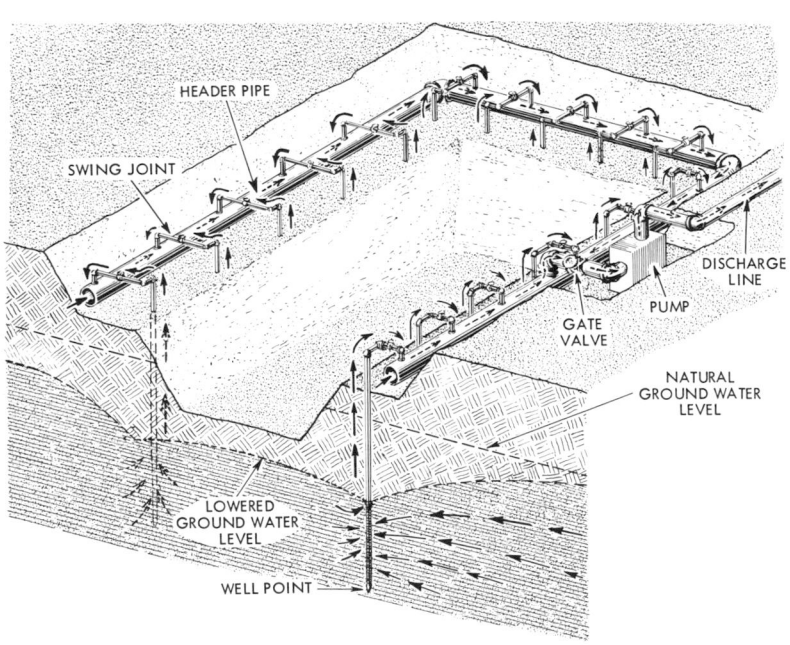

Fig. 5-8. The well point system of dewatering.

water table by means of the well point pump. Fig. 5-8 shows how such a system works.

Ground water is a great problem and unless the water is properly controlled it can cause sand excavations to be much more costly than rock excavations. Improper handling of ground water can slow the construction work and even bring it to a halt. Engineers realize that removing ground water is costly and when preparing an estimate, they utilize the of the footing and then up on a slope." This is done because soft soil cannot stand vertically; it might slide into the excavation.

Soil Swell. Soil, in its natural state, is generally closely compacted. When this soil is disturbed by excavating, and loaded into dump trucks or piled, it increases its volume due to *swell*. Such swelling can occur to the extent of 50 percent. Table II shows the average increase in volume for various soils.

TABLE II. SOIL SWELL

SOIL	BEFORE EXCAVATION	AFTER EXCAVATION
EARTH CLAY.....................	1	1 1/4
SAND AND GRAVEL..............	1	1 1/2
BROKEN STONES................	1	1 1/2
FREE STONE	1	1 1/4
ROCK..........................	1	1 1/2

best available advice on the practical and economical methods of dewatering. It is advisable, should the contractor discover water beneath his proposed excavation, to contact reputable dewatering contractors to bid on and dewater the building site.

Payline. The payline establishes the exact amount of excavation for which the contractor is responsible and for which he will be paid. The way to accurately compute the amount of excavation is outlined in some specifications as follows: "The payline shall be 2 feet from the top edge of the footing out, and then vertically up." This, of course, would not apply to soft soil. In soft soil the payline statement would read "two feet from the top edge

Neat Cut. A vertical-walled excavation is called a neat cut. It is made if, in the opinion of the contractor, the soil is well compacted and the excavation shallow enough to eliminate the possibility of slides or cave-ins.

Slope Cut. The term slope cut is applied to an inclined or sloped wall excavation. The slope cut is used for deep excavations in soft, shifting soil. The angle of a slope cut is determined by the angle of repose for soil in the area.

Backfill. Backfilling is the process by which the excavator replaces the earth around the sub-structure or foundation. In backfilling, it is necessary for the estimator to make sure the ground doesn't contain large boulders or extremely large masses of earth or debris. These would increase the cost of backfilling since it would require more cautious methods of pushing back the earth.

Compaction. Compaction is attempting to place the disturbed earth back into its original compact condition. This means trying to remove the air spaces so that the particles are as closely packed as possible. This is usually achieved by wetting the earth while it is being backfilled and then continually going over it with a large piece of heavy equipment or a vibrator tamper.

Rough Grading. The excavation specifications will probably call for rough grading of the site. Rough grading is done over that portion of the site which is not occupied by the building, roads, or walks. This is done before spreading of top soil and *finish grading.*

Excavating, Filling, and Rough Grading. In regard to this section of the specifications the contractor for general construction (and in some cases the excavator) must closely follow the written specifications.

Topsoil. Excavating for a building requires many types of work. For example, before actual excavating can begin, it is profitable to remove the existing brush and small trees. If there is any topsoil on the site, it is financially desirable to scrape as much of it as possible and stockpile it for future use; if there is an excess of topsoil, it can be sold to landscape gardeners or other builders. Topsoil is usually spread 6 to 12 inches deep.

If the building site is lower than the proposed grade or the proposed finished grade of the building, it is necessary for the estimator to calculate the amount of fill which is required to bring the land around the building up to the proposed grade. If the development, building, or structure is to be near existing structures, it might be necessary to shore, brace, and protect the adjacent structures with either temporary shoring or permanent supports, in which case they would require underpinning. To avoid injury to men and equipment, the excavating contractor must shore all areas where excessive vibrations (due to traffic and movement of equipment) might affect the side banks of the excavation.

Some of the different types of excavation for common structures are as follows:

1. Excavating and/or filling for all foundations of the building to the bottom of the piers, footings, wall beams, grade beams, and buttresses.

2. Excavating of the cellar, interior beams, and walls.

3. Excavating for all pits except housetrap pits which are usually estimated by the plumbing contractor.

4. Excavation for ramps, ramp walls, exterior stairs and walls, areaways, or exterior and interior retaining walls as detailed in the plans is another distinct part of the excavating contract which the excavator must study carefully since (in most cases) the special excavation will be in confined areas.

5. Excavation for the boiler stack and incinerator foundations are two

special excavations. The boiler stack and incinerator foundations are usually the deepest part of the structure and might have to sustain extreme loads. The excavator must be careful not to excavate too deeply because filling in with loose soil will reduce the load-bearing capacity of the soil.

6. Backfilling around all wall beams, slabs on the ground, foundation and interior walls, and wherever specified in the drawings or written specifications.

7. Excavation for utility supports for heating and sanitation lines, storm sewers, house sewers, as well as for water, fire, and gas lines is sometimes considered as a special excavation under a separate part of the contract— the heating and plumbing section. Regardless of responsibility, the general contractor and estimator should read and understand what is required by the excavation specifications for this particular portion of the work.

There are many other things which might be specified under the excavation section:

8. If there are large trees they are usually mentioned in the specifications. The method of removal and disposal of the trees can vary significantly. For example, in some areas you may cut the trees into small logs and bank them on the site. In other areas, you may cut the trees into small logs and use them as a portion of your backfill on the general site. In yet other areas, you might be required to cut the trees into small logs and transport them to a public or private rubbish dump. Each of these removal methods

is expensive; the estimator and/or excavating contractor must therefore visit the site, understand the local requirements, and price the job in accordance with the methods required for the removal of the trees.

If the trees are to remain as a permanent part of the site, it is usually required that the excavating contractor protect these trees. Protection might be in the form of tying the roots, tying the trees with guy wires, building fences around the trees, marking them, blocking them out, or even pruning the trees. Regardless of what is necessary to protect the trees, the estimator must be sure to calculate its cost.

9. As previously discussed, the excavating of a building is divided into general and special excavation. General excavation can be done with equipment which is readily available and accessible to the construction site. Special equipment and hand excavation is considered as special excavation. In quantity take-offs for excavation, it is necessary to estimate and check-off each item separately since general excavation will usually cost less than special excavation.

10. Excavation estimates for a site are frequently approximate estimates because it is seldom possible to determine the exact geological conditions at the site. Regardless of the number of test borings, the spacing between bores is usually so great that what lays between the borings is seldom known for sure. In such cases there are certain items which the excavation estimator must submit as *unit prices*. (Unit prices are stated as price

per yard, per hour, per ton, or per truckload.) These items are considered as special excavation. For example, the removal of an old concrete foundation, where the plans for the demolished building no longer exist, must be taken as special excavation and the contractor must be given a unit price. Without a unit price, the estimate is a gamble to all concerned parties.

11. The removal of rock is special excavation because it is not always possible to determine how large or deep the boulders are at the construction site. When removing a boulder, it is not always possible to break it—leaving a portion below the grade line. The complete removal of rocks can be very expensive. The estimator should either set a unit price for rock excavation or the owner or contractor should give the company a unit price. The unit price is based upon the amount of rock by tonnage, boulder, truckload, or pieces, depending on the area and accessibility of the site.

12. Removal of underground water is one of the most costly and important items. You cannot work under or around water without special equipment. Dewatering is a special excavation item and an allowance is made to the contractor for pumping or installing a wellpoint system.

After the hole is dug and the concrete workers, the steel workers, and the other tradesmen who work below grade level have finished their work, it is necessary to backfill. For a large building, you must not only backfill from the exterior walls to the sidewalk but also to the pier frames of the structure. The cellar slabs must be backfilled and leveled. The estimator can easily see that the cost of backfilling the outside of a structure is usually less per yard than backfilling the interior or the interior foundation walls, where the working radius of the equipment is shortened considerably because of the pillars, pilasters, columns, and other parts of the sub-structure.

Now you are almost ready to do an excavating estimate. You are aware of what is involved in an excavating estimate and should see the need for carefully reading the excavation portion of the specifications to acquaint yourself with the various jobs and items. After acquainting yourself with these jobs and items, you should list them in the proper sequence, on an estimating sheet, and price each one. Then estimate for the various pieces of equipment.

First you will learn how the estimate may be done in several ways. Then the use of excavating equipment will be explained in more detail to give you a better understanding of haulage, the cost of loading material, and the cost of the machinery itself. This does not mean a useable dollar-and-cents value for the job because costs vary greatly among areas. However, the time that each piece of equipment takes to do a given job varies little, and these figures are therefore accurate and can be applied to most jobs.

QUANTITY TAKE-OFFS

Profiles. Prior to the actual excava-

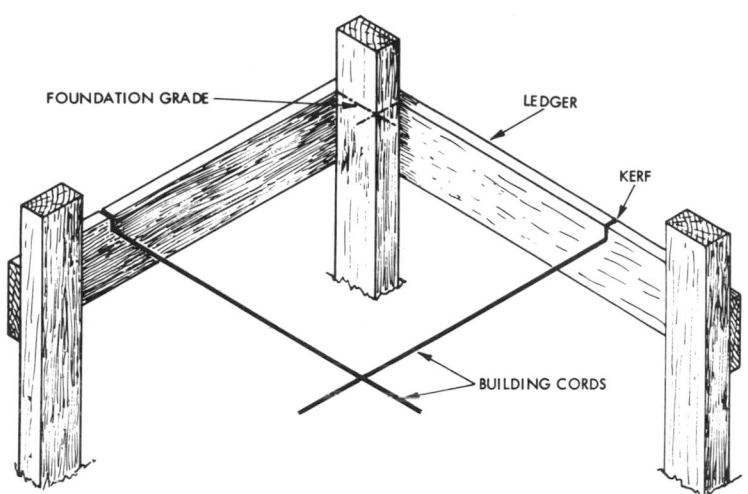

Fig. 5-9. Batter boards are used to mark the boundaries of the construction site.

tion, the area is staked out or *profiled* by one of several methods, depending on the size of the structure. On a very small excavation, a 6-8-10 triangle is used at each corner, and on larger jobs a transit is used to establish the lines. After the lines are established they are transformed to *batter boards,* Fig. 5-9, which are 1″ × 6″ boards nailed at right angles in a horizontal position to three 2″ × 6″ boards. The batter boards thus nailed form a corner in the shape of an L. Batter boards can also be used at the sides of the cut without the L leg to establish a line for special interior foundations and walls. The main purpose of the batter boards is to maintain excavation lines, foundation lines, and building lines.

Ordinarily so little lumber or other material is required that the contractor supplies it without consideration in the quantity take-off or estimate. On exceptionally large jobs where a great many 2 × 4's and planking are required, the amount of lumber is estimated by actual count of the pieces. Since batter boards are reuseable, the small material cost is not considered.

Gridding. Suppose that an airport is to be constructed on land which varies in elevation. The large portion of land which is needed might have hills, valleys, and gullies. The terrain must be cut (excavated) or filled until the final grade is uniform.

To estimate this job you must use the topographical site map and divide it into small squares, or *grids.* This operation is called *gridding* or *grilling.*

The grids are usually drawn on an overlay sheet of tracing paper. The size of the grids is determined by the nature of the terrain. If the terrain is gradually sloped, the grids might represent squares of 100 feet on each side.

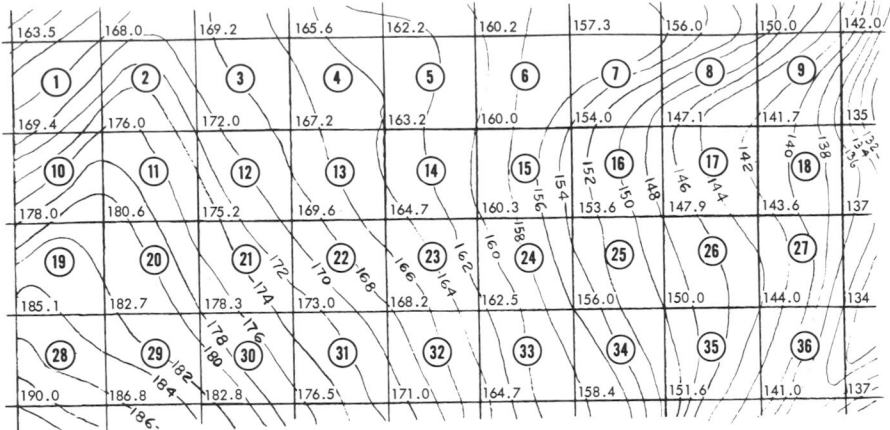

Fig. 5-10. An example of gridding. The contour plan is divided into equal areas.

When the terrain is very irregular the grid size is decreased to 25 feet square; this gives a more accurate picture of the cut and fill needed for the particular site.

In Fig. 5-10, assume that the length and width of each grid is 100 feet; thus each grid is 100′ × 100′, or 10,000 square feet. The approximate elevation at each corner is established by using the nearest contour line. (These elevations are given on the drawing.)

Look at Grid 32 and note which contour lines pass close to the corners of the grid. The corners are approximately at elevations of 168.2′, 162.5′, 171.0′, and 164.7′. adding together these four elevations and then dividing by four gives the mean or average elevation of the grid:

$$\frac{666.4'}{4} = 166.6'$$

The final grade is to be 162.0′, therefore, the existing grade for this grid is at an average of 4.6′ above the final grade. The excavators must cut 4.6 × 10,000, or 46,000 cubic feet from this particular grid.

Fig. 5-11 shows the cut (or fill) required for each grid. The differences of the totals for the cut and the fill indicate how much fill (if any) must be brought onto the site. In this case, you would have to haul 26,400 cubic feet, or 978 cubic yards of fill.

The excavated earth is stockpiled on the site and represents a further cost. The cost of moving this earth around the site is added to the cost of the additional fill. The unit costs and the total cost depend on the type, the size, and the efficiency of the equipment used for this job.

General Excavation. A quantity take-off for the excavation of a house foundation (in two different types of soil) will help you to understand better the method of computing general excavation yardage.

GENERAL EXCAVATION

QUANTITY TAKE-OFF — AIRPORT SITE

Nc	Average Elevation ft.	Avg. Cut ft/sq.ft.	Quantity cu.ft.	Avg. Fill ft/sq.ft.	Quantity cu.ft.	Total Cut cu.ft.	Total Fill cu.ft.	Difference cu.yds.
	0 FT. GRADE LEVEL = 162.0'							
1	169.22	7.22	72,000					
2	171.30	9.30	93,000					
3	168.50	6.50	65,000					
4	164.55	2.55	25,000					
5	161.40			0.60	6,000			
6	157.87			4.13	41,300			
7	153.60			8.40	84,000			
8	148.70			13.30	133,000			
9	142.17			19.83	198,300			
10	176.00	14.00	140,000					
11	175.95	13.95	139,500					
12	171.00	9.00	90,000					
13	166.17	4.17	41,700					
14	162.05	0.05	500					
15	156.97			5.03	50,300			
16	150.65			11.35	113,500			
17	145.07			16.93	169,300			
18	139.33			22.67	226,700			
19	181.60	19.60	196,000					
20	179.20	17.20	172,000					
21	174.02	12.02	120,200					
22	168.88	6.88	68,800					
23	163.92	1.92	19,200					
24	158.10			3.90	39,000			
25	151.87			10.13	101,300			
26	146.34			15.89	155,300			
27	139.65			22.35	223,500			
28	186.15	24.15	241,500					
29	182.65	20.65	206,500					
30	177.65	15.65	156,600					
31	172.17	10.17	101,700					
32	166.60	4.60	46,000					
33	160.40			1.60	16,000			
34	154.00			8.00	80,000			
35	146.65			15.35	153,500			
36	139.00			23.00	230,000			
						1,995,600	2,022,000	978

Fig. 5-11. A quantity take-off for the general excavation of an airport site.

The house foundation is to be 40 feet long, 20 feet wide, and 5 feet deep for both examples. The footings extend 6 inches beyond the foundation. *Example 1*. The foundation is in light soil and the payline is 2 feet out from the edge of the footing and then up on a 1:1 slope. Figs. 5-12 to 15 illustrate this simple excavation. Since a sloped payline is required, it is necessary to use a formula for the volume of a frustum of a right pyramid:

$$V = a/3 \, (A_1 + A_2 + \sqrt{A_1 \times A_2})$$
$$or \; V = a/6 \, (A_1 + A_2 + 4M)$$
$$\text{where } A_1 = \text{bottom area}$$
$$A_2 = \text{top area}$$
$$M = \text{mid-section area}$$

The second formula will be used for the volume calculations; thus the areas of the bottom, top, and the midsection must be found. Since the footings extend 6 inches beyond the foundation, 1 foot must be added to the length and to the width. Figs. 5-14 and 5-15 are cross-sectional views of this excavation. The payline is 2 feet beyond the footings, so 4 feet must be added to both the length and width.

Bottom length =
  $40.0' + 1.0' + 4.0' = 45.0'$

Bottom width =
  $20.0' + 1.0' + 4.0' = 25.0'$

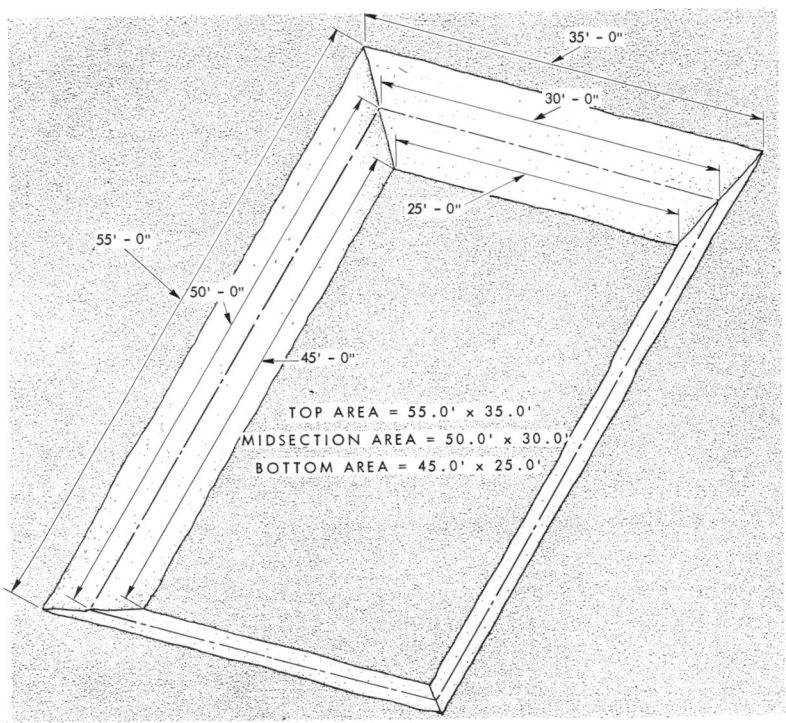

TOP AREA = 55.0' x 35.0'
MIDSECTION AREA = 50.0' x 30.0'
BOTTOM AREA = 45.0' x 25.0'

Fig. 5-12. A pictorial view of an excavation.

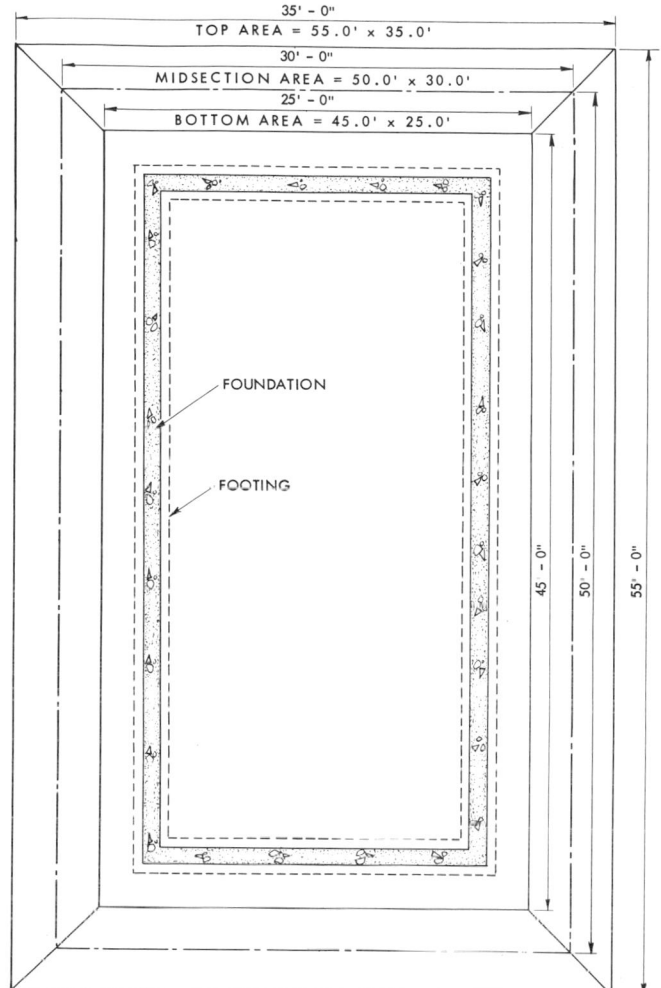

Fig. 5-13. Plan view of an excavation.

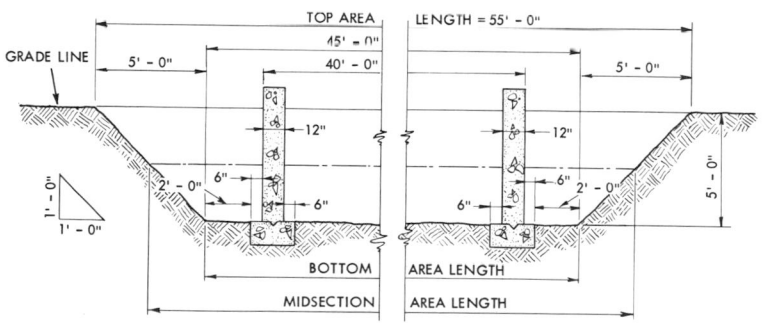

Fig. 5-14. Cross-sectional view of excavation's length.

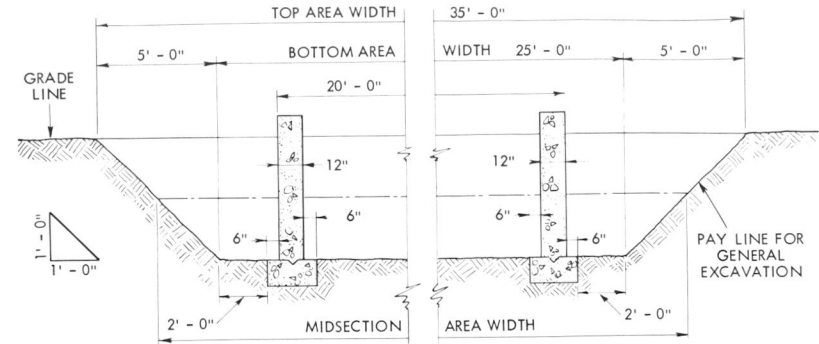

Fig. 5-15. Cross-sectional view of excavation's width.

$A_1 = 25.0' \times 45.0' = 1,125.0$ sq. ft.

Since the soil is on a slope, the length and width of the top area are the bottom dimensions increased by the horizontal dimensions of the slope. A 1:1 slope means that for every foot gained horizontally there is one foot of depth. The depth here is 5 feet so the top area dimensions have 5 feet added at each end:

Top length $= 45.0' + 10.0' = 55.0'$
Top width $= 25.0' + 10.0' = 35.0'$
$A_2 = 35.0' \times 55.0' = 1,925.0$ sq. ft.

The mid-section area is obtained by finding the average of the top and bottom dimensions to get the average length and width, and then multiplying the averages:

Av. length $= \dfrac{45.0' + 55.0'}{2} = 50.0'$

Av. width $= \dfrac{25.0' + 35.0'}{2} = 30.0'$

$M = 50.0' \times 30.0' = 1,500.0$ sq. ft.

Substituting in the formula:

$V = a/6 (A_1 + A_2 + 4M)$
$= 5/6 (1,125.0 + 1,925 + 6,000)$

$= 7,541.67$ cu. ft. or 279.32 cu. yds.

Example 2. The excavation is in heavy soil with the payline also 2 feet beyond the footings but vertically up (*neat cut*). Since the cut is vertical, the area of the top, bottom, and middle of the cut are equal, and the volume is found from the simple volume formula as follows:

$V = 1 \times w \times a$
where $1 =$ length
$w =$ width
$a =$ altitude or depth

From the previous problem, the width is known to be $25.0'$ and the length is $45.0'$. The depth was given as $5.0'$. The volume is found by substituting these figures in the formula:

$V = 45.0' \times 25.0' \times 5.0'$
$= 5,625.0$ cu. ft. or
208.33 cu. yds.

Note that there is an additional 71 cu. yds. of excavation when the walls of the excavation are sloped.

Special Excavation. Fig. 5-16 illus-

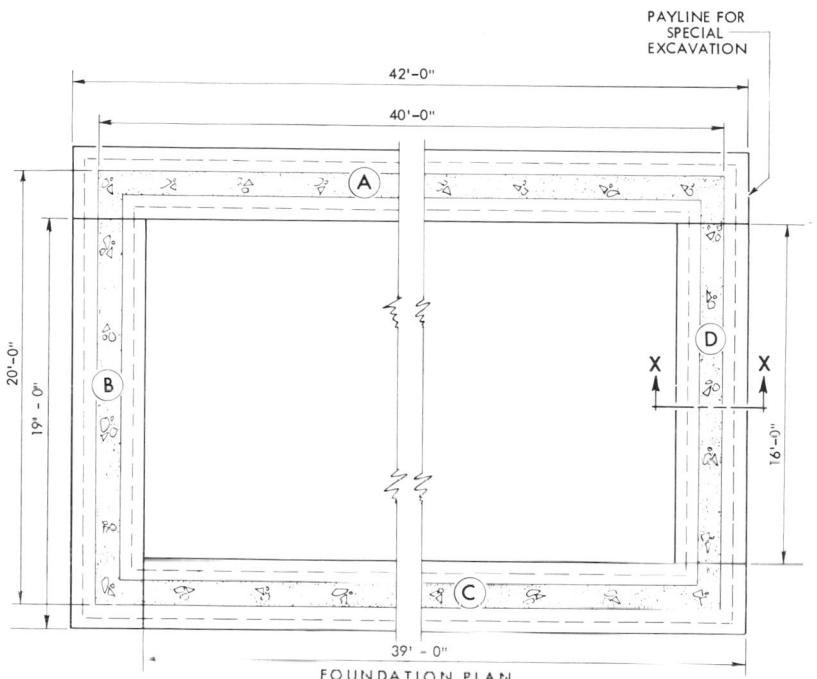

FOUNDATION PLAN

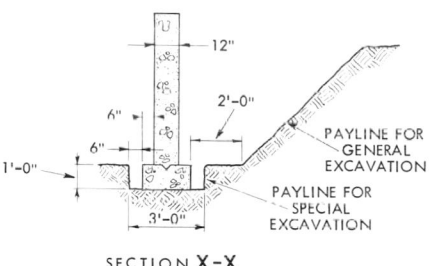

SECTION X-X

SPECIAL EXCAVATION - FOOTINGS

FOOTING	LENGTH	WIDTH	DEPTH	CU. FT.
A	42.0'	3.0'	1.0'	126.0
B	19.0'	3.0'	1.0'	57.0
C	39.0'	3.0'	1.0'	117.0
D	16.0'	3.0'	1.0'	48.0
			TOTAL	348.0 = 12.9 cu. yds.

Fig. 5-16. The method of calculating special excavation of footings.

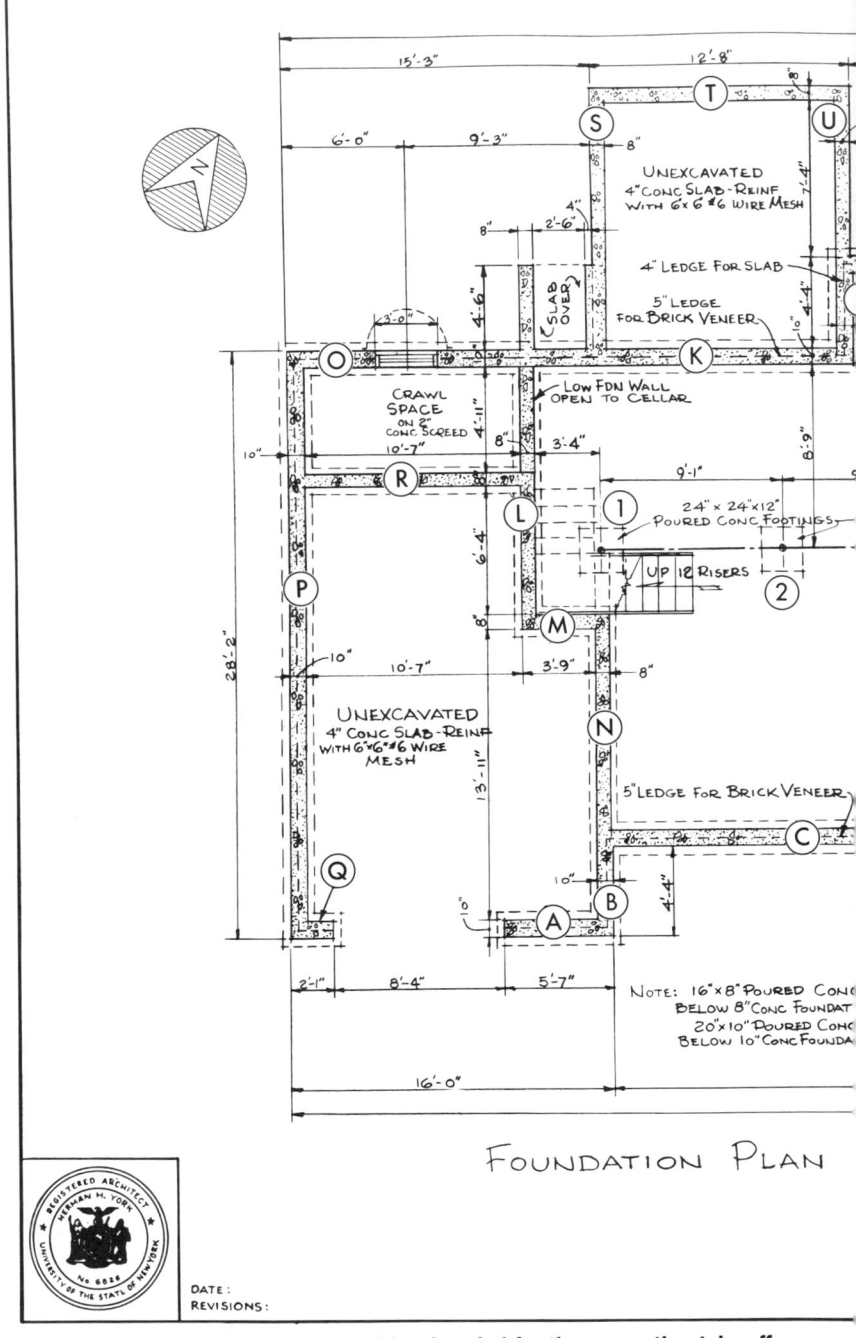

Fig. 5-17. House Plan A coded for the excavation take-off.

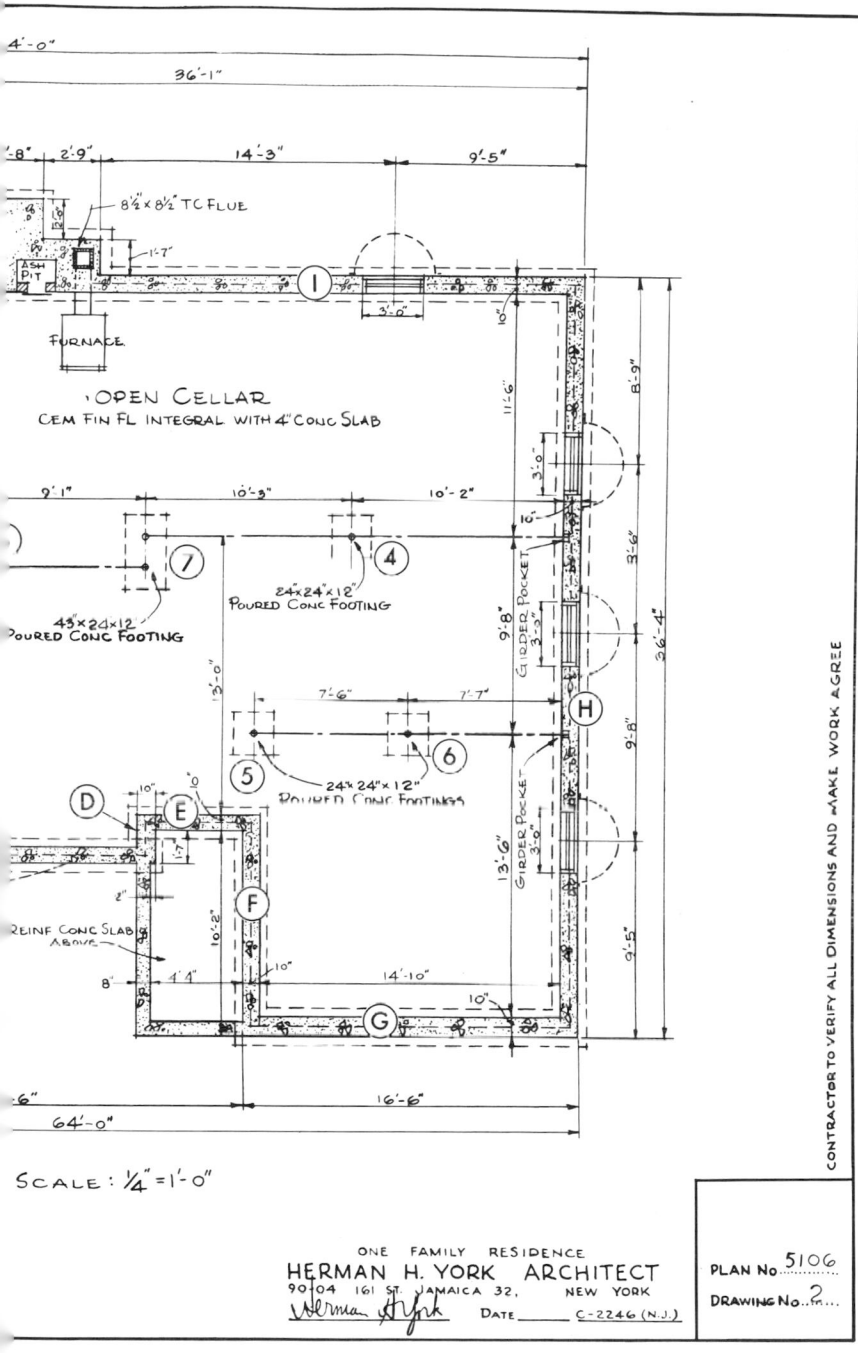

SCALE: $\frac{1}{4}" = 1'-0"$

ONE FAMILY RESIDENCE
HERMAN H. YORK ARCHITECT
90-04 161 ST. JAMAICA 32, NEW YORK
Herman H. York   DATE_____   C-2246 (N.J.)

PLAN No. 5106
DRAWING No. 2

trates how the special excavation is calculated for the footings in the previous example. The payline for special excavation (in this case) is 6″ on either side of the foundation footing; this leaves space for the placing and bracing of concrete forms.

The backfill for a foundation footing is figured as the difference between the volume of concrete needed for the footing and the volume of the total footing excavation.

To avoid duplicating the calculations at the corners of the excavation, the take-off is divided into four parts, A, B, C, and D. The volume of each part is found and then the parts are added together to obtain the total excavation for the footings.

Excavation for wall trenches without footings (such as for cheek walls) is calculated as in the following example.

Example 3. Assume that the wall is to be 4′-0″ below grade level. The length is 10′-0″ and the width is 6″. Allow 2′-0″ on all sides for the payline. How much soil must be excavated?

Solution: Use the formula for volume.

$$V = 14.0' \times 4.5' \times 4.0'$$
$$= 252.0 \text{ cu. ft.}$$
$$= 9.33 \text{ cu. yds.}$$

Quantity Take-off for House Plans.
Fig. 5-17 is the foundation plan of HOUSE PLAN A. The walls have been coded for a detailed excavation take-off. The complete set of plans at the end of Chapter 2 should be referred to for other views of the foundation walls and footings.

The base of the payline for *general excavation* is 2′-0″ from the base of the north, south, and east footings, and 2′-0″ west of the base of the footing at wall *A*. The slope is 1:1.

The *chimney footing* is taken as an additional part of the general excavation. The volume of the removed earth is calculated as the volume of a wedge:

$$V = \frac{1}{2} \times l \times w \times a$$

The *cheek walls* (under the porch) are considered as *special excavation*. Walls *S* and *U* are included as part of the general excavation, so the excavation is figured as the volume of a wedge with wall *T* as the length—plus 1.0′ at each end.

Fig. 5-18 itemizes the general and special excavation for HOUSE PLAN A.

The step footing is a special excavation and is usually excavated with its cross-section in the shape of a trapezoid. The volume is calculated as a parallelepiped (see Chapter 4) whose base is equal to the mean of the trapezoid's top and bottom bases, b_1 and b_2.

The step footing for the crawl space under the lavatory and laundry room is within the boundaries of the general excavation.

LABOR COSTS

When estimating labor costs, it is necessary to know how much work a man can do in a given time — usually an hour.

Wages vary greatly among different areas, but the *amount* of work which a man can accomplish in a given

		Length l ft.	Width w ft.	Depth a ft.		Volume V cu. ft.	V cu. yds.
		EXCAVATION				SHEET 1	
		QUANTITY TAKE-OFF				*HOUSE PLAN A*	
GENERAL EXCAVATION							
	MASS EXCAVATION				$a = 6' - 4''$, SLOPE = 1:1		
	$A_1 = 58.42' \times 41.17' = 2,405.15$ SQ. FT.						
	$A = 71.08' \times 53.83' = 3,826.24$ SQ. FT.						
	$4M = 4 (64.76 \times 47.50) = 12,304.4$ SQ. FT.						
	VOL = $\frac{a}{6}(A_1 + A + 4M)$					19,555.26	724.27
	CHIMNEY FOOTING						
		7.42	3.0	3.0			
	V = $\frac{a}{2}(l \times w)$					33.39	1.24
						TOTAL	725.51
SPECIAL EXCAVATION							
	WALL FOOTINGS						
	A	7.42	3.67	0.83		22.66	
	B	4.33				13.19	
	C	27.33				83.25	
	D	1.58				4.81	
	E	5.00				15.23	
	F	10.17				30.98	
	G	15.67				47.23	
	H	35.50				108.14	
	I	35.92				109.42	
	J	4.33				13.19	
	K	15.50				47.21	
	L	12.58	3.33	0.67		28.07	
	M	3.75	3.33	0.67		8.37	
	N	7.83	3.33	0.67		17.47	
	O	11.42	3.67	2.50		104.78	
	P	27.33	3.67	2.50		250.35	
	Q	1.25	3.67	2.50		11.47	
	R	8.83	3.33	2.50		73.61	
						990.33	36.68
	CHEEK WALLS						
	T	14.67	4.00	3.00			
	V = $\frac{a}{2}(l \times w)$					88.02	3.26

Fig. 5-18-A. The quantity take-off for the excavation of House Plan A.

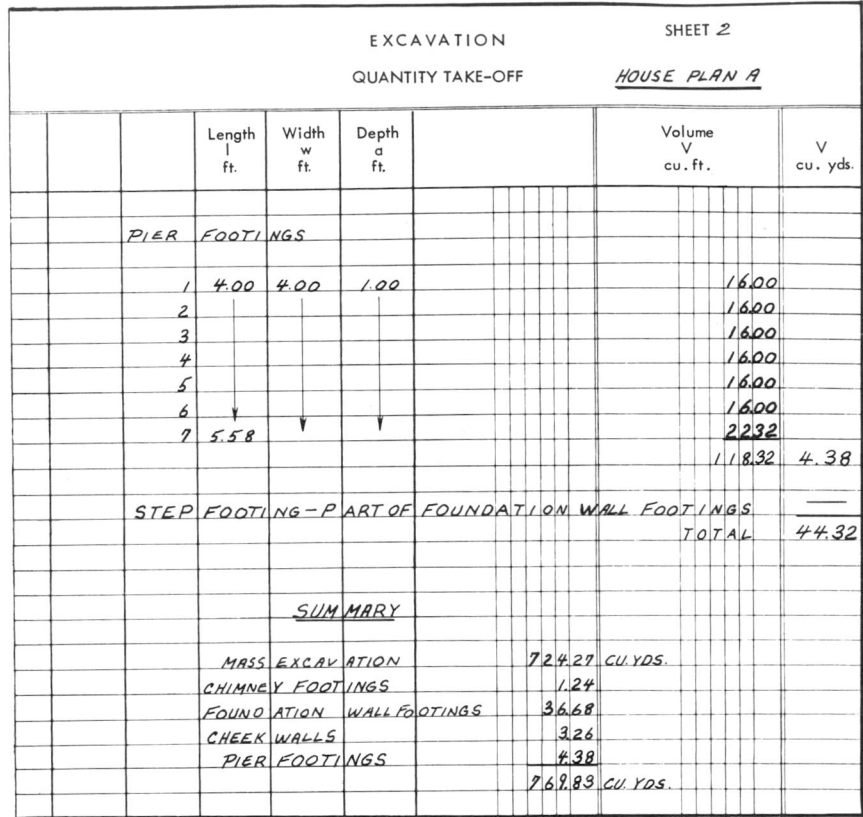

		Length l ft.	Width w ft.	Depth a ft.		Volume V cu. ft.	V cu. yds.
	EXCAVATION					SHEET 2	
	QUANTITY TAKE-OFF					*HOUSE PLAN A*	
	PIER FOOTINGS						
	1	4.00	4.00	1.00		16.00	
	2					16.00	
	3					16.00	
	4					16.00	
	5					16.00	
	6					16.00	
	7	5.58				22.32	
						118.32	4.38
	STEP FOOTING—PART OF FOUNDATION WALL FOOTINGS					TOTAL	44.32
	SUMMARY						
	MASS EXCAVATION					724.27 CU. YDS.	
	CHIMNEY FOOTINGS					1.24	
	FOUNDATION WALL FOOTINGS					36.68	
	CHEEK WALLS					3.26	
	PIER FOOTINGS					4.38	
						769.83 CU. YDS.	

Fig. 5-18-B. The quantity take-off for the excavation of House Plan A.

period of time varies little throughout the country.

For ordinary residences of 5 to 8 rooms two men should be able to erect the profiles in about 3 hours. For houses having considerably more area than usual, 4 hours labor might be necessary.

Special excavation with hand tools is usually figured as the amount of earth which a man can shovel in approximately one hour. One man digging in soft ground can remove one cubic yard of earth with a wheelbarrow in approximately one hour. This is true for both footing excavations and wall trenches.

When backfilling, one man can shovel about 24 cubic yards of earth in 7 hours (loose or semi-hard earth such as sand, gravel, etc.).

Labor costs for spreading and tamping fine gravel or cinders is usually given in a combined material and labor price per square inch or square foot (*unit cost*) by the supplier.

EQUIPMENT COSTS

The estimator should be able to determine the cost of owning and operating power equipment.* There are many factors which must be considered when calculating the hourly cost of owning and operating equipment and the cost per yard—the true measure of equipment performance.

Owning costs are the "fixed costs" which are always present regardless of whether or not the equipment is in operation. These include depreciation, interest, insurance, and taxes.

Operating costs are "variable costs." These include fuel and lubricants, repairs, and tires.

Operator's wages must be added to the owning and operating costs to obtain a true figure for hourly costs of operation. When renting equipment, the operator's salary is often included in the hourly rate. An example of this is the cost of renting a truck. Trucks are usually rented by the day. A 10 cubic yard water level truck, which can carry 12.5 yards of heaped earth, can be rented as a unit complete with driver for a fixed cost per day.

Overhead should not be considered with the owning and operating costs. This item includes indirect costs such as supervision, transportation, storage, etc., and should be placed under *General Operations*.

Trencher. This is a specialized piece of equipment that is used to dig trenches for water lines, gas lines and various other utility installations. This

*Material in this section courtesy of Caterpillar Tractor Co.

piece of equipment can only work efficiently in moderately soft ground, free of large boulders which might have a tendency to jamb the trenching wheel. The trencher can dig a trench as wide as is required but the depth of the trench is limited to the radius of the wheel. The trencher can be compared to a miniature ferris wheel.

The cost of excavation using the trencher is determined by the cost of rental of the machine and the type of ground. It is difficult to give the amount of excavation that a machine can perform due to the fact that there are various sizes of machines. As a guide, it can be assumed that a trencher can trench about 400 cubic yards of moderately soft ground in one day.

In order to get the unit cost per cubic yard of trenching, compute the following costs:

1. Cost of machine rental per day

2. Cost of the operator per day

3. Total cubic yards that can be excavated in one day (8 hours)

4. Divide the cost of the machine and the operator by the amount of excavation of earth and get the unit cost of excavation per cubic yard.

Example. The machine and operator

$$\frac{\$200}{400 \text{ cubic yards}}$$

Excavated material = $.50 per cu. yd. To the figure of $.50 must be added the various costs of operation such as insurance, fees, compensation, welfare, etc.

Backhoe. This piece of equipment may also be used for trenching. This machine can trench in hard, well-compacted soil and also in a partial rock terrain. This machine is almost identical to a power shovel with the exception that the power shovel digs away from the machine and the backhoe digs toward the machine. The cost of excavation using a backhoe depends on the size of the bucket and the type of ground that has to be excavated. Once this factor is known, we can apply the same formula as used for the trencher.

Cost ÷ Amount = unit cost of excavation per cubic yard.

Bulldozer. The bulldozer is merely a tractor mounted on crawlers or wheels with a front end attachment, either a cutting or scraping blade or a bucket. A bulldozer has the ability to dig, push, accumulate and mound material. If the bulldozer is equipped with a hydraulic lift blade it is commonly called a payloader. The bulldozer or payloader is used to cut or dig in soft earth and is generally used to make shallow cuts. It is not used for deep excavations or for digging rock or hard material. This type of excavation is left to the power shovel. Bulldozers as other pieces of excavation equipment come in various sizes and digging capacities.

In order to get a unit cost the formula for the trench may be applied.

As a general rule of thumb, a DC-4 operating efficiently can move about 220 cubic yards in a day (8 hours).

Power Shovels. When trucks are power shovel loaded, it is necessary to estimate the number of *hauling units* (vehicles) required to keep the shovel working efficiently. For estimating purposes, it is first assumed that the shovel operates at an ideal rate. The number of required hauling units is

TABLE III. ESTIMATED HOURLY PRODUCTION OF POWER SHOVELS AT 100% EFFICIENCY.

MATERIAL CLASS	$\frac{3}{8}$	$\frac{1}{2}$	$\frac{3}{4}$	1	$1\frac{1}{4}$	$1\frac{1}{2}$	$1\frac{3}{4}$	2	$2\frac{1}{2}$	$2\frac{3}{4}$	3	$3\frac{1}{2}$	4	$4\frac{1}{2}$	5	$5\frac{1}{2}$	6	$6\frac{1}{2}$
Moist loam or Sandy clay	85	115	165	205	250	285	320	355	405	435	454	525	580	635	685	740	795	840
Sand and gravel	80	110	155	200	230	270	300	330	390	420	450	505	555	600	645	695	740	785
Common earth	70	95	135	175	210	240	270	300	350	380	405	455	510	560	605	645	685	725
Clay, tough, hard	50	75	110	145	180	210	235	265	310	335	360	405	450	490	530	570	605	640
Rock, well-blasted	40	60	95	125	155	180	205	230	275	300	320	365	410	455	500	540	575	610
Common with rock	30	50	80	105	130	155	180	200	245	270	290	335	380	420	460	500	540	575
Clay, wet & sticky	25	40	70	95	120	145	165	185	230	250	270	310	345	385	420	455	490	520
Rock, poorly blasted	15	25	50	75	95	115	140	160	195	215	235	270	305	340	375	410	440	470

SHOVEL DIPPER SIZES IN CUBIC YARDS

NOTE: The above table is based on cubic yards (bank measure) per hour, 90-degree swing, best digging depth, material loaded into hauling units, and no delays.

determined by the following formula:

Hauling Units =
$$\frac{\text{shovel production at } 100\% \text{ efficiency}}{\text{hauling unit production}}$$

Note: *Bank-cubic yards* refers to the earth's volume before excavation. After being dug, the earth will usually swell and occupy a greater volume. See Table III.

Equipment Cost Records. Fig. 5-19

is a sample estimate for a tractor with an attached scraper. Its purpose is to determine the hourly cost of operating a typical piece of excavation and grading equipment.

Check list for submitting a bid for excavation:

1. Type of excavation equipment
2. Cost of operation of the excavation equipment

Cost Estimating Sheet

MACHINE & MODEL *DW 20-Series* G-*Tractor* H.P. _____ 320

ATTACHMENTS *No. 27 Cable Control* TIRES SIZES 29.5 29 (28 ply) Front, 14.0 24 (14 ply) Rear

FOB FACTORY PRICE *$52,905.00* FREIGHT _ _ _ _ TOTAL *$52,905.00*

CONDITIONS *Average*

OWNING COST:

Depreciation @ . ./*10,000*. . . . . . Hours. . . *4,114*. . . . . $*4.38*

Interest, Insurance, Taxes @ *$0.03* /$1000. . . . ./*59*

Total Owning Cost *5.97* *5.97*

OPERATING COST:

Fuels and Lubricants

Diesel Fuel. . *9.0*. . . . gph x. *$ 0.28* /gal 2.52

Gasoline 0.03

Lubricating Oil. *0.19*. . gph x. *$ 1.00* /gal 0.19

Transmission Oil. *0.04* gph x. *$ 0.70* /gal 0.03

Hydraulic Oil. . *0.04*. . gph x. *$ 1.00* /gal 0.04

Filters 0.04

Grease . . *0.15*. . . . . lbs/hr x *$ 0.20* /lb. 0.03

Repairs Incl. Labor. . . *90%*. . . . of Depreciation *4.38 X 0.9* 3.95

Tires @ . *45,000*. . Miles @. . ./*10*. . M.P.H. . *4,500*. . . Hours

. *10,135*. . . . Replacement Cost . *10,135*/*4500* 2.26

Other . . . . _____

Total Operating Cost. *7.73* *7.73*

Total Hourly Owning and Operating Cost (Excl. Operator). *13.60*

Operator's Wages *4.50*

Total Hourly Owning and Operating Cost *$18.36*

Fig. 5-19. Sample cost estimating sheet for a tractor and scraper.

3. Cost of the operator for the excavation equipment

4. How many days the piece of excavation equipment will be on the job

5. How many trucks will be needed to haul away the excess earth or rock

6. How many days the trucks will be needed

7. Cost of the operators for the trucks

8. The purchase of fill if necessary

9. Type of equipment used to backfill

10. How many days the backfill operation will take

11. Cost of the operation of the equipment used for backfilling

12. Cost of the operator for the backfilling equipment

13. Hand labor that may be required in the excavation operation

14. Labor that may be needed for tamping or miscellaneous cleaning

15. Supervision of the equipment and the laborers

16. Cost of surveying team to survey and stake out the line for the excavation

17. Insurance, fees and permits that may be required

18. Profit

19. Overhead.

It must be taken into consideration that in some cases there will be a fee for dumping the excess of earth at a local dump or a fill site. On the other hand, there may be a need in the area for fill in which case the fill may be able to be sold and in some cases the person buying the fill might supply the trucks needed for the hauling operation.

Questions and Problems

1. How is the amount of materials for profiles figured?

2. How is the amount of excavation for a basement figured when the surface is level?

3. How is the amount of backfill figured for a foundation footing?

4. How is the amount of excavation figured for wall trenches when the wall doesn't have a footing (such as a cheek wall)?

5. How much excavation is figured for a pier footing with the following dimensions: 3.0′ × 3.0′ × 1.0′? Allow a 1.0′ payline on all sides. (Answer: 25.0 cu. ft.)

6. Why must soil swell be considered by the estimator?

7. How is the total amount of backfill calculated for a job site?

8. A house occupies 1,500 sq. ft. of a 3,000 sq. ft. site. The garage and walks occupy 350 sq. ft. The remainder of the lot is to be covered with topsoil to a depth of 6″. How much topsoil is needed?

(Answer: 21.3 cu. yds.)

9. Foundation wall footings for a house are to be backfilled by hand. One man can shovel about 24 cubic yards of earth per work day. How much earth can 2 men backfill in 2 days?

(Answer: 96 cu. yds.)

10. How many trucks can a half-yard payloader fill in 7 hours if it is able to load one truck in 12 minutes?

(Answer: 35)

11. Why are excavation costs higher when hard earth is encountered?

See Appendix D, page 470, for Hourly Estimates for:
Excavations, Backfills, Concrete Walls and Slabs: Excavation; Backfilling; Ditching; Sewers and Drains; Footers (Excavating).

6

CONCRETE

~~~~~~

Almost all buildings have some form of concrete foundation and many have walls and floors of concrete. This chapter discusses common and recommended types of concrete footings, foundations, floors, and walls and explains how their costs are estimated. Some small buildings are built on posts or blocks, but these are not common and are not discussed in this chapter.

CONCRETE AND MORTAR MIXES

Concrete. In laboratory testing, the proportioning of sand and cement is done by weight. In theory, this is the way it should be done. The volume of a given weight of concrete will be different if it is packed loosely or thrown in a pile from the same weight packed tightly. The same is true of sand. Portland cement is standardized; it is always delivered from the manufacturer in 94-lb. bags (1 cubic foot). Portland cement will increase in volume from 10 to 30 percent merely by being dumped from the bag into a pile or from being shoveled into a measuring box. On the construction site, the cement should be measured in the original container as it comes from the manufacturer. This gives each container the same or nearly the same weight. Sand and stone, of course, must be measured loose as it is thrown in the measuring boxes. Sand will not vary quite as much in its weight-to-volume as does portland cement, but the dryness of the sand does have an effect on the volume of sand. Loose, dry sand occupies less space than does wet sand.

The *absolute* volume of concrete is easily explained by the following experiment. The materials needed are readily available and it is recommended that you do the experiment.

Take four quart-size jars and a one-gallon jar and fill the quart jars:

1 with beads   1 with flour
1 with marbles   1 with water

First pour the marbles into the gallon jar. Then add the beads and shake well. You can see that there is less than two quarts volume in the gallon jar.

This is because the *voids (spaces)* between the marbles have been occupied by the beads—so you have a denser mixture, but the volume is less.

Now pour the flour into the gallon jar. Again shake the jar until the three ingredients are thoroughly mixed. By measuring, you can see that now you have less than the three quarts of ingredients which were added. Finally add the quart of water. This too will find its way into the spaces between the other ingredients.

Thus from the original four quarts of ingredients, you get a volume that is much less in volume than the expected gallon. This loss is caused by the displacement of the air spaces between the marbles, beads, and flour.

The same effect is encountered when mixing concrete. In the experiment above, the marbles represent (and act like) the coarse aggregate; the beads represent sand; the flour represents cement.

Concrete is measured in cubic yards. Although sidewalks are specified in square yards or square feet, their thickness is also given so that the volume of concrete can be calculated.

Ready-Mixed Concrete. Most concrete used in building construction is ready-mixed. The concrete is delivered to the construction site in special trucks which dump it directly into forms or into a hopper for distribution by wheelbarrows, mechanical concrete *buggies,* or crane-hoisted buckets.

Cinder Concrete. Cinder concrete is made in much the same manner as ordinary concrete except that cinders are used in place of stone as aggregate.

Cinder concrete can be used for fireproofing and for short reinforced floor and roof slabs. It's principal advantage is its lightness in weight. The cinders should be from well burned hard coal, reasonably free of sulphides. Excessive sulphides cause reinforcing steel bars to rust.

Lightweight Concrete. The aggregate consists of heat-expanded minerals, which have better strength properties than cinders.

Reinforced Concrete. Concrete which has steel bars or mesh embedded in it is stronger than plain concrete. This concrete is obviously more expensive than plain concrete because of the additional cost of the steel bars or mesh and the tie wire. The labor cost per cubic yard of concrete will also be greater.

Prestressed Concrete. In recent years, prestressed concrete structural members have been used more frequently in building construction. The American Concrete Institute defines prestressed concrete as, "Concrete whose stresses resulting from external loadings are counterbalanced by prestressing reinforcement placed in the structure." Some of the concrete structural members made by this method are double and single tees, channel sections, beams, hollow slabs, and flat planks.

On a large construction project where many prestressed concrete members are of identical size, the cost of each will be lowered.

Weight of Concrete. Contrary to common belief, the weight of concrete varies considerably. The weight usually is between 105 and 160 pounds per cubic foot, depending on the materials used. The type of concrete generally used weighs about 145 pounds per cubic foot. This figure is an average which the estimator may use safely.

Cinder concrete, however, weighs between 75 and 115 pounds per cubic foot.

Cement may be specified at 4.00, 3.80, etc., cubic feet to the barrel. If it is specified at 4.00 cubic feet per barrel, then 1 barrel weighs 376 pounds and contains 4 bags. Each bag weighs 94 pounds and occupies a cubic foot.

Proportions. For ordinary work, the concrete ratio is stated in the specifications. The ratio is usually given in the form 1:2:4:8. This ratio means 1 part cement, 2 parts sand (or fine aggregate), 4 parts stone (or coarse aggregate), and 8 parts water. The ratio will vary according to the architect's purposes and local conditions.

Sometimes the ratio is given in a shorter form, such as 1:2:4. In this case the amount of water is not stated in the specifications and the contractor must use his own judgment. The amount of water varies with the type of structure, degree of exposure, and compressive strength of the concrete.

The voids in ordinary broken stone are somewhat less than half of the volume, and it is common practice to use one-half as much sand as the volume of the broken stone. The proportion of cement is varied according to the strength required in the structure, and according to the desire to economize.

On this principle you have the familiar ratios 1:2:4, 1:2½:5, 1:3:6, and 1:4:8. Note that in each of these cases the ratio of the sand to the broken stone is a constant, and the ratio of the cement alone is variable.

Tables I and II give the ingredients in 1 cubic yard of concrete at various mixes, and the volume factors of various mixes.

Ideal Conditions in Preparing the Mix. The general principle to be adopted in mixing concrete is that the amount of water used should be only enough to crystallize the cement paste; the amount of paste should be no more than is sufficient to fill the voids between the particles of sand; and the mixture thus produced should be only enough to fill the voids between the broken stones. If this ideal could be realized, the total volume of mixed concrete would exactly equal the total volume of the stone involved.

No matter how thoroughly the ingredients are mixed—and no matter how careful the mixer—some of the particles of cement will get between the grains of sand and the volume of the mixture will be greater than the volume of the sand. The grains of sand will get between the smaller stones and the smaller stones will get between the larger stones and separate them.

Actual Conditions. Because perfect mixing is highly impractical, the amount of water used in mixing concrete is always a little excessive. The cement paste is also usually more than is necessary to fill the spaces between the sand particles. The amount of *mortar* (cement, sand, and water), therefore, is also considerably in excess of the amount required for filling the spaces between the stones. Even allowing some excess in all particulars, however, there is much variation in the percentage of voids in the sand as well as the stone. Because of this variation, an experimental determination of the voids must be made. Even in the best concrete work there is a periodic re-testing of the mixture. For less careful work, the proportions ordinarily adopted in practice are considered sufficiently accurate.

Concrete Problem. Compute the amount of materials needed to obtain 15 cubic yards of concrete. Use a 1:3:7:8 ratio and the *absolute volume* method of finding the amount of concrete.

Solution: The absolute volume method of finding the amount of concrete is very accurate. The 8 in the ratio means there are to be 8 gallons of water to every sack of cement.

Each of the materials used to make the concrete mix has an absolute volume constant. The product of the ratios and the constants are computed and then added together. The total gives the amount of concrete obtained from each bag.

Absolute Volume Constants

cement	0.485
sand	0.567
aggregate	0.567
water	0.134

The sand and aggregate have the same constant because their absolute volumes are the same, even though they differ in size and shape.

$$1 \times 0.485 = 0.485$$
$$3 \times 0.567 = 1.701$$
$$7 \times 0.567 = 3.969$$
$$8 \times 0.134 = 1.072$$

7.227 cu. ft. concrete per bag cement

Since the problem is given in terms of cubic yards, you must divide the number of cubic feet per bag into the number of cubic feet per yard in order to obtain the required number of bags per cubic yard.

$$7.227 \overline{)27.00} \quad 3.7 \text{ bags per cu. yd.}$$

Multiplying each of the ratios by 3.7 gives the amount of each material needed for one cubic yard of concrete:

cement: $1 \times 3.7 = 3.7$ bags
sand: $3 \times 3.7 = 11.1$ cu. ft.
aggr: $7 \times 3.7 = 25.9$ cu. ft.
water: $8 \times 3.7 = 29.6$ U.S. gallons

To find the amount of materials needed for 15 cu. yds. of concrete, multiply by 15:

cement: $15 \times 3.7 = 55.5$ bags
sand: $15 \times 11.1 = 166.5$ cu. ft.
aggr: $15 \times 25.9 = 388.5$ cu. ft.
water: $15 \times 29.6 = 444.0$ U.S. gals.

If each bag weighs 94 pounds and if water weighs 8.35 lb./gal., the weight of the materials is as follows: cement, 5,217 lbs.; sand, 15,651 lbs.; aggregates, 36,519 lbs.; and water, 3,707 lbs.

Mortar. Mortar is a mixture of sand, cement, and water. The ratio of cement to sand (by volume), doesn't exceed 1:3 for most applications. See Table II.

TABLE I. MATERIALS REQUIRED FOR 1 CUBIC

Mixture	3/4" GRAVEL		
	BARRELS CEMENT	CUBIC YARD SAND	CUBIC YARD STONE
Mortar			
1:1 1/2	3.61	0.80	USING
1:2'	3.02	0.90	"
1:2 1/2	2.60	0.96	"
1:3	2.28	1.01	"
Concrete			
1:1:2	2.30	0.35	0.74
1:1 1/2:3	1.71	0.39	0.78
1:1 3/4:2 3/4	1.75	0.43	0.75
1:2:3	1.54	0.47	0.73
1:2:3 1/2	1.44	0.44	0.77
1:2:4	1.34	0.41	0.81
1:2 1/2:4	1.24	0.47	0.75
1:2 1/2:4 1/2	1.16	0.44	0.80
1:2 1/2:5	1.10	0.42	0.83
1:3:4	1.15	0.52	0.72
1:3:5	1.03	0.47	0.78
1:3:6	0.92	0.42	0.84

TABLE II. VOLUME FACTORS OF

KIND OF CONCRETE WORK	VOLUME CEMENT BAGS	MIX BY JOB DAMP SAND CU. FT.	MATERIA STONE GRAVE CU. FT
FOOTINGS, HEAVY FOUNDATIONS	1	3.75	5
WATERTIGHT CONCRETE FOR CELLAR WALLS AND WALLS ABOVE GROUND	1	2.5	3.5
DRIVEWAYS)) ONE COURSE FLOORS, WALKS)	1	2.5	3
DRIVEWAYS)) TWO COURSE FLOORS, WALKS)	1 1	TOP 2 BASE 2.5	0 4
PAVEMENTS	1	2.2	3.5
WATERTIGHT CONCRETE FOR TANKS, CISTERNS AND PRECAST UNITS (PILES, POSTS, THIN REINFORCED SLABS, ETC.)	1	2	3
HEAVY DUTY FLOORS	1	1.25	2

YARD OF MORTAR AND CONCRETE

	1" STONE DUST OUT			2 1/2" STONE DUST OUT		
BARRELS CEMENT	CUBIC YARD SAND	CUBIC YARD STONE	BARRELS CEMENT	CUBIC YARD SAND	CUBIC YARD STONE
VERY	FINE	SAND	3.87	0.86	
"	"	"	3.21	0.95	
"	"	"	2.74	1.01	
"	"	"	2.39	1.06	
2.57	0.39	0.78	2.63	0.40	0.80
1.85	0.42	0.84	1.90	0.43	0.87
1.85	0.47	0.80	1.93	0.46	0.84
1.70	0.52	0.77	1.73	0.53	0.79
1.57	0.48	0.83	1.61	0.49	0.85
1.46	0.44	0.89	1.48	0.45	0.90
1.35	0.52	0.82	1.38	0.53	0.84
1.27	0.48	0.87	1.29	0.49	0.88
1.15	0.46	0.91	1.21	0.46	0.92
1.26	0.58	0.77	1.28	0.58	0.78
1.11	0.51	0.85	1.14	0.52	0.87
1.01	0.46	0.92	1.02	0.47	0.93

VARIOUS MIXES

WORK- ABILITY OR CON- SIST- ENCY	A ONE BAG BATCH MAKES THIS VOLUME OF CONCRETE CU. FT.	TOTAL WATER PER BAG GALLONS	MATERIALS FOR ONE CUBIC YARD OF CONCRETE		
			CEMENT BAGS	SAND CU. FT.	STONE GRAVEL CU. FT.
STIFF	6.2	8.00	4.3	16.3	21.7
MEDIUM	4.5	6.00	6.0	15.0	21.0
STIFF	4.1	5.50	6.5	16.3	19.5
STIFF	2.14	...	12.6	25.2	...
STIFF	2.8	6.00	5.7	14.2	22.8
STIFF	4.2	5.25	6.4	14.1	22.4
MEDIUM	3.8	5.00	7.1	14.2	21.3
WET	3.9	5.75	6.9	13.8	20.7
STIFF	2.8		9.8	12.3	19.6

TABLE III. BARRELS OF PORTLAND CEMENT PER CUBIC YARD OF MORTAR

(VOIDS IN SAND BEING 45 PERCENT, AND 1 BBL. CEMENT YIELDING 3.4 CUBIC FEET OF CEMENT PASTE.)

Proportions of Cement to Sand	1:1	1:1.5	1:2	1:2.5	1:3	1:4
	Bbls.	Bbls.	Bbls.	Bbls.	Bbls.	Bbls.
Bbl. specified to be 3.5 cu. ft. ...	4.62	3.80	3.25	2.84	2.35	1.76
Bbl. specified to be 3.8 cu. ft. ...	4.32	3.61	3.10	2.72	2.16	1.62
Bbl. specified to be 4.0 cu. ft. ...	4.19	3.46	3.00	2.64	2.05	1.54
Bbl. specified to be 4.4 cu. ft. ...	3.94	3.34	2.90	2.57	1.86	1.40
Cu. yds. sand per cu.yd. mortar...	0.6	0.8	0.9	1.0	1.0	1.0

TABLE IV. BARRELS OF PORTLAND CEMENT PER CUBIC YARD OF MORTAR

(VOIDS IN SAND BEING 35 PERCENT AND 1 BBL. CEMENT YIELDING 3.65 CU. FT. OF CEMENT PASTE.)

Proportion of Cement to Sand	1:1	1:1.5	1:2	1:2.5	1:3	1:4
	Bbls.	Bbls.	Bbls.	Bbls.	Bbls.	Bbls.
Bbl. specified to be 3.5 cu. ft. ...	4.22	3.49	2.97	2.57	2.28	1.76
Bbl. specified to be 3.8 cu. ft. ...	4.09	3.33	2.81	2.45	2.16	1.62
Bbl. specified to be 4.0 cu. ft. ...	4.00	3.24	2.73	2.36	2.08	1.54
Bbl. specified to be 4.4 cu. ft. ...	3.81	3.07	2.57	2.27	2.00	1.40
Cu. yds. sand per cu. yd. mortar..	0.6	0.7	0.8	0.9	1.0	1.0

Table III shows barrels of cement per cubic yard of mortar when the voids in the sand are 45% of its volume and 1 barrel of cement yields 3.4 cubic feet of cement paste. Table IV shows the same thing when the voids are 35% of its volume and the yield is 3.65 cubic feet of cement paste.

Such mortar is commonly used as topping for concrete floors.

Fireproofing. Experience and tests have shown that the fire-resisting qualities of concrete are greater than those of any other known type of building material. During a fire, the temperature may reach 1,900° F and injure the concrete to a depth of ¾-inch but the body of the concrete is not affected.

Two inches of concrete will safely protect an I beam. Reinforced concrete beams and girders should have a clear thickness of 1½ inches of concrete outside the steel on the sides and 2 inches on the bottom. Slabs should have at least 1 inch below the bars, and columns should have 2 inches.

Fig. 6-1 shows a typical steel beam cross section and the fireproofing concrete surrounding it. The concrete is sometimes reinforced by heavy wires wound around the column. This also helps to maintain the bond between the concrete and steel.

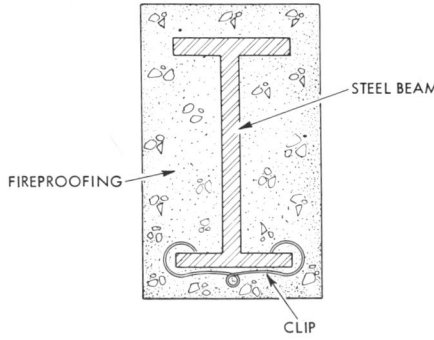

Fig. 6-1. Cross section of fireproof steel beam. The clip helps to keep the concrete bonded to the steel.

Theory. Portland cement concrete is made of sand, stone, cement paste, and water—none of these materials are combustible and therefore portland cement concrete is used as fireproofing material. Furthermore, the finished concrete is porous and retains 12 to 18 percent of its water, thus it will not readily transfer heat. This water is chemically combined and is not given off at the boiling point. On heating, a part of the water is given off at 500°F, but dehydration does not take place until 900°F is reached. The mass is kept for a long time at comparatively low temperature by the vaporization of water which absorbs heat. A steel beam imbedded in concrete is thus cooled by water vaporization in the surrounding concrete.

Resistance to the passage of heat is offered by the porosity of concrete. Air is a poor conductor, and an air space is an efficient protection against conduction. The outside of the concrete may reach a high temperature; but the heat only slowly and imperfectly penetrates the mass, and reaches the steel so gradually that it is carried off by the metal as fast as it is supplied.

Porosity of Cinder Concrete. Porous substances, such as asbestos, are always used as heat-insulating material. For this reason, cinder concrete, being highly porous, is a much better insulator than a dense concrete made of sand and gravel or stone.

FOOTINGS, FOUNDATIONS AND FLOORS

Footings. Footings distribute the load transmitted to them, by foundations, over a large ground area. The reason for this is that a narrow or pointed object will pierce the ground more readily than a blunt or broad object. A stick of wood, for example, which has a point one inch square will pierce the ground more readily than a stick of wood which has a two inch square point. This same principle applies to foundations.

Look at the foundation shown in Fig. 6-2. This foundation is 12 inches thick and 20 feet long. In other words,

Fig. 6-2. Cross section of foundation wall and footing.

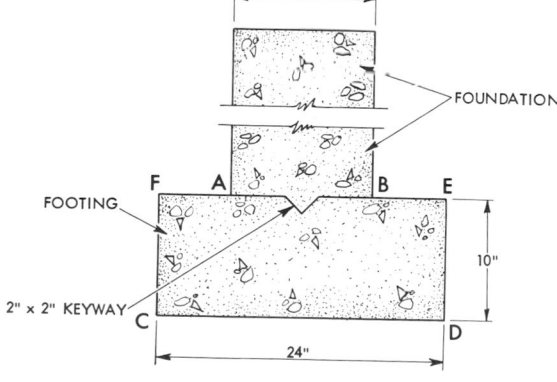

the foundation is 20 square feet (1 ft. × 20 ft. = 20 sq. ft.). Now suppose that the foundation will support 2,000 pounds per square foot. The total supportable load for this foundation then will be 2,000 × 20 = 40,000 lbs.

Remember that the various kinds of soil have varying load-bearing capacities. One kind might easily support 1,500 pounds per square foot while another might support 2,000 pounds per square foot. If the soil can support a maximum of 1,500 pounds per square foot, then that kind of soil has a capacity of 1,500 pounds per sq. ft.

If the foundation in Fig. 6-2 carries 2,000 pounds per square foot and it rests on soil which can only carry 1,500 pounds per square foot, the soil will be overloaded and the foun-

$$\frac{40,000 \text{ lbs.}}{40 \text{ sq. ft.}} = 1,000 \text{ lbs. per sq. ft.}$$

This is below the limit of the soil's load-bearing capacity.

Shape of the Footing. A footing should have a perfectly flat edge in contact with the soil (*CD* in Fig. 6-2). This edge should also be horizontal to prevent sliding.

Composition of Footings. Although in rare instances, stone, blocks, and other materials are used, footings are usually made of concrete. Concrete is the most desirable material for two reasons: (1) when dry, the concrete becomes one solid piece—this helps to avoid the possibility of settling; (2) concrete requires less labor than other kinds of footings—thus lowering the labor cost.

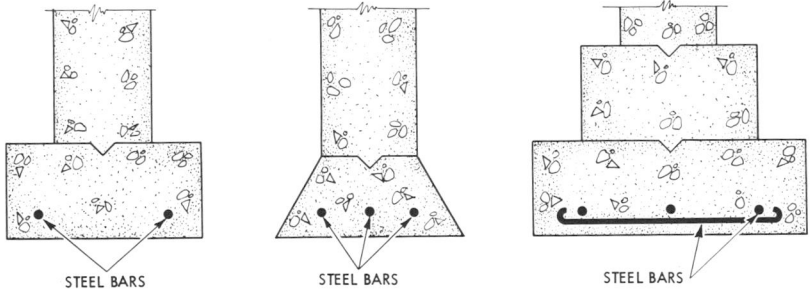

Fig. 6-3. Steel reinforcing bars used in various types of footings.

dation will settle. Thus, a footing is needed to distribute the load over a wider area. The footing shown in Fig. 6-2 is 24 inches wide and 20 feet long, or 40 square feet (2 ft. × 20 ft. = 40 sq. ft.). Thus, the 40,000 pound foundation is now distributed over 40 square feet. Now the soil is not overloaded since:

Steel reinforcing bars or rods, Fig. 6-3, add to the strength of the footings and are often used. The written specifications and the working drawings give the size, spacing, and other necessary data about the reinforcing steel.

Footing Forms. A footing should have a constant area in its rectangular cross section. To assure this, wood

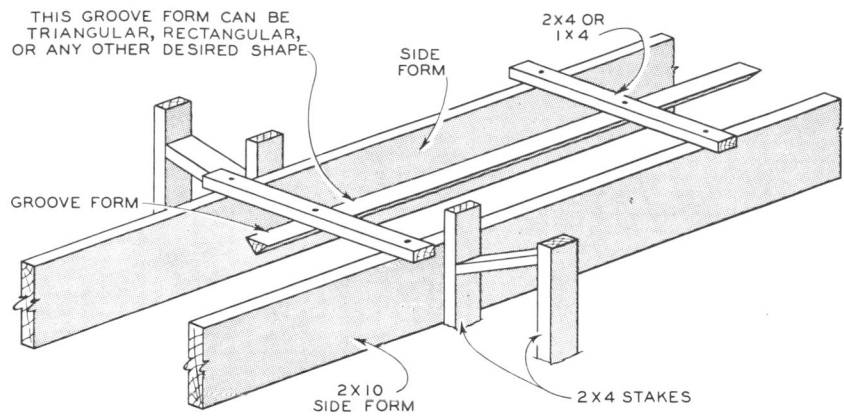

Fig. 6-4. Typical formwork for a concrete footing.

forms are sometimes necessary—especially where the soil is loose or sandy or will not remain vertical (line *DE* in Fig. 6-2). When wood forms are not used, the soil must be wetted so that it doesn't absorb water from the wet concrete. Fig. 6-4 shows typical formwork for concrete footings.

Column Footings. Inside areas of a building are usually supported by interior partitions. The partitions are in turn supported by I beams which run along the basement ceilings. At one or

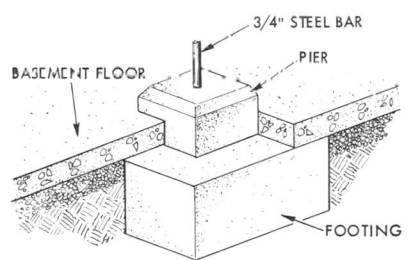

Fig. 6-5. Column footing with concrete pier. The pier keeps a wood post above the floor level to protect it from moisture. The steel bar holds the post in place.

more places these I beams are supported by columns. These columns will each require a footing, Fig. 6-5. The size of the footing depends upon the total load transmitted by the column and the load capacity of the soil.

Foundations. Foundations, Fig. 6-6, are of four basic types: full basement, which consists of masonry walls resting on footings and includes a basement floor. The surface type excludes the floor and has a crawl space. The slab or slab-at-grade type consists of concrete laid directly over a gravel pad on the ground. Pier type foundations support the structure on several posts or piers.

More and more homes in all parts of the country are being built using the slab-at-grade floor. See Fig. 6-7. Certain precautions must be made in order that the floor be satisfactory.

The Small Homes Council suggests the following: the earth around the house must be graded so that the water will drain away properly. The entire

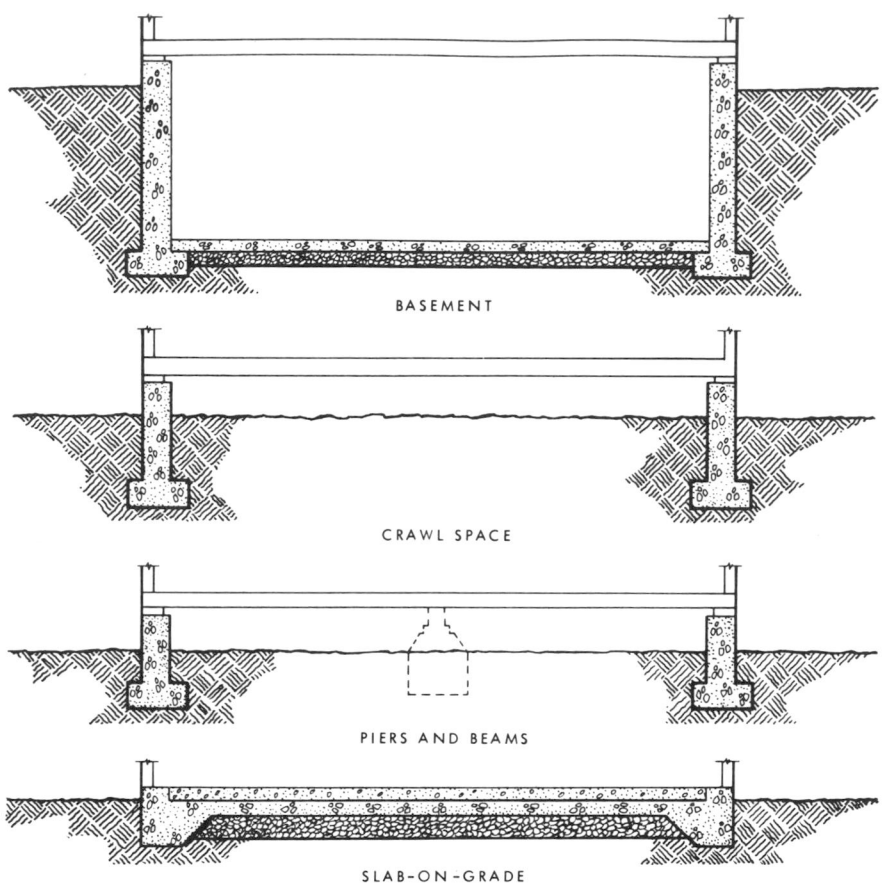

BASEMENT

CRAWL SPACE

PIERS AND BEAMS

SLAB-ON-GRADE

Fig. 6-6. Basic types of foundations.

area where the floor will be laid should be covered with 4 inches of washed gravel or crushed rock in order to reduce the capillary rise of moisture. A membrane should be provided over the gravel strong enough to resist puncturing when the concrete is placed. This membrane serves as a vapor barrier to keep moisture from the ground. Polyethylene film, asphaltum board ⅛-inch thick, or reinforced duplex paper with asphaltum center

may be used. Overlap paper 4 inches. One additional problem to solve in cold climates is heat loss. The heat loss is primarily around the perimeter of the house, and to counteract the loss and prevent condensation resulting from the cold floor, edge insulation is required. Two-inch-thick rigid waterproof insulation extending 2 feet from the walls is suggested. Where panel heat is used in the floor, the insulation should cover the entire floor area.

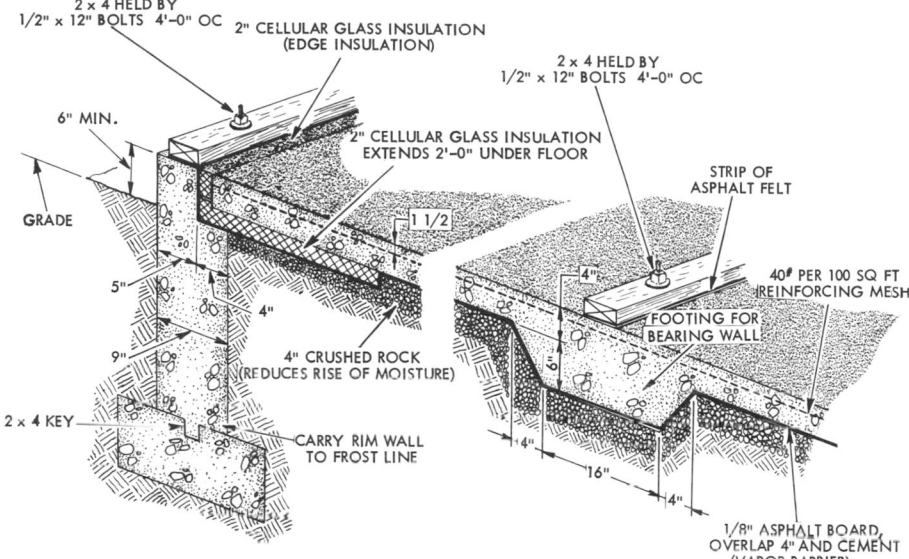

Fig. 6-7. A concrete floor laid on the ground requires a vapor barrier, and in cold climates must have edge insulation.

The footing and foundation wall will be formed in the manner used locally. In some areas the foundation wall is omitted entirely and a simple perimeter support is constructed instead by merely thickenening the edge of the slab. See Fig. 6-8.

In some regions of the South, footings are laid for both the exterior wall and the bearing partition by making trenches in the firm earth. Reinforcing rods are used in the footings and cement block is used for the foundation wall. A polyethylene film is spread

Fig. 6-8. A simple perimeter support will serve for light construction in warm climates.

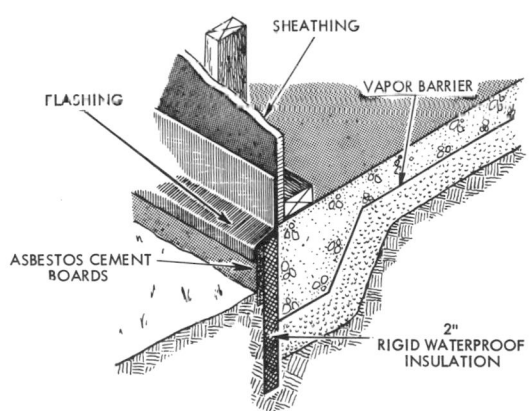

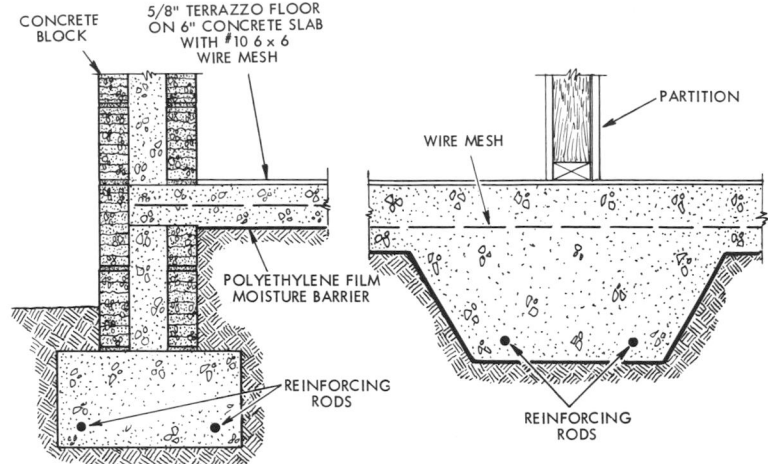

Fig. 6-9. Detail shows method used in warm climates for houses made with concrete block exterior walls and terrazzo floors.

over the earth where the slab is to be laid. The floor slab will be finished with terrazzo and in order to make it as rigid as possible, wire mesh is imbedded in the concrete. See Fig. 6-9.

In some areas where the soil is too unstable to permit the use of conventional footings and foundation walls, grade beams may be used. See Fig. 6-10. (Grade beams are continuous beams running around the house perimeter; they rest on piers.) Holes are dug at the perimeter of the building, a maximum of 8 feet apart, to a depth sufficient to bring them to solid soil well below the frost line. Concrete is poured into the holes, or into shells made for that purpose, to form piles. Forms are made to contain the grade beams which rest on the piles. A steel rod (a dowel) serves to position the grade beam. Horizontal reinforcing

rods add strength to the grade beams. This construction may be used with slab-on-ground foundations and for houses with crawl spaces.

Anchor bolts are embedded in the top of the foundation walls before the concrete has set. These bolts hold wood *sills* securely in place on top of the wall. (Wood sills are discussed in Chapter 8.)

Waterproofing. In areas where the soil does not provide adequate natural drainage, it is advisable to waterproof foundations. As shown in Fig. 6-11, asphalt or some other waterproofing agent, such as polyethylene film, can be used. The agent to be used will be on the drawing or in the specifications. In addition to the waterproofing agent, it is helpful to place loosely connected drainage tiles as shown in Fig. 6-11. This is an especially good practice here, since much of the backfill is gravel and

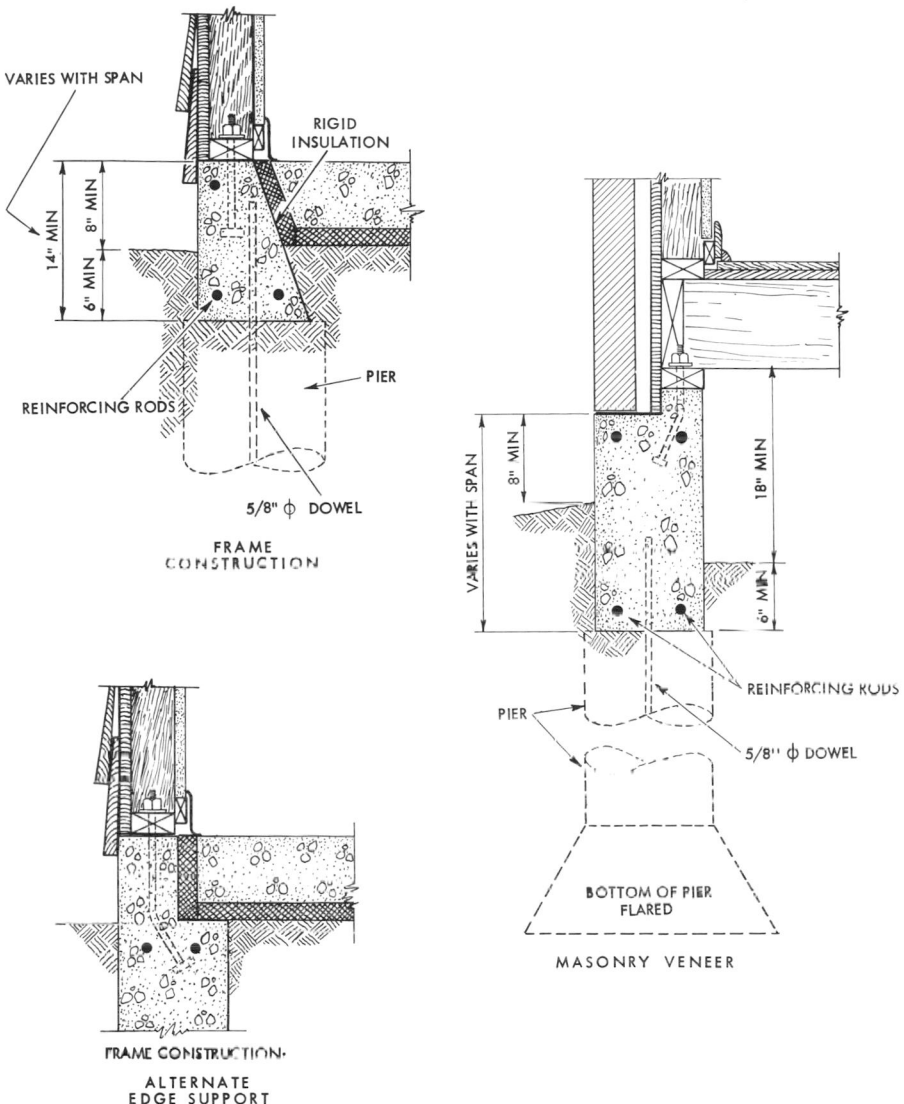

Fig. 6-10. Examples of grade beam and pier construction.

small rocks; seepage water can find its way through this type of material to the drain tile. The tile is connected to a sewer or to some other natural means of disposal. Exceptionally wet conditions may require a sump pump.

There are a number of ways to waterproof a foundation. The estimator should therefore check carefully the drawings and specifications for each building.

Cheek Walls. These are foundation

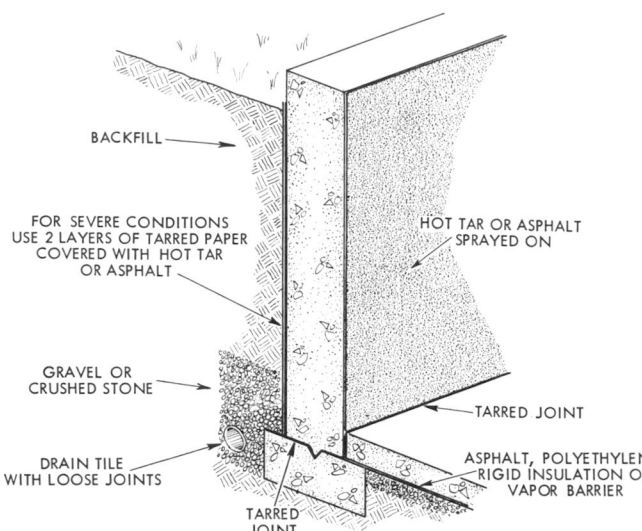

BACKFILL

FOR SEVERE CONDITIONS
USE 2 LAYERS OF TARRED PAPER
COVERED WITH HOT TAR
OR ASPHALT

HOT TAR OR ASPHALT
SPRAYED ON

GRAVEL OR
CRUSHED STONE

TARRED JOINT

DRAIN TILE
WITH LOOSE JOINTS

ASPHALT, POLYETHYLENE,
RIGID INSULATION OR
VAPOR BARRIER

TARRED
JOINT

Fig. 6-11. Methods of waterproofing and insulating foundations and floor slabs.

walls which support light structures and structures which impose relatively little load, such as porches, areaways, and entrance steps. Such structures rarely need footings. The term derives from the smoothing of the face of the wall after the concrete has set.

Floors. Concrete floors generally vary from 3 to 8 inches in thickness. The specifications and drawings will indicate how thick the floor is to be. As indicated in Fig. 6-11, the concrete foundation floor is generally laid over a well-tamped layer of gravel, crushed stone, or cinders. Sometimes a cement topping ($\frac{1}{4}$ to 1-inch thick) is specified for the concrete base.

Floors are sometimes constructed with metal or plastic pans, Fig. 6-12. These create a flat slab on the top surface and a ceiling below which looks like a waffle—hence they are often called "waffle-slab" systems.

Another type of floor system uses a

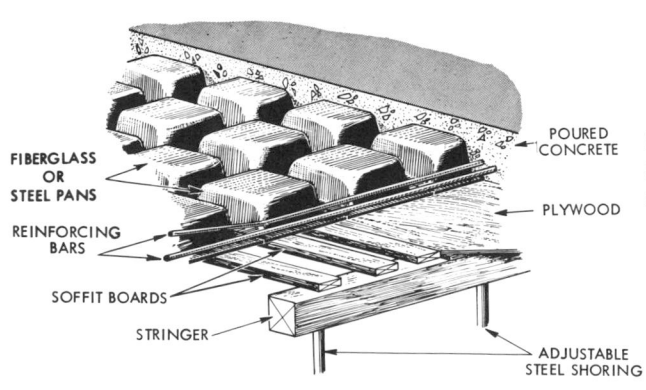

FIBERGLASS
OR
STEEL PANS

POURED
CONCRETE

REINFORCING
BARS

PLYWOOD

SOFFIT BOARDS

STRINGER

ADJUSTABLE
STEEL SHORING

Fig. 6-12. Metal pans used for concrete floor construction.

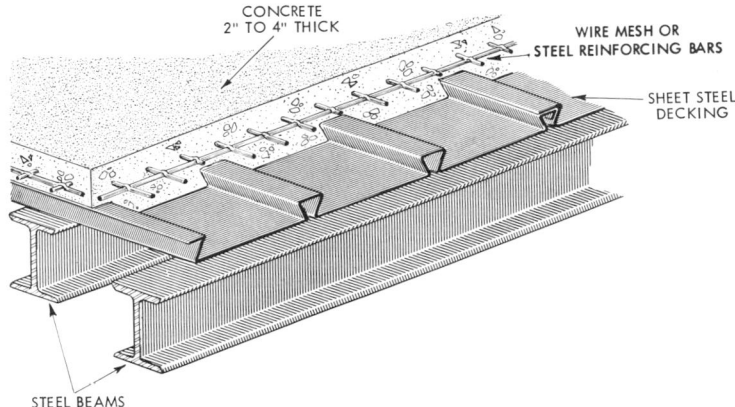

Fig. 6-13. Reinforced concrete floor slab with metal decking.

layer of wire mesh over a metal decking which rests upon the structural steel. See Fig. 6-13. A thin concrete slab (2 to 4 inches thick) is formed by pouring concrete over the mesh and metal decking.

Still another system uses the *reinforced* and *prestressed concrete plank method* to construct the floors.

BEAMS, COLUMNS AND STAIRS

Beams. Beams are used for horizontal support of floors and roofs. They are usually rectangular or 'I'-shaped in cross section. Concrete beams are usually an integral part of the floor which they support. Fig. 6-14 shows cross sections of some concrete beams.

Spandrel Beams, of reinforced concrete, span openings in the outer walls of buildings. They are integral parts of the floor slab and must be able to support masonry. The depth of the outside face of the beam is on the drawings or in the specifications.

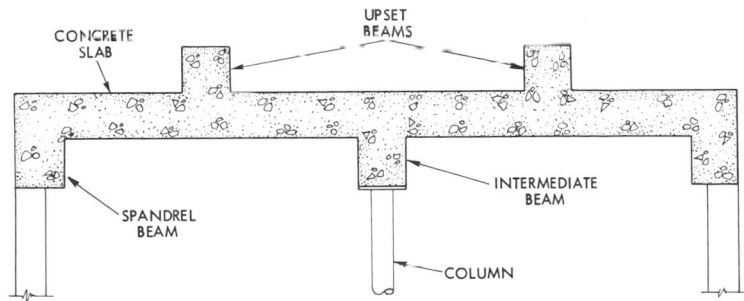

Fig. 6-14. Concrete beams as integral parts of a concrete slab.

Spandrel inserts are placed within the forms for the spandrel beams and project from the spandrel beam face. These inserts take the bolts which hold up the lintels which support the brickwork on the building face.

Intermediate beams also form and support a part of the floor slab but are not on the outside edge of the building.

Free-standing beams span an open space and frame into slabs only at the beam ends. An example of this is a beam across an elevator shaft.

Upset beams are found on the roofs of high-rise buildings. They form the underside supports for water towers and air conditioning units.

Columns. Columns are vertical supports which support beams which in turn support floor slabs. Sometimes a column directly supports a floor slab.

Lally columns are steel pipes filled with concrete and set on plates on top of footings, Fig. 6-15. They are used to support steel beams (S or W), and are bolted to the beams.

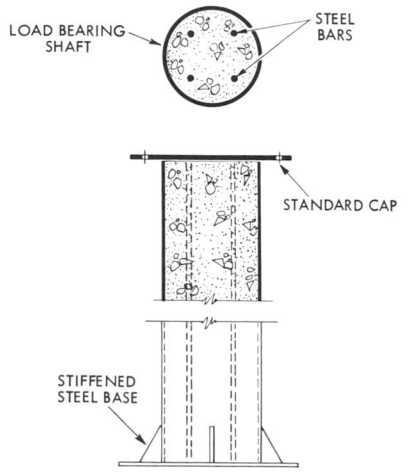

Fig. 6-15. A lally column with reinforcing steel bars.

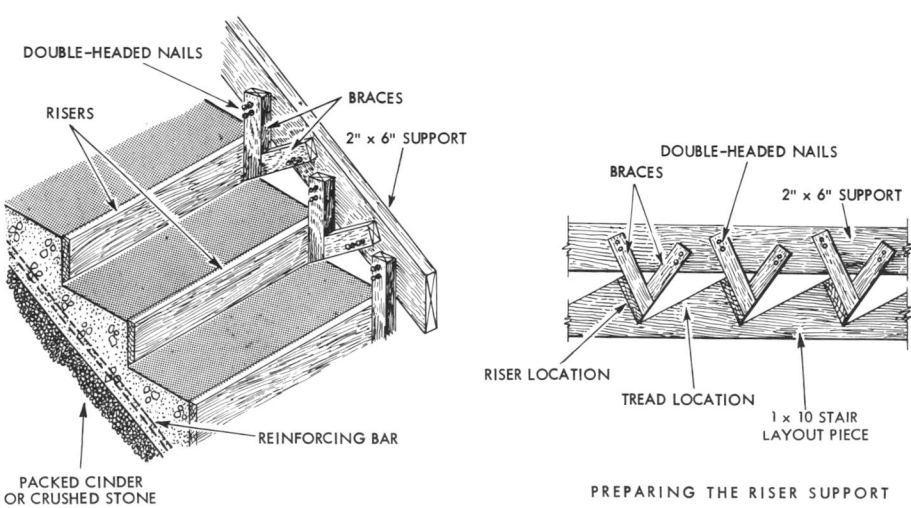

Fig. 6-16. Concrete stairs. The stairs are formed by risers held in place with braces.

Dovetail slots are metal strips placed inside the columns to form slots on which hang the brick tie inserts.

Stairs. Fig. 6-16 shows a typical concrete stairway. The concrete is poured into forms which are removed after the concrete has hardened. Steel reinforcing bars are embedded in the concrete to increase its load-bearing capacity.

FORMWORK

Concrete formwork is usually figured separately for each section of the concrete work; "foundation walls," "interior column footings," "cheek walls," etc., are some of the separate listings for formwork.

When a basement floor is poured, the wall acts as the form, thereby elimi-nating any need for framing members around the edges of the slab. However, when an *exterior slab* is poured, (such as a stair platform, a terrace, or a patio), formwork is necessary for the sides.

If one side of the slab is abutting against the foundation wall, forms will not be necessary on that side of the slab. The face of the slab is dowelled to the wall by iron ties which protrude through the foundation wall. When the slab is resting on cheek walls or spanning steel members, as in a porch or terrace, it is necessary to have forms under the slab. In all of these cases, careful consideration must be given to the formwork before the estimator can begin his calculations.

Panel forms. Panel forms are the

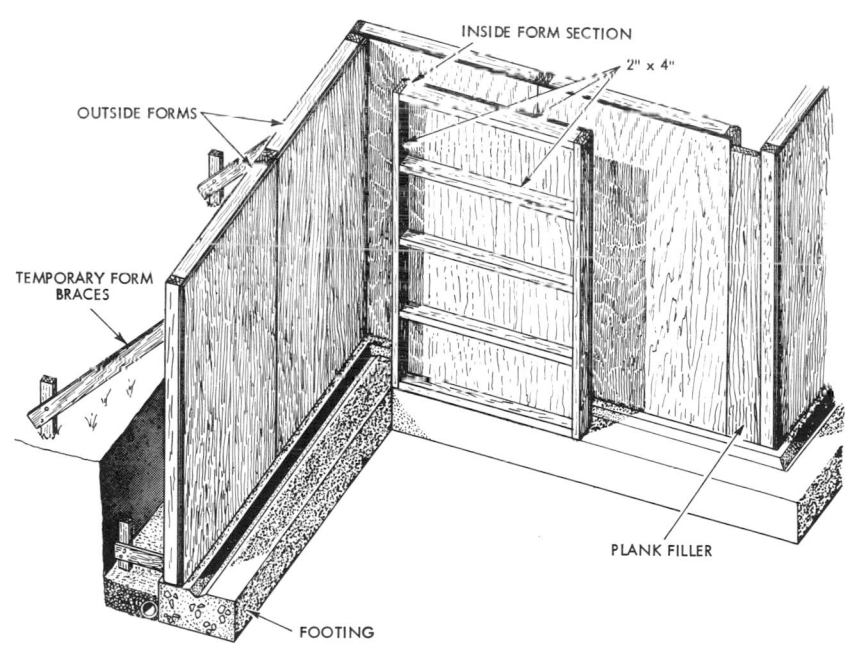

Fig. 6-17. Wood panel forms can be re-used many times.

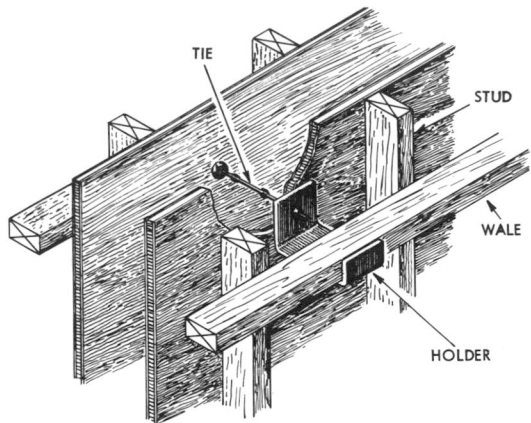

Fig. 6-18. Walers are held against the studs by patented form ties and holders. (Allenform Corporation)

most common forms in use today and consist of 4' × 8' plywood panels, 2' × 8' panels or 4' square panels, Fig. 6-17. These smaller sizes allow the panels to be quickly moved about by the carpenters.

The panels are made of either ⅝" or ¾" sheathing grade plywood cut to the desired dimensions and treated with oil or a commercial sealant.

Vertical bracing studs hold the panels in place, Fig. 6-18. The bracing is doubled along the plywood seams. Usually the studs are set 2'0" *on centers,* o.c. This means that the distance from the center of one stud to the center of another is 2'-0".

The *walers* (2" x 4" horizontal bracers along the walls) are laid across the vertical studs at about 16" o.c. spacing, Fig. 6-18.

Pan Forms. Pan forms are a series of interlocking metal panels which are set up to form the desired wall for the form. They are usually 2' × 2' sections which are wired or locked together. Pan forms are rapidly replacing

panel forms in the small house field because of their ease in handling and because they can be re-used.

The preferred bracing for pan forms is horizontal walers set at 2'-0" o.c. In long, continuous walls, vertical bracing is required to prevent the forms from buckling; otherwise the walers keep the smaller pan forms intact.

Angular Braces. Angular braces help to keep the foundation wall form vertical. They run at a 45° angle to the walls from the earth in front of the forms, Fig. 6-19. *Stakes* are driven into the ground to secure them.

Kickers. Kickers are placed along the ground perpendicular to the wall and secured to the brace at the stake, Fig. 6-19. They aid in bracing the bottom of the forms and prevent buckling or kicking out.

Spacing Bars and Spreaders. Spacing bars or rods and spreaders are not usually the concern of the estimator, but he should be familiar with their use. They are usually made of scrap

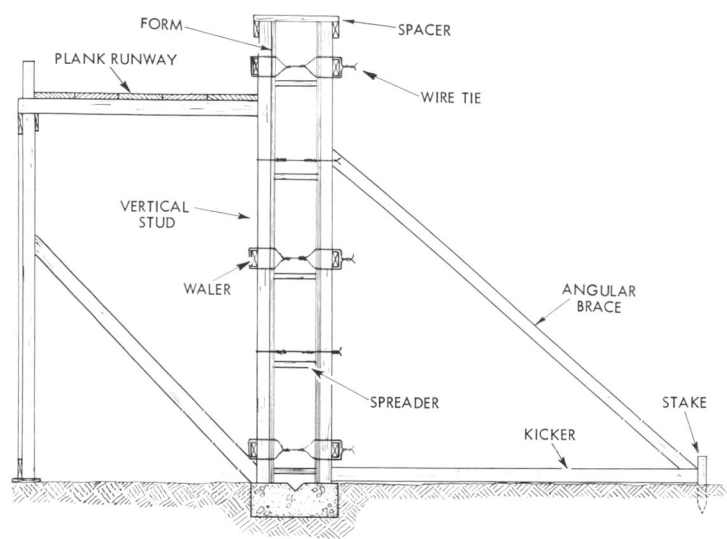

Fig. 6-19. Angular braces and kickers shown in formwork cross section.

lumber and are cut to fit over the tops of the forms — *spacers*, or inside the forms — *spreaders*. See Fig. 6-19. They are placed at varying intervals so as to be most effective without affecting the strength of the concrete.

The spreaders are one-inch square; their length is the width of the foundation wall. These spreaders are removed as the concrete is poured and they are no longer needed to keep the forms apart.

Ties. Ties are metal or plastic wires which are placed through the width of the foundation walls and secured to the outside ends of the forms before the concrete is poured. They prevent the buckling out of the forms

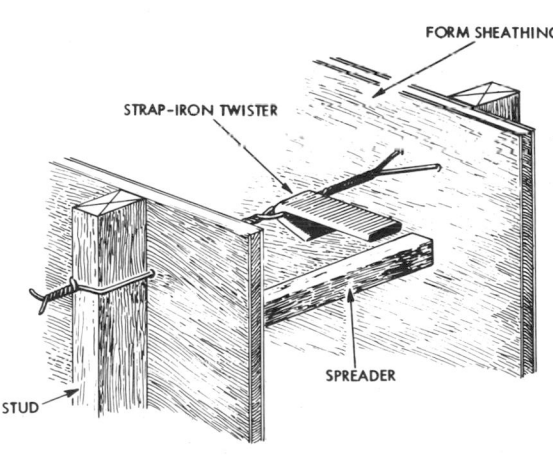

Fig. 6-20. Wire ties.

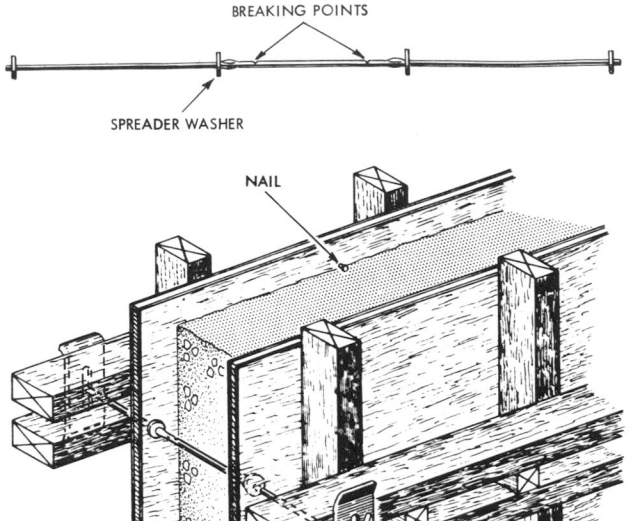

BREAKING POINTS

SPREADER WASHER

NAIL

CLAMP

Fig. 6-21. Snap ties.

under the pressure of the poured concrete. These ties are placed throughout the forms at specified intervals and clamped to the outside braces, Fig. 6-20.

Snap ties are special ties which are twisted and snapped off when the forms are removed, leaving the ties embedded about ½ inch inside the foundation wall, Fig. 6-21.

When the forms are ready to be stripped, the clamps are struck a blow which releases the wales. The forms are then freed from the wall and slipped from the ties. When the end of the tie is twisted, it breaks off inside the wall at the "break back" point. This leaves a hole in the wall which is patched with grout. Some ties are equipped with wood or plastic cones

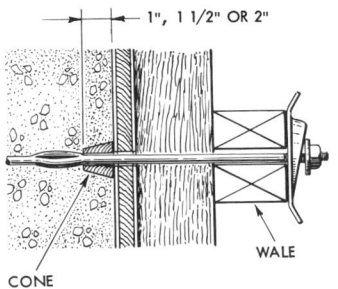

1", 1 1/2" OR 2"

WALE

CONE

Fig. 6-22. Some ties are made with wood or plastic cones which are removed from the wall, leaving a smooth conical hole for patching. (Richmond Screw Anchor Co., Inc.)

which are removed by twisting with a special wrench, leaving a clean conical hole that may be patched neatly. See Fig. 6-22. Snap ties are made for different wall thicknesses and are designed to take different loads. They should be placed at proper intervals, depending on the problems, with wales arranged to keep the forms in line. Bracing can be kept to a minimum because the ties are effective in retaining the concrete in the forms.

SIDEWALKS AND DRIVEWAYS

Sidewalks are concrete slabs and, as such, provision must be made for adequate drainage. At least a 2-inch layer of granular material, thoroughly tamped (subgrade), is required beneath the concrete. Often the surface of the sidewalk is slanted slightly so that water runs off the top.

After the forms have been placed and the subgrade is cleaned free of debris and dampened, concrete is poured in amounts small enough to be leveled quickly. Care must be taken not to overwork the concrete as this causes many problems later such as scaling and dusting.

To allow for initial contraction and later expansion to prevent unnecessary cracking and to avoid uneven settling, sidewalks must have joints.

Expansion joints, (also called isolation joints) are needed where sidewalks meet curbs, crosswalks, buildings or other fixed objects.

Control joints or transverse joints are also needed. These are grooves about 1/5 the thickness of the con-

Fig. 6-23. Hand tooling control joints in sidewalk. (Portland Cement Association)

crete. They may be tooled by hand (Fig. 6-23) or cut with a power saw at four or five foot intervals. Control joints allow for horizontal movement of the concrete. Cracking will then occur at these joints rather than on the surface of the slab.

Driveways are laid in a manner similar to sidewalks but they are usually of reinforced concrete to allow for greater loads. The center is slightly arched to permit water to run off, and they are thicker and wider than sidewalks. Control joints should be placed approximately 10 feet apart in drives.

Finishing Concrete. Various ways of finishing the poured concrete are possible. Finishing operations depend on the final effect desired.

Screeding. This is the process of leveling the top of the slab with a templet having a straight or curved

Fig. 6-24. The concrete is levelled by drawing a templet over guide strips. This operation is called screeding.

edge. The templet is drawn over wood or metal guide strips called screeds. See Fig. 6-24.

Floating. To obtain a surface smoother than that resulting from screeding but rough enough for a good foothold, a floating operation is employed. Floats may be of wood, rubber, or metal and may be attached to the end of a long handle as shown in Fig. 6-25.

Troweling. This is necessary to cover any rough spots or holes in the face of the concrete. Such items as noticeable honeycombs or the impression of the snap ties must be patched. On slabs, troweling follows floating and results in a smoother surface.

Rubbing. Usually, any surface im-

Fig. 6-25. Floating to smooth concrete surface. (Portland Cement Association)

pressions of the formwork that appear on the concrete after the forms are stripped must be removed. Rubbing is the method used to smooth these rough surfaces.

Special Effects. A variety of special effects on slab surfaces may be obtained. Scoring, brooming, and impressions in the face of the concrete are only some of the ways of finishing the surface.

QUANTITY TAKE-OFFS

Steel. Steel beams, columns, and *lintels* (framing over windows) are being used in many small homes. Sometimes they are used alone as in HOUSE PLAN A, but often they are used with reinforced concrete.

Reinforcing steel is estimated by the pound or ton (2,000 lbs.). To find the amount of reinforcing steel bars needed for the structure, first find the number of pieces and lengths of each size bar. Then multiply by the weight per linear foot. A 5% waste factor is usually added: this is sufficient to take care of waste and overlap.

Steel bars must be wired in place (horizontally and vertically) in retaining walls, cantilever girder frames, foundation walls, etc. *Tie wire* must therefore be taken into consideration in the quantity take-off.

Footings. The amount of concrete needed for footings is estimated by adding the volumes of each section of footing, each section being one of the foundation walls. Where corners exist, you must be careful not to overlap

the lengths and thereby increase the quantity of concrete.

Example 1. Find the amount of concrete needed for the wall footings of the house foundation in Fig. 6-26. (Note that the cheek walls have no footings.)

Solution: Set up a table as follows:

Width = 2.0 Height = 1.0'

Footing	Length	Volume
A	41.0'	82.0 cu. ft.
B	19.0'	38.0
C	39.0'	78.0
D	17.0'	34.0
	116.0'	232.0 cu. ft. or 8.6 cu. yds.

Example 2. Find the amount of concrete needed for the chimney footings in Fig. 6-26.

Solution: The chimney footing is 3.0' by 2.5'; the depth is 1.0'. This is a simple volume problem.

$$V = 3.0' \times 2.5' \times 1.0' = 7.5 \text{ cu. ft.}$$

Example 3. Find the volume of concrete needed for the column footings in Fig. 6-26.

Solution: There are two column footings and they are equal in size.

$$V = 2' \times 2' \times 1' = 4 \text{ cu. ft.}$$
for each footing
$$2 \times 4 = 8 \text{ (cu. ft.)}$$

Foundation Walls. The amount of concrete needed for foundation walls is estimated in the same manner as the footings.

Example 4. Find the volume of concrete needed for the foundation walls of the house in Fig. 6-26.

Solution: Use a table as in Example 1.

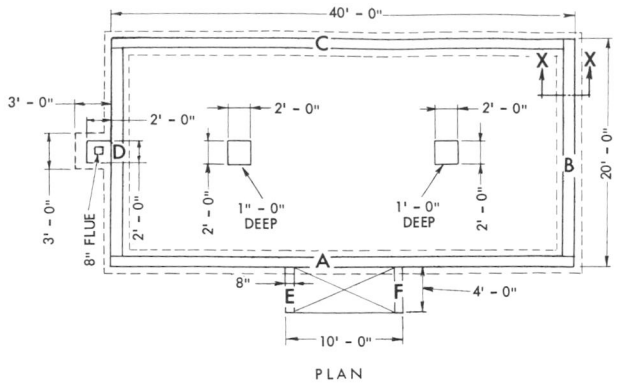

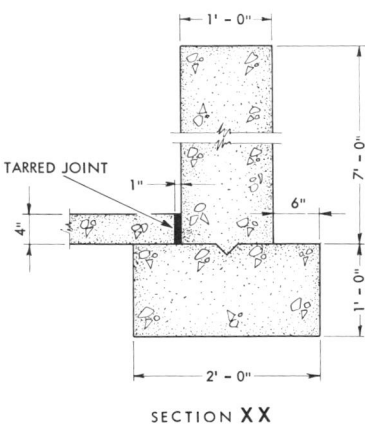

Fig. 6-26. Plan and section view of a house foundation.

Width = 1.0'   Height = 7.0'

Wall	Length	Volume
A	40.0'	280.0 cu. ft.
B	19.0'	133.0
C	39.0'	273.0
D	18.0'	126.0
	116.0'	812.0 cu. ft. or 30.1 cu. yds.

Note that the total length of the foundation wall is equal to the total length of the wall footings (116.0'). Since the center lines of the footings and walls are the same, their lengths should be the same. This serves as a useful check on your calculations.

Example 5. Find the volume of concrete needed for the chimney foundation in Fig. 6-26. The chimney foundation is 7.0' deep.

Solution: Find the total volume of the chimney foundation and subtract the volume of the flue.

Chimney foundation volume =
    2.0' × 2.0' × 7.0' = 28.0 cu. ft.

Flue volume =
$$0.67' \times 0.67' \times 7.0' = 3.14 \text{ cu. ft.}$$
$$V = 28.0 - 3.14 = 24.86 \text{ cu. ft.}$$

Example 6. Find the volume of concrete needed for the porch cheek walls. The cheek walls are 3.0' deep.

Solution: The cheek walls are figured in the same manner as the foundation walls.

Length = 4.0' Width = 0.67' Height = 3.0'

Wall	Volume
E	8.0 cu. ft.
F	8.0
	16.0 cu. ft.

Floor Slabs. The concrete for all floor slabs is figured as the volume of the particular slab; the volumes of large openings such as stairwells are deducted from the floor volume.

Example 7. Find the volume of concrete needed for the basement floor slab in Fig. 6-26.

Solution: The volume is the area of the basement (minus the area of the column footings) multiplied by the slab thickness.

Area of basement =
$$38.0' \times 180' = 684.0 \text{ sq. ft.}$$
Area of columns =
$$2 \times 2.0' \times 2.0' = 8.0 \text{ sq. ft.}$$
Slab area =
$$684.0 - 8.0 = 676.0 \text{ sq. ft.}$$
Slab volume =
$$676.0 \times 0.33 = 223 \text{ cu. ft.,}$$
$$\text{or } 8.26 \text{ cu. yds.}$$

Example 8. Find the volume of concrete needed for the porch slab in Fig. 6-26.

Solution: This problem is figured as a simple volume problem.
$$10.0' \times 4.0' \times 0.33' = 13.2 \text{ cu. ft.}$$

Note that the floor slab in Fig. 6-26 has a 1″ space around it to allow for expansion. This *expansion joint* is usually filled with tar or other expansion material. The amount of material needed is found by finding the area which comes in contact with the slab.

Example 9. Find the amount of expansion material needed for the foundation in Fig. 6-26.

Solution: Find the inside perimeter of the foundation wall.

$$38.0' + 38.0' + 18.0' + 18.0' = 112.0'$$

The joint is to be only 1″ wide and 4″ deep, so the volume of needed material is found by converting these dimensions to feet:
$$0.08' \times 0.33' \times 112.0' = 2.96 \text{ cu. ft.}$$

Beams. The concrete for a beam is computed as the volume of the beam, or that portion apart from the floor slab. The lengths of beams with identical cross-sectional area are added together before finding the total beam volume.

Example 10. Find the amount of concrete needed for the spandrel beams in Fig. 6-27.

Solution: There are 2 beams which are 8.0' long and 2 beams which are 18.0' long, each framing into columns at the corners of the structure. The beams have identical cross-sectional areas.

$$2 \times 8.0' = 16.0' \quad 2 \times 18.0' = 36.0'$$
$$16.0' + 36.0' = 52.0'$$

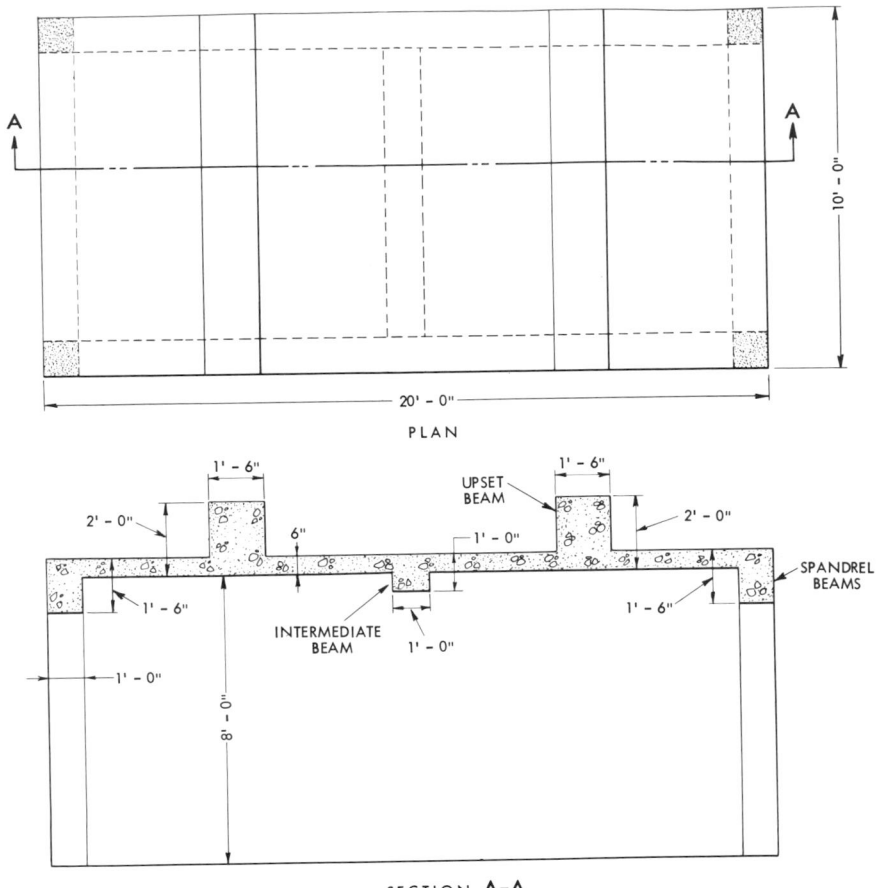

SECTION A-A

Fig. 6-27. Concrete beams, slabs, and columns.

The beams are 1.0′ wide and 1.0′ deep (after 6″ for the floor slab), or 1 sq. ft. in cross-sectional area.

$$1.0 \times 52.0 = 52.0 \text{ (cu. ft.)},$$
$$\text{or } 1.93 \text{ cu. yds.}$$

Example 11. Find the amount of concrete needed for the intermediate beam in Fig. 6-27.

Solution: The beam is 8.0′ long and 1.0′ wide. The depth is 6″ (1′-0″ −0′-6″ = 0′-6″ or 0.5′).

$$8.0' \times 1.0' \times 0.5' = 4.0 \text{ cu. ft.}$$
$$\text{or } 0.15 \text{ cu. yds.}$$

Example 12. Find the amount of concrete needed for the upset beams in Fig. 6-27.

Solution: There are 2 upset beams of equal size.

$$2 \times 10.0' \times 1.5' \times 1.5' = 45.0$$
$$\text{cu. ft. or } 1.67 \text{ cu. yds.}$$

The volume of the slab in Fig. 6-27 is found by multiplying its surface area by its thickness:

$$10.0' \times 20.0' \times 0.5' = 100.0 \text{ cu. ft.,}$$
or 3.7 cu. yds.

Columns. The volume of concrete needed for columns is usually computed by multiplying the cross-sectional area by the height which is taken as the total distance from floor to floor. Thus in Fig. 6-27 the heights of the columns are each 8'-6". The 6" thickness of the floor slab is neglected since the concrete will probably *shrink* as it hardens.

Example 13. Find the amount of concrete needed for the columns in Fig. 6-27.

Solution: There are 4 columns of equal height and cross-sectional area.

$$4 \times 8.5' \times 1.0' \times 1.0' = 34.0 \text{ cu. ft.}$$
or 1.2 cu. yds.

Concrete Constant. When finding the amount of concrete needed for a column, it is convenient to find first the average cross-sectional area. The number thus obtained is called the *concrete constant.* When multiplied by the height of the column, the concrete constant will give the volume of the column.

Example 14. Find the concrete constant and volume of a column which has a square base 2'-0" on each side, and whose height is 9'-0".

Solution: First find the area of the base.

$$\text{Area} = 2.0' \times 2.0' = 4.0 \text{ sq. ft.}$$

The concrete constant is 4.

Volume $= 4 \times 9.0 = 36.0$ (cu. ft.), or 1⅓ cu. yds.

Stairs. The volume of stairs is figured by multiplying the number of treads by the width of the stairway. You are generally safe to assume 1 cu. ft. of concrete per linear foot of tread.

Precast treads are figured by the piece or by the linear foot.

Example 15. Find the amount of concrete needed for a concrete stairway with 10 treads each 3'-6" wide.

Solution: Multiply the number of treads by the width of the stairway.

$$10 \times 3.5' = 35.0'$$

Assuming 1 cu. ft. per linear foot of tread, 35.0 cu. ft. of concrete is needed.

Concrete Fill. The estimator must consider the concrete fill necessary for concrete floors, stairs, and roofs.

Floor and roof fills are best estimated using cubic yards (instead of square feet). Since roofs usually slope to a drain, their thicknesses will vary. When using the square foot (of surface area) method to find the concrete, the *average* thickness of the roof is used in the calculations.

Fills for stairs and platforms are figured by the square foot of tread area in relation to the thickness, or by the linear foot (of treads).

ESTIMATING FORMWORK

Each class of formwork is figured separately. In general, formwork is figured in *square feet of contact area, (SFCA)*.

The following sections first discuss

the amount of formwork needed for each class of construction—footings, foundations, columns, etc. Then the bracing and ties are discussed separately.

Footings. The two main types of footings which the estimator figures are for foundations and for columns.

Foundation footings. These footings must have formwork on both sides. The lengths of the inside and outside forms are multiplied by the depth of the footing to obtain the *SFCA*.

Example 15. Find the *SFCA* of the foundation footings in Fig. 6-28. The footings are 1.0′ deep.

Depth = 1.0′

Outside Forms

Wall	Length	SFCA
A	41.0′	41.0
B	21.0′	21.0
C	41.0′	41.0
D	21.0′	21.0
		124.0

Inside Forms

Wall	Length	SFCA
A	37.0′	37.0
B	17.0′	17.0
C	37.0′	37.0
D	17.0′	17.0
		108.0

Solution: Multiply the length of each footing by its depth and add together the areas.

Footing Forms:

$$124.0 + 108.0 = 232.0 \text{ (sq. ft.)}$$

Column footings. Formwork for the interior column footings is figured in the same manner as the foundation footings. Since column footings are usually solid, only an outside form will be required.

The notation *F4S* means that forms are necessary on 4 sides.

Example 16. Find the amount of formwork required for the two column footings in Fig. 6-28.

Solution: The two forms are identical so the area of one is found and

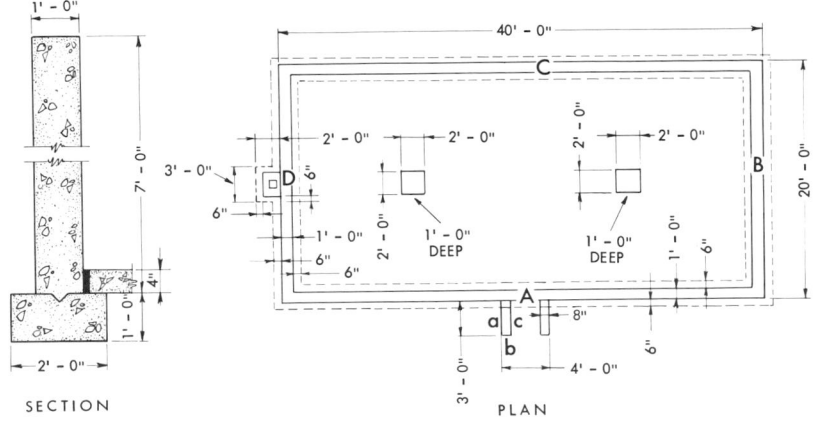

Fig. 6-28. House foundation plan for formwork take-off.

doubled. The area is the perimeter of the footing times the depth of the footing.

$$Perimeter = 2' + 2' + 2' + 2' = 8'$$
$$Area = 8' \times 1' = 8 \text{ sq. ft.}$$
$$Total\ area = 2 \times 8 = 16 \text{ (sq. ft.)}$$

Walls. Concrete foundation walls are sometimes grouped by the estimator, for convenience, into foundation walls with footings and those without footings (cheek walls).

Foundation walls are figured in the same manner as the foundation footings.

Example 17. Find the *SFCA* of the foundation wall in Fig. 6-28.

Solution: Set up tables for the outside and inside faces.

Depth = 7.0'

Outside Forms

Wall	Length	SFCA
A	40.0'	280.0
B	20.0'	140.0
C	40.0'	280.0
D	20.0'	140.0
		840.0

Inside Forms

Wall	Length	SFCA
A	38.0'	266.0
B	18.0'	126.0
C	38.0	266.0
D	18.0	126.0
		784.0

Total area =
$$840.0 + 784.0 = 1624.0 \text{ (sq. ft.)}$$

Note that there is no deduction for window openings in the amount of wall formwork. This allowance is made in the framing of the forms as an interior form fitting between the walls, Fig. 6-17.

The estimator doesn't allow for the *lap joints* at each corner of the walls. This extra inch at the butt-ends of the formwork is negligible and is only figured in the carpenter's layout.

Cheek walls. The forms for cheek walls are figured in the same manner as the foundation walls.

Example 18. Find the *SFCA* of the cheek walls in Fig. 6-28. The walls are 3.0' deep.

Solution: The two walls are identical so you need only find the area of one wall and then double the area.

Side	Length	Depth	SFCA
a	3.0'	3.0'	9.0
b	0.67'	3.0'	2.0
c	3.0'	3.0'	9.0
			20.0

Total area =
$$2 \times 20.0 = 40.0 \text{ (sq. ft.)}$$

Floor Slabs. Formwork for slabs depends on the type of slab to be poured and whether or not the sides of the slabs are to be open or framed.

If the slab is to be a *monolithic* pour, i.e., poured as part of the walls, then part of the slab formwork will be an extension of the exterior wall formwork and figured as such.

The formwork for slabs is usually figured as the *SFCA* of the underside of the slab.

Example 19. Find the amount of formwork required for the slab which goes over the cheek walls in Fig. 6-28. (The slab is not monolithic, i.e., it is poured separately from the cheek walls.) The depth of the slab is 3″. (Remember that 3″ = 0.25′.)

Solution: The formwork is figured in a manner similar to that of the column footings. The perimeter of the slab times its depth will give the *SFCA* of the sides of the slab.

Perimeter =
    3.0′ + 4.0′ + 3.0′ = 10.0 ft.
Area of sides =
    10.0′ × 0.25′ = 2.5 sq. ft.
Underside area =
    3.0′ × 2.66′ = 7.98 sq. ft.
Total area =
    2.5 + 7.98 = 10.48 (sq. ft.)

Beams. Formwork for beams is figured as the *SFCA* of open surface, beam sides, and beam bottom. Thus a spandrel beam has formwork on three sides—the outside beam side includes slab edge form. An upset beam has formwork only on its two sides, and an intermediate beam has formwork on three sides (beam sides and beam bottom).

Example 20. Find the amount of formwork needed for the spandrel beams in Fig. 6-27.

Solution: Each beam has three open sides. For each beam, add together the widths of these three open sides; multiply by the total length of the beams of that size. (Since there are two sets of two beams each—all with identical widths—it is only necessary to multiply the lengths of each size by two.) Add together the lengths.

Width of beams =
    1.5′ + 1.0′ + 1.0′ = 3.5′

Lengths: 2 × 8.0′ = 16.0′
         2 × 18.0′ = 36.0′
                    ───────
                     52.0′

Total area =
    3.5′ × 52.0′ = 182.0 sq. ft.

Example 21. Find the amount of formwork needed for the upset beams in Fig. 6-27.

Solution: The beams are identical so the area of one need only be multiplied by 2. Note that these beams do not need formwork on the top or bottom.

Area = 2 × 10.0′ × (1.5′ + 1.5′)
     = 60.0 sq. ft.

Columns. Each side of a column has formwork. Thus the *lateral* or *face area* must be calculated. If the average perimeter (or diameter) of the column is known, then it is easy to find the lateral area; it is the perimeter (or circumference) times the height of the column.

The perimeter is also called the *form constant.* Thus the form constant for columns with a perimeter of 2.0′ is 2. The form constant times the column width will give the lateral area of a column.

Example 22. Find the amount of formwork needed for the columns in Fig. 6-27.

Solution: The four columns are of equal size. Find the form constant and multiply it by four times the length of one column.

Perimeter = 1′ + 1′ + 1′ + 1′ = 4′
    Form constant is 4
Area = 4 × 4 × 8 = 128 (sq. ft.)

Stairs. The *SFCA* of stairs is figured as the areas (in square feet) of the underside, ends, and risers of the stairs. (See Fig. 6-16.)

Angular Braces and Kickers. An angular brace forms the hypotenuse of a right triangle with the form as one leg and a kicker as the other leg. Since the angular brace is usually at a 45° angle, the legs of the triangle are equal. If the length of one of the three elements of a 45° right triangle is known, then the other elements can be found.

Example 23. Find the length of an angular brace which is at a 45° angle to the top of a 7.0′ form.

Solution: Since the form is one leg of a 45° right triangle, the kicker is also 7.0′ and the Right Triangle Law can be used to find the length of the angular brace (the hypotenuse).

$$c^2 = a^2 + b^2$$
$$= 7^2 + 7^2$$
$$= 98 \text{ sq. ft.}$$

c = 9.9′, length of the angular brace.

Snap Ties. Snap ties are figured as a specified number per square foot of wall area.

Example 24. Find the number of snap ties needed for the foundation walls in Fig. 6-28. One tie is to be used for each 9 square feet of wall area.

Solution: Find the outside area of each wall and divide by 9.

Wall	Outside Area	No. of Ties
A	40.0′ x 7.0′	31
B	20.0′ x 7.0′	16
C	40.0′ x 7.0′	31
D	20.0′ x 7.0′	16
		94

Vertical Bracing. Vertical bracing studs for the walls and column footings are figured as the length of the wall divided by the on-center distances of the studs. If there are two studs at each position, then the amount of studs is doubled.

Example 25. Find the number of pieces needed for the vertical braces on the outside walls of Fig. 6-28. There are two braces at each on-center distance. Each of the studs is 8.0′ long.

Solution: If the centers are two feet apart, then divide each wall length by two feet and double your answer to get the required number of braces.

Wall	Length : OC Distance	No. of Pieces
A	40 ÷ 2	40
B	20 ÷ 2	20
C	40 ÷ 2	40
D	20 ÷ 2	20
		120

Total brace lengths =
$$120 \times 8.0′ = 960.0′$$

Walers. The walers or horizontal braces which run the length of the wall are usually spaced 16″ apart, o.c. distance. The first few rows of walers may be doubled because of the greater pressure at the bottom of the forms. The greater the height of the form, the greater will be the number of double rows of walers.

In Fig. 6-28 the number of walers needed is computed as follows:

No. of walers

$$= \frac{\text{Height of wall}}{\text{On-center distance}} + 1$$

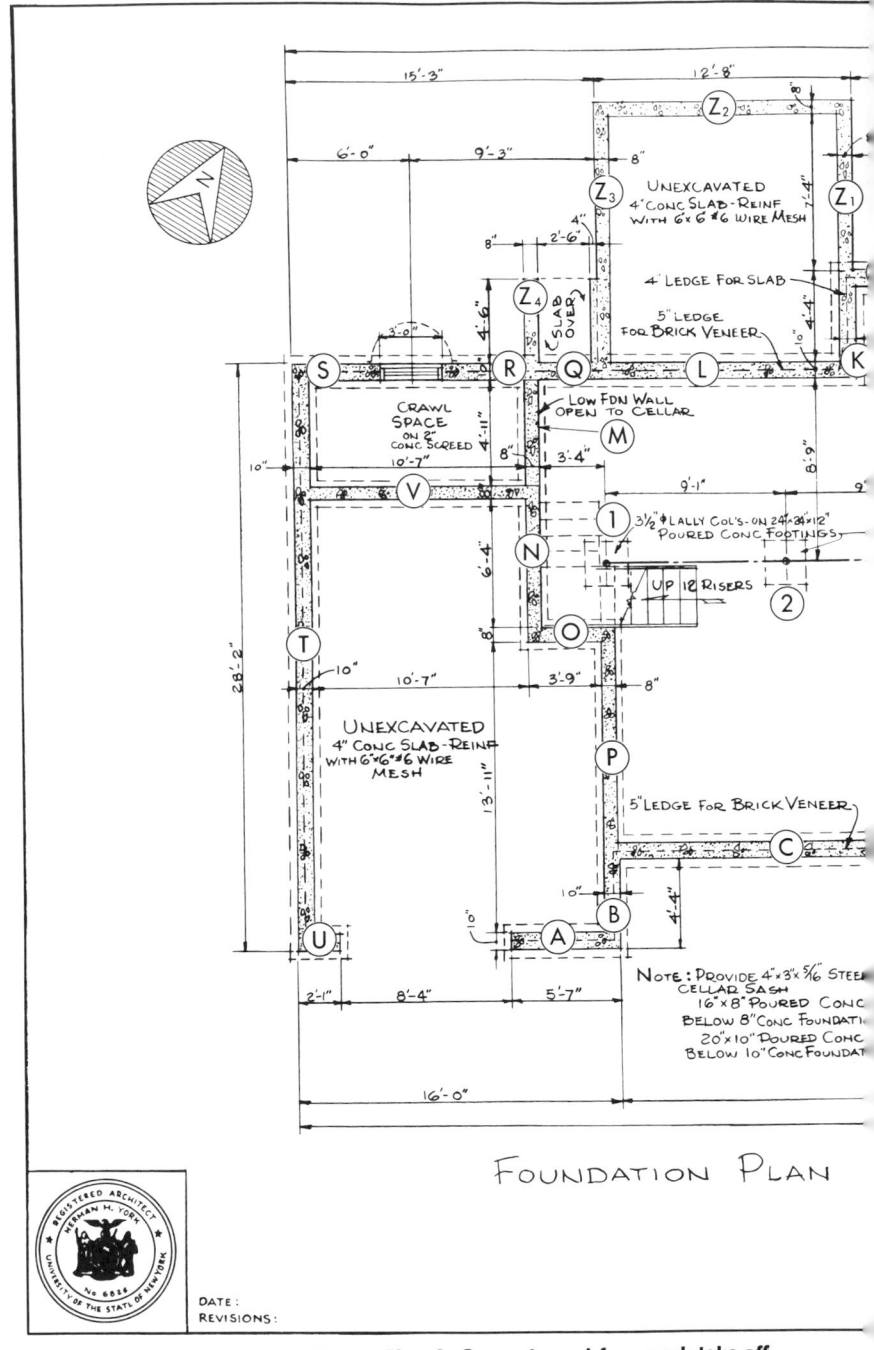

Fig. 6-29. House Plan A. Concrete and formwork take-off.

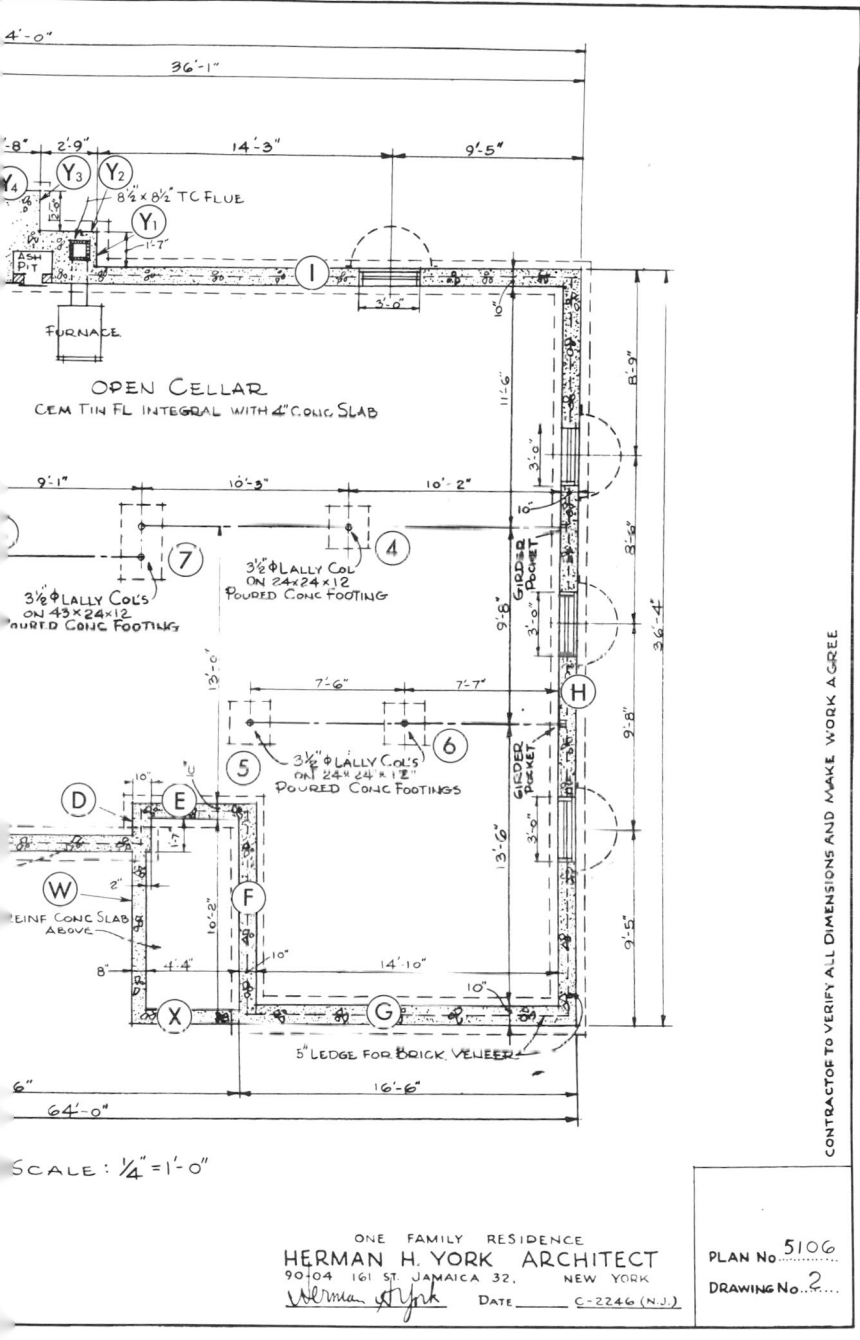

OPEN CELLAR
CEM TIN FL INTEGRAL WITH 4" CONC SLAB

FURNACE

8½" × 8½" TC FLUE

ASH PIT

3½"∅ LALLY COL
ON 24×24×12
POURED CONC FOOTING

3½"∅ LALLY COL'S
ON 43×24×12
POURED CONC FOOTING

3½"∅ LALLY COL'S
ON 24"×24"×12"
POURED CONC FOOTINGS

GIRDER POCKET

GIRDER POCKET

REINF CONC SLAB
ABOVE

5" LEDGE FOR BRICK VENEER

CONTRACTOR TO VERIFY ALL DIMENSIONS AND MAKE WORK AGREE

SCALE: ¼" = 1'-0"

ONE FAMILY RESIDENCE
HERMAN H. YORK ARCHITECT
90-04 161 ST. JAMAICA 32, NEW YORK

Date _____ C-2246 (N.J.)

PLAN No. 5106
DRAWING No. 2

$$= \frac{7.0'}{133'} + 1$$

$$= 6.26 \text{ or approximately}$$

6 walers

Outside Wall	Waler Length	Linear Feet
A	42'	126
B	22'	66
C	42'	126
D	221	66
		384 ft.

This means that each wall has 6 walers running the length of the wall. In addition to these 6 walers, each wall will have 3 extra walers for additional bracing near the bottom of the form.

Example 26. Find the linear footage of the walers needed for the outside foundation walls in Fig. 6-28.

Solution: As shown earlier, each wall will have walers running its entire length; multiply each wall length by 9.

Outside Wall	Linear Ft. (+ 1' overhang)	Total Feet
A	41'	369
B	21'	189
C	41'	369
D	21'	189
		1116 ft.

Foundation footings also must be braced so that the form will hold its shape after pouring. This is best done by using horizontal walers along the length of the footings. For a footing 1' deep, a double waler along the base and a single waler along the top of the form should be adequate.

Example 27. Find the number of linear feet needed for the outside walers of the foundation footings in Fig. 6-28.

Solution: Each foundation footing form has a double waler along the base and a single waler at the top— 3 walers running the length of the footing form. Therefore, multiply the length of each form by 3. (There is a 1' overlap at each end.)

Example 28. Find the number of linear feet needed for the vertical stakes for the outside footing forms in Fig. 6-28. Each stake is 4' long and 2' apart, o.c.

Solution: Dividing the length of each wall by the distance between stakes will give the number of stakes needed for each wall.

Wall	Length ÷ 2 ft.	Linear Ft.
A	21	84
B	11	44
C	21	84
D	11	44
		256 ft.

Quantity take-off for House Plan A

Figs. 6-30 to 6-37 contain detailed quantity take-offs for the concrete and formwork needed for the foundation of HOUSE PLAN A. Fig. 6-29 is used for this take-off. In general, the walls are labeled counter-clockwise. (Use the detailed drawings at the end of Chapter 2 for further clarification).

Concrete. The take-off for the foundation and cheek walls is taken as the volume of the walls less the volume of five windows—labeled "Outs" on the estimating sheet.

A 5″ wide ledge for brick veneer is at the top of the exterior foundation walls. The ledge extends 6″ below the

CONCRETE

QUANTITY TAKE-OFF

SHEET 1

HOUSE PLAN A

Wall Code	Length l ft.	Width w ft.	Height h ft.	Area A sq.ft.	Volume V cu.ft.	V cu. yds.
EXTERIOR FOUNDATION WALLS						
A	5.58	0.83	7.00		31.25	
B	4.33				24.25	
C	27.33				153.05	
D	1.58				8.85	
E	4.33				24.25	
F	11.00				61.39	
G	14.83				83.05	
H	36.33				203.45	
I	22.83				127.64	
J	7.83				43.85	
K	4.33				24.25	
L,Q	15.50		7.00		86.80	
R	5.83		5.00		24.19	
S	6.00		4.00		19.92	
T	27.33		4.00		90.74	
U	1.25	0.83	4.00		4.16	
	196.22				1011.08	
OUTS (WINDOWS)	3.00	0.83	1.67		-20.83	
				TOTAL	990.25	37.—
INTERIOR FOUNDATION WALLS						
M,N,O,P	26.00	0.67	7.00		121.94	
V	10.58	0.67	7.00		28.35	
	36.58				150.29	6.—
EXTERIOR CHEEK WALLS						
Z4	4.50	0.67	7.00		21.11	
W	8.58	0.67	7.00		40.24	
X	4.33	0.67	7.00		20.31	
Z1	7.33	0.67	4.00		19.64	
Z2	12.67	0.67	4.00		33.96	
Z3	7.17	0.67	4.00		19.22	
Z3	4.50	1.00	4.00		18.00	
					172.48	7.—
CHIMNEY FOOTING						
	6.25	5.25	1.00		32.81	
	2.75	2.00	1.00		-5.50	
					27.31	1.—

Fig. 6-30-A. House Plan A. Concrete take-off.

		CONCRETE				SHEET 2	
		QUANTITY TAKE-OFF				HOUSE PLAN A	

Wall Code	Length l ft.	Width w ft.	Height h ft.	Area A sq.ft.	Volume V cu.ft.	V cu. yds.
	CHIMNEY STACK					
	2.67	4.42	7.00		82.60	
	2.75	2.42	7.00		46.62	
FLUE	0.71	0.71	3.00		1.51	
ASH PIT	1.50	1.83	1.00		2.75	
					124.96	5—
	EXTERIOR WALL FOOTINGS					
A	6.42	1.67	0.83			
B	4.33					
C	27.33					
D	—					
E	6.67					
F	10.17					
G	15.67					
H	35.50					
I	22.00					
J	7.67					
K	4.33					
L,Q	15.08					
R	5.75					
S	6.42					
T	27.33					
U	1.25	1.67	0.83			
	195.92				272.33	11—
	INTERIOR WALL FOOTINGS					
M,N,O,P	25.50	1.33	0.67		22.70	
V	9.83	1.33	0.67		8.76	
	35.33				31.46	2—
	PIER FOOTINGS					
1 TO 6	2.00	2.00	1.00		24.00	
7	3.58	2.00	1.00		7.17	
					31.17	2—
	CELLAR SLAB					
	14.83	34.67	0.33	514.16		
	5.83	24.58		142.84		
	26.50	26.58		704.37		
	3.75	11.92		44.70		
	4.33	12.16	0.33	-52.65		
				1,353.42	446.63	17—

Fig. 6-30-B. House Plan A. Concrete take-off.

CONCRETE

QUANTITY TAKE-OFF

SHEET 3

HOUSE PLAN A

Wall Code	Length l ft.	Width w ft.	Height h ft.	Area A sq.ft.	Volume V cu.ft.	V cu. yds.
	MISC.	CONCRETE		SLABS		
FRONT ENT.	10.17	5.00	0.33	50.85	17 —	
REAR P.	12.67	12.33	↓	156.22	52 —	
REAR ENT.	4.50	3.50		15.75	5 —	
GARAGE	13.92	14.33	↓	199.47	66 —	
"	7.00	10.58	0.33	74.06	25 —	
CRAWL SPACE	10.58	4.92	0.17	52.06	9 —	
				548.41	174 —	7 —
					TOTAL CONCRETE	95 —

Fig. 6-30-C. House Plan A. Concrete take-off.

top of the wall. This volume is subtracted from the total volume of each exterior foundation wall.

The additional concrete for walls *R*, *S*, and *V* is the volume of concrete of the stepped wall below the top of the normal footing.

The additional concrete for the cheek walls is the volume of a wedge of concrete. This wedge starts at the bottom of wall Z_2 and slopes to the top of the foundation walls.

The concrete for the chimney is taken as the volume of the chimney less the volumes of the flue stack and ash pit.

The cellar slab is divided into 4 parts for the quantity take-off. Working from right to left, the area of each part is calculated and added together. Any area outside the boundary of the cellar slab is then subtracted.

The volumes are converted to whole cubic yards and in almost all cases the amount of concrete is greater than the volume of the wall, column, or slab. This "extra" concrete can be counted as part of the wastage.

Formwork. Only the quantity of formwork, bracing, and ties is computed here. Any estimate made from these *minimum* figures would, of course, have to consider waste, transportation, and labor costs.

						SHEET 1
		FORMWORK				
		QUANTITY TAKE-OFF				*HOUSE PLAN A*

Wall Code	Length l ft.-in.	Length l ft.	Height h ft.-in.	Height h ft.	Area A sq.ft.
WALLS—OUTSIDE FACE					
A	5'-7"	5.58	7'-4"	7.33	40.90
B	4'-4"	4.33			31.74
C	26'-6"	26.50			194.25
D	1'-7"	1.58			11.58
E	4'-2"	4.17			30.57
F	9'-6"	9.50			69.64
G	16'-6"	16.50			120.95
H	36'-4"	36.33			266.30
I	23'-8"	23.67			173.50
Y₁	1'-7"	1.58			11.58
Y₂	2'-9"	2.75			20.16
Y₃	2'-0"	2.00			14.66
Y₄	2'-8"	2.67			19.57
Y₅	3'-7"	3.58			26.24
J	7'-0"	7.00	↓	↓	51.31
Z₁	8'-0"	8.00	7'-4"	7.33	58.64
Z₂	12'-8"	12.67	4'-0"	4.00	50.68
Z₃	12'-4"	12.33	7'-4"	7.33	90.38
Q	2'-6"	2.50	7'-4"	7.33	18.33
Z₄	9'-8"	9.67	7'-4"	7.33	70.88
R	5'-9"	5.75	4'-8"	4.67	26.85
S	6'-0"	6.00	2'-8"	2.67	16.02
T	28'-2"	28.17	2'-8"	2.67	75.21
U	2'-1"	2.08	2'-8"	2.67	5.55
W	8'-7"	8.58	7'-4"	7.33	62.89
X	5'-0"	5.00	7'-4"	7.33	36.65
					1,595.03
WALLS—INSIDE FACE					
C	26'-8"	26.50	7'-4"	7.33	194.25
D	1'-7"	1.58			11.58
E	5'-10"	5.83			42.74
F	10'-2"	10.17			74.55
G	14'-10"	14.83			108.70
H	34'-8"	34.67			254.13
I,J	35'-1"	35.08	↓	↓	257.14
K	4'-4"	4.33			31.74
L,Q	15'-8"	15.67	7'-4"	7.33	114.86
M	4'-11"	4.92	4'-8"	4.67	22.98
N	7'-0"	7.00	7'-4"	7.33	51.31
O	3'-9"	3.75	7'-4"	7.33	27.49
P	9'-10"	9.83	7'-4"	7.33	72.05
					1,263.52

Fig. 6-31-A. House Plan A. Formwork take-off.

| | | FORMWORK | | | | SHEET 2 |
| | | QUANTITY TAKE-OFF | | | | *HOUSE PLAN A* |

Wall Code	Length l ft.–in.	Length l ft.	Height h ft–in.	Height h ft.	Area A sq.ft.	
	WALLS—UNEXCAVATED INSIDE					
A	4'-9"	4.75	7'-4"	7.33	34.82	
P,B	13'-11"	13.92	7'-4"	7.33	102.03	
O	3'-9"	3.75	7'-4"	7.33	27.49	
N	7'-0"	7.00	7'-4"	7.33	51.31	
V	10'-7"	10.58	2'-8"	2.67	28.25	
T	20'-11"	20.92	2'-8"	2.67	55.86	
U	1'-3"	1.25	2'-8"	2.67	3.34	
BUTTENDS	1'-8"	1.67	2'-8"	2.67	4.46	
M	4'-11"	4.92	2'-8"	2.67	13.14	
R	4'-11"	4.92	4'-8"	4.67	22.97	
T	4'-11"	4.92	2'-8"	2.67	13.14	
L	11'-4"	11.33	7'-4"	7.33	83.05	
Z₁	11'-8"	11.67	7'-4"	7.33	85.54	
Z₂	11'-4"	11.33	4'-0"	4.00	45.32	
Z₃	11'-8"	11.67	7'-4"	7.33	85.54	
W	7'-11"	7.92	7'-4"	7.33	58.05	
X	4'-4"	4.33	7'-4"	7.33	31.74	
F	10'-2"	10.17	7'-4"	7.33	74.55	
S	6'-8"	6.67	2'-8"	2.67	17.81	
					838.41	
	FOOTINGS—OUTSIDE FACE					
A	6'-5"	6.42	10"	0.83	5.33	
B	4'-4"	4.33			3.59	
C	27'-4"	27.33			22.68	
D	1'-7"	1.58			1.31	
E	3'-6"	3.50			2.91	
F	10'-2"	10.17			8.44	
G	17'-4"	17.33			14.38	
H	37'-2"	37.17			30.85	
I	23'-8"	23.67			19.65	
Y₁	1'-7"	1.58			1.31	
Y₂	2'-9"	2.75			2.28	
Y₃	2'-0"	2.00			1.66	
Y₄	3'-6"	3.50			2.91	
Y₅	3'-7"	3.58			2.97	
J	7'-8"	7.67			6.37	
K	4'-4"	4.33			3.59	
L,Q	15'-1"	15.08			12.52	
R	5'-9"	5.75			4.77	
S	6'-5"	6.42			5.33	
T	29'-0"	29.00	10"	0.83	24.07	
					176.92	

Fig. 6-31-B. House Plan A. Formwork take-off.

| | | FORMWORK | | | | SHEET 3 | | |
| | | QUANTITY TAKE-OFF | | | | HOUSE PLAN A | | |

Wall Code	Length l ft.–in.	Length l ft.	Height h ft.–in.	Height h ft.	Area A sq.ft.	
	FOOTINGS–INSIDE FACE					
A	4'-9"	4.75	10"	0.83		3.94
B,P	13'-2"	13.17	10"	0.83		10.93
O	3'-9"	3.75	8"	0.67		2.51
N	6'-8"	6.67	8"	0.67		4.47
V	9'-10"	9.83	8"	0.67		6.59
T	20'-2"	20.17	10"	0.83		16.74
U	1'-3"	1.25				1.04
BUTT ENDS	3'-4"	3.33				2.76
C	25'-9"	25.75				21.37
D	1'-8"	1.67				1.37
E	6'-8"	6.67				5.54
F	10'-2"	10.17				8.44
G	14'-0"	14.00				11.62
H	33'-10"	33.83				28.08
I,J	34'-3"	34.25				28.43
K	4'-4"	4.33				3.59
L,Q	15'-11"	15.92	10"	0.83		13.21
M	4'-6"	4.50	8"	0.67		3.00
N	6'-8"	6.67				4.47
O	3'-9"	3.75				2.50
R	5'-5"	5.42	10"	0.83		4.50
S	4'-9"	4.75	10"	0.83		3.94
T	4'-2"	4.17	10"	0.83		3.46
						192.50
	INTERIOR COLUMN FOOTINGS					
8 COL.	8'-0"	8.0	1'-0"	1.0		64 —
	(FORM CONSTANT = 8)					
	TOTAL FORMWORK				4,130.38	

Fig. 6-31-C. House Plan A. Formwork take-off.

					HORIZONTAL WALERS			SHEET 4	

2' – 0" OC— 6 walers per 7' – 4" wall
wall length + 6" overhang at free ends

QUANTITY TAKE-OFF

HOUSE PLAN A

Wall Code	Waler Length ft.-in.	Waler Length ft.	No. of Walers		Total Length ft.	
	FOUNDATION—EXT. FACE					
A	6'-7"	6.58	6—		39.48	
B	4'-10"	4.83			28.98	
C	26'-6"	26.50			159—	
D	2'-1"	2.08			12.48	
E	4'-2"	4.17			25.02	
F	9'-6"	9.50			57.00	
G	17'-6"	17.50			105.00	
H	37'-4"	37.33			223.99	
I	24'-2"	24.17			145.02	
Y1	2'-1"	2.08			12.48	
Y2	3'-3"	3.25			19.50	
Y3	2'-6"	2.50			15—	
Y4	3'-8"	3.67			22.02	
Y5	4'-1"	4.08			24.48	
J	7'-0"	7.00	↓		42—	
Z1	8'-6"	8.50	6—		51.—	
Z2	13'-8"	13.67	4—		54.68	
Z3	12'-10"	12.83	6—		76.98	
Q	2'-6"	2.50	6—		15.—	
Z4	11'-8"	11.67	6—		70.—	
R	5'-9"	5.75	5—		28.75	
S	6'5"	6.42	4—		25.68	
T	29'-2"	29.17	3—		87.51	
U	3'-1"	3.08	3—		9.24	
W	9'-1"	4.08	6—		54.48	
X	6'-0"	6.00	6—		36—	
					1,440.76	
	INSIDE FACES—EXC. WALLS					
C	26'-8"	26.67	6—		160.02	
D	2'-1"	2.08			12.48	
E	6'-10"	6.83			40.98	
F	10'-8"	10.67			64.02	
G	14'-10"	14.83			88.98	
H	34'-8"	34.67			208.02	
I, J	35'-1"	35.08			210.48	
K	4'-10"	4.83			28.98	
L, Q	15'-11"	15.92			95.52	
M	4'-11"	4.92			29.52	
N	7'-0"	7.00			42—	
O	4'-3"	4.25	↓		25.50	
P	10'-9"	10.75	6—		64.50	
					1071—	

Fig. 6-32-A. House Plan A. Horizontal walers.

				HORIZONTAL WALERS		SHEET 5	
2' - 0" OC— 6 walers per 7' - 4" wall wall length + 6" overhang at free ends				QUANTITY TAKE–OFF		HOUSE PLAN A	
Wall Code	Waler Length ft.-in.	Waler Length ft.	No. of Walers		Total Length ft.		
	UNEXCAVATED INSIDE						
A	5'-3"	5.25	6			31.50	
P, B	13'-11"	13.92	6			83.52	
O	4'-3"	4.25	6			25.50	
N	7'-6"	7.50	6			45.—	
V	10'-7"	10.58	3			31.74	
T	20'-11"	20.92	3			62.76	
U	1'-9"	1.75	3			5.25	
V	10'-7"	10.58	3			31.74	
M	4'-11"	4.92	3			14.76	
R	6'-5"	6.42	4			25.68	
T	4'-11"	4.92	3			14.76	
L	11'-4"	11.33	6			67.98	
Z₁	11'-8"	11.67	6			70.02	
Z₂	11'-4"	11.33	3			33.99	
Z₃	11'-8"	11.67	6			70.02	
W	9'-6"	9.50	6			57.—	
X	4'-4"	4.33	6			25.98	
S	5'-2"	5.17	3			15.51	
BUTT END	1'-10"	1.83	3			5.49	
"	1'-10"	1.83	6			10.98	
						729.18	
				TOTAL		3240.94	

Fig. 6-32-B. House Plan A. Horizontal walers.

Wall Code	Wall Length ft.	No. of Studs	Stud Length ft.		Total Length ft.	
	OUTSIDE FORMS					
A	5.58	3 —	8 —		24 —	
B	4.33	2			16	
C	26.50	13			104	
D	1.58	2			16	
E	4.17	2			16	
F	9.50	5			40	
G	16.50	8			64	
H	36.33	18			144	
I	23.67	12			96	
Y_1	1.58	2			16	
Y_2	2.75	2			16	
Y_3	2.00	2			16	
Y_4	2.67	2			16	
Y_5	3.58	2			16	
J	7.00	4			32	
Z_1	8.00	4 —	8 —		32	
Z_2	12.67	6 —	5 —		30	
Z_3	12.33	6 —	5 —		30	
Q	2.50	2 —	5 —		10	
Z_4	9.67	5 —	5 —		25	
R	5.75	3 —	5 —		15	
S	6.00	3 —	4 —		12	
1	28.17	14	4		56	
U	2.08	2 —	4 —		8	
W	8.58	4 —	4 —		16	
X	5.00	3 —	4 —		12	
		131 —			878 —	

VERTICAL STUDS
SHEET 6
2' – 0" OC— minimum of 2 per wall
QUANTITY TAKE-OFF
HOUSE PLAN A

Fig. 6-33-A. House Plan A. Vertical studs.

				VERTICAL STUDS		SHEET 7	
2' – 0" OC— minimum of 2 per wall				QUANTITY TAKE-OFF		*HOUSE PLAN A*	

Wall Code	Wall Length ft.	No. of Studs	Stud Length ft.		Total Length ft.		
	INSIDE FACE						
C	26.50	13	8—		104—		
D	1.58	2			16		
E	5.83	3			24		
F	10.17	5			40		
G	14.83	8			64		
H	34.67	17			136		
I,J	35.08	17			136		
K	4.33	2			16		
L,Q	15.67	8			64		
M	4.92	2			16		
N	7.00	4			32		
O	3.75	2			16		
P	9.58	5	8—		40—		
		88			704—		
	UNEXC INSIDE						
A	4.75	3	8—		24—		
P,B	13.92	7	8		56		
O	3.75	2	8		16		
N	7.00	4	8		32		
V	10.58	5	3		15		
T	20.92	10	3		30		
U	1.25	2	3		6		
BUTTEND	0.83	2	3		6		
"	0.83	2	8		16		
V	10.58	5	3		15		
M	4.92	3	3		9		
R	5.75	3	5		15		
T	4.92	3	3		9		
L	11.33	6	8		48		
Z1	11.67	6	8		48		
Z2	11.33	6	5		30		
Z3	11.67	6	8		48		
W	9.50	5	8		40		
X	4.33	2	8		16		
S	5.17	3	3—		9—		
		85			488—		

Fig. 6-33-B. House Plan A. Vertical studs.

				WALERS–FOOTINGS		SHEET 8	
3 per length, + 6" overlap at free ends				QUANTITY TAKE-OFF		*HOUSE PLAN A*	
Wall Code	Waler Length ft.-in.	Waler Length ft.	No. of Walers		Total Length ft.		
	FOOTINGS—OUTSIDE FACE						
A	7'-5"	7.42	3—		22.26		
B	4'-10"	4.83			14.49		
C	27'-4"	27.33			81.99		
D	2'-1"	2.08			6.24		
E	3'-9"	3.75			11.25		
F	10'-8"	10.67			32.01		
G	18'-4"	18.33			54.99		
H	38'-2"	38.17			114.50		
I	24'-2"	24.17			72.50		
Y1	2'-1"	2.08			6.25		
Y2	3'-3"	3.25			9.75		
Y3	2'-6"	2.50			7.50		
Y4	4'-6"	4.50			13.50		
Y5	4'-1"	4.08			12.25		
J	8'-2"	8.17			24.50		
K	4'-10"	4.83			14.49		
L, Q	15'-7"	15.58			46.77		
R	6'-3"	6.25			18.75		
S	7'-1"	7.08			21.24		
T	31'-0"	31.00	3—		93.00		
					678.23		
	INTERIOR COLUMNS						
8 COLS.	8'-0"	8.0	3		192—		

Fig. 6-34-A. House Plan A. Footing walers.

				WALERS–FOOTINGS		SHEET 9		
			3 per length, + 6" overlap at free ends	QUANTITY TAKE-OFF		*HOUSE PLAN A*		

Wall Code	Waler Length ft.–in.	Waler Length ft.	No. of Walers			Total Length ft.		
		FOOTINGS–INSIDE FACE						
A	5'–3"	5.25	3 —			15.75		
P,B	13'–2"	13.17				39.51		
O	4'–3"	4.25				12.75		
N	7'–6"	7.50				22.50		
V	9'–10"	9.83				29.49		
T	20'–2"	20.17				60.50		
U	2'–2"	2.17				6.50		
BUTT ENDS	5'–4"	5.33				16.00		
C	25'–8"	25.67				77.00		
D	2'–1"	2.08				6.24		
E	7'–8"	7.67				23.00		
F	10'–8"	10.67				32.00		
G	14'–0"	14.00				42.00		
H	33'–10"	33.83				101.49		
I,J	34'–3"	34.25				102.75		
K	4'–10"	4.83				14.50		
L,Q	16'–11"	16.92				50.75		
M	4'–6"	4.50				13.50		
N	7'–6"	7.50				22.50		
O	4'–3"	4.25				12.75		
P	11'–1"	11.09				30.00		
V	9'–10"	9.83				29.49		
M	4'–2"	4.17				12.51		
R	5'–4"	5.33				16.00		
S	4'–9"	4.75				14.25		
T	4'–2"	4.17	3 —			12.50		
						816.23		

Fig. 6-34-B. House Plan A. Footing walers.

			VERTICAL BRACING-FOOTINGS		SHEET *10*		
4' – 0" OC; minimum of 2 per wall			QUANTITY TAKE-OFF		*HOUSE PLAN A*		

Wall Code	Wall Length ft.	No. of Stakes	Stake Length ft.		Total Length ft.		
	FOOTING STAKES—INSIDE FACE						
A	4.75	2	4'		8 —		
P,B	13.08	3			12		
O	3.75	2			8		
N	7.00	2			8		
V	9.75	2			8		
T	20.75	5			20		
U	1.67	2			8		
BOTTENS	3.33	2			8		
C	25.17	6			24		
D	.83	2			8		
E	6.67	2			8		
F	10.17	3			12		
G	14.00	4			16		
H	33.83	9			36		
I,J	34.25	9			36		
K	4.33	2			8		
L,Q	15.17	4			16		
M	4.50	2			8		
N	6.67	2			8		
O	3.75	2			8		
P	9.50	2			8		
V	9.88	2			8		
M	4.17	2			8		
R	5.75	2			8		
S	6.50	2	↓		8 ↓		
T	4.17	2	4'		8 – –		
					316 —		

Fig. 6-35-A. House Plan A. Vertical bracing for footings.

				VERTICAL BRACING-FOOTINGS		SHEET 11
4' – 0" OC; minimum of 2 per wall				QUANTITY TAKE-OFF		*HOUSE PLAN A*

Wall Code	Wall Length ft.	No. of Stakes	Stake Length ft.		Total Length ft.	
				FOOTING STAKES–OUTSIDE FACE		
A	6.42	2	4'		8	
B	4.33	2			8	
C	26.83	7			28	
D	1.58	2			8	
E	3.50	2			8	
F	10.17	3			12	
G	17.33	4			16	
H	37.17	9			36	
I	23.67	6			24	
Y_1	1.58	2			8	
Y_2	2.75	2			8	
Y_3	2.00	2			8	
Y_4	3.50	2			8	
Y_5	3.58	2			8	
J	7.67	2			8	
K	4.33	2			8	
L,Q	16.42	4			16	
R	5.75	2			8	
S	2.75	2			8	
T	29.00	7	4'		28	
					264	
				FOOTING STAKES–COLUMNS		
8 COL.		8	4'		256	
				TOTAL STAKES	836	

Fig. 6-35-B. House Plan A. Vertical bracing for footings.

		TIES, BRACES AND KICKERS				SHEET *12*			
		QUANTITY TAKE-OFF				*HOUSE PLAN A*			
						ft.			
SNAP TIES									
(ONE TIE EVERY 4 SQ.FT.)									
	4,130.38÷4=1,033 SNAP TIES								
ANGULAR BRACES									
(ONE −10'-0" BRACE EVERY FOURTH VERTICAL STUD)									
	OUTSIDE STUDS÷4 = 13 1/4								
		33×10.0'					330 −		
	INSIDE STUDS÷4 = 17 3/4								
		44×10.0'					440 −		
					TOTAL		770 −		
KICKERS									
(ONE −3'-0" KICKER TO EVERY BRACE)									
		77×3.0'					231 −		

Fig. 6-36. House Plan A. Ties, braces, and kickers.

		Item				Length l ft.		Area A sq.ft.		Volume V cu.yds.
		CONCRETE								
		EXTERIOR FOUNDATION WALLS								37
		INTERIOR FOUNDATION WALLS								6
		EXTERIOR CHEEK WALLS								7
		CHIMNEY FOOTING								1
		CHIMNEY STACK								5
		COLUMN FOOTINGS								2
		WALL FOOTINGS								13
		CELLAR								17
		MISC. SLABS								7
										95
		FORMS								
		EXTERIOR FOUNDATION WALLS						2,858.55		
		INTERIOR FOUNDATION WALLS						838.41		
		WALL FOOTINGS						369.42		
		COLUMN FOOTINGS						64.—		
								4,130.38		
	SNAP TIES—1,033									
	ANGULAR BRACES-77					770.—				
	KICKERS-77					231.—				
	WALERS					3,240.94				

Fig. 6-37. House Plan A. Concrete and formwork summary.

ESTIMATING LABOR

The average time it takes a worker to do his part of the job will vary with the local conditions, but the variance should be small except for extreme conditions.

Forms and concrete equipment must be hauled onto the site before work is begun. This is an item of overhead expense. Some contractors add what they think the cost in time will total.

One carpenter can erect about 300 square feet of forms in 8 hours. Usually he has a helper whose time is figured as one-half of the carpenter's.

A man can mix and place about 110 cubic feet of concrete in about 7 hours. This includes charging the mixture and wheeling the concrete to the forms. For long hauls, some extra time should be allowed.

Concrete forms can be removed and piled (ready to be moved) by one man at the rate of 700 square feet in 8 hours.

The surface of the concrete must be rubbed when it is desired to remove all form marks. One man can rub about 100 square feet per hour.

One cement-finisher with a laborer-helper can *top dress* about 425 square feet of sidewalk in about 7 hours. For basement floors, 600 square feet can be done in the same time.

See Appendix D, pages 470 to 474, for Hourly Estimates for:

Excavations, Backfills, Concrete Walls and Slabs: Footers (Placing); Foundations (Concrete); Foundations (Masonry); Concrete (Walls); Concrete Slab Construction; Insulating Concrete; Concrete Slab Pouring;

Concrete Flatwork, Steps: Walks, Driveways and Patios; Steps;

Coping, Lintels, Beams, Columns: Concrete Beams, Lintels; Concrete Columns.

Questions and Problems

1. How many bags of cement are needed for the 95 cu. yds. of concrete foundation used in House Plan A? The concrete mixture is 1:2:4:8. Use the absolute volume method. (See the "Concrete Problem" in the section on Concrete and Mortar Mixes.)

   (Answer: 517 bags)

2. It is estimated that the following numbers and sizes of reinforcing bars will be needed for concrete reinforcement:
   20 #5 bars, each 20 ft. long
     at 1.043 lbs./lin. ft.
   20 #6 bars, each 15 ft. long
     at 1.502 lbs./lin. ft.
   Find the total amount of steel needed for this job.

   (Answer: 867.8 lbs.)

3. Find the concrete constant and volume of a concrete column with a circular base, 1' in diameter and 10' high. The cross-sectional area is constant.

   (Answer: 31.4 cu. ft.)

4. Find the volume of concrete needed for a concrete stairway with 12 treads which are each 4' wide.

   (Answer: 1.8 cu. yds.)

5. Find the amount of formwork needed for the intermediate beam in Fig. 6-27.

   (Answer: 16.0 sq. ft.)

6. Find the form constant and **SFCA** of the 2 columns in Fig. 6-28, each of the columns being 8.0' high. (Ans.: 8,128 sq. ft.)

7. Find the length of board needed for 20 kickers if the angular braces are 7' long. (Assume the braces are at a 45° angle.)

   (Answer: 99 ft.)

8. Snap ties are to be used every 12 square feet in foundation walls whose outside areas total 1,500 square feet. How many snap ties are needed?

   (Answer: 125)

9. Find the amount of lumber needed for the vertical bracing studs on the inside walls of Fig. 6-28. The on-center distances are 2' and there are two 8' braces at each on-center distance. (Answer: 896 ft.)

10. Find the linear footage of the walers needed for the inside foundation walls of Fig. 6-28. Each wall has 9 walers running its entire length.

    (Answer: 1,008 ft.)

7

FRAMING-PART I
WALLS AND FLOORS

In order to make it easier to understand, the material on framing has been divided into two chapters. This first section deals with Wall Framing and Floor Framing.

Most residential buildings are constructed in a certain order: excavation will necessarily come before concrete work which must precede framing and millwork. Although the order of construction may vary, you should learn to think of the operations in a definite order. This is especially true of framing and millwork where it is easy to overlook small but costly items.

WALL FRAMING

Outside Wall Framing. Plans often do not include framing drawings for outside walls. When they do, the estimator's job is simplified. When they do not, the estimator must calculate the number of 2 × 4 studs needed for the wall and around the window and door openings.

Fig. 7-1 is a picture of typical outside wall framing. It is a simple matter to calculate the number of 2" × 4" studs required because they are either on 16- or 24-inch centers. A typical corner post is shown in Fig. 7-2.

The pictorial views of *Western* and *ballon* framing, Figs. 7-3 and 7-4, will help you to visualize the outside wall framing details.

The following materials and methods of construction should be carefully studied and understood by the estimator. The framing illustrations are detailed so that terms such as *sill, sole,* and *plate* will not cause confusion when in the specifications.

Western or Platform Framing. Fig. 7-3 is a pictorial view of Western or *platform* framing. The letters *A* through *F* are near the load-bearing walls. For Western framing, most of the studs will have been pre-cut to the desired lengths at the mill. Most of the cutting in the field is for sills, soles, plates, and headers.

Fig. 7-1. Typical outside wall framing.

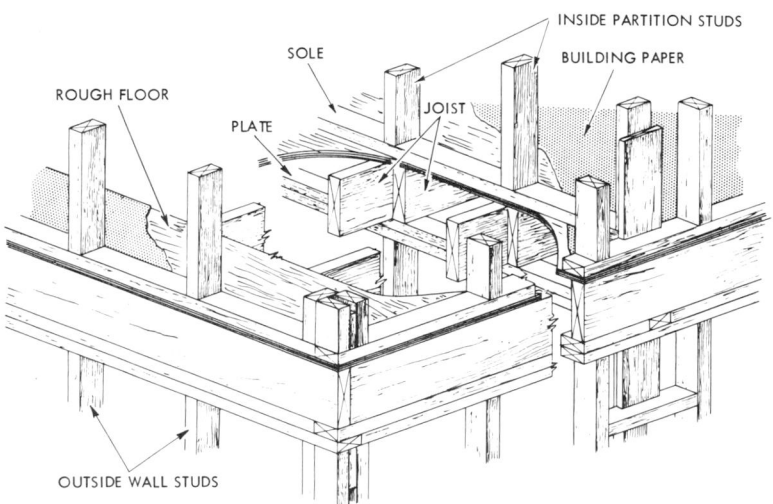

Fig. 7-2. Construction at second floor for Western or platform framing. Note the three-piece corner post.

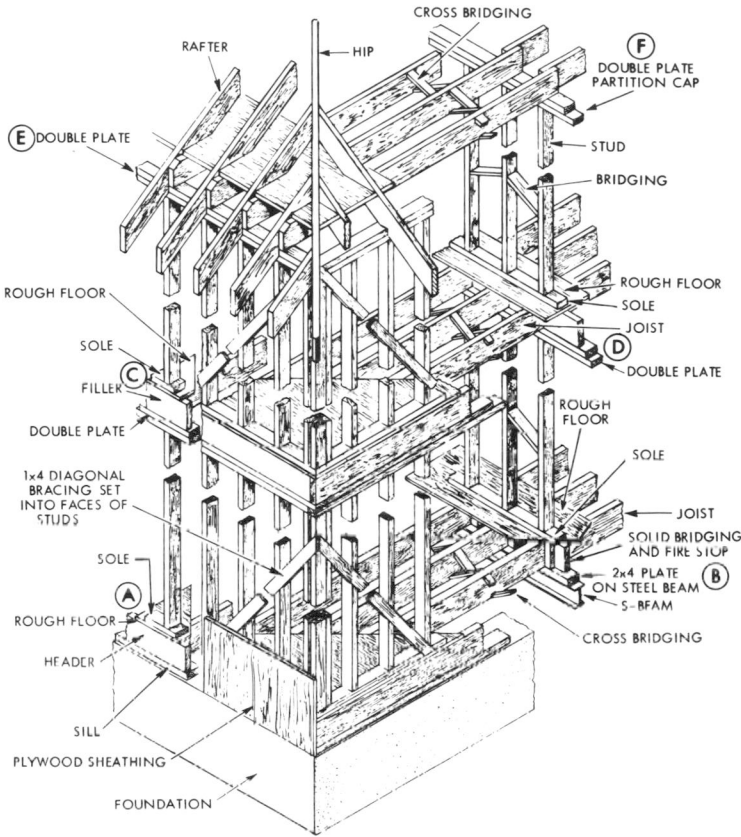

Fig. 7-3. Western or platform framing.

Balloon Framing. Fig. 7-4 is a pictorial view of balloon framing. This method of construction is not as popular as the simpler Western frame partly because it does not lend itself to either prefabrication or pre-cutting. Dry lumber suitable for the long, two-story studs used in this type of framing is expensive.

Plank and Beam Framing. Fig. 7-5 is a pictorial view of plank and beam framing. This method of construction uses a few large members to replace the many small members used in typical wood framing. The floor and roof planks serve as structural (that is, load-bearing) members and must be well-seasoned high-grade lumber. Very often these planks will be left exposed and serve as the ceiling for the rooms below. Note that No. 1 common grade lumber is usually the minimum

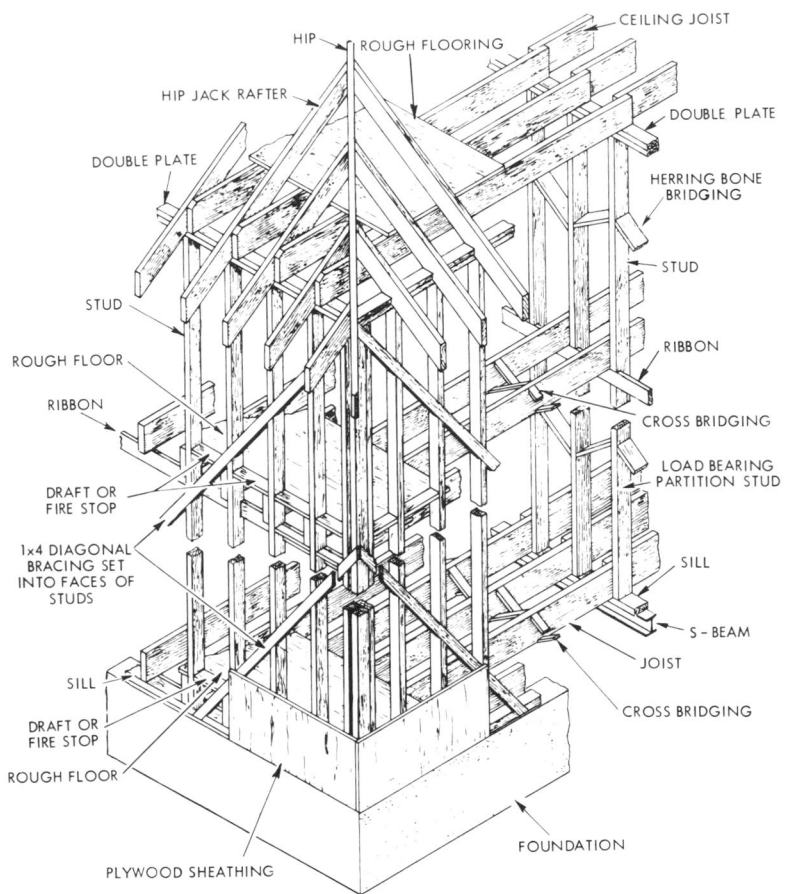

Fig. 7-4. Balloon framing.

acceptable quality. See Appendix A for Grading of Lumber and the new Softwood Standards.

Sheathing. After the frame is constructed it is covered with sheathing. The sheathing helps to insulate the building while providing a better base for siding, and exterior trim. See Fig. 7-6.

Plywood is commonly used for sheathing. Lumber (usually tongued-and-grooved or shiplapped) or fiber insulation board may also be used.

The most common size of plywood and insulation board is $4' \times 8'$ panels. Insulation board is about $\frac{3}{4}''$ thick and plywood is usually $\frac{3}{8}''$ to $\frac{5}{8}''$ thick, depending on the stud spacing and structural factors.

In both the Western and balloon frames, the sheathing is nailed directly to the outside edge of the stud. Either

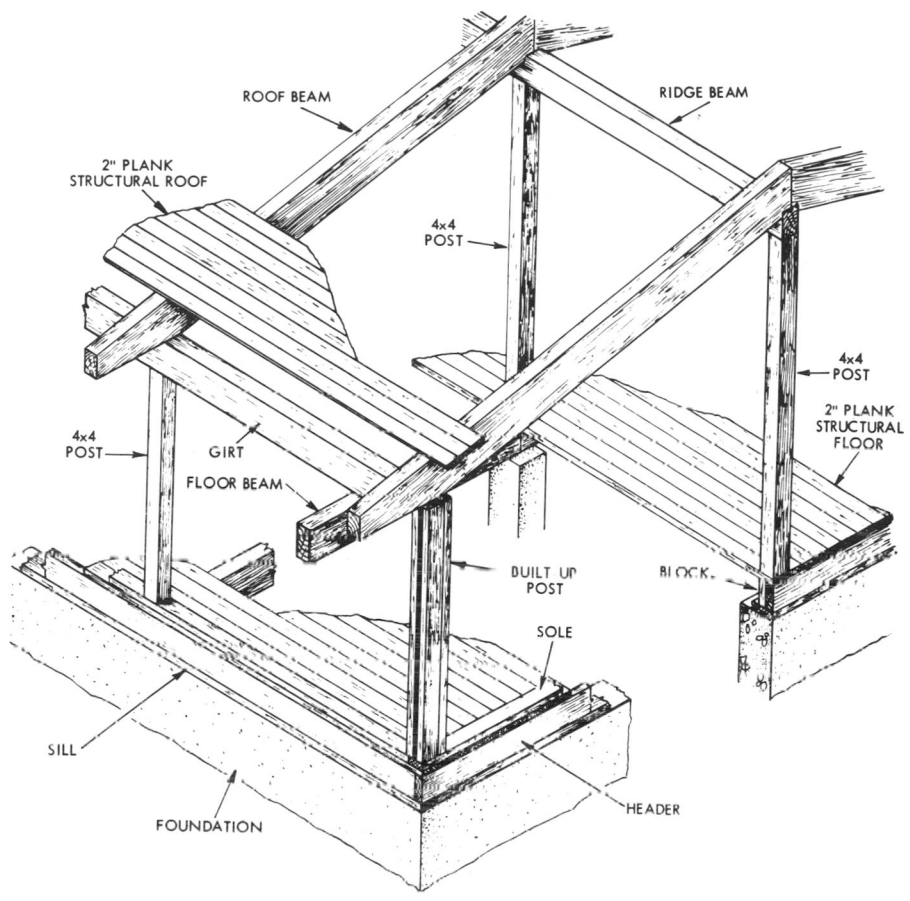

Fig. 7-5. Plank and beam framing.

boards or panels may be used depending on the specifications and local building codes.

Building paper or *felt paper* is customarily nailed to the wood or plywood sheathing and the exterior finish (such as shingles or siding) is nailed over the paper. Building paper is not needed when insulation board sheathing is used.

Siding. Fig. 7-6 shows three different types of siding. Clapboards may be obtained as shown, or with a rabbet so that the back will fit tightly against the wall. Plywood siding is available in both tapered and flat shapes. Prefinished hardboard and metal siding are becoming increasingly popular.

Boards are used for siding by nailing them in a vertical position flat on the wall and then covering the joint with another narrow strip called a bat-

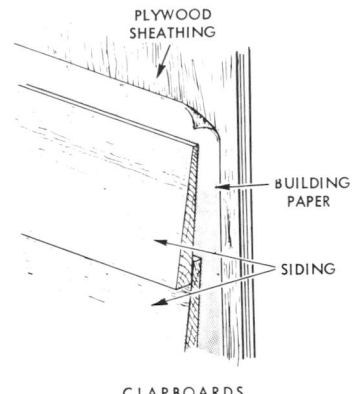

CLAPBOARDS

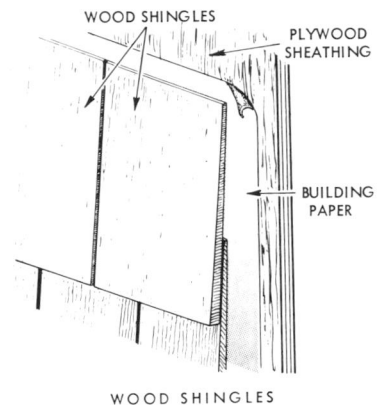

WOOD SHINGLES

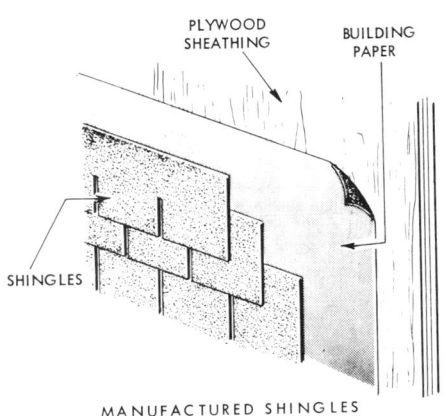

MANUFACTURED SHINGLES

Fig. 7-6. Types of siding.

ten. This is called board and batten siding.

Rough cedar channel siding is used quite extensively today. It has a ship-lap joint with the tongue longer than the rabbet and is installed vertically, showing a recess (channel) at each joint.

Wood shingles shown in Fig. 7-6, *center,* are from 20" to 24" long, ½" to $9/16$" thick at the butt, and of negligible thickness at the opposite end. Some roofing shingles are of random widths varying from 2½" to 14" or 16". Dimension shingles (for walls) are sawed to uniform widths of 4", 5", or 6".

Cement-asbestos, asphalt, and other types of manufactured shingles, Fig. 7-6, *bottom,* vary in size. The specifications should either state the size or give enough information so that the size can be accurately determined. This is important for the labor estimate as the sizes of the shingles will affect the cost.

Bracing. Specifications may require bracing for the rough framing. Figs. 7-3 and 7-4 show one type on the outside walls called diagonal bracing. 1 × 4 or 1 × 6 inch boards are cut into the studs and nailed fast.

Walls and Partitions. Figs. 7-7 and 7-8 show details of outside wall framing. Fig. 7-9 shows details of interior wall framing. The walls are assembled on the floor and tipped up. See Fig. 7-1. The soles are then nailed to the floor.

Interior Framing. In frame buildings the interior walls or partitions are framed using 2" × 4" vertical studs,

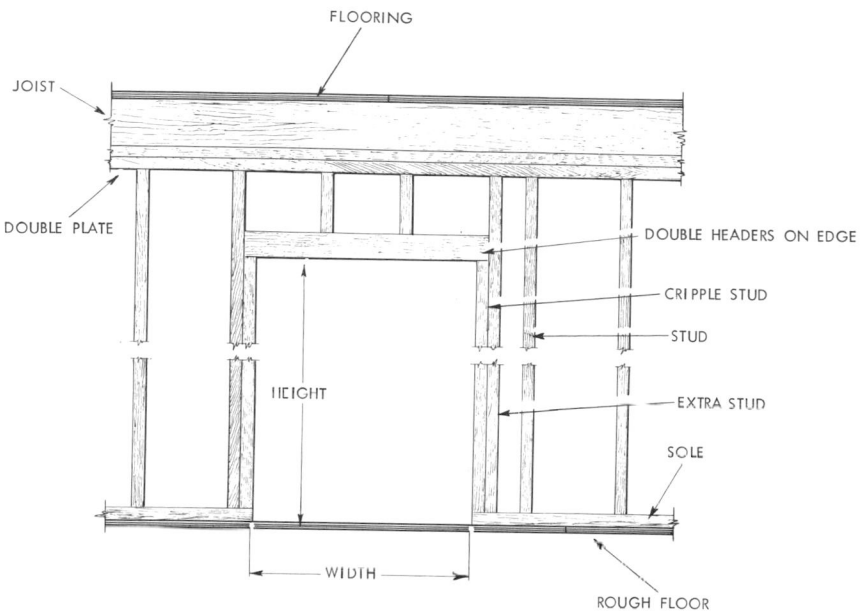

Fig. 7-7. The rough opening for a door in a Western or platform framed house.

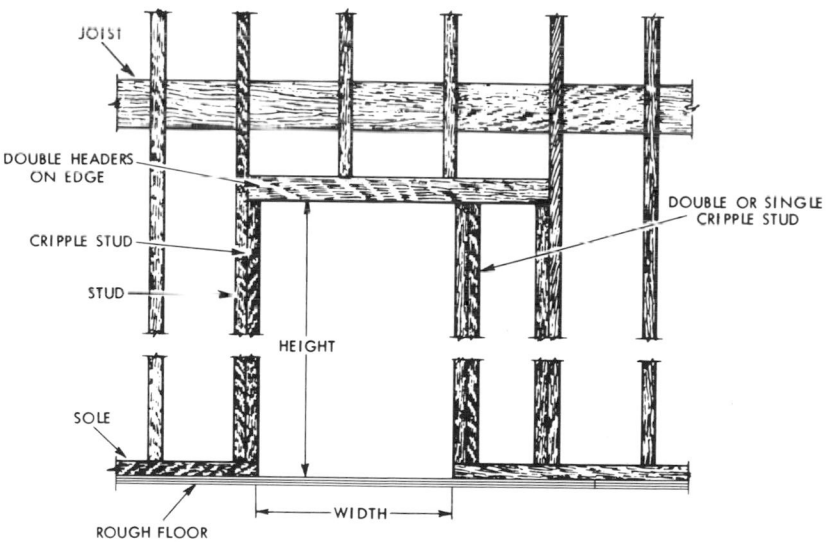

Fig. 7-8. The rough opening for a door in a balloon framed house is framed with a long header.

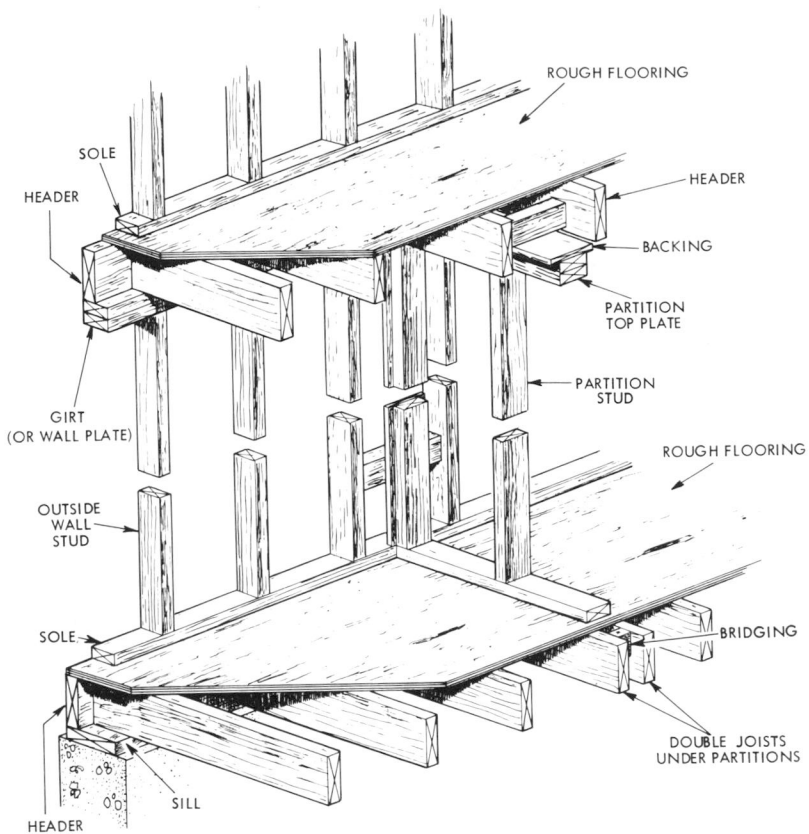

Fig. 7-9. Interior partition framing for a house with Western or platform framing. The joists are doubled under the partition.

usually spaced 16″ o.c. (on centers). Lathing and plaster, or drywall covering is then applied.

Sometimes an interior wall is framed with 2 × 6 studs in order to have more space between the walls. (See Fig. 7-21*B*.) This is required for walls which have soil pipes in them, or where extra wall strength is desired. In balloon framing the studs extend down to the girder or sill. If no girder comes under a bearing wall or partition, the joists directly under the wall or partition are doubled, as shown in Fig. 7-10.

Insulated Walls. Insulation is necessary for both comfort and to keep fuel costs reasonable. Some insulation materials also have structural value.

Fig 7-10 illustrates a common insulating practice. *Rigid* insulation is substituted for sheathing. The insulation is delivered to the job in 4 × 8 foot sheets. These sheets are nailed directly to the studs and provide some rigid-

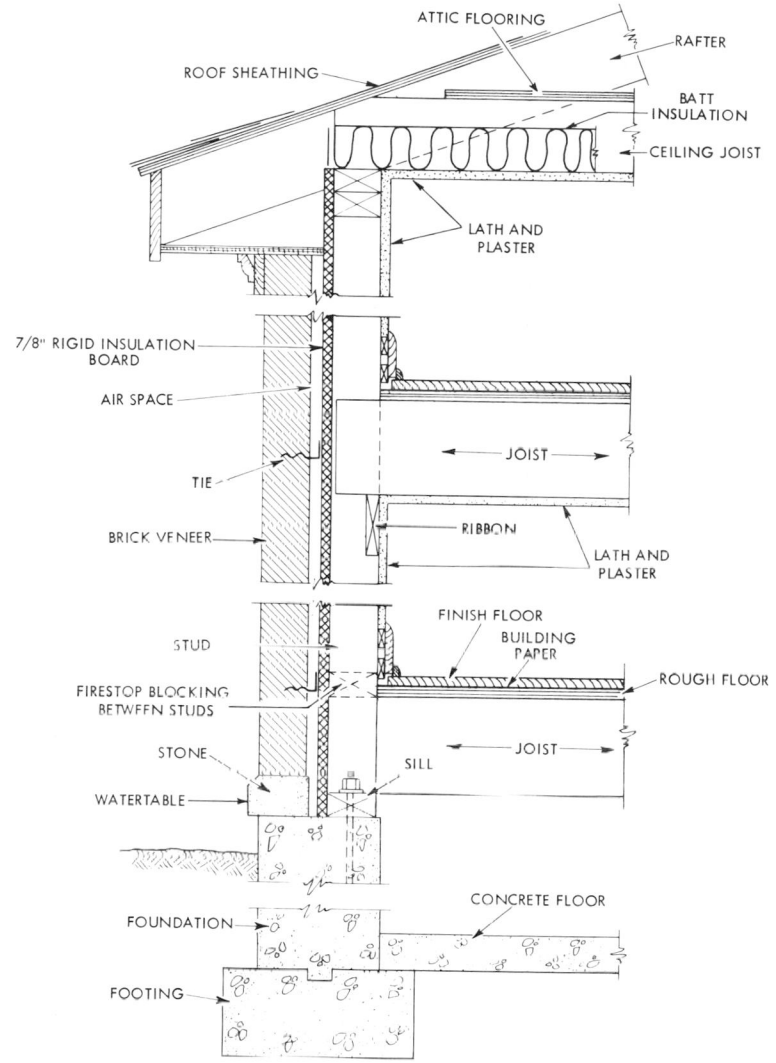

ATTIC FLOORING

RAFTER

ROOF SHEATHING

BATT INSULATION

CEILING JOIST

LATH AND PLASTER

7/8" RIGID INSULATION BOARD

AIR SPACE

TIE

JOIST

BRICK VENEER

RIBBON

LATH AND PLASTER

FINISH FLOOR

BUILDING PAPER

STUD

ROUGH FLOOR

FIRESTOP BLOCKING BETWEEN STUDS

STONE

SILL

JOIST

WATERTABLE

CONCRETE FLOOR

FOUNDATION

FOOTING

Fig. 7-10. Rigid insulation is used in walls, floors, and ceilings (Balloon Section).

ity to the structure. Sometimes gypsum lath is specified for plaster backing. The sheets can be installed rapidly and are easily handled. The air space between the bricks and sheathing also helps to insulate the house.

The undersides of the first-floor joists (Fig. 7-10) may have rigid insulation applied to them. This produces a desirable basement ceiling and reduces heat transmission between the basement and the first floor.

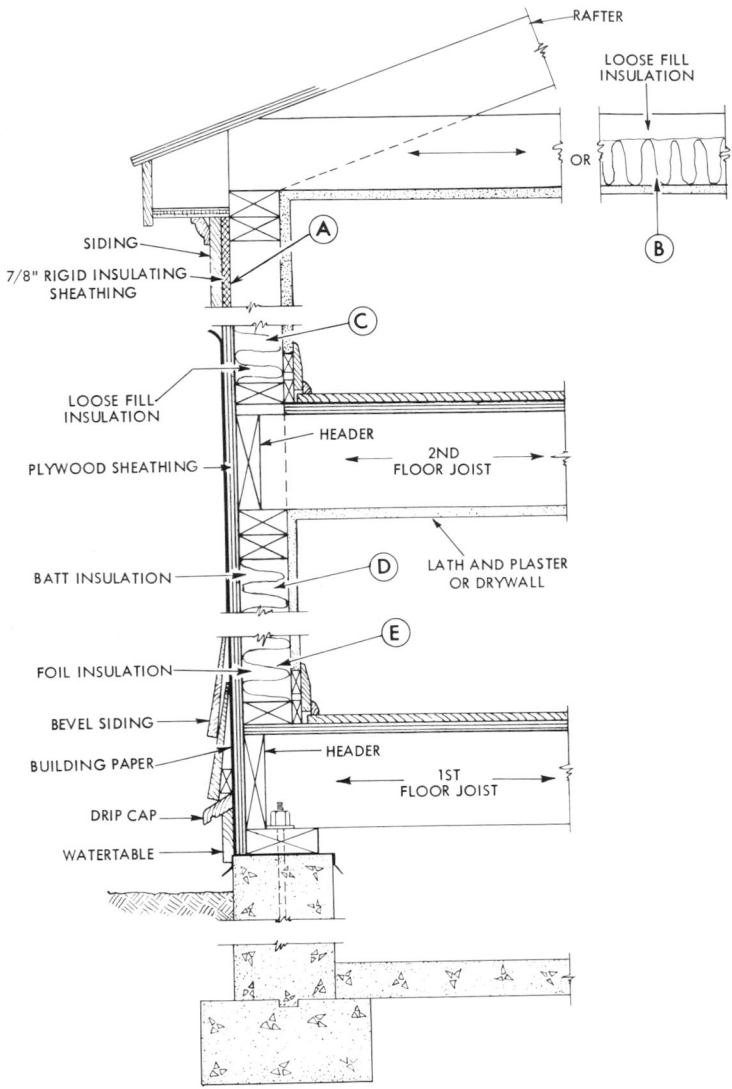

Fig. 7-11. Various types of insulation in a Western or platform framed wall.

A full frame wall, such as Fig. 7-11, can be insulated in many ways. At *A* is the same type of rigid insulation as in Fig. 7-10. In this case, siding is nailed to the studs through the insulated sheathing.

At *B,* between attic joists, *loose fill* insulation is used as a method of protecting the second-floor rooms from the varying summer and winter temperatures of the attic.

At *C* loose fill is suggested for insulation although it is not generally used on new work. The sheathing can be either plywood or rigid insulation board. Rigid insulation board will give maximum insulation value.

At *D,* insulating *batts* (or *bats*) are shown in place between the studs. These are made to fit between studs spaced 16″ o.c.

At *E,* the use of *aluminum foil* insulation is illustrated.

The insulating materials shown in Fig. 7-11 can be applied to stucco and stone veneer walls as well as brick

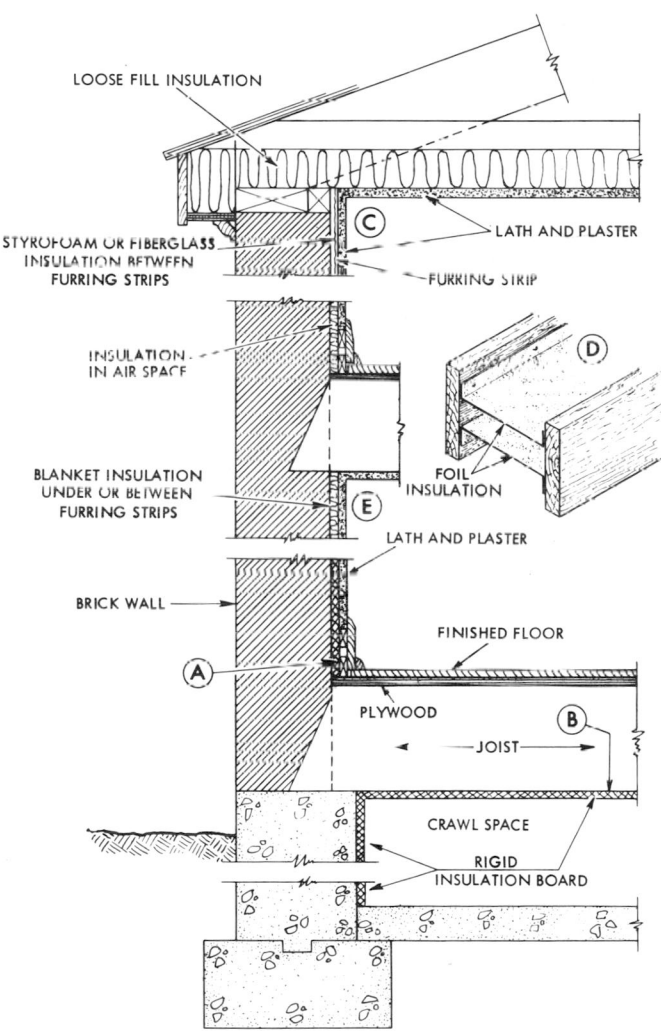

Fig. 7-12. A solid brick wall with various kinds of insulation.

veneer and all frame walls. Several different combinations of insulating materials are possible.

Solid brick walls, Fig. 7-12, can be insulated by leaving an open air space as shown at *A,* or using mineral wool or other blanket insulation as shown at *E* or styrofoam or fiberglass insulation as shown at *C.*

Rigid insulation can be used under the joists as shown at *B.* Aluminum foil insulation may be used between joists as shown at *D.*

Fig. 7-13 shows the method of insulating concrete walls. It is similar to that employed for brick walls. Rigid insulation can be *furred* out from the concrete surface and used as backing

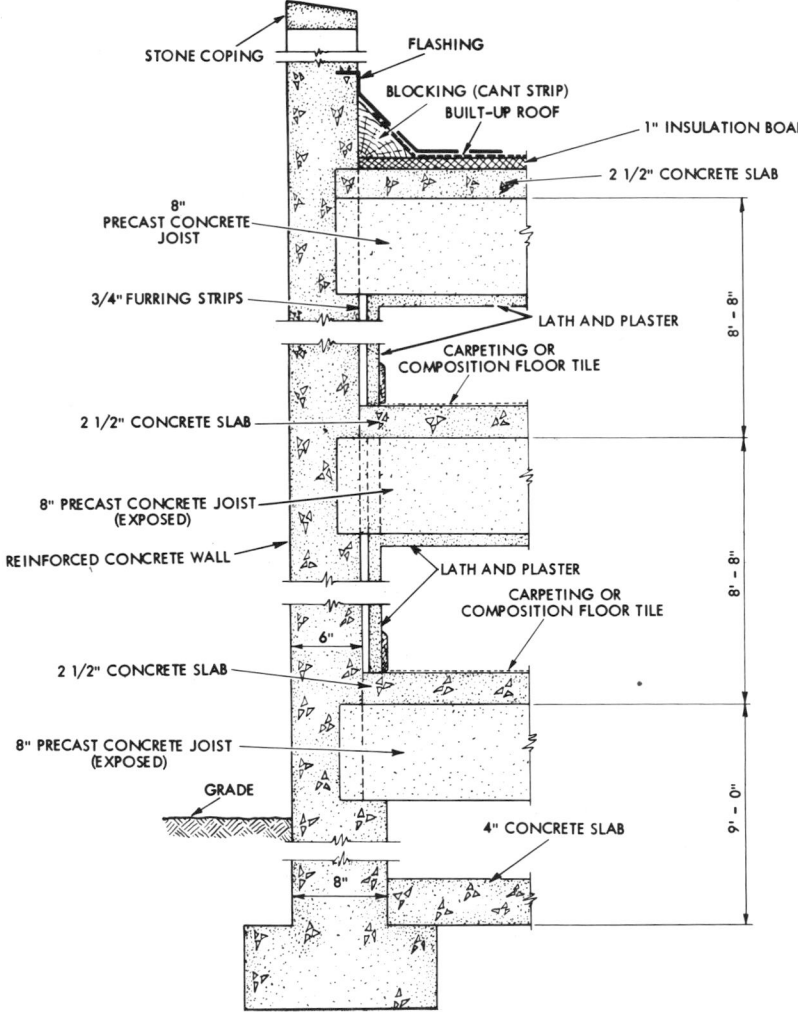

Fig. 7-13. Insulation for concrete walls, joists, and slabs.

for lath. Fiberglass insulation or some other form of blanket insulation is placed between the furring strips. Aluminum foil can also be used between the furring strips.

In Fig. 7-14, where concrete blocks are used for wall construction, loose fill or rock wool and rigid insulation can be used as shown.

Steel walls are of many types and

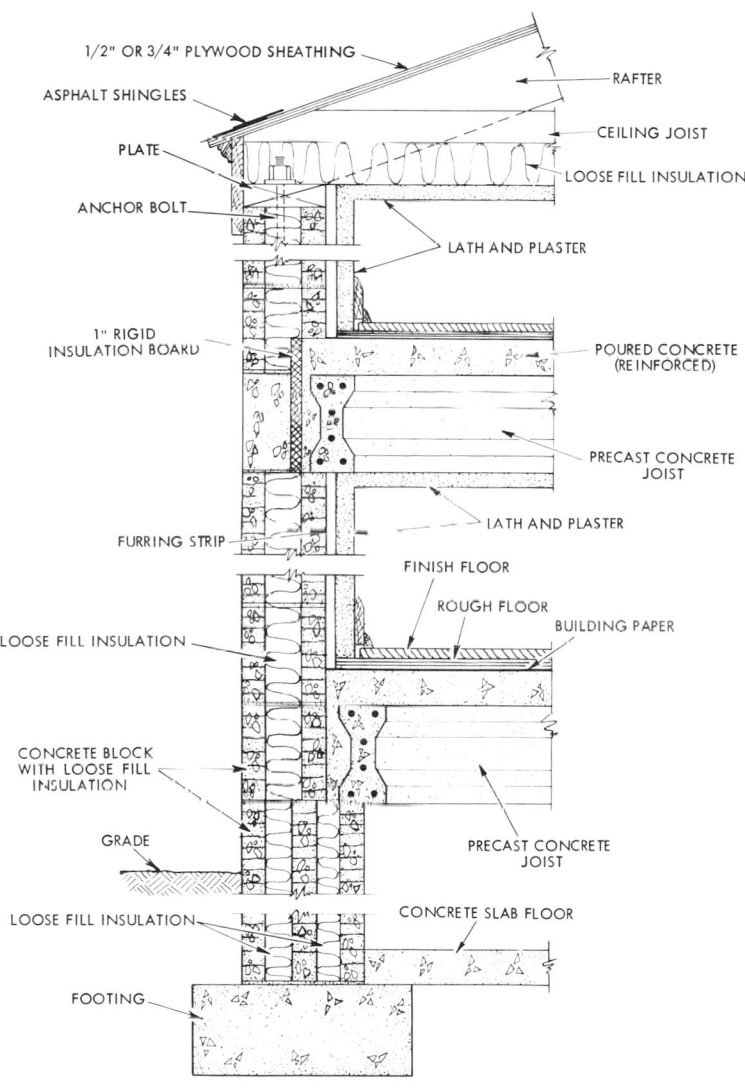

Fig. 7-14. Insulation for concrete block walls.

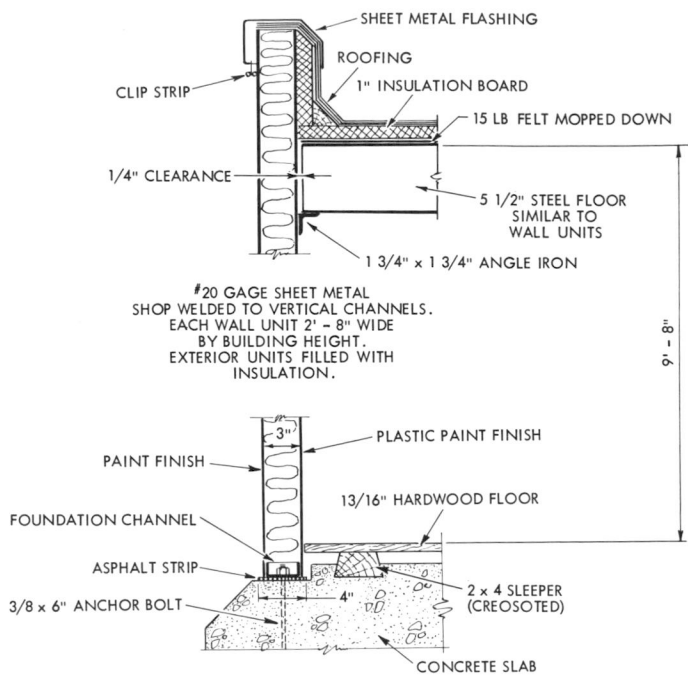

Fig. 7-15. Insulation for steel walls.

require various forms of insulation. The wall in Fig. 7-15 is relatively thick, permitting the use of rock wool or similar types of insulating material. Rigid insulation may be necessary for other types of steel walls. The estimator should therefore study the wall section drawings to better visualize the areas or volumes which the insulation will occupy.

Basement Partitions. These are generally of wood or brick or concrete block with wood being the most popular.

For basement recreation rooms the partitions are built of 2 × 4's and then plastered or covered with drywall board on one or both sides. Furring is used if plaster is to be applied to foundation walls. Concrete, concrete blocks, and tile are often used depending on the other structural features of the building. In any case, the symbols or written instructions should plainly indicate wall specifications and details.

Soundproofing. Soundproofing of the type shown in Fig. 7-16 requires double the usual number of studs, with quilt insulation between rows, and almost double the usual amount of labor is required.

Modular Planning. There is a tendency today toward pre-built panels which are made either on the site or elsewhere and which are assembled to

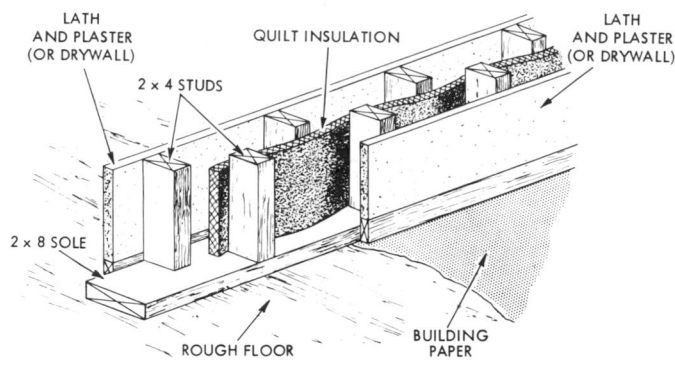

Fig. 7-16. Soundproofing a frame wall.

make the walls and partitions of the house. A whole new concept of planning buildings and manufacturing materials to fit the needs of modular design was necessary to make the idea possible. It was necessary first of all to decide on a common unit of measure which is called a module. This idea

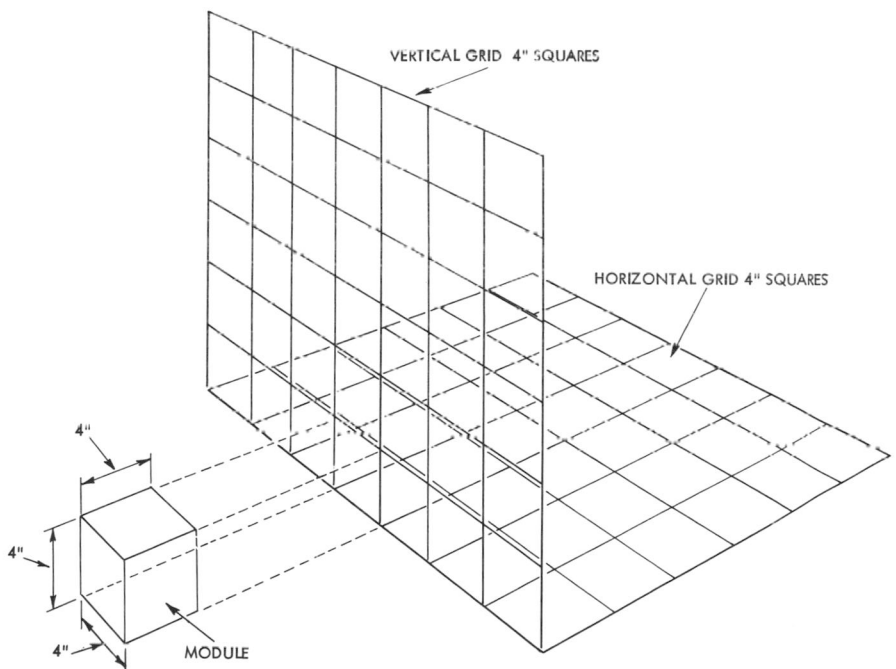

Fig. 7-17. A grid using 4-inch squares is used to lay out plan and elevation drawings. The module itself is a 4-inch cube.

Fig. 7-18. Windows and doors are made using modular dimensions. Western or platform framing permits the assembly of walls in a horizontal position. (Douglas Fir Plywood Assoc.)

is based on the use of a standard grid which is divided into 4-inch squares. The grid is three dimensional so that the module—the unit of measurement —is actually a 4-inch cube. See Fig. 7-17. All dimensions are based on multiples of 4; for example, $16'' \times 32'' \times 8'\text{-}0''$.

The first step in achieving modular coordination is accomplished by the architect. He lays out his plan over a sheet which is divided into squares representing 4-inch units (modules).

Modular coordination, however, still could not be attained unless the manufacturer produced materials of appropriate dimensions. Lumber, sheathing, panels, windows, doors, etc., are produced by many manufacturers to be adaptable to modular construction. See Fig. 7-18. Bricks, concrete blocks, and other masonry materials, however, must be made to allow room for the bonding material.

The modular idea, when applied to the construction of building components, uses multiples of 16, 24 and 48 inches wherever possible. These are all multiples of the 4-inch module. Modular layout demands careful work on the part of the designer of the building in order that the elements may fit properly within the grid and be in correct relation to the adjoining building elements.

FLOOR FRAMING

In frame buildings, the floor joists are wood members specified as 2 × 8's, 2 × 10's, 2 × 12's, etc., spaced at intervals of 12″, 16″, or 24″ o.c. The lumber is sold in lengths which are multiples of two—such as 6′, 8′, 10′, and 12′. A good designer tries to keep this fact in mind so that he makes the maximum use of the material. The length of lumber is an important item to the estimator.

Fig. 7-19 shows a girder being used as a joist support. Girders are W-

beams (formerly called wide flange) or S-beams (formerly called I-beams), solid timber, or several 2 × 8's, 2 × 10's, etc., spiked together. The joists can rest directly upon the girder, or they can bear on a 2 × 6 as shown in the illustration.

Joists can be fitted between the flanges of an S-beam girder as shown in Fig. 7-20, in which case the estimator must consider such additional items as *iron dogs* and *bolts*. The added amounts of material and labor when the joists are cut out to fit the S-beam must also be considered.

Sometimes, the architect will supply framing plans, in which case the estimator can easily count the number of joists and note the lengths, sizes, and

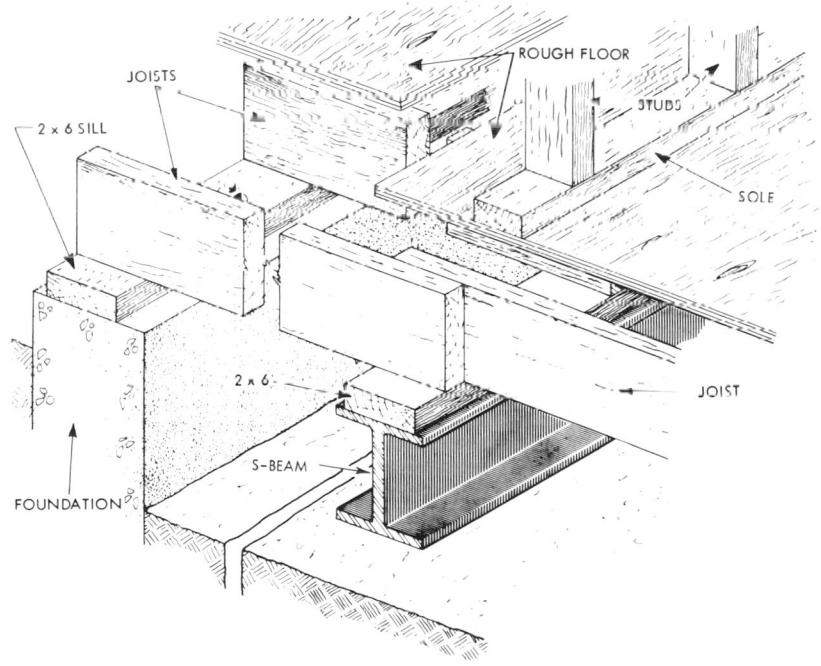

Fig. 7-19. Two ways of supporting joists.

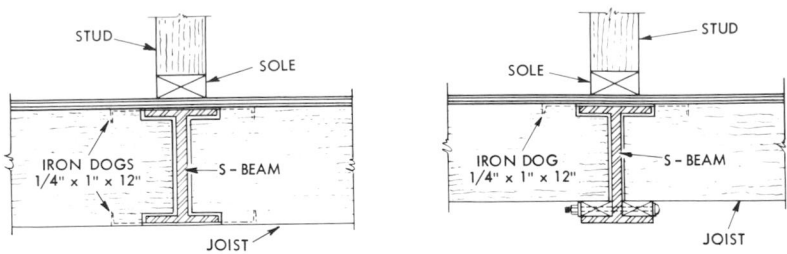

Fig. 7-20. Wood joists fitted between the flanges of an S-beam.

spacing. If framing plans are not supplied, the estimator must determine the sizes and spaces from the specifications and then calculate the amount needed according to the dimensions shown on the floor plans. There is *double* framing around all openings such as chimneys, fireplace hearths, and stair wells, Fig. 7-21-*A*.

Whenever a partition on the first or second floor occurs which is parallel to the joists underneath, the joists are doubled. Joists under bathrooms present special problems to be solved by the carpenter to provide passage of pipes. See Fig. 7-21-*B*.

Where the outside walls are brick, the joists must be beveled at the ends; this increases labor costs. Anchor bolts also add extra material and labor costs.

Sometimes concrete or steel joists are specified together with a concrete floor. In such a case, material and form costs must be studied carefully. If framing plans are supplied, they are used for quantity take-offs.

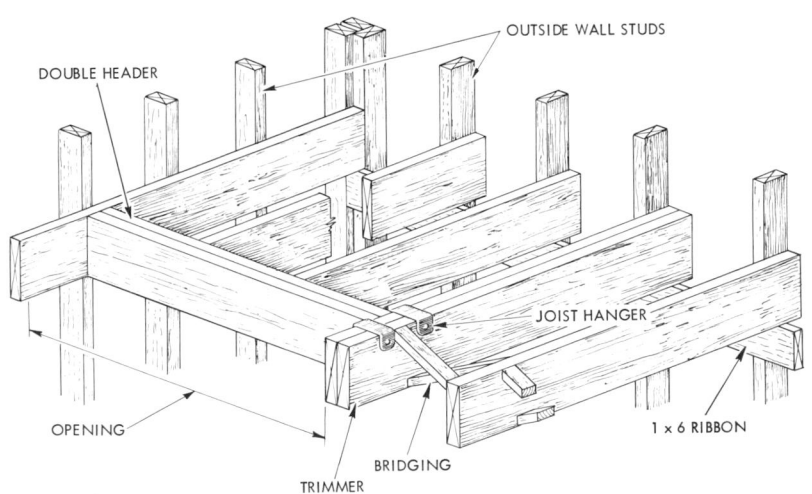

Fig. 7-21-A. Double framing around a floor opening.

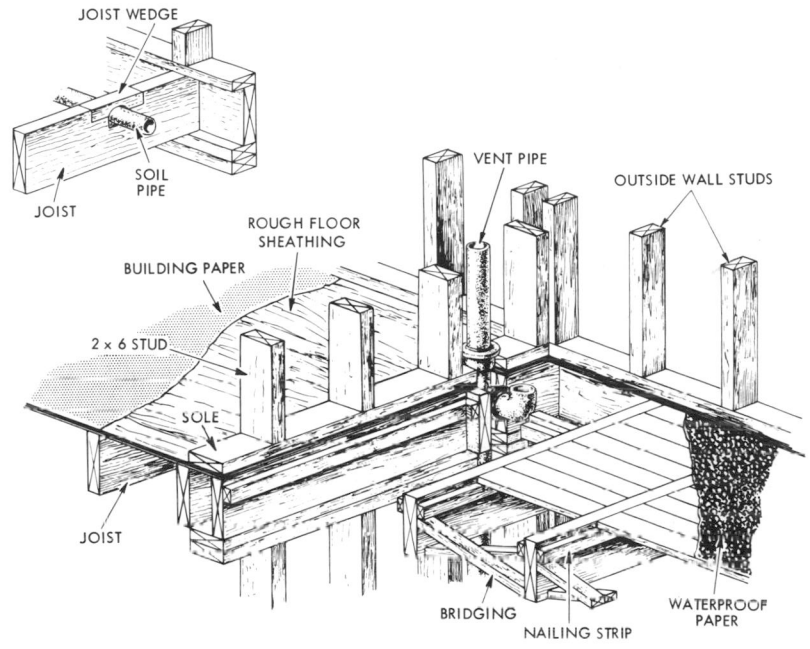

Fig. 7-21-B. Bathroom joists and wall framing.

Wood Floors. The rough flooring is laid over the joists. This flooring consists of plywood panels. Tongue-and-groove boards or 1 × 4 or 1 × 6 boards may be used.

Fig. 7-22 is a detailed drawing of

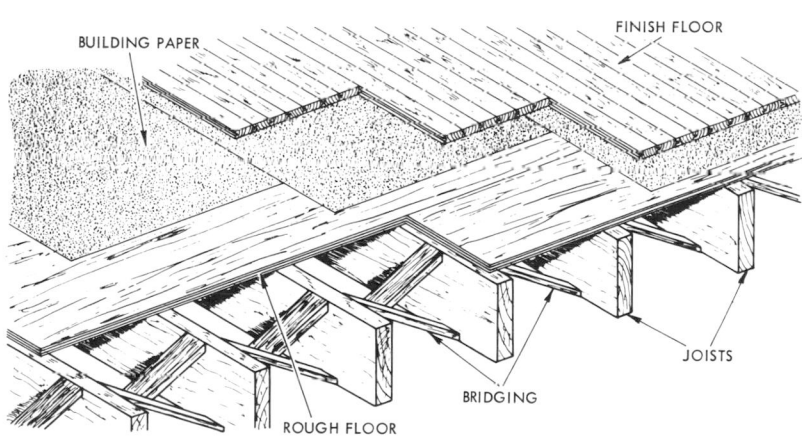

Fig. 7-22. Typical wood floor construction.

joists, bridging, rough floor, and finish floor. Rough flooring is sometimes put on diagonally and nailed directly to the joists. In many cases, 1 × 2 wood furring strips are laid on the subfloor at 12 or 16″ centers to provide a space for the electrical conduit. The finish floor is nailed to the furring strips.

The finish floors are not laid until the building is practically complete. Finish flooring is laid over the rough flooring after it has been overlaid with building paper. The laying of finish flooring involves considerably more time than rough flooring because more careful matching and fitting is required.

Different kinds and qualities of lumber are used for the finish flooring. Generally the rooms are floored with oak, maple, or similar wood.

In new homes where wall-to-wall carpeting or linoleum is always to be used, plywood, not pine, is used for flooring. If pine were used an underlayment of fiberboard (Masonite, etc.) would be advisable. Irregularities in a pine floor will show up in the linoleum and will wear out a carpet.

Where concrete floors are indicated, wood *sleepers* may be embedded in the concrete before it sets. See Fig. 7-23. The sleepers are *creosoted* or otherwise treated to resist deterioration. They extend above the concrete a specified distance and provide an insulating air space.

If wall-to-wall carpeting is laid over concrete floors, wood flooring is not always necessary. The surface of the concrete is ground smooth and a heavy layer of felt is cemented to it. The carpet is laid over the felt.

Sound Insulation. This item adds greatly to both material and labor costs, so it is an important item to the estimator. Fig. 7-24-*top,* is a cross section of an ordinary ceiling-floor without soundproofing. Various methods of soundproofing are shown at center and bottom. The most expensive is shown at lower left; here double the usual number of joists is required.

ESTIMATING MATERIAL

Wood and Lumber. The lumber take-off should show each piece of material, location, number of pieces, board-feet, and linear footage. The take-off is then given to the contractor, who reviews it and then forwards it to the lumber yard. At the lumber yard, the wood is selected and fabricated in accordance with the take-off, and then delivered

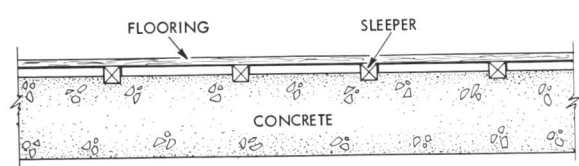

Fig. 7-23. Wood floors can be secured to concrete slabs by placing sleepers in the wet concrete.

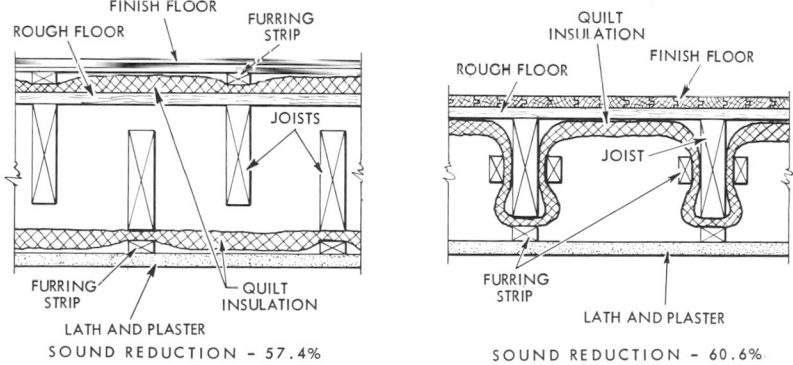

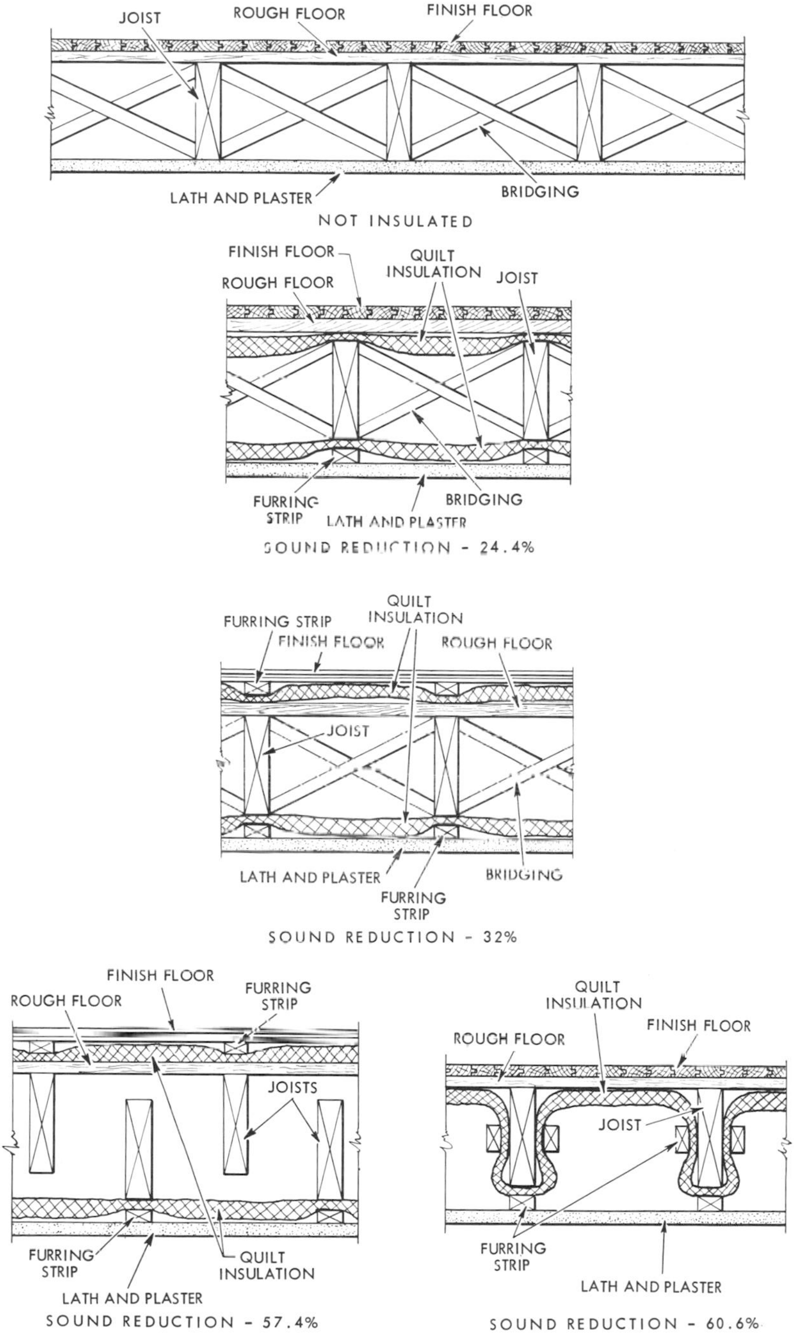

Fig. 7-24. Methods of providing sound insulation in ceilings and floors.

to the building site as needed, over a period of time. (This is done to lessen the possibility of theft, destruction, and the improper use of lumber which is not immediately required by the carpenters.)

Copies of the final estimate are given to the carpenter and contractor who, by using the plans, can locate each piece of framing material on the job. You can readily understand the importance of a correct quantity take-off which shows numbers and locations of materials.

Lumber Ordering. The two main classifications of wood are hardwood and softwood. Hardwood is generally used for fine finish work and softwood is used for framing and sheathing.

Examples of hardwood are birch, maple, oak, walnut, mahogany, gum, beech, and poplar.

Examples of softwoods are pine, spruce, hemlock, and fir.

Hardwood is purchased in various lengths and widths. Softwood is bought in standard sizes. A 2 × 8 plank, 10'-0" long is actually 1½" × 7¼" × 10'-0" long, but is ordered as a 2 × 8.

Most lumber is calculated and bought by the board-foot. Board measure is explained in Chapter 4.

A board foot is 1-inch thick, 12 inches wide and 1 foot long. The idea can be used for any size piece of wood of any length. The formula is:

$$\text{board feet} =$$

$$\frac{\text{thickness in inches} \times \text{width in inches} \times \text{length in feet}}{12}$$

A 2 × 4, 10 feet long, would be figured as follows:

$$\frac{2 \times 4 \times 10}{12} = \frac{80}{12}$$

or 6⅔ board feet.

Dressed-and-Matched Lumber. 1 × 6 D & M lumber is calculated as the area to be covered plus 20%. Deduct all openings.

Paneling. When interior walls are not plastered or left with a finished concrete surface they are usually finished with drywall (gypsum wallboard), board paneling or plywood paneling.

Wood paneling is figured as the area to be covered plus 10% for waste. All calculations are in square feet. The boards are ordered in lengths to fit the space with the least amount of waste. Plywood paneling is ordered in 4 × 7, 4 × 8, or 4 × 10 foot sheets.

Plywood. Plywood for general purposes of wall and roof sheathing or rough flooring is also estimated in square feet and ordered either in square feet or by the panel. The contractor should order the sizes which can be used with a minimum of waste. Careful planning is necessary in order to specify which sizes cut to the best advantage.

Wall and Partition Framing. This section is concerned mainly with the lumber and insulating material.

Sills are usually secured to the concrete foundation wall with anchor bolts. The bolts are spaced approximately 8'-0" o.c. The section detail on the plans should show the size of the sill. In Fig. 7-25 the sill is made of

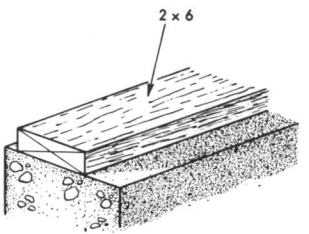

Fig. 7-25. A typical sill on top of a foundation wall.

cedure is then to find the lengths of the studs, the distances between studs, and the number of studs; then calculate the required board feet.

For a balloon frame residence, you can usually assume one stud for every foot. This takes care of double studs and wastage. Thus if the outside walls are 200′ long, you will need 200 studs. The stud lengths are equal to the dis-

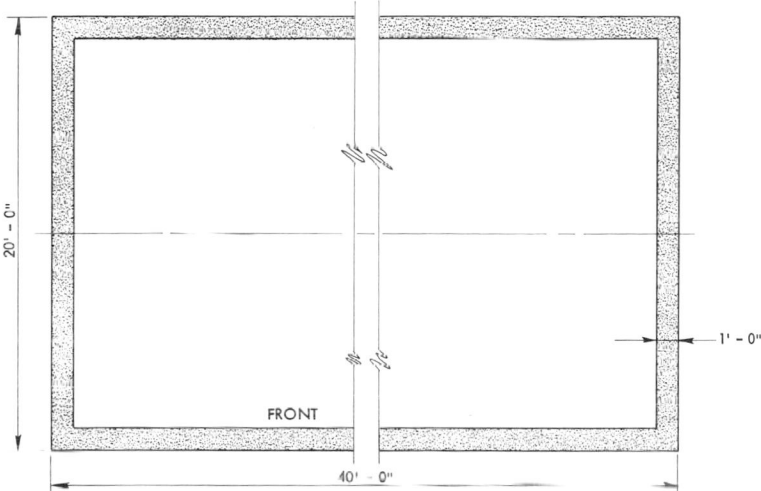

Fig. 7-26. Plan view of a foundation wall.

1 piece of 2 × 6. The foundation in Fig. 7-26 needs approximately 120 linear feet of 2 × 6 inch planks, in this case 120 board feet.

Headers. Headers are usually the same size planks as the floor joists. If 2 × 10 joists are used, the headers are also 2 × 10 inches (1½″ × 9¼″ actual size).

Outside Wall Studs. Calculate the perimeter of the building, that is, the lengths of the outside walls. The pro-

tance between the bottom plate (sole) and the top plate.

For a Western frame residence, the procedure is similar. The Western frame, however, has 2 sets of studs— one for the first floor and one for the second floor. (If the house is to be only one story high, only one set is necessary.) If the distance around the residence is 150′, you would assume one stud per foot, or 150 studs. If 2 sets are required, you would assume 300 studs. The lengths of each set are equal to the distance between the sole and the plate for each one.

Wall Plates. Plates are figured separately from the studs, even though they are usually the same size — 2 × 4. Measure the linear footage or calculate the lengths from the dimensions on the working drawings. Some plates are single 2 × 4's and others are double 2 × 4's. Where double, simply multiply by two.

The platform (or Western) frame requires three horizontal members: a sole at the bottom and two plates at the top. For this type of frame, the linear footage of one plate is multiplied by 3; this gives the total linear footage of the sole and top plates.

For the balloon frame, the linear footage is multiplied by 2 because there is only the upper plate (two 2 × 4's) to be considered. (See Fig. 7-4.)

The balloon frame requires a 1 × 4 *ribbon* at the second-floor level. (In Fig. 7-4 the ribbon is just below the second-floor joists.) To find the length of ribbon, measure the linear footage (or use the given dimensions) of the two outside walls which carry the ends of the joists.

Sheathing. Measure the length of walls requiring sheathing and multiply by the height to which the sheathing will be applied. Then deduct the areas of all openings such as windows and doors. Where boards are applied diagonally, add 30% for waste. Allow 20% where boards are placed horizontally. Sometimes the combined areas requiring sheathing must be figured individually because of irregular shapes. The lengths of all areas can either be scaled or found from the dimensions. The heights should be found in a similar manner.

Insulating rigid sheathing is calculated by square feet. Allow about 6% additional for waste. Deduct all openings such as windows and doors. A carpenter can help to cut wastage by figuring the most economical sheets to order. The sheets are 4'-0" wide and come in lengths of 6', 8', 10', and 12'.

Insulation. Batt insulation for frame walls is figured in square feet. Calculate the square footage of surface area to be insulated and deduct all openings such as windows and doors. The material comes in batts which are 1", 2", 3" or 4" thick and made to fit between studs spaced on 16- or 24-inch centers. The batts are 24, 48, or 96 inches long or may be obtained in blanket form in a continuous roll.

The calculation of side wall areas is simple and is done in the same manner as for sheathing.

Quilt insulation is also figured by the square foot. Calculate the area to be insulated and deduct all openings. No wastage need be considered. The material is in rolls 48" wide and ½" or 1" thick. Area calculations are the same as batt insulation. (For brick walls the process is the same as for frame walls: 1"-thick blanket insulation is used on brick walls. It is stapled to the furring strips and fills the space between the brick and plaster.)

Loose wool insulation is bought in bags. Each bag covers 15 square feet, 4" thick. Therefore, calculate the area to be insulated and divide by 15 to determine the number of bags. Deduct all openings. Areas are calculated as explained for batt insulation.

Wall and Partition Sheathing. Boards for walls and partitions are calculated by finding the total surface

ing, and adding 30% for ½ × 6, 25% for ½ × 8, and 20% for ¾ × 10. The additional amount takes care of lapping and waste. Deduct for all openings. Table I can be used for other sizes and exposures.

TABLE I. BEVELED SIDING

SIZE (INCHES)	EXPOSED INCHES	TO AREA TO BE COVERED ADD	SIZE (INCHES)	EXPOSED INCHES	TO AREA TO BE COVERED ADD
1/2 x 4	3 1/4	1/4	1/2 x 5	3 3/4	1/3
1/2 x 4	3	1/3	1/2 x 6	5 1/4	1/5
1/2 x 4	2 3/4	1/2	1/2 x 6	5	9/40
1/2 x 5	4 1/4	1/5	1/2 x 6	4 3/4	1/4
1/2 x 5	4	1/4	1/2 x 8	6 1/2	2/5

area and deducting the openings. Then add 20% for wastage.

Partition Studs. The lengths of interior wall studs depend on the room height. These figures can be obtained from the elevation drawings. Studs (2 × 4's) are purchased in lengths which are multiples of two and also in 9' lengths. They may be ordered to fit the exact length required. Two or more short studs are obtained from one stud.

When finding the stud dimensions, do not include the floor and joist thicknesses or heights.

After the heights have been determined, the linear footage of each partition height can be scaled, allowing one stud per foot (when the spacing is 16" o.c.).

Studs in inside walls of dormers are counted with interior studs. Exterior dormer wall studs are counted with exterior studs.

Siding and Felt. The amount of siding is calculated by combining the area of all surfaces requiring sid-

The amount of felt is determined by adding 10% to the required area. This felt is laid between the sheathing and the siding. Felt for side walls usually comes in rolls containing 432 square feet.

Floors. The main areas of concern under this heading are joists, bridging, subflooring, finish flooring, and floor felt.

Joists. The positions of all floor joists should be indicated on the plans. For example, HOUSE PLAN A has double-headed arrows. These arrows indicate floor joists and are generally accompanied by a notation giving their size and spacing. The direction of the arrows indicates the direction the floor joists are to run. If no arrow is shown, the estimator should check the section view of the blueprints or ask the architect to find out the direction of the joists. Remember that the joists must seat or bear at least 4" at each bearing point. Thus if the length of a room is given as 12'-8", and the arrow indicates that the joists are running the long way, you must add 4" for bearing at each end. This gives 12'-8" + 4" + 4", or 13'-4" as the required joist length. Since joists are bought in lengths corresponding to multiples of two, you would have to buy floor joists 14'-0" long.

Sometimes architects include several framing plans with their drawings. The joists for all floors are indicated on these plans by dash- or double lines in their proper positions. In such cases the estimator need only count the number of joists and determine the various lengths according to standard lengths which are multiples of 2'-0".

If no framing plans are included, there are several methods which can be used. The rule for estimating joists spaced 16" o.c. is shown in the following example.

Example 2. Find the number of joists required along the rear of Fig. 7-26. The joist spacing is 16" o.c.

Solution: Divide the length (40'-0") by 1.33 (or 4/3) and add 1.

$$\frac{40.0'}{1.33'} = 30$$

$$30 + 1 = 31 \text{ joists}$$

(If each of the joists is 10' long, then 310 linear feet of board is needed.)

For estimating joists spaced 12" o.c., the length is divided by 12", or 1'. Then add 1.

For estimating joists spaced 24" o.c., you divide the length by 2 and then add 1.

Tables, such as Table II, may also be used to find the number of joists required. Look under the first column

TABLE II.
NUMBER OF WOOD JOISTS REQUIRED
FOR ANY FLOOR AND SPACING

Length of Building	Spacing of Joists		
	12"	16"	24"
6	7	6	4
7	8	6	5
8	9	7	5
9	10	8	6
10	11	9	6
11	12	9	7
12	13	10	7
13	14	11	8
14	15	12	8
15	16	12	9
16	17	13	9
17	18	14	10
18	19	15	10
19	20	15	11
20	21	16	11
21	22	17	12
22	23	18	12
23	24	18	13
24	25	19	13
25	26	20	14
26	27	21	14
27	28	21	15
28	29	22	15
29	30	23	16
30	31	24	16
31	32	24	17
32	33	25	17
33	34	26	18
34	35	27	18
35	36	27	19
36	37	28	19
37	38	29	20
38	39	30	20
39	40	30	21
40	41	31	21

One joist has been added to each of the above quantities to take care of extra joist required at end of span.

Add for doubling joists under all partitions.

until you come to the length. Then read across the spacing columns until you come to the spacing required by your job. There you will find the number of joists needed. For example, for a length of 20 ft. with joists spaced 16 inches on center, you need 16 joists.

Bridging. Each joist normally requires two pieces of bridging. To determine the total linear footage of required bridging, you must determine the length of an average piece of bridging. Each piece of bridging forms the hypotenuse of a right triangle with the height of the joist as one leg and the distance between joists as the other

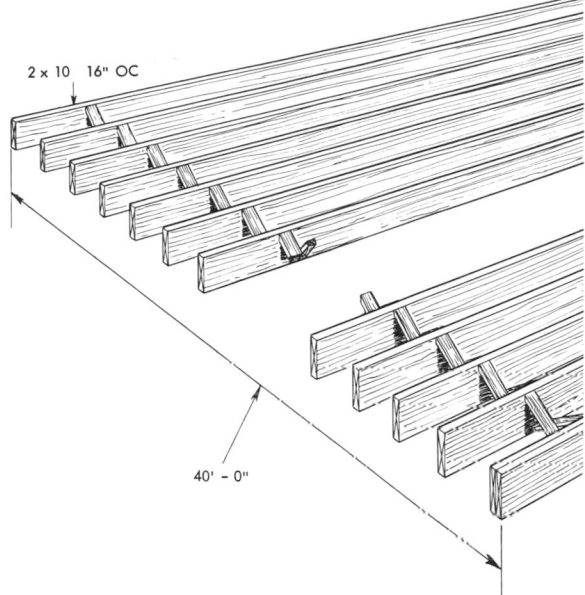

Fig. 7-27. Joists and bridging.

leg. Thus if 2 × 10 joists are used, the height of the joist is 10″. If the joists are spaced 12″ o.c., the other leg of the triangle is 10″. (Remember that the o.c. distance is greater than the actual distance between joists.)

Example 3. Find the length of board needed for bridging of the house foundation shown in Fig. 7-26. The joists are 2 × 10's spaced 16″ o.c. There are two sets of bridging per run. (See Fig. 7-27.)

Solution: First find the length of a single piece of bridging.

$$10^2 + 14^2 = 296$$

$$\sqrt{296} = 17.2, \text{ or about } 18'' \text{ each}$$

$$\text{No. sets} = \frac{40.0'}{1.33'} = 30 \text{ sets per run}$$

$$30 \times 2 = 60 \text{ pieces}$$

Since there are 2 runs, 120 pieces, each 18″ long are required.

$$120 \times 18'' = 2,160, \text{ or } 180 \text{ lin. ft.}$$

If metal bridging is to be used, only the number of required sets need be found.

Subflooring. The rough floor is calculated as the area of the entire floor minus such large openings as stairwells. The openings are figured roughly, being sure to allow too little rather than too much area for them. The flooring, as shown by the typical wall section in Fig. 7-28, does not always extend to the sheathing, Fig. 7-28, *right*. Thus the dimensions to the outside of the building cannot always be used. For irregularly shaped floors it is good idea to divide the surface into rectangles, squares, etc., to simplify area calculations.

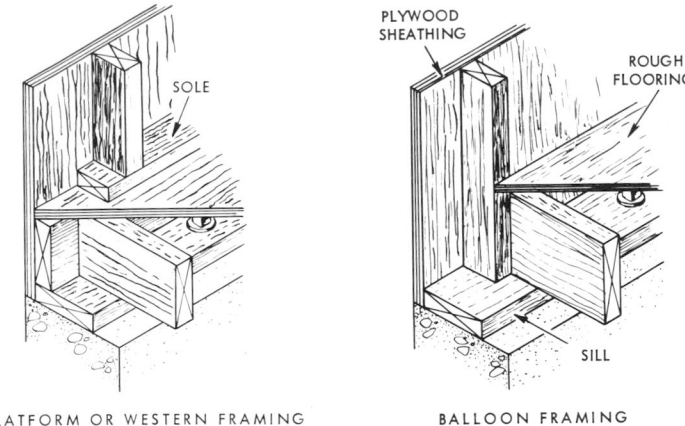

PLATFORM OR WESTERN FRAMING   BALLOON FRAMING

Fig. 7-28. Rough flooring extends (*left*) to the sheathing in platform or Western Framing and (*right*) to the inner face of studs in balloon framing.

If the subflooring is specified as diagonally applied, add 30% to the calculated area. If not diagonally applied, add 20% to the area.

Finish Flooring. For flooring which is ¾″ × 2″ or ¾″ × 2¼″ in thickness and width: Find area of each floor surface requiring finish flooring, then add together these areas; add 33% for waste.

For 1″ × 4″ fir or pine add 25% to the total area.

Drop Flooring for Tile. Drop floor-ing, for tile, requires additional material. This material is figured the same as for ordinary subflooring. Add 20% for waste. The 1 × 2 or 2 × 2 cleats are measured by assuming two per joist. They are figured in linear feet and ordered in lengths which can be used with a minimum of waste.

Floor Felt. This item is calculated in square feet. Find the area of the floor, deduct the areas of openings, and add 10%. One roll of felt usually contains 500 square feet.

ESTIMATING LABOR

See Appendix D, pages 474 to 476, 478, and 476 to 477, for Hourly Estimates for:

Framing, Sheathing, Decking: Wood Floor Joists; Steel Bar Joists; Steel Beams; Concrete Joists; Precast and Prestressed Concrete Floor Beams; Studding; Ceiling Joists; Sheathing; Bar Rib Lath; Steel Decking; Paper Backed Wire Mesh;

Siding: Wood Siding; Asphalt Siding; Composition Siding; Asbestos Siding; Shingle Backer; Paper, Felt, Etc.; Aluminum Siding; Galvanized Steel Siding; Stucco;

Subfloors, Finished Floors: Subfloors on Wood Joists; Insulation; Underlayment.

Questions and Problems

See **Hourly Estimates** in Appendix D to answer the following:

1. Find the number of joists required for the foundation shown in Fig. 7-26 if the joist spacing is 12″ o.c.

   (Answer: 41)

2. Find how many hours to allow for a skilled craftsman to complete 100 sq. yds. of stucco wall with troweled finish.

   (Answer: 28.75 hours)

8

FRAMING-PART II ROOFS, STRUCTURAL OPENINGS, AND DETAILS

~~~~~~

This chapter contains more material on framing. Roofs and structural openings are discussed in depth. Finishing operations, millwork and trim are also touched upon. Of course, appropriate estimating procedures are also covered.

Note: The actual layout of a roof and the layout of each rafter is generally left to experienced carpenters. It can be a very complicated process depending on how complex the roof shape is. The estimator is concerned primarily with (1) finding the length of the rafters and (2) finding the number he needs. He orders lengths long enough to provide an opportunity for the carpenter to make the necessary cuts. He does not have to know how to lay out rafters to estimate the material needed.

The framing square is a tool which contains tables which will give short cuts in finding rafter lengths with the least amount of mathematics. It will be explained under the topic of *Roof Rafter Lengths* in this chapter.

ROOFS

Fig. 8-1 shows a typical roof framing diagram. The *rafters* are shown in their proper positions and from such a diagram the estimator can easily count the number of rafters, *ridge pieces,* etc. The lengths of the pieces must be calculated unless they are in-dicated on the drawing. Generally, the size of the rafters and the spacings are given. The specifications or details should indicate the *pitch* or the rise and run. The pitch is the ratio of roof height (rise) to roof width (span). With this information, in addition to the dimensions on the plans, the lengths can be computed.

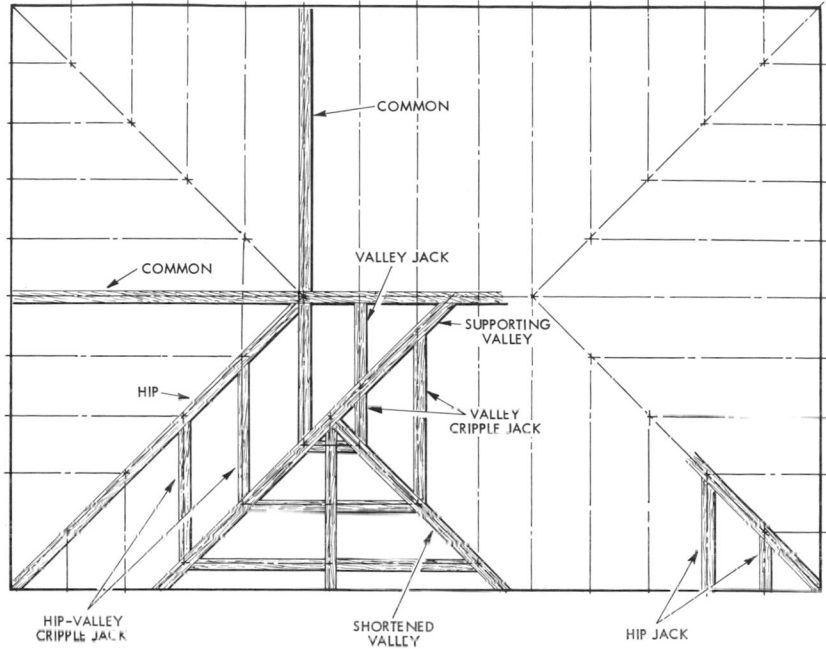

Fig. 8-1. The roof structure is made up of several types of rafters.

One way to find the length of the plank needed for the rafter is to calculate the hypotenuse of a right triangle with the level distance the total run and the vertical distance the total rise.

The overhang extending beyond the wall plate is a separate triangle. See Fig. 8-2. Note: this procedure will be discussed under *Estimating Material* later in this chapter.

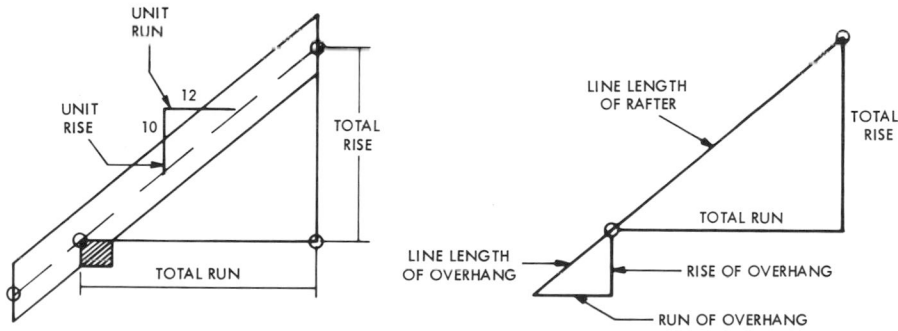

Fig. 8-2. Finding the length of the rafter plank.

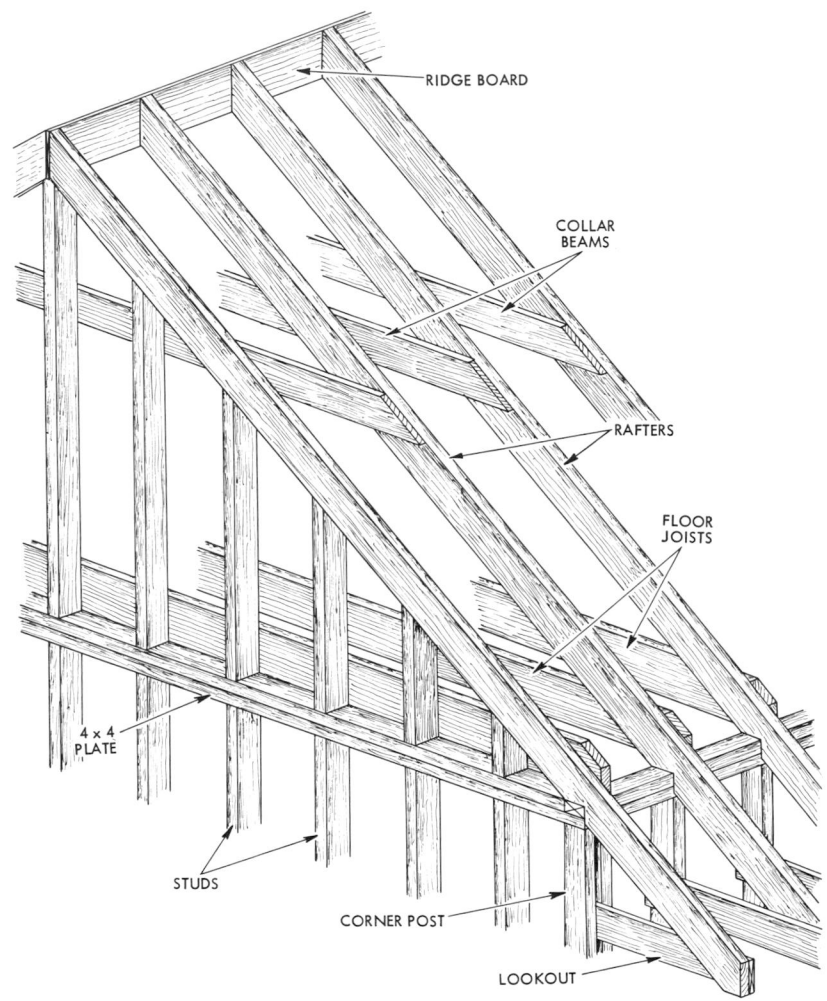

Fig. 8-3. Gable roof construction.

Planks should be somewhat longer than the calculated length to provide for the angle cuts at both ends.

Figs. 8-3 and 8-4 show two common roof types. From these illustrations you can visualize the necessary roof construction materials.

For a gambrel roof (Fig. 8-4) two lengths of rafters must be calculated.

The *ridge pieces* and *purlins* are usually 2-inch material wide enough to provide full bearing for the members which rest against them.

Fig. 8-5 shows nailed truss rafters used as roof framing for a concrete block-wall house. The block walls support the entire roof load. No bearing partitions are needed for the structure.

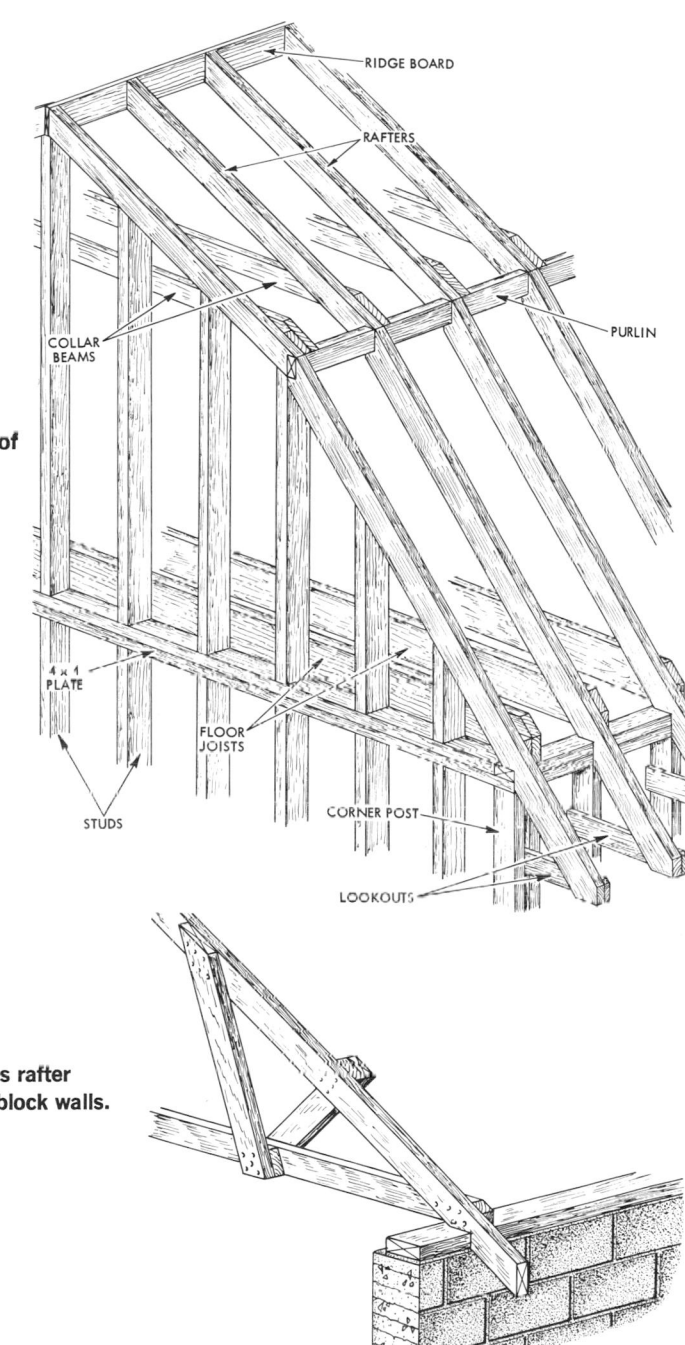

Fig. 8-4. Gambrel roof construction.

Fig. 8-5. Nailed truss rafter used with concrete block walls.

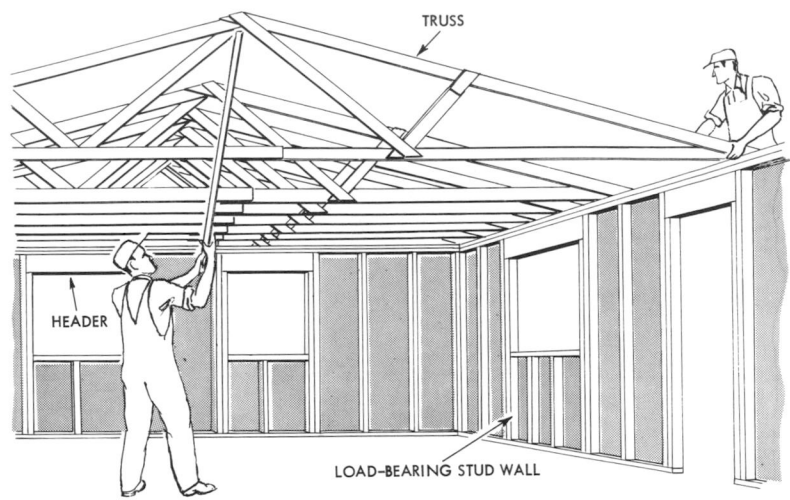

Fig. 8-6. Prefabricated trusses are swung into place, aligned, and fastened quickly. (University of Illinois, Small Homes Council)

The type of roof truss shown in Fig. 8-5 has been widely used in recent years for small house construction.

The use of roof trusses eliminates the need for bearing partitions and allows the house to be built as one large room. Floors are laid and ceiling hung or applied from wall to wall before the partitions are installed.

Fig. 8-6 shows the ease with which

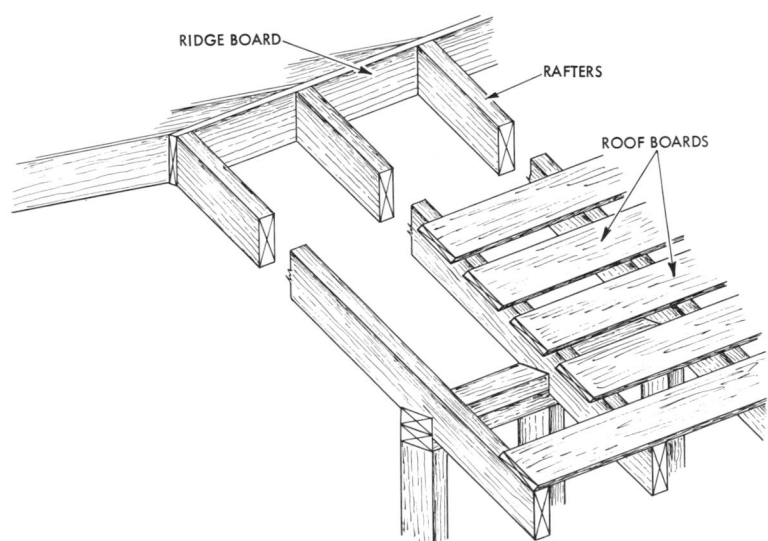

Fig. 8-7. Open-spaced boards are used with wood shingles.

trusses can be installed. The trusses are usually prefabricated in an assembly plant and then shipped to the construction site. They can be assembled in several ways. The simplest are nailed together; the larger, heavier trusses are bolted together. Many use a combination of nails and glue.

Special hardware is often necessary to attach the trusses to the wall plate. A careful study of all drawings and specs must be made before estimating a job calling for roof trusses.

Roof Boards. Fig. 8-7 shows a simple gable roof. The roof boards are placed in either of two ways: they can either be spaced as shown, with at least 2″ between boards, or they can be close together as in Fig. 8-8. The open or spaced roof board method is generally used in connection with wood shingles, although some other roofing methods use spaced boards. When cement asbestos or asphalt saturated felt shingles are used, the roof boards are laid close together. There

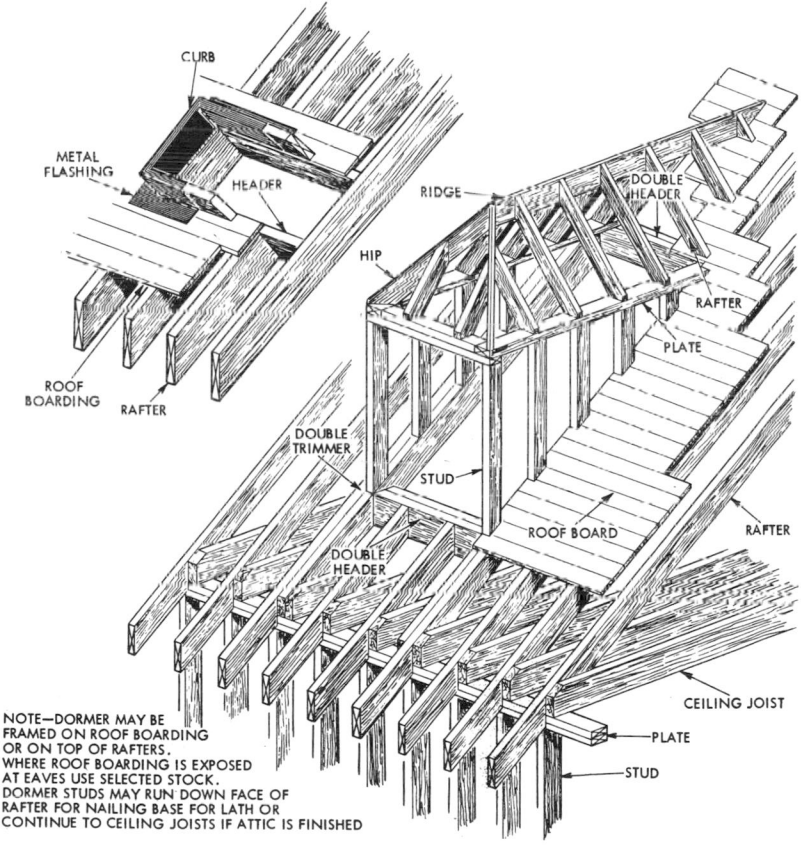

Fig. 8-8. Framing for hipped dormer and roof opening on a gable roof.

is a considerable difference between the two in both material and labor.

Nominal 1-inch material is used in widths varying from 4 to 10 inches.

Dormers. Figs. 8-8 and 8-9 show typical dormers. These drawings should be carefully studied to obtain a clear concept of each of the members used in dormer construction. For example, in Fig. 8-9 the common rafters are cut off and held in position by a double header. A header also supports the rafters above the dormer.

The corner posts of the dormer (Fig. 8-9) are 4 × 4's or two 2 × 4's spiked together. Sometimes one or more special details are specified, such as the tying pieces, shown by the detail drawing in Fig. 8-9. (Fig. 8-8 shows the construction of a hipped dormer, with the end of the ridge resting on sheathing.)

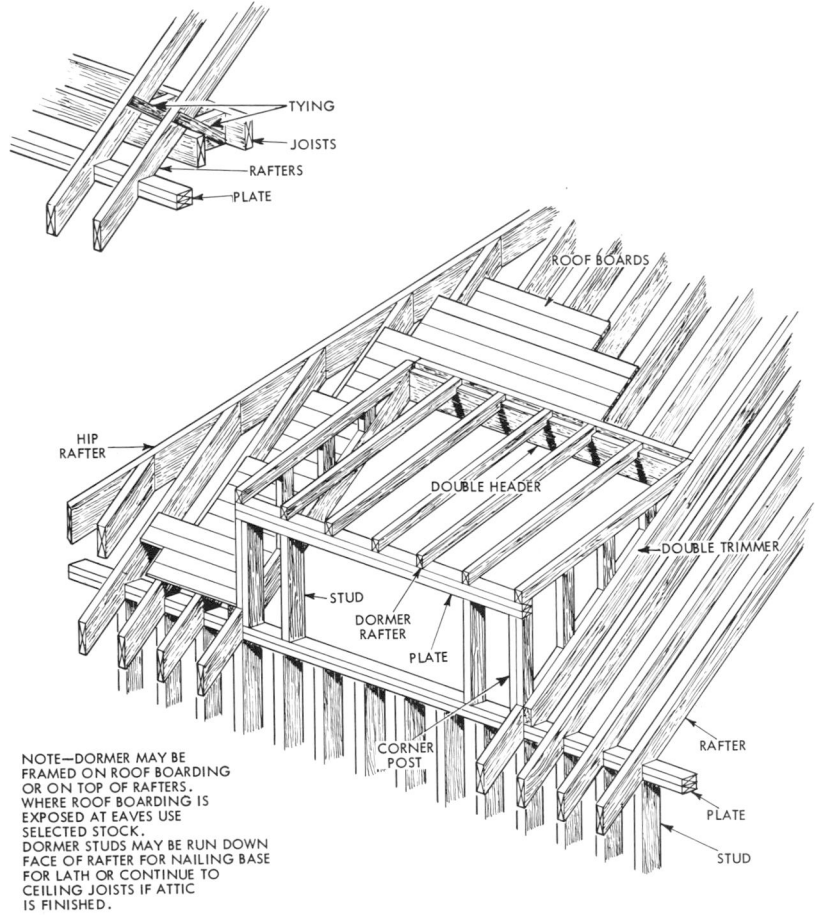

NOTE—DORMER MAY BE FRAMED ON ROOF BOARDING OR ON TOP OF RAFTERS. WHERE ROOF BOARDING IS EXPOSED AT EAVES USE SELECTED STOCK. DORMER STUDS MAY BE RUN DOWN FACE OF RAFTER FOR NAILING BASE FOR LATH OR CONTINUE TO CEILING JOISTS IF ATTIC IS FINISHED.

Fig. 8-9. Shed dormers require doubled studs and plates at the top, bottom, and corners.

The lengths of the various members of the dormer assembly must be calculated or estimated from the working drawings. Enough dimensions can be obtained from the drawings (from actual dimensions or by scaling) for close calculation of the various lengths.

The sides of dormers are covered with regular sheathing; the roofs are covered with the same types of roof sheathing used on the main roof area. Dormer construction is costly because of the labor involved in cutting, beveling, and fitting the large number of members.

Cornices and Sills. Typical cornices for frame and brick walls are shown in Fig. 8-10. The top plate is composed of two 2 × 4's and the rafters are notched to fit over it.

Where slate or tile roofing is used, added strength must be provided by using larger rafters or spacing them closer.

Fig. 8-11 shows water tables for frame and stucco walls. Fig. 8-12 is a frame wall section showing the water table. The section view of the working drawings will show the type of water table. It is often omitted. The lowest piece of siding serves as a drip.

Fig. 8-13 shows one recommended method for installing *flashings* on a flat roof connected to a brick wall. The important part of this detail, a *raggle block* with a *mitered* corner, is purchased specifically for this purpose.

An older method of roof topping (and flashing) is shown in Fig. 8-14. Fig. 8-14, right, shows the building paper (or flashing) nailed to the masonry using a nailed strip.

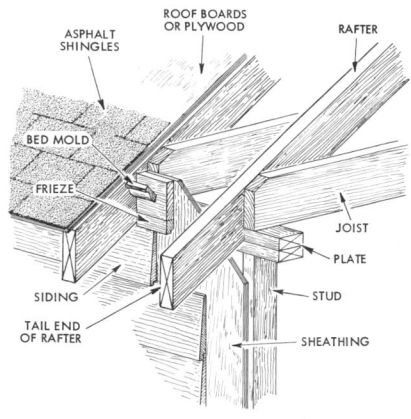

OPEN CORNICE

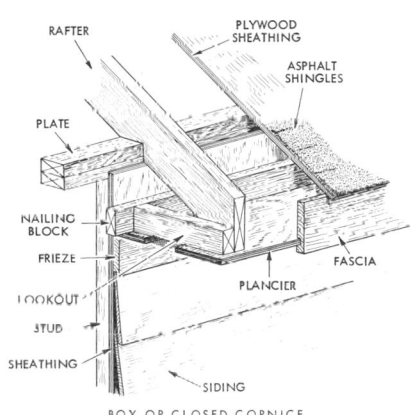

BOX OR CLOSED CORNICE

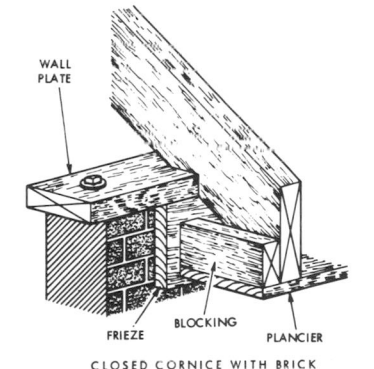

CLOSED CORNICE WITH BRICK

Fig. 8-10. Typical cornices for frame and brick walls.

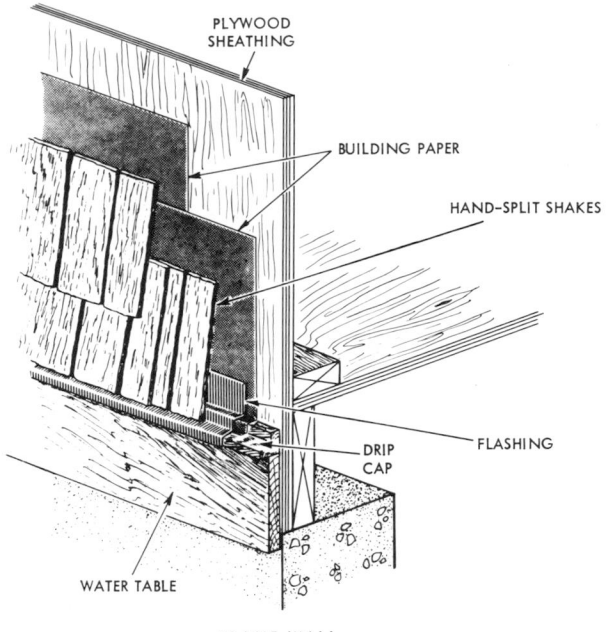

FRAME WALL

STUCCO WALL

Fig. 8-11. Water tables used in frame and stucco walls.

Methods of constructing and flashing flat roofs on masonry and concrete walls vary widely. If a concrete slab roof is poured, the insulation and roof-ing are applied as explained for frame roofs, and flashing is installed in a similar manner.

Roof Insulation. Aluminum foil is

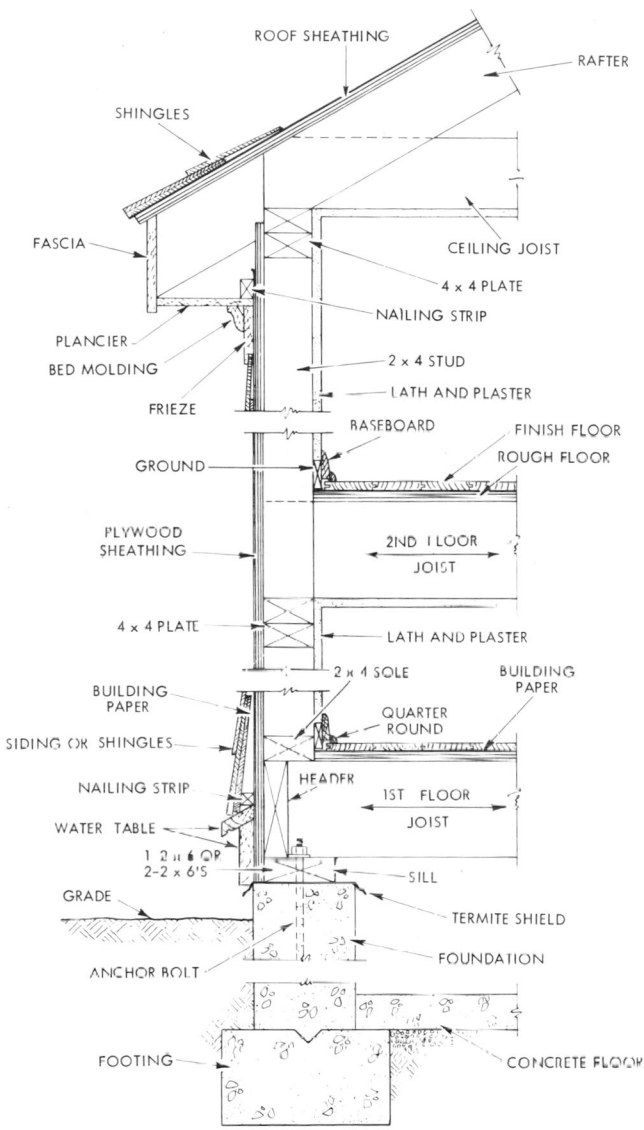

Fig. 8-12. A section through a frame wall which shows details of the cornice and water table.

commonly used as insulating material in frame roofs and is well adapted here because of its thinness and ability to reflect rather than hold heat. Fig. 8-15 shows a typical installation.

Rigid insulation is commonly used

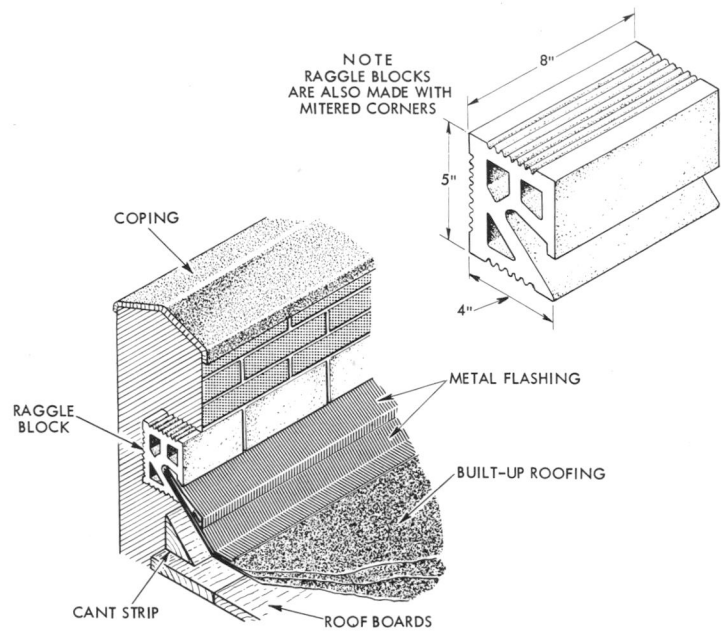

Fig. 8-13. Sheet metal flashing installed on flat roof.

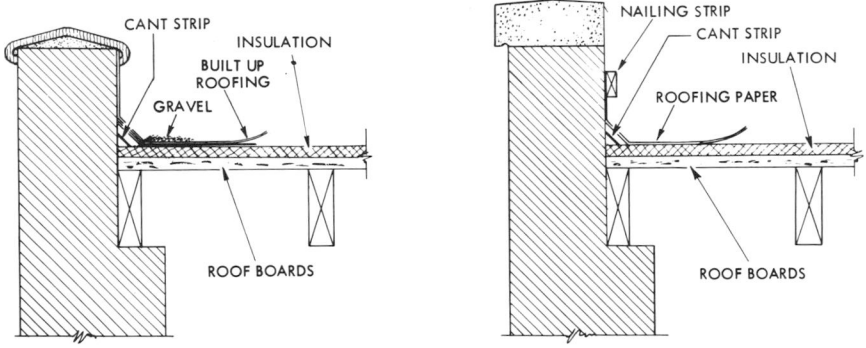

Fig. 8-14. An older method of installing roofing and flashing.

in frame roofs as shown in Fig. 8-16. Batt and blanket insulation can be used on sloped roofs.

Red Cedar Shingle and Hand Split Shakes. Cedar shingles and shakes have exceptional strength in proportion to weight, a low ratio of expansion and contraction with changes in moisture content and a high impermeability to liquids.

Shingles are sawed to a finish dimensional size.

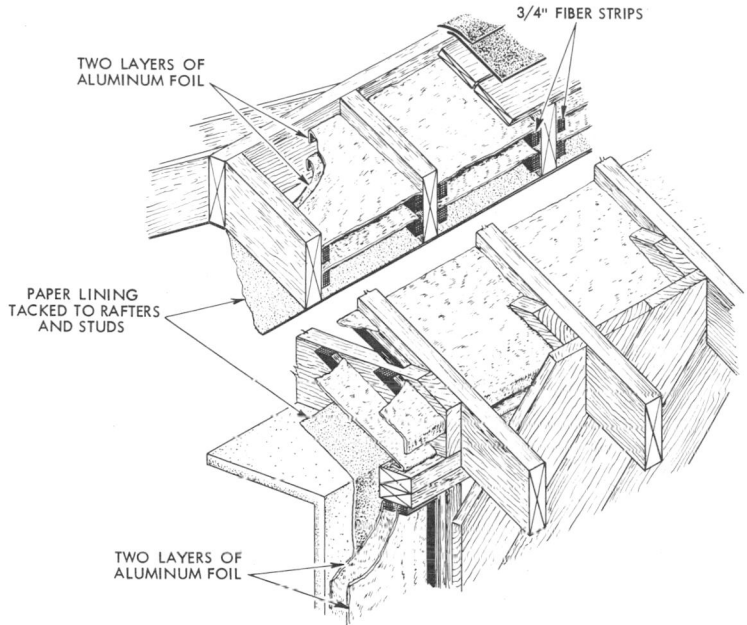

Fig. 8-15. Aluminum foil used as roof Insulation.

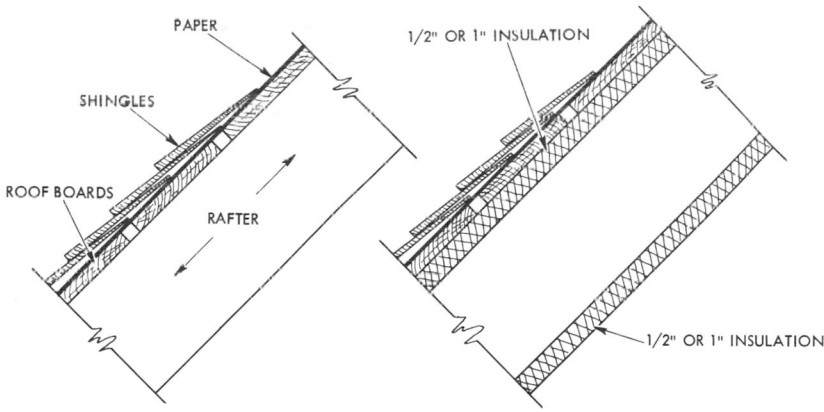

Fig. 8-16. Rigid insulation is also used on frame roofs.

Shakes are hand-split resulting in random widths.

Shingles and shakes are generally used for roofs in a natural or unstained state.

Nails should be considered here

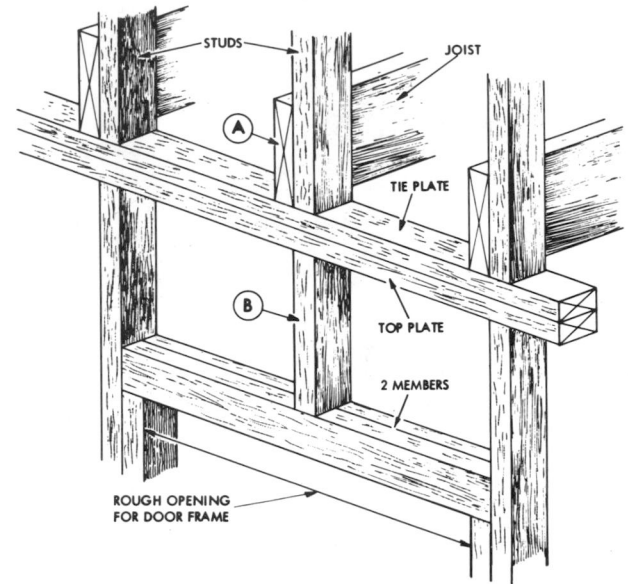

Fig. 8-17. Framing above a door opening.

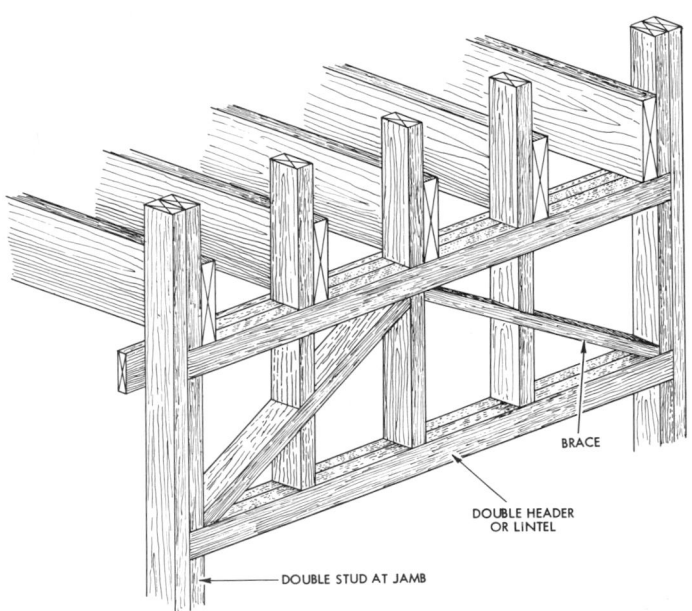

Fig. 8-18. Braces are sometimes used to strengthen a door opening.

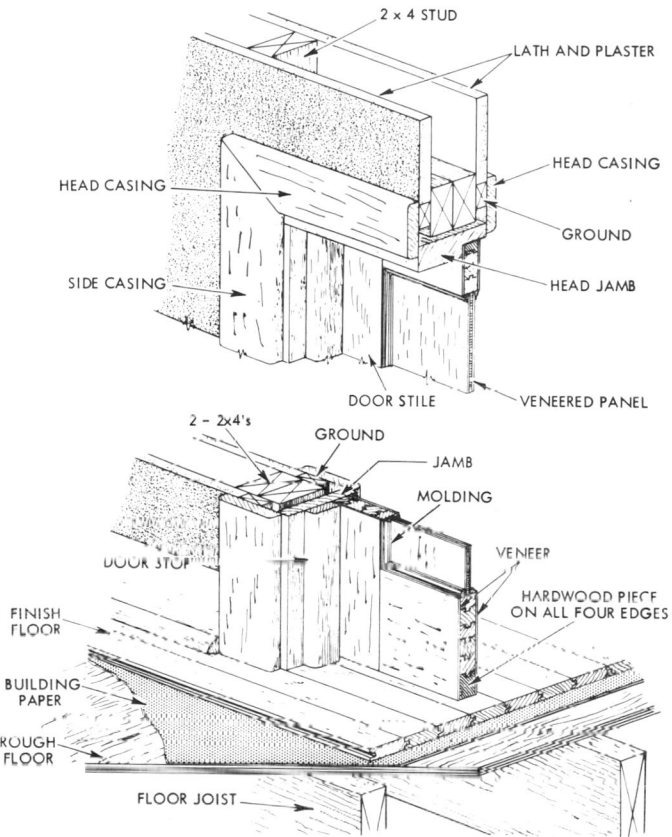

Fig. 8-19. Interior door framing.

also. When installing cedar shingles or hand split shakes only rust resistant (hot dipped zinc) or aluminum nails should be used because of their exposure to the weather.

STRUCTURAL OPENINGS

Doors. If studs must be eliminated for a door opening in a load-bearing partition, the framing is done as in Fig. 8-17. Two members are doubled and set on edge over the door opening.

They in turn bear on an extra set of 2×4's added to the vertical framing. In this manner the joist *A* and stud *B* have a solid bearing.

Fig. 8-18 is another method of framing a door opening. The braces strengthen the framework.

The framing around a door is done at the same time as the wall and partition framing. Fig. 8-19 shows framing of a typical interior door.

The door frame (the jambs) are usually made in a mill. The door must be

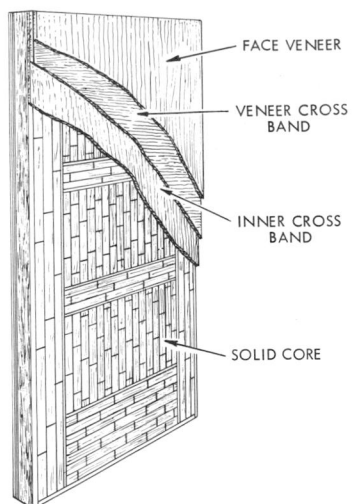

Fig. 8-20. Solid Core Flush Door.

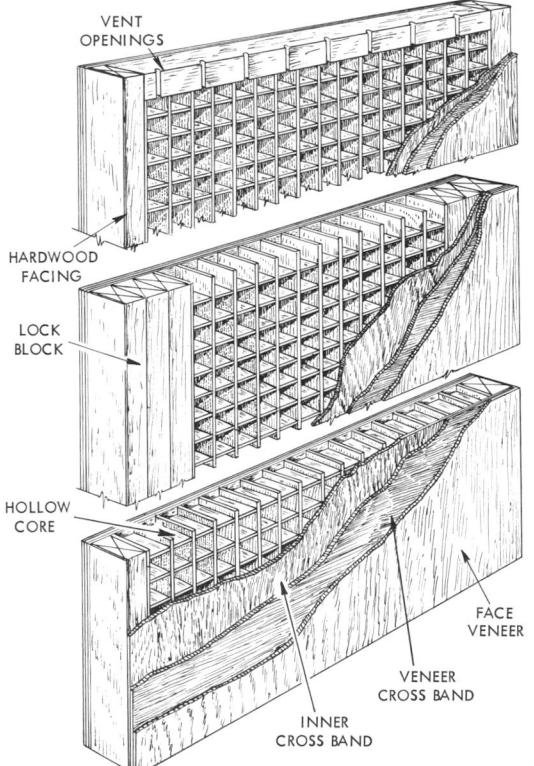

Fig. 8-21. Hollow Core Flush Door.

fitted, the hardware installed, the stops nailed on and the casings cut and fitted to the jamb.

The door itself is delivered to the job ready to be installed. It is therefore considered as millwork. Hinges and locks are separate items and are usually listed as hardware.

Pre-hung doors are delivered with the door hung in the frame ready to put in place in the wall with the casings cut and fitted to the jamb.

Two main classes of doors are the flush door and the panel door. Flush doors have one large flat surface on each side. Panel doors are made of vertical members called *stiles,* and horizontal members called *rails.* Panels which have raised surfaces on them are held between stiles and rails. Both flush and panel doors are used as interior and exterior doors. Exterior doors should be 1¾ inches thick.

Two types of flush doors are manufactured, the solid core and the hollow core. Solid core doors are usually used as exterior doors. They are heavier and can withstand weather better than hollow core doors. The solid core door is made up of many small soft wood blocks which are glued together. Three layers of veneer are glued to each face of the core. The edges are finished with softwood or hardwood. See Fig. 8-20. Hollow core doors (Fig. 8-21) are lightweight and quite satisfactory for inside use. They are made up of strips of wood, forming a frame and a core of interlocking strips. There are several different ways of arranging the parts of the core. Inside doors are usually 1⅜″ thick.

Windows. The rough opening members for windows are arranged in a manner similar to those for doors. See Fig. 8-17, also previous chapter, Figs. 7-7 and 7-8.

There are three basic types of windows, depending on how they operate. In some windows the sash is held in a fixed position; in others the sash slides vertically or horizontally, and in still others the sash is hinged to swing in or out. See Fig. 8-22.

The introduction of large insulating glass units has made the so-called picture window possible, and the desire on the part of owners to have an unobstructed view from inside the building has made it very popular. It is often used in combination with other types of sash, or with louvered ventilators within the same frame.

Double-Hung Frames. The double-hung window is considered most practical in cold climates although every other type is used also. The two sashes are arranged in grooves so as to slide past one another in a vertical direction. The horizontal sliding window uses the same principle, except that the grooves or tracks are horizontal. Both types can be provided with spring channels which permit the easy removal of the sash for cleaning and painting.

The basic construction of the double-hung window frame is shown in Fig. 8-23. (Note: It will help to understand how the section views are drawn by looking at the elevation view, Fig. 8-23, and an isometric view of another type of window, Fig. 8-28, which is a different type but similar in

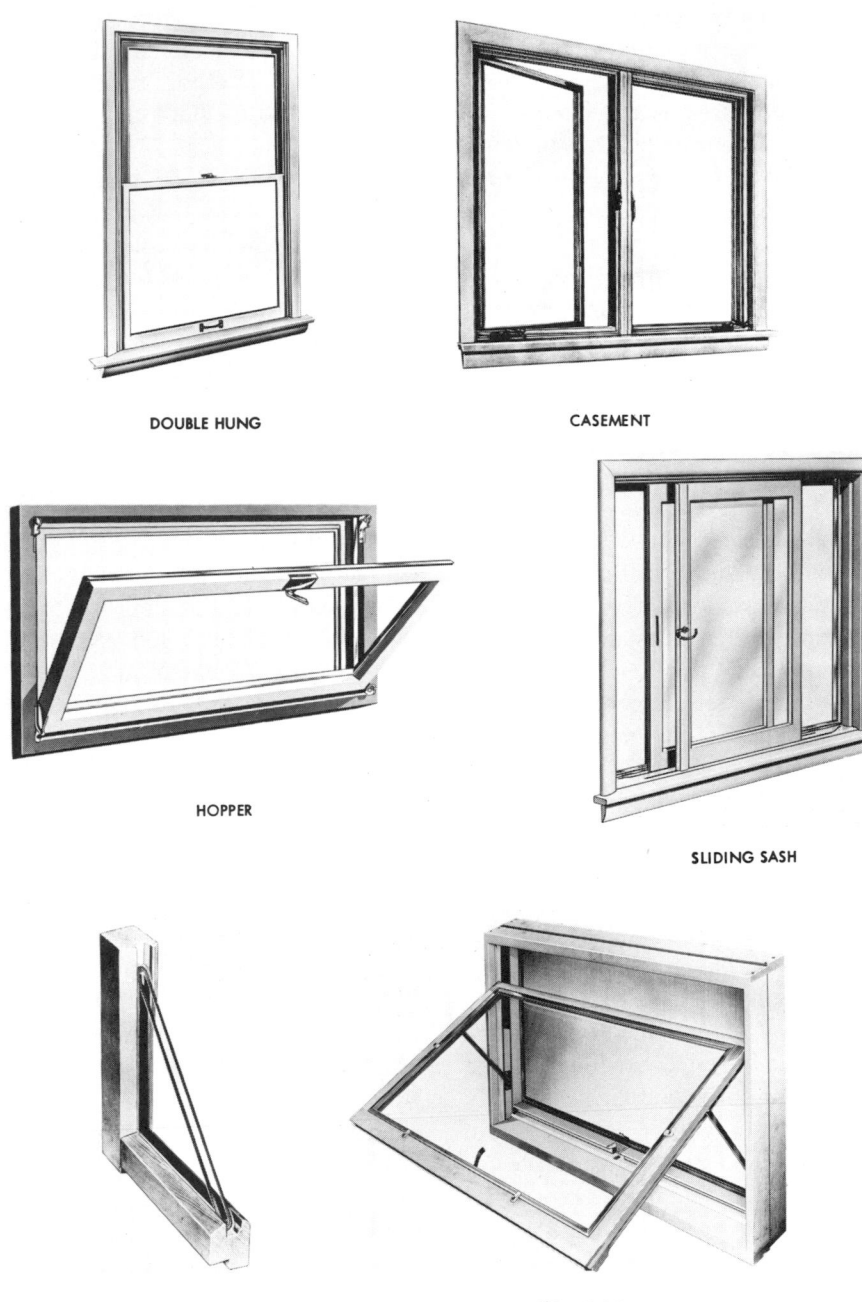

DOUBLE HUNG

CASEMENT

HOPPER

SLIDING SASH

INSULATING GLASS

AWNING WINDOW

Fig. 8-22. Windows commonly used for residential construction.

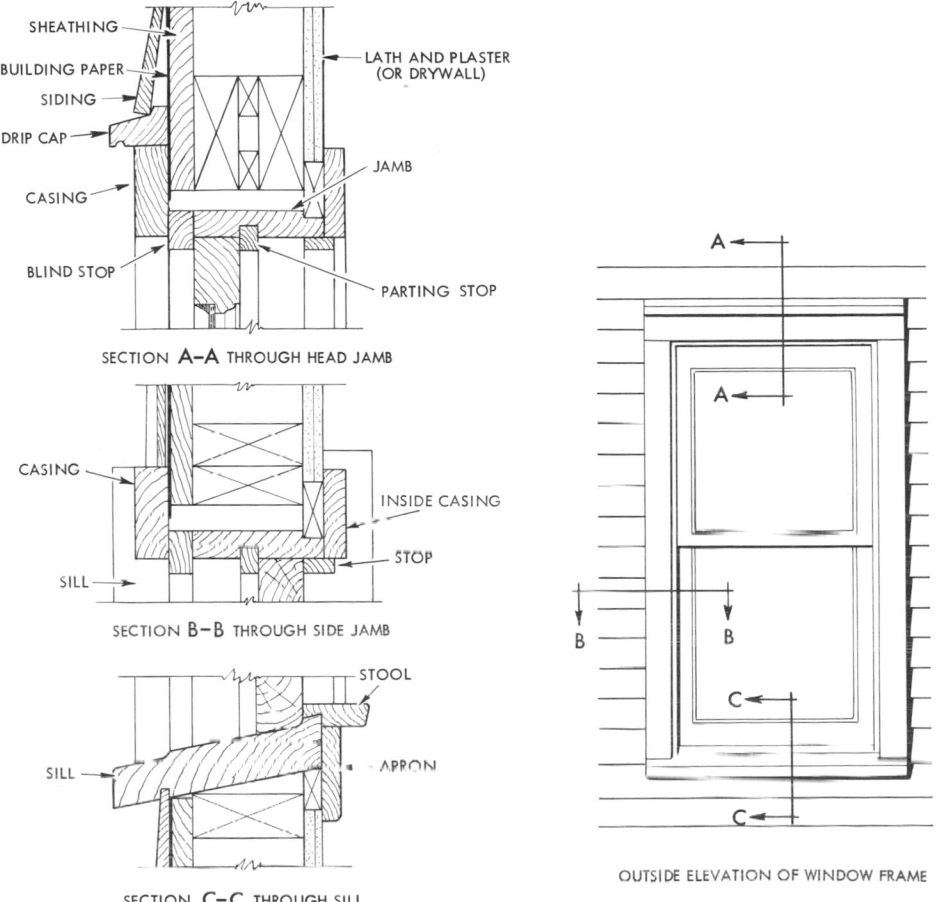

SECTION **A–A** THROUGH HEAD JAMB

SECTION **B–B** THROUGH SIDE JAMB

SECTION **C–C** THROUGH SILL

OUTSIDE ELEVATION OF WINDOW FRAME

Fig. 8-23. A simplified sectional view of a wood double-hung window gives the names of the parts and shows how the parts are related.

some respects.) The jambs are ¾ of an inch in thickness. They have a parting stop to separate the top from the bottom sash and also a blind stop which produces a groove in the frame in which the window slides. The outside casings are of the same width and thickness as those of the other openings in the building. The joints between frame parts are usually butt joints. On the better-built frames, tongue-and-groove joints are used.

Windows in pairs or multiples are separated by *mullions* which may be a box construction or a single 1½ inch structural member with jambs fastened to each side. (A box construction would require two jamb pieces and an exterior or an interior casing built to form a box.)

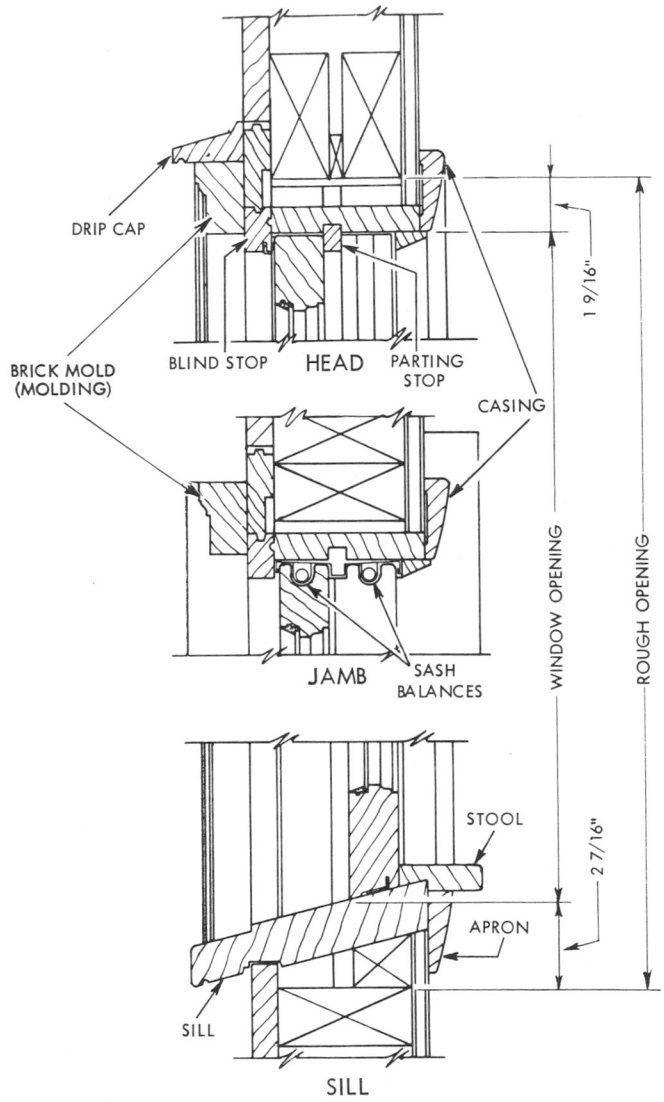

Fig. 8-24. A wood double-hung window and wood frame in frame construction. (Rock Island Millwork Co.)

Manufacturers of windows have developed features which make them more weathertight, easier to install and maintain, and interchangeable so that the same window frame can be used in frame construction, with brick ve-

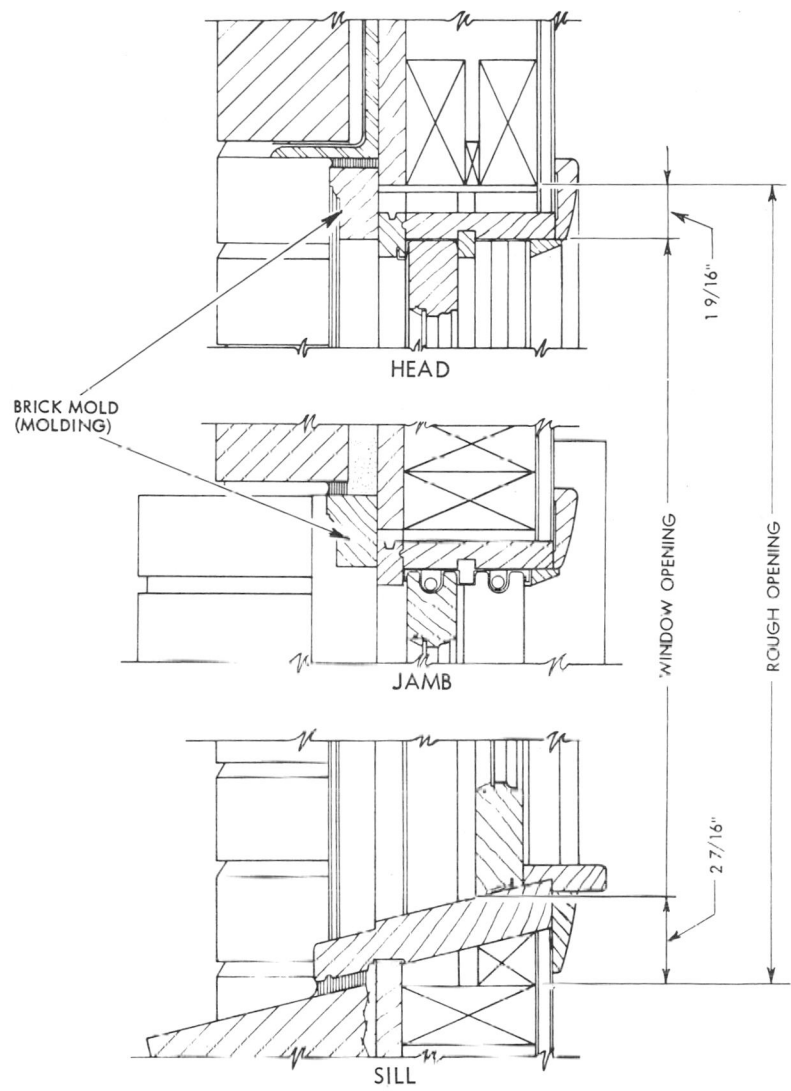

BRICK MOLD
(MOLDING)

HEAD

JAMB

SILL

4 7/8" FRAME WALL WITH BRICK VENEER
3/4" SHEATHING - 1/2" DRY WALL

Fig. 8-25. The wood window and frame shown in Fig. 8-44 is used in a brick veneer wall with only slight modifications. (Rock Island Millwork Co.)

neer, or in a masonry wall with only minor modifications. Fig. 8-24 shows a typical package window for use in a frame wall. Instead of outside casing a brick mold (molding) is used. A spring-type sash balance is shown in the section through the jamb. The window can be ordered for drywall (gypsum board) or lath and plaster application. (Drywall is usually ½ inch thick whereas lath and plaster is ¾ inch thick.)

Fig. 8-25 shows the same basic frame used in a brick veneer wall. The

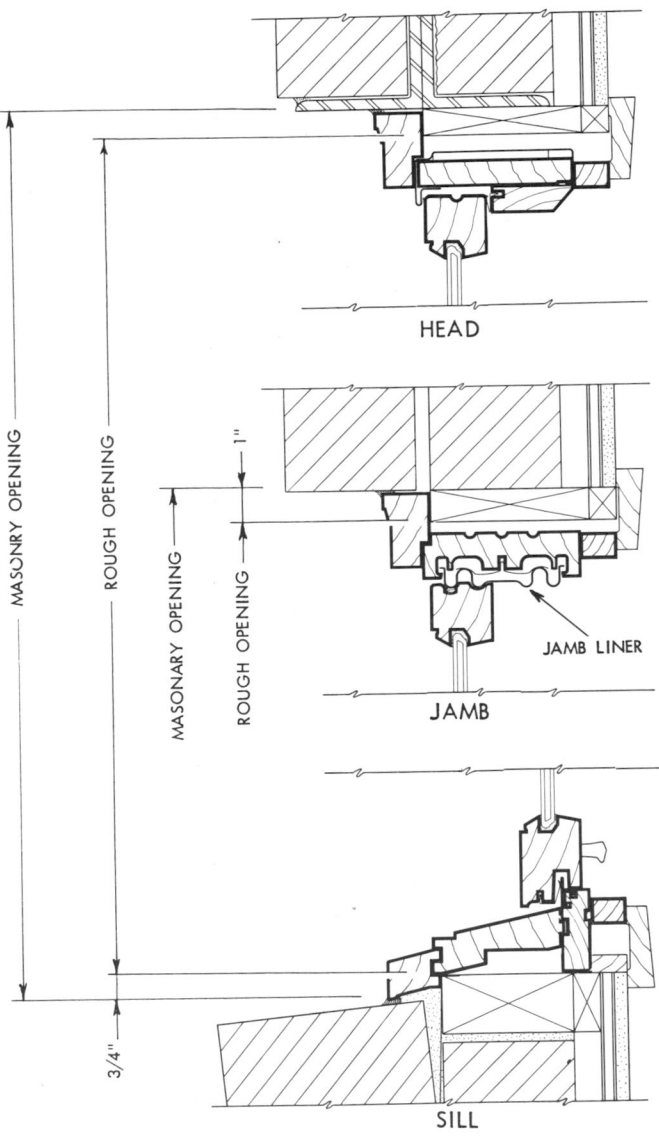

Fig. 8-26. A package wood window frame which can be used on either a masonry or frame wall. Note absence of blind stops, stool and apron. (Rolscreen Co.)

brick mold (molding) is placed away from the masonry a short space to provide for calking. The window frame in a brick veneer wall is fastened to the frame part of the wall. Fig. 8-26 shows a wood window made by another manufacturer which is also universal in application. In this case it is applied to a brick wall. The section through the head shows steel lintels supporting the masonry above the window. The window is designed to eliminate the blind stop entirely. (See Figs. 8-23 and 8-25 for a blind stop.) When drywall is used instead of lath and plaster, the block at the inside of the jamb is cut to the proper width. The section through the jamb shows a combined jamb and vinyl-spring jamb liner which serves (1) to hold the window in position as it is raised and lowered to any height, (2) to permit the window to be removed for washing, and (3) to provide a weather seal. The section through the sill shows the weatherstripping arrangement and the absence of the usual stool and apron.

Casement Windows. Casement windows (Fig. 8-22) are the most common windows that swing on hinges. Generally they swing out and are operated by a crank device.

The construction and thickness of materials used in casement window frames made in a local mill is often similar to that used in exterior door frames with rabbeted jambs. If the same jambs are used as those of the double-hung windows, the allowances for fitting will be the same as for a double-hung window with sash balances.

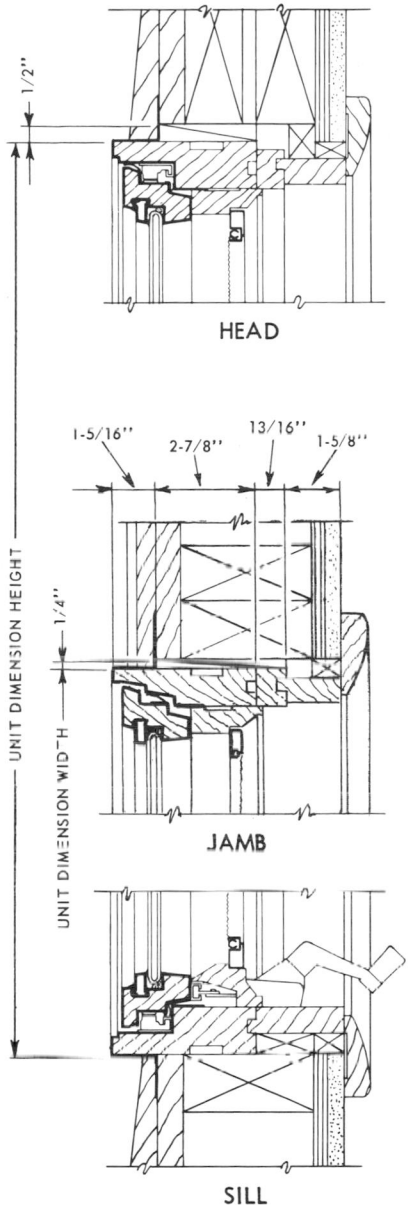

Fig. 8-27. A package casement window sealed in rigid vinyl plastic. Note the absence of exterior casings. The window is operated by a roto-operator through a gear arrangement so as not to interfere with the inside screen. (Andersen Corp.)

Large manufacturers who specialize in windows have introduced many innovations, which should be appreciated by the carpenter and the home owner. Fig. 8-27 shows a package outswinging casement window with many such features. It is glazed with welded insulating window glass. The sash has a wood core completely sealed in rigid vinyl plastic. It is also used on frame parts to form flashing on all four sides of the window, including the sill. There is no exterior casing. The jamb extends beyond the face of the wall to form a thin projection. The siding is cut to fit against it and is calked. Spring-tension vinyl weatherstripping provides a weather seal. The window is operated by a roto-operator through a gear arrangement so as not to interfere with the screen which is on the inside of the window. The stool and apron on the inside of the window have been omitted.

Awning and Hopper Windows. These resemble casement windows in most details. See Fig. 8-22, and Fig. 8-28. The hinges are located at the top of awning windows and roto-operators or bar lock devices are arranged so that the windows can be opened without interfering with the screens. Hopper windows do not need roto-operators because they swing in and the screen is on the outside. They often make up the bottom portion of large combination frames.

Patent Sash Balances. There are several patent sash balance devices on the market which have been accepted as satisfactory by home builders. Some

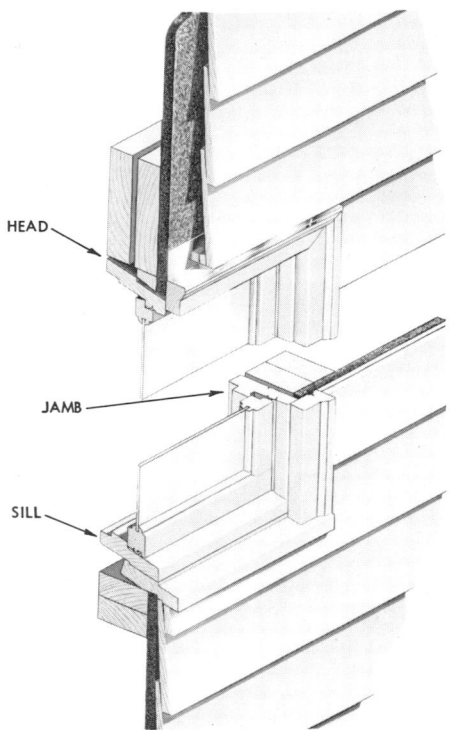

Fig. 8-28. An isometric view of an awning type window shows how the section through the head, jamb and sill are taken. (Andersen Corp.)

of them operate on the principle of a spring which winds up as the sash is raised, thus providing tension. Some of them operate on a friction principle. Some of them are adjustable so that the tension on the spring counter-balances the weight of the sash. They are installed quickly. Because they eliminate the need for the space for window weights, the rough opening need be no more than that required for a casement window, thus permitting the use of narrow trim. The sash balance and weatherstrip shown in Fig. 8-29 is a

Fig. 8-29. Sash balances use springs to counterbalance the weight of the sash. (Zegers Inc.)

metal jamb liner fastened to the jamb, replacing the usual wood parting stop and containing springs which permit the sash to remain open in any position. This type combines the value of a sash balance with a weatherstrip feature.

Several manufacturers have developed windows with a "take out" sash. One of the jambs is equipped with a metal or plastic jamb liner that rests against springs, which compress to permit the sash to be removed. A few extra precautions are needed such as: (1) setting the frames so that the jambs remain plumb and straight, (2) trimming the window so that the jamb liner is not bound but can move when the

jamb is pressed, and (3) painting so that the spring feature is not hampered. The result is a smooth operating window which can be removed for cleaning. See Figs. 8-30 and 8-31.

Lintels. For both windows and doors in masonry walls, some form of lintel is necessary. Sometimes wood members are used, but generally angle irons set back-to-back are used.

Window frames are rarely made on the job. They are purchased complete from the mills and delivered completely assembled. The labor cost occurs in the placing of the frame in position in the wall. Window trim is purchased separately and cut and fitted on the job.

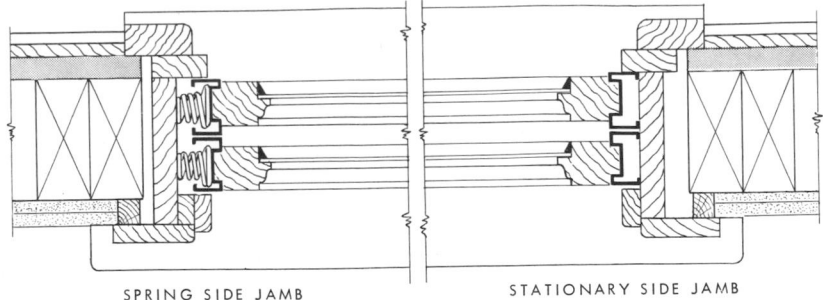

Fig. 8-30. Plan view through jambs. Removable windows have a metal jamb which compresses to permit the removal of the sash. (R. O. W. Window Corp.)

SPRING SIDE JAMB STATIONARY SIDE JAMB

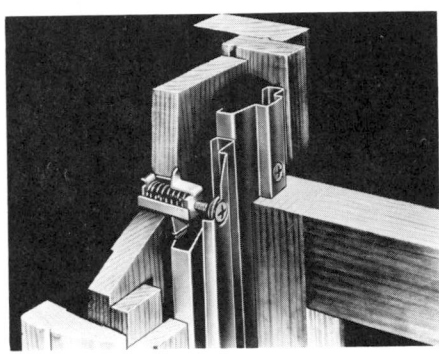

Fig. 8-31. Detail view of adjustable side jamb. (ARB Window Sales Co.)

Sash balances, friction holding devices and other window controls are usually estimated as hardware. (Pulleys, ropes and sashweights are rarely used except in old work.)

Window frames for masonry walls are slightly different in construction, but like windows for frame walls, they are delivered to the job ready to be installed. The sash and glass may be delivered to the job installed in the frame. In this case, the sash, sash balances and glass are all in place.

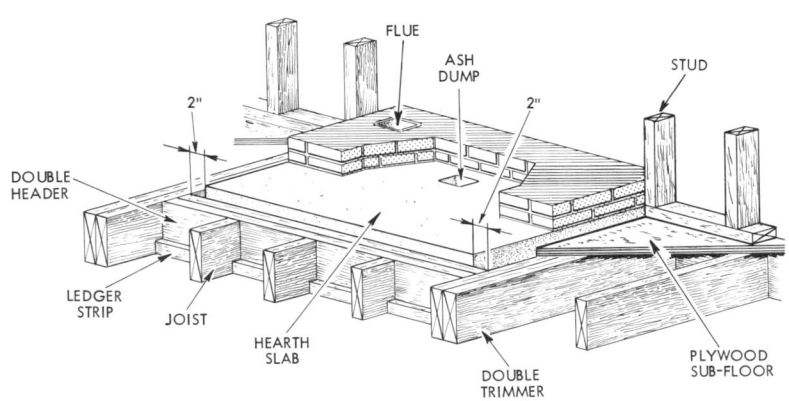

Fig. 8-32. A 2″ clearance must also be provided around the sides of the fireplace.

Fireplace Framing. Framing members, flooring, roof members, and walls must be free of the chimney. A minimum space of 2″ must always be provided. Non-combustible insulation is inserted between the framing members and the chimney masonry. The fireplace itself always has a 2″ clearance with insulation on both sides and the back. See Fig. 8-32. If the exterior covering of the dwelling is wood siding, shingles, etc., the junction of the wood and brick must be calked.

Framing around the chimney placed on an exterior wall is accomplished by two double trimmers resting on the wall plate (Fig. 8-33). A double header is spiked between the trimmers to carry the regular floor joists. A 2″ clearance with insulation is provided

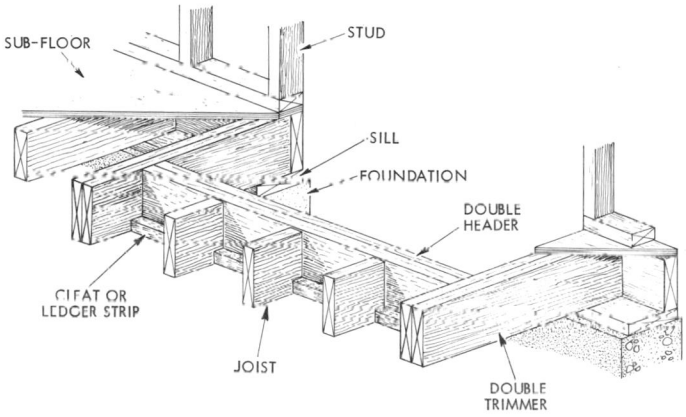

Fig. 8-33. Framing is doubled around a chimney placed on an outside wall.

Fig. 8-34. A 2″ clearance must be provided between the chimney and the roof framing members. This space is filled with non-combustible material.

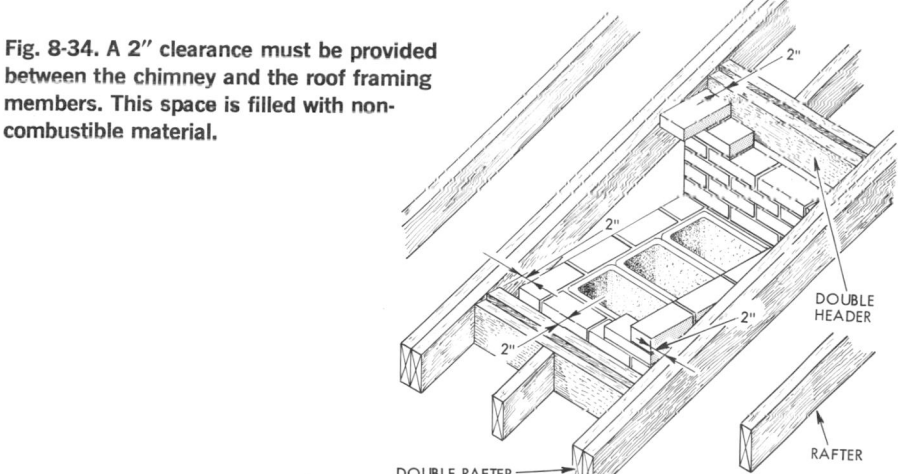

only on the sides of an exterior fireplace.

The wall opening for a fireplace is framed exactly like a door opening with the additional tolerances for the fireplace material in lieu of the door frame or buck.

The opening in the roof is framed in a similar manner with double rafters and trimmers. Fig. 8-34 shows roof framing details around the chimney.

Stairs. This is a difficult item of construction for the inexperienced estimator to visualize, unless he has prior experience with stairs and their framing. The estimator should be able to visualize every detail of a stairway before he does a quantity take-off. Stairway framing is largely composed of walls or partitions which form rooms. Typical stair members, however, are important enough to deserve the estimator's close attention.

Note Fig. 8-35. There are three sep-

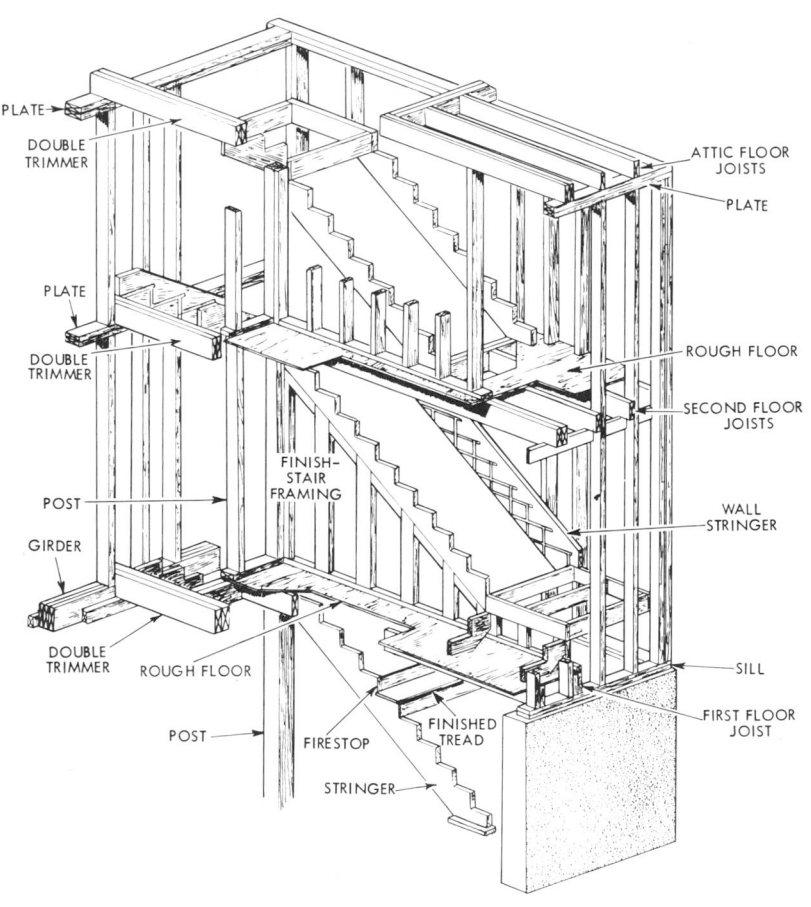

Fig. 8-35. Stairway framing.

arate stairways leading from (1) the basement to the first floor, (2) the first to the second floor, and (3) the second floor to the attic. The basement and attic stairways each have one turn; the second-floor stairway has two turns.

The basement stairs are the simplest in design and construction. They consist of two *stringers*. The stringers are cut from either a 2 × 10 or a 2 × 12 plank. The *treads* (rests) and the *risers* (vertical members between treads) are carefully fitted. Such stairways generally have a railing made from 2 × 2's fastened to the wall. Plates may be used at the bottom of the stringers but can be ignored in the quantity take-off.

The stairway to the second floor in Fig. 8-35 is more complicated. This stairway consists of a wall stringer and 2 or 3 other stringers. (Only the out-side stringers are shown.) The platforms for all three stairways are framed of 2 × 4's or 2 × 8's and floored in the same manner as the first floor.

The attic stairway is usually of simple construction. In Fig. 8-35 two stringers are used in each leg of the stairs with a platform between them. Note that double joists surround the stairwell.

The stairway going to the second floor is open on one side, Fig. 8-36, and has a closed (or housed) stringer on the wall side, Fig. 8-37. The stair parts are usually constructed in a mill.

Instead of the stairs shown going to the attic in Fig. 8-35, folding stairs might have been used. Such stairways are usually factory made and either partially or completely assembled.

Fig. 8-36 shows the open edge of a stairway in which the stringer (or car-

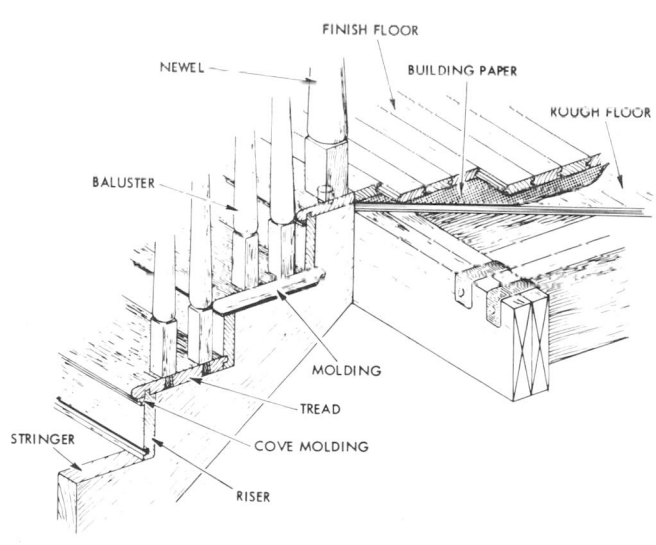

Fig. 8-36. The open edge of an inside stairway.

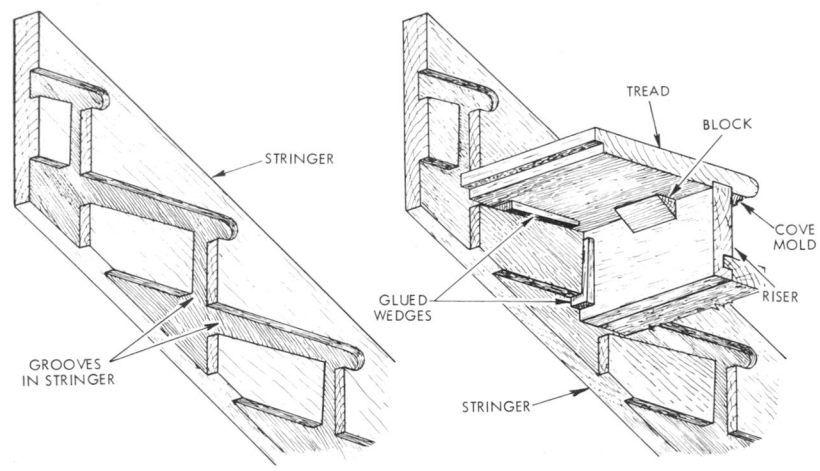

Fig. 8-37. Stringers and parts of a stairway. A closed stringer is routed out to receive the treads, risers and wedges.

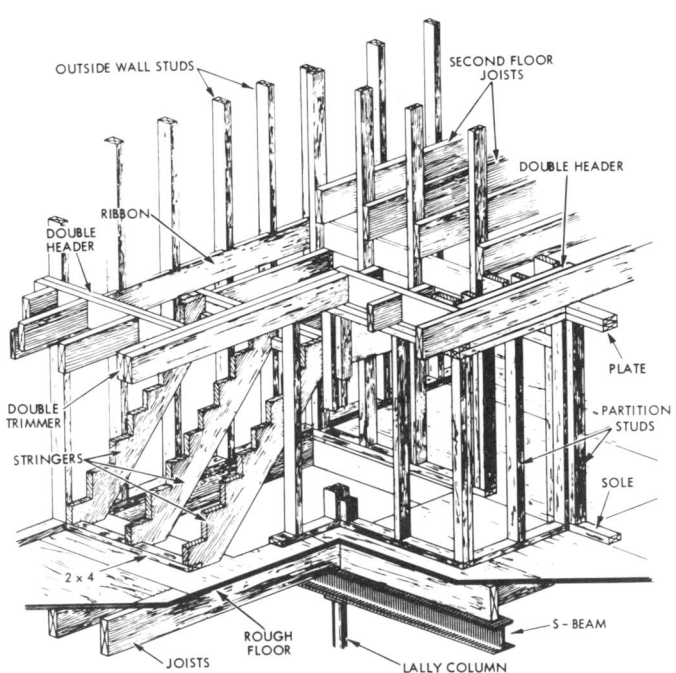

Fig. 8-38. A 3-stringer stairway and stairwell details.

riage) and the treads and risers are shown in detail. The trim for the stringer is also shown. Railings and balusters can vary greatly in design.

Fig. 8-38 shows a three-stringer stairway and adds more details to the framing in and around stairwells.

The working drawings will contain lumber dimensions—length and size. The *run* of a stairway is the horizontal projection. Knowing this length and the *rise* (distance between landings or height of each leg), the stringer lengths can be approximated. The lumber size depends on the riser and tread dimensions.

Dimensions for platforms are also given on the working drawings. Lumber for risers and basement and attic treads is generally specified as to thickness and kind of wood.

Often the contractor does no more than build the stringers (for main stairways) and surrounding frame and platforms. In such cases, the stairs together with trim, railings, and balusters are supplied by a mill, ready to be installed. Generally, only the cheaper residential designs call for contractor-built main stairways.

DETAILS

Grounds. When lath and plaster is used for wall finish, grounds are necessary. Fig. 8-12 shows small rectangular strips or grounds next to the 2 × 4 sole at the first and second floors. (Grounds may also be required around windows.) These are pieces of wood about ¾" to 1½" wide. They are nailed to the sole or the studs and run

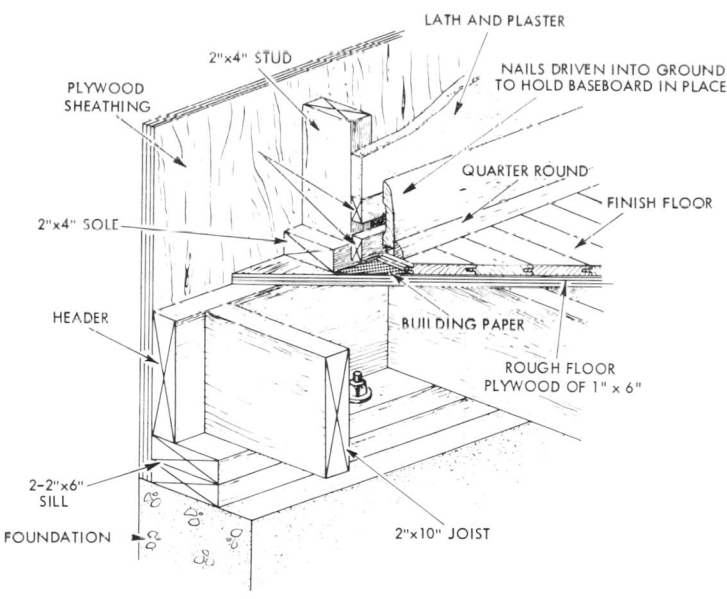

Fig. 8-39. Grounds installed in a wood frame building.

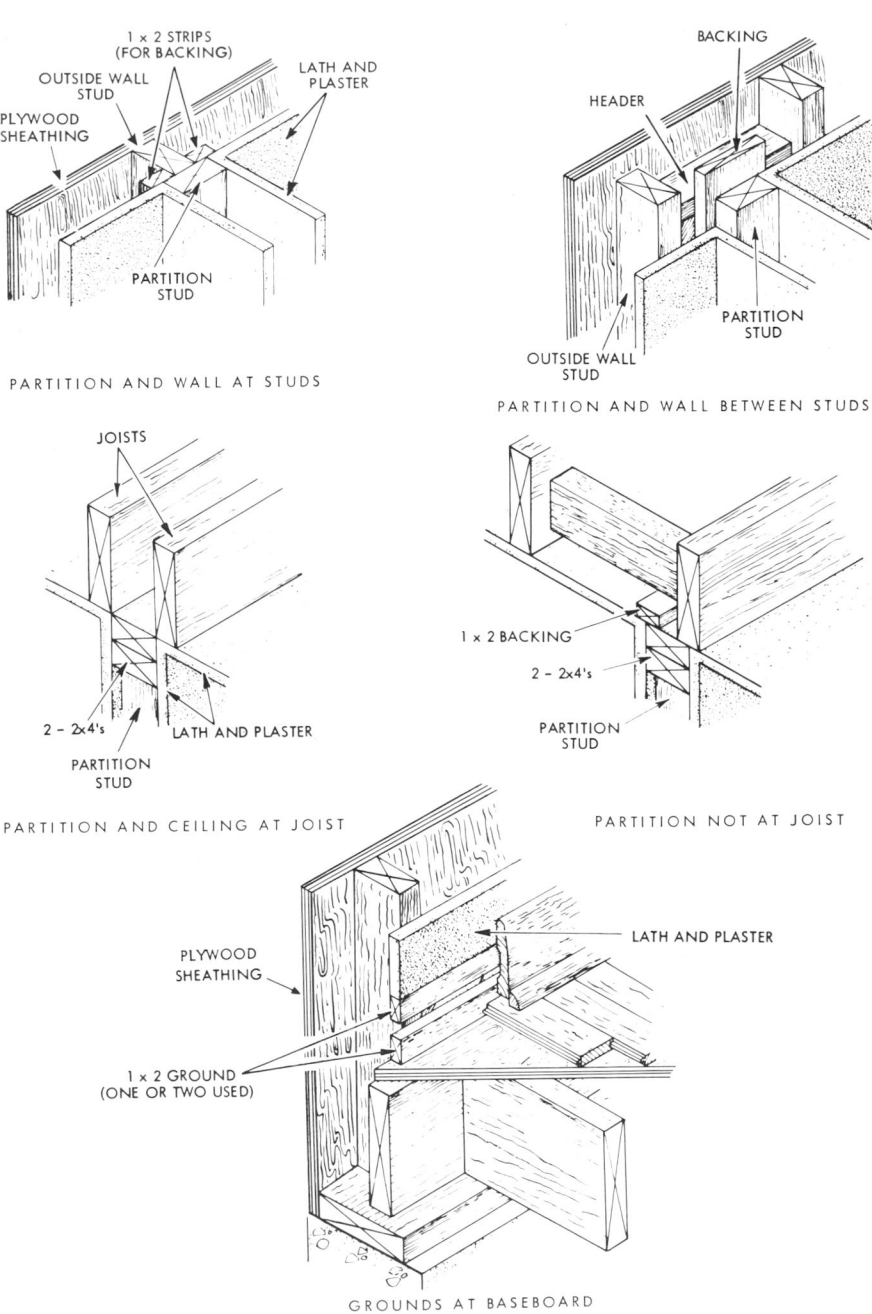

Fig. 8-40. Intersections of partitions. Backing is shown at top, left and right, and center, right. Grounds are shown at bottom.

along the floor all around the rooms. The lath is brought up to the ground. The plaster is applied to the ground and flush with the surface of the ground. When baseboards are being installed, the grounds form a nailing surface. An enlarged view is shown in Fig. 8-39.

If *chair rails* (wooden molding at chair-back height) are to be installed on one or more walls, grounds should be nailed to the studs under the place designated for the rail.

Fig. 8-40 shows various places where grounds or backing are used. Top, *left,* shows two pieces of 1 × 2 used for backing where an inside partition meets an outside wall exactly at a stud location. At top, *right,* the partition meets the outside wall between studs. Center, *left,* shows a partition reaching the ceiling at joist locations. No backing is required. When the partition is off from the joist location as at center, right, a piece of backing is required. Bottom shows the use of grounds at the baseboard. Grounds will also be needed if a wood cornice is used at the ceiling.

Grounds can require significant material and labor, so this item should be examined carefully.

Fireproofing. Where buildings are constructed of flammable materials, such as wood, fireproofing is impossible unless some form of fire-resisting paint or chemical is applied to all wood parts. *Firestopping,* however, can be built into the walls as shown in Fig. 8-41.

The main purpose of firestopping is to prevent the spread of fire. For ex-ample, if fire originates in the basement of a frame residence, the hot smoke, gases, and flame have a tendency to escape and spread up through the walls. The walls form a chimney unless they are built to obstruct the sweep of fire, smoke, and gases.

Chimneys themselves can be fire hazards unless some fireproofing material separates them from adjoining combustible surfaces.

Many insulating materials are designated fireproof or *slow-burning.* For example, most of the wool and blanket types are made of non-combustible materials or they are treated to make them fireproof. The rigid types burn slowly; they smolder but do not burst into flame. Thus a well-insulated structure is also well fireproofed, or at least made fire retarding.

Concrete structures are considered fireproof. Steel structures are generally protected from intense heat by concrete fireproofing. Light steel members used in residential construction will not long withstand fire and intense heat, so many designers specify large amounts of non-inflammable material to serve the dual purpose of insulating and firestopping.

Special Details and Millwork. There are many special details encountered in wood construction.

Cabinets. Kitchen cabinets and special cabinets of all kinds are delivered to the job ready to install in most cases. Shelves for linen closets, pantry shelves, etc., are usually cut and fitted on the job.

Storage Areas. Storage spaces, especially in attics, are not generally fin-

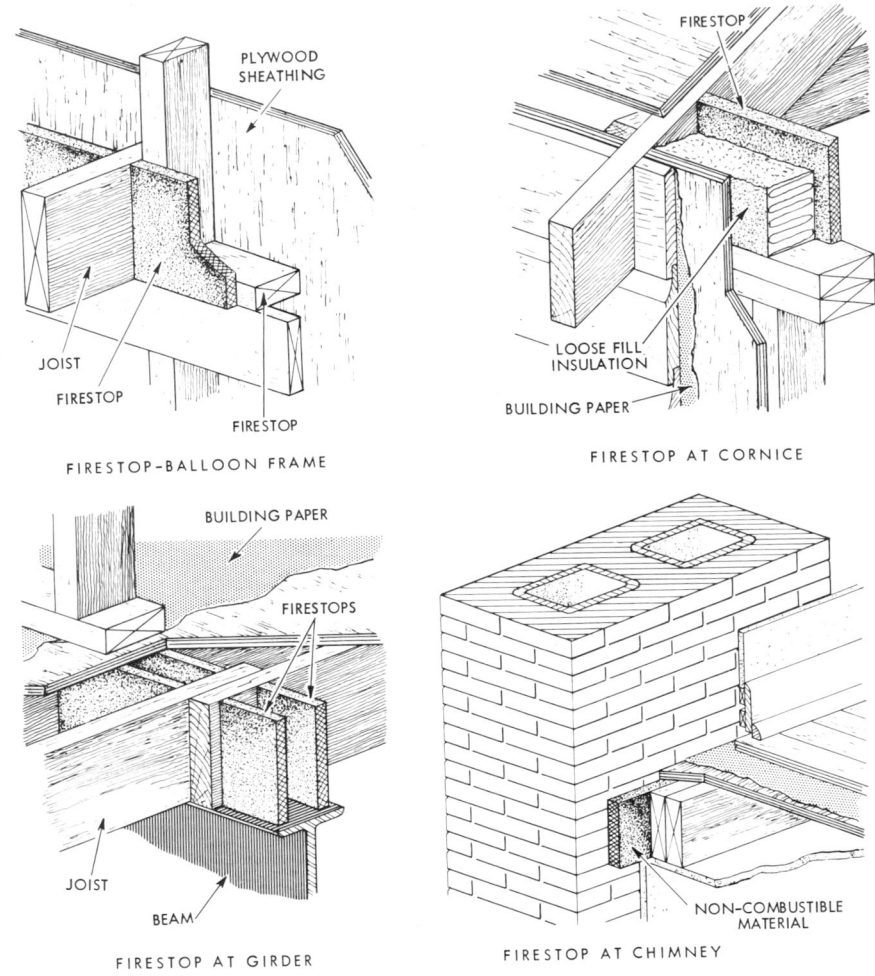

PLYWOOD
SHEATHING

JOIST

FIRESTOP

FIRESTOP

FIRESTOP-BALLOON FRAME

FIRESTOP

FIRESTOP

LOOSE FILL
INSULATION

BUILDING PAPER

FIRESTOP AT CORNICE

BUILDING PAPER

FIRESTOPS

JOIST

BEAM

FIRESTOP AT GIRDER

NON-COMBUSTIBLE
MATERIAL

FIRESTOP AT CHIMNEY

Fig. 8-41. Firestopping can be built into wood walls.

ished. The specifications should cover any special work to be done.

Closet Accessories. The coat rods and shelves in clothes closets are cut and fitted on the job.

Trim. This includes such items as baseboards, coves, and casings. Trim is important to the estimator because of the material and labor costs involved. The various items of trim differ greatly in material and size. See Fig. 8-42. All of it, however, is purchased in straight lengths and cut and fitted as required.

Picture moldings and chair rails are delivered to the job ready-formed and in lengths which have to be cut and fitted. Mantel details may be made on the job or purchased ready to fit into place. Stair rails, balusters, tread nos-

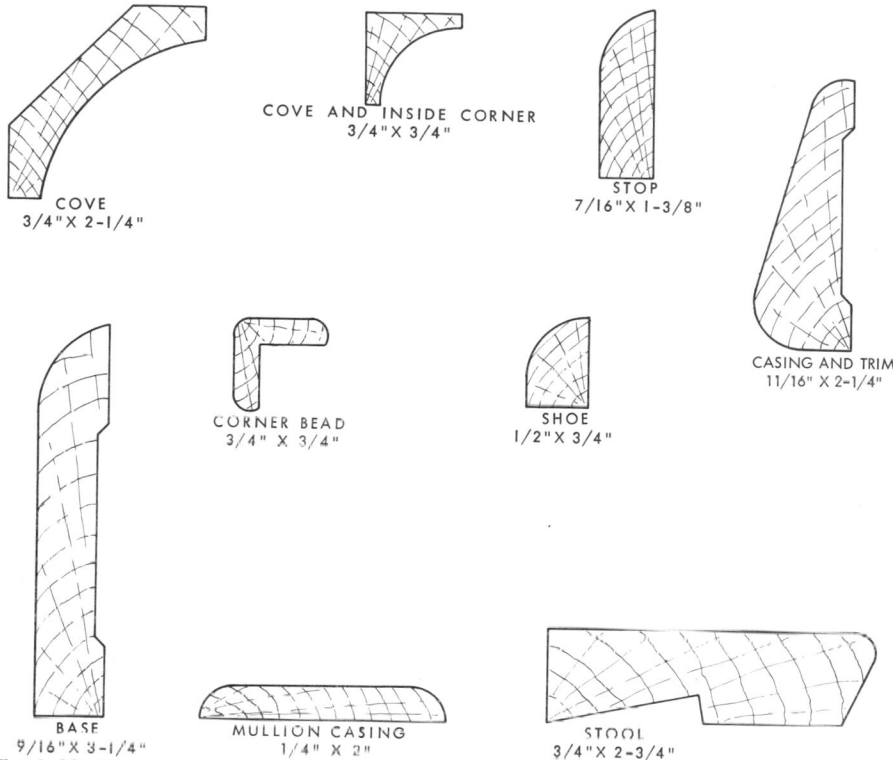

Fig. 8-42. Moldings for general use have been reduced to a few simple shapes. (Georgia Pacific Corp.)

ing, etc., are purchased ready to cut, fit, and install. Few trim details are actually made on the job. Stone trim used around fireplaces, as window and door sills, etc., is purchased in correct sizes and shapes. Shutters are purchased complete and ready to hang.

ESTIMATING MATERIAL

The biggest worry of an estimator is to unintentionally omit a necessary item of lumber when taking off a lumber list. His best safeguard is to use a check list. When he thinks he has all the necessary material listed, he should check the items with a check list like the one shown in Table I.

Roofs. In addition to the roof boards, there are various types of rafters, beams, insulation, shingles, and cornices associated with roof takeoffs.

Roof Area. The area of any roof, regardless of its shape and no matter how it may be divided, is calculated as follows: Determine the exact area of the horizontal projection of the roof, that is, the floor area beneath the roof. Then add for the roof pitch the percentage specified in Table II. To the results thus obtained, add the hori-

TABLE I. LUMBER CHECK LIST.

SIZE	LENGTH	DESCRIPTION	PRICE	AMOUNT
		Outside Plates		
		Outside Studs		
		Floor Joists		
		Ceiling Joists		
		Roof Joists (Rafters)		
		Hip Rafters		
		Ridge Pole		
		Lookouts		
		Cant Strips		
		Inside Plates		
		Inside Studs		
		Lintels (Door and Window)		
		Subfloor and Finish Floor		
		Roof Sheathing		
		Wall Sheathing		
		Siding and Corners		
		Furring Strips		
		Sleepers (Floor Strips)		
		Grounds (Plaster Part. Only)		
		Fascia and Plancier (Soffit)		
		Insulation		
		Asphalt Felt		
		Roof Shingles--Hip and Ridge		
		Bridging, Collar Beams		
		Ribbons		
		Diagonal Braces		
		Stair Material--		
		Stringers		
		Treads		
		Risers		
		Railing		
		Porch Material--		
		Posts and Beams		
		Floor and Joists		
		Steps		
		Railings		
		Nails		
		Rough Hardware		

TABLE II. ROOF AREAS OF PITCHED ROOFS		
Pitch	Increase of Area Over Flat Roof	Multiplication Factor
1/4	12%	1.12
1/3	20%	1.20
3/8	25%	1.25
1/2	42%	1.42
5/8	60%	1.60
3/4	80%	1.80
7/8	101%	2.01

zontal cornice projection. This is a sufficiently accurate method of finding roof areas for most purposes.

If the length of a common rafter is known, the roof area can be accurately determined by multiplying the rafter length by the total length of the roof over which the rafter spans.

Example 1. Find the area of the roof of a 1/3 pitch hip-roof building, 30' by 30'.

Solution: Find the area of the horizontal projection of the roof and add 20% (or multiply by 1.20). Refer to Table II.

$$A = 30' \times 30' = 900 \text{ sq. ft.}$$
$$900 \times 1.20 = 1080$$
$$A = 1080 \text{ sq. ft.}$$

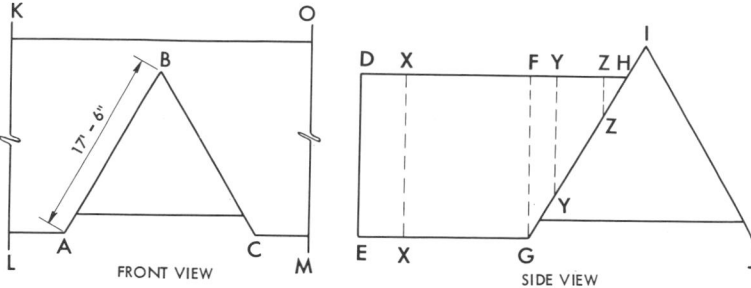

Fig. 8-43. A two-section sloped roof.

(Note: This includes all dormers but excludes cornice projections.)

Roof Boards. To find the required number of roof boards, first find the area of the roof, and then add 20%. Deduct dormer openings and other similar openings. Include all roof *surfaces* such as main roof, dormer, etc. No deductions for small openings such as chimneys need be considered.

If roof boards are to be laid 2" apart, figure the exact area of the surface to be boarded. Do not add anything for waste.

Roof Rafter Lengths. The required lengths of roof rafters are determined from the working drawings by either scaling or calculating from the given dimensions. Roof rafters 12'-0" long must be ordered for a roof which requires 10'-9" rafters because rafters, like other dimension lumber, come in lengths which are multiples of two.

Fig. 8-43 shows a front view of roof section *ABC* intersecting roof section *KOML*. In the side view, section *ABC* becomes *DHGE*. To determine the rafter lengths required for roof section *ABC* you must scale the true length of the longest rafters, which are the *com-*

mon rafters. These rafters are used in the area *DFGE* in the side view. The true length of rafters such as *XX* cannot be scaled from the side view. You must use the front view and scale the line *AB* or *BC*. The rafters in the area *DFGE* are the same length as *BC*. *Example 2.* Find the length of the common rafters in roof section *DFGE* in Fig. 8-43.

Solution: The lengths are equal to *BC*. Since the length of *BC* is the same as *AB* and the length of *AB* is 17'-6", the rafters in section *DFGE* are 17'-6". The closest available length is 18'-0".

For the roof area *FHG* in Fig. 8-43 the same 18'-0" lengths are ordered even though the rafters gradually become shorter. The amount sawed off an 18'-0" rafter to make rafter *YY* leaves enough for rafter *ZZ* so that there is little waste.

Fig. 8-44 shows a common gable roof. This is the type of roof which is often used on a rectangular-shaped house. To find the length of rafter *XX* (side view), you measure its length on the front view, because this is the only place where the true length can be measured. Thus if either *AB* or *BC*

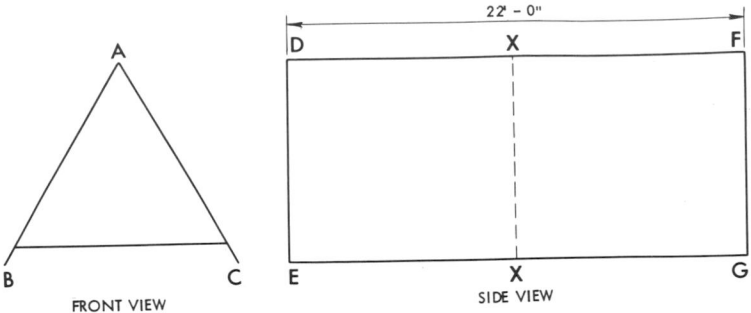

Fig. 8-44. A common sloped roof.

(front view) measures 15'-0" you know the length of rafter XX is 15'-0". (You would order 16'-0" lengths.)

Fig. 8-45 shows another common roof type. The top view shows the ridges J and K. Suppose that you desire to find the length of rafter XX. To do so you must find the length of MQ on the front view. MQ is the same length as MR and is found to be 16'-10". Therefore, 18'-0" rafters will be ordered for XX and all common rafters in the areas $HFWZ$ and $ZWGI$. For areas FWE and EWG you would

also order 18'-0" lengths which will be cut into shorter lengths.

To determine the length of rafter YY in Fig. 8-45 you will need an end view of roof section $DABC$. Since the working drawings usually show all four outside wall dimensions, all true lengths can be measured.

To obtain the line length of a rafter when you know the unit rise per foot and the run of the roof (which is ½ the width of the building) you will find the framing table on the framing square very useful. Fig. 8-46 shows a

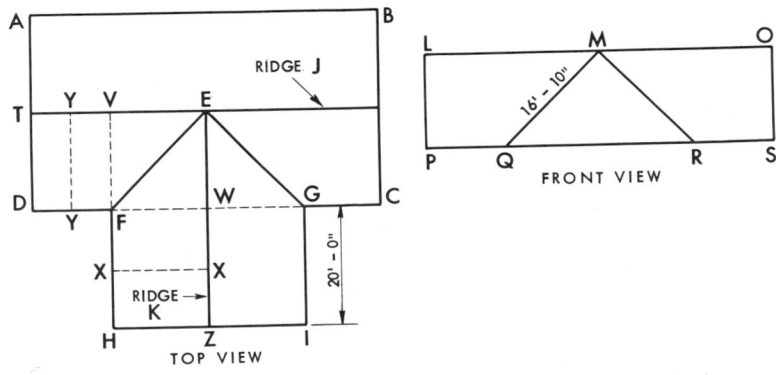

Fig. 8-45. Determining rafter lengths from roof drawings.

LENGTH OF MAIN RAFTERS PER FOOT OF RUN	21.63	18.44	17.69
LENGTH OF HIP OR VALLEY RAFTERS PER FOOT OF RUN	24.74	22.00	21.38
DIFFERENCE IN LENGTH OF JACKS - 16 INCHES ON CENTERS	28.84	24.585	23.588
DIFFERENCES IN LENGTH OF JACKS - 2 FEET ON CENTERS	43.27	36.38	35.38
SIDE CUT OF JACKS		6-11/16	7-13/16	8-1/8
SIDE CUT OF HIP OR VALLEY		8-1/4	9-3/8	9-5/8

Fig. 8-46. Unit-length rafter table on face of framing square.

line of numbers which reads *Length of Main Rafters per Foot of Run*. Fig. 8-47 shows how this is used to find the unit length for a roof with a slope of 8 inches rise per foot of run. You look under the 8-inch mark to find the number you want. If the run were 5'-0" you multiply 14.42 by 5 to get 72.10 inches, which is equal to 6'-0⅛" for the length of the common rafter.

Valley Rafters. Another common problem is finding the lengths of valley rafters such as *FE* and *EG* in Fig. 8-45. Valley rafters are longer than common rafters. There is no place where the true lengths of these rafters can be scaled quickly.

Fig. 8-48 illustrates a method for finding the true lengths of the valley rafters. The top view (plan) is drawn directly above a front view (elevation). Using a radius equal to *AB,* draw the arc *AC.* (The center of the arc is at *B.)* From point *C* drop line *CE* to the elevation view. Then connect *E* to *G.*

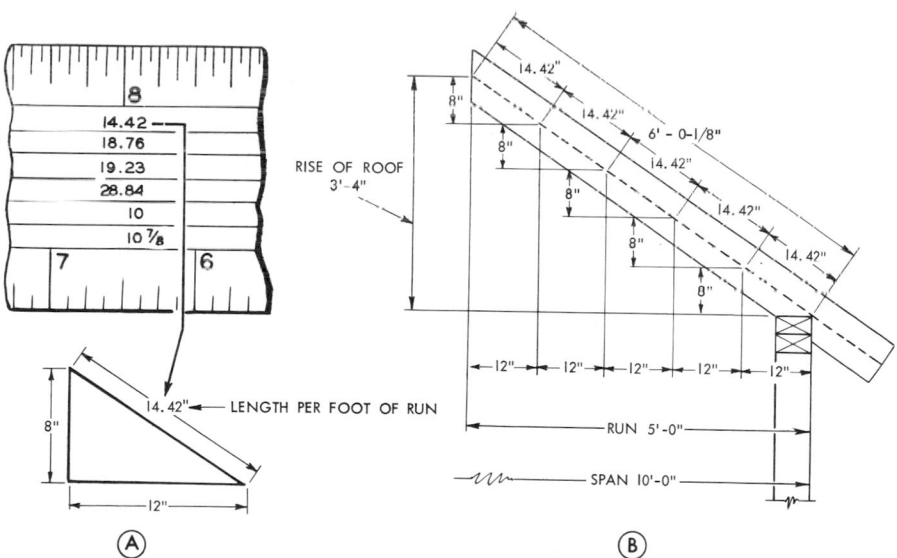

Fig. 8-47. Unit length of common rafter for 8-inch unit rise.

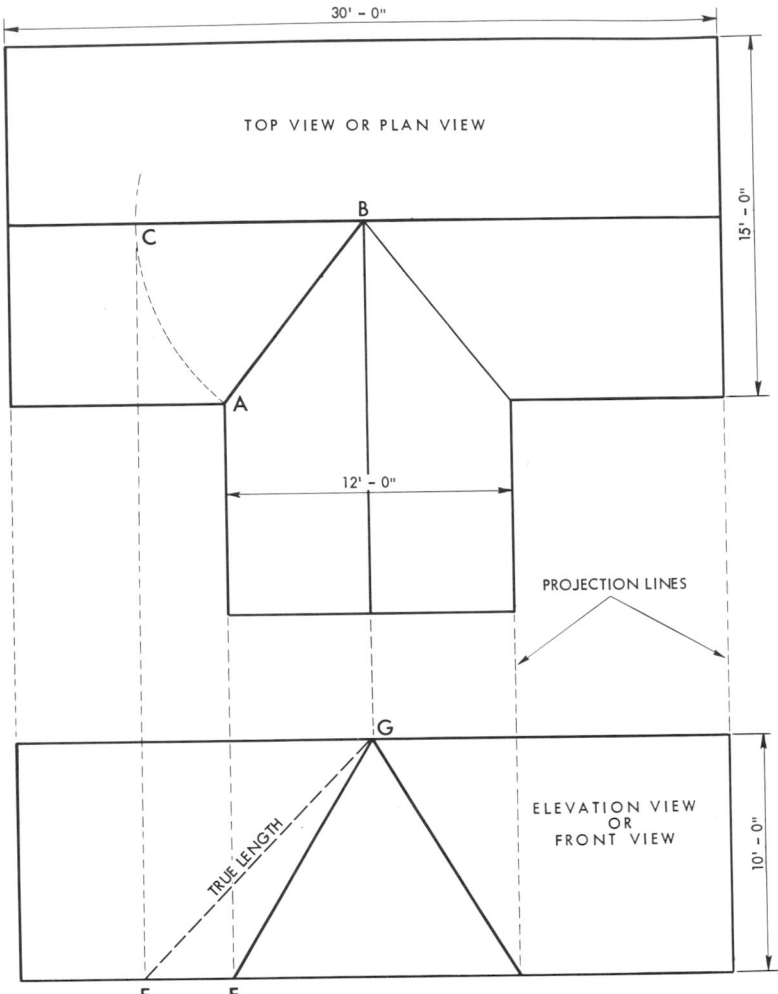

Fig. 8-48. A method for finding true lengths of valley rafters.

The line *EG* is the true length of the valley rafter. The difference between lines *FG* and *EG* is the difference between the true and apparent lengths.

Fig. 8-49 illustrates a roof which has two roof sections of unequal widths but identical pitch. Since the two roofs have the same pitch, the val-

ley rafter *CF* intersects the ridge of the main roof section at the point *F,* which is apparently just as far from the point *A* as is the point *C*. In other words, distance *AF* is equal to the distance *AC* when measured on a horizontal plane. The distance *AC,* however, is the *run* or horizontal projection of the

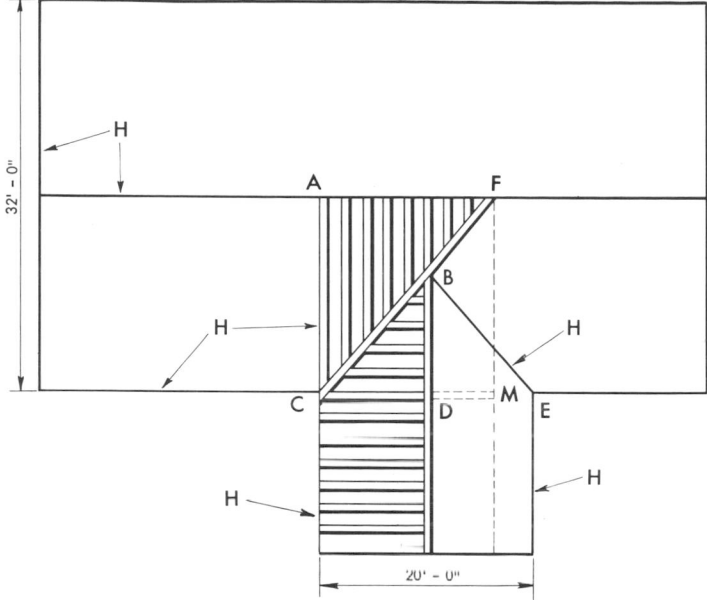

Fig. 8-49. Two roof sections with the same pitch.

common rafter in the main roof. Since the main roof is 32' wide, the distance *AC* is 16' and the distance *AF* is also 16'. If the two roof sections had been of identical width, the dotted line *FM* would have been the ridge of the smaller roof section; but since this roof is only 20' wide, the line *BD* is the ridge of the narrower roof and intersects the valley rafter at *B*. The valley rafter is continued up to the main ridge at *F* and is supported there by the main ridge, thus giving support to the inner end, *B*, of the secondary ridge.

The point *F* is the same distance measured vertically above the level of the top of the wall plate as is the point *A*, and this distance is the *rise* of the common rafter *AC*. Now suppose that the slope of the main roof is 10" to the foot (For every foot in the run of rafter *AC*, this rafter and the main roof surface rise 10".) Since the span of the main roof is 32' and the half-span is 16', the total rise is 16 × 10", or 160". This is also the rise of the valley rafter *CF*.

The two roof sections are of the same pitch so a run of 16 × 12" along *AC* (from *C* to *A*) corresponds to a run of 16 × 12" along *AF* (from *A* to *F*) which also corresponds to a total run of 16 × ? along the valley rafter from *C* to *F*. Since the two distances *AC* and *AF* are equal (measured horizontally), it appears that the unit run of *CF* is equal to the length of the hypotenuse of a right triangle, the other two sides each being 12".

$$12^2 + 12^2 = CF^2$$

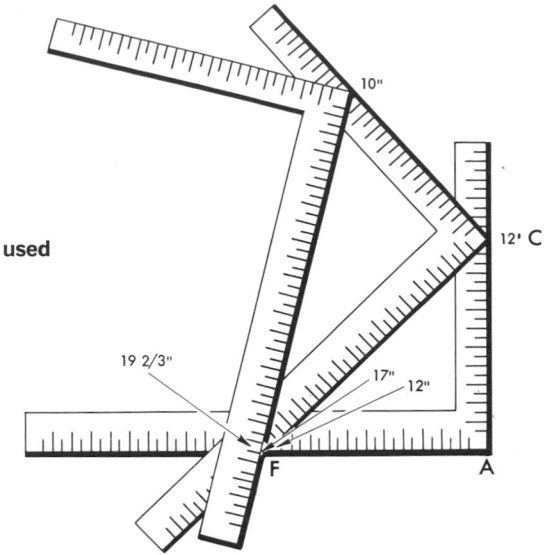

Fig. 8-50. A framing square can be used to find valley rafter lengths.

$$CF = \sqrt{288} = 17''$$

Thus the question mark in the equation for the total run $16 \times ?$ is replaced by $17''$ and $16 \times 17''$, or $272''$ is the total run of the valley rafter.

Knowing the rise $(160'')$ and the run of the valley rafter enables you to calculate its length. The length is the hypotenuse of the right triangle with legs equal to $160''$ and $272''$.

$$\sqrt{(160)^2 + (272)^2} = 315.5'', \text{ or}$$

$26'\text{-}3\frac{1}{2}''$

Using the framing square, Fig. 8-50, a quick method of finding the lengths of valley rafters is as follows: Find the $12''$ mark on the tongue and $12''$ mark on the blade; with a rule or another framing square measure the distance between these marks as shown in Fig. 8-50. The distance is almost exactly $17''$. This is the unit run.

When setting the framing square at $10''$ (for the unit rise) and $17''$ (for the unit run) and measuring between the two points, a distance of $19\frac{2}{3}''$ is found. This is the unit line length on the rafter. There are 16 units of run on the common rafter, thus

$$16 \times 19\frac{2}{3} = 315\frac{1}{2}'' \text{ or } 26'\text{-}3\frac{1}{2}''$$

This is the actual length.

The quickest way to find the length of a valley or hip rafter is to use a table which is stamped on the side of the framing square which gives the *Length of Hip or Valley per Foot of Run* of the common rafter. The number for this case is 19.70.

$$19.70 \times 16 = 315\frac{1}{2}'' \text{ or } 26'\text{-}3\frac{1}{2}''.$$

The number 16 is the number of one foot units of run on the common rafter on this roof.

Fig. 8-51 shows an example of figuring the line length of a hip rafter with an 8-inch rise per foot of run for the common rafter and a $5'\text{-}0''$ total length for the common rafter.

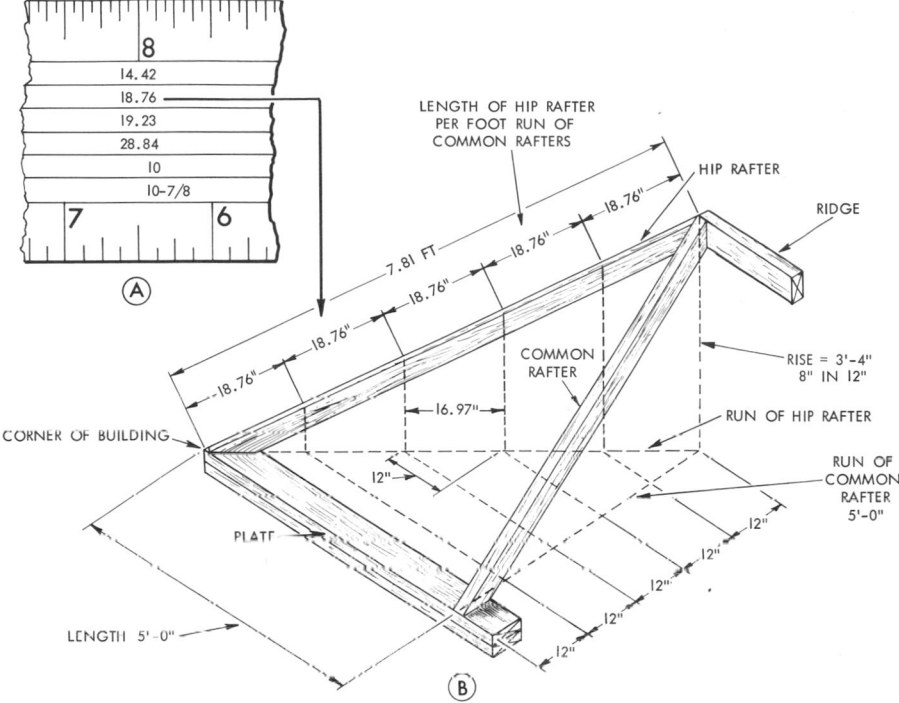

Fig. 8-51. Unit length of hip or valley rafter for 8-inch unit rise.

Quantity of Rafters. For simple roofs, such as Fig. 8-44, the process of finding the quantity of rafters is simple. If the rafters are spaced 16″ o.c., as they generally are, you *divide the roof length by 1.33′ (or 16″) and add 1—or multiply by 0.75′ and add 1.* For rafters with 12″ spacing allow one rafter per foot (or 12″) and add 1. *Example 3.* Find the number of rafters needed for the 22′-0″ roof length in Fig. 8-44. The spacing is 16″ o.c.

Solution:

$$\frac{22.0'}{1.33'} = 16.5 \quad \text{(call it 17)}$$

$$17 + 1 = 18$$

There are two sides to the roof in Fig. 8-44, so 2 × 18, or 36 rafters are required.

For roofs such as Fig. 8-45 the process is more laborious but just as simple to figure. For example, the area *HZEF* is composed of common rafters between lines *HZ* and *FW*. The triangular section *FWE* is composed of *jack rafters.* For the area between *HZ* and *FW* the rule for 16″ o.c. rafters can be used.

$$\frac{20.0'}{1.33'} = 15$$

$$15 + 1 = 16 \text{ rafters}$$

For the *triangular* area *FWE,* an-

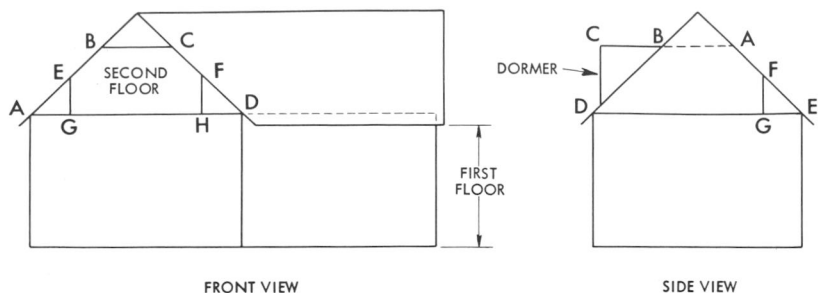

FRONT VIEW   SIDE VIEW

Fig. 8-52. Sometimes the roof forms part of the second floor ceiling.

other rule is used: For 16″ o.c. spacing, multiply the length by 0.5 and add 1. For 12″ o.c. spacing, multiply the length by 0.75 and add 1.

In a similar manner all other roof areas in Fig. 8-45 can be calculated for rafters. All dormers and other areas requiring rafters are computed in the same manner.

Extra rafters must be provided for valleys and hips; these usually are counted by studying the roof plans or elevations.

A *collar beam* is often required between rafters to give added strength particularly for snow loads. See line *BC* in Fig. 8-52. The length can be found by measuring across the end of the roof elevation at the height required above the floor.

Roof Insulation. The common roof type, Fig. 8-44, doesn't require special calculations when applying insulation because the area is easily determined.

For roofs such as that shown in Fig. 8-45 the area calculations are more difficult because of the triangular surfaces such as *EWG* and *VFE*. Divide the roof surface into easily calculable shapes (see Chapter 4) and the calculation of areas will be simplified.

Sometimes roofs are encountered where the second-floor rooms have broken ceilings, Fig. 8-52, left. Here the roof forms part of the ceiling for the second floor. The lines *EB* and *CF* are parts of the roof. If the blueprints include a section view through the room, the length of lines *GE, EB, BC, CF,* and *FH* can be measured. Insulation would be placed to line these spaces. Multiplying the lengths at each part by the length of the building would give the area of insulation required.

Fig. 8-52, right, shows a typical dormer. Here the ceiling of the room becomes *CA,* and *CD* and *FG* are side walls. Areas are calculated by scaling each part.

In full two-story houses the attic floor is sometimes insulated instead of the roof. It is then only a matter of calculating the attic floor area.

Roofing Material. In order to estimate the proper quantity of roofing material to be used, you will have to: first, determine the pitch of the roof; second, find the required exposure to be used on the roof; third, calculate the number of shingles or shakes required. In some areas, due to the weather conditions, up to three layers of shingles may be needed.

Shingles are usually bought by the *square* or in *bundles.* There are four bundles in one square (100 square feet).

For shingles on a roof slope of 4″ to 12″ and steeper, the standard exposures are 5″ for a 16″ shingle, 5½″ for an 18″ shingle and 7½″ for a 24″ shingle.

For *hand split shakes,* generally a 7½″ exposure is recommended for 18″ shakes, 10″ exposure for 24″ shakes and 13″ exposure for 32″ shakes. The minimum roof slope for handsplit shakes is 4″ rise per foot of run.

To estimate the amount of roofing shingles for the following roof pitches, add the appropriate percentage to the total roof area to be covered.

4″ in 12″ add 5% to area
5″ in 12″ add 8½% to area
6″ in 12″ add 12% to area
8″ in 12″ add 20% to area

Once the amount of roofing material has been determined, you must then make allowances for doubling courses at eaves and starter courses. One square of hand split shakes will provide about 120 lineal feet of starter course. Allow about one square of cedar shingles and two squares of hand split shakes for every 100 lineal feet of valley.

Nails. At standard exposure, it takes approximately 2½ lbs of nails per square of both cedar shingles and hand split shakes.

Labor. The cost of labor must then be determined by the estimator. The approximate application of cedar shingles and hand split shakes takes a carpenter approximately one hour to install one square.

For slate, tile, and composition shingles first calculate the area, then add 20% for slate, 18% for tile, and 5% for composition shingles.

Cornices. For cornices, you multiply the total linear footage by the width of frieze, fascia, and plancier (bottom board).

Example 4. A house has 60′ of cornice, and the plancier width is 12″, the frieze 8″, and the fascia 4″. Find the board measure necessary for this cornice.

Solution:

$$\text{Soffit or Plancier} \quad \frac{12 \times 1 \times 60}{12} = \frac{60 \text{ bd. ft.}}{1 \times 12}$$

$$\text{Frieze} \quad \frac{8 \times 1 \times 60}{12} = \frac{40 \text{ bd. ft.}}{1 \times 8}$$

$$\text{Fascia} \quad \frac{4 \times 1 \times 60}{12} = \frac{20 \text{ bd. ft.}}{1 \times 4}$$

Doors. These are designated *exterior* or *interior* because they are usually two distinct types. The exterior doors are usually thicker. Doors are listed in a *Door Schedule* (below):

TABLE III. EXTERIOR DOOR SCHEDULE

Size	Description	Amount
3′ - 0″ x 6′ - 8″ x 1 3/4″	Flush Solid Core	1
2′ - 8″ x 6′ - 8″ x 1 3/4″	3 Lite 2 Panel	1

Screen and storm doors are generally factory made and are furnished complete with hardware.

Door Casings. These are bought by sides which include two side casings and one head casing for each side of each doorway. The material list in the door schedule or quantity take-off should show the sizes of doors and sides.

Door Stops. Stops are bought by the set which consists of two side pieces and one head piece.

Windows. If the windows are not listed in a *Window Schedule,* you must list them as in the following:

TABLE IV. WINDOW SCHEDULE

Size	Description	Amount
3' - 0" x 4' - 0"	6 over 1 DH	1
2 (3' - 0" x 4' - 0")	gang of 2 DH 6 over 1	1
2 (3' - 0" x 4' - 0") + (4' - 0" x 4' - 0")	gang of 3 DH 6 over 1	1

Windows ordered in this manner are usually delivered complete with sash, frame, trim, casing, mullions, aprons, and stools.

Packaged windows are ordered by type and catalog number and are designated by light (glass) sizes.

Chapter 13 explains in detail how glass is estimated.

Screens and Storm Windows. Screens are usually either a part of or sold to accompany "packaged" windows. Metal combination screen and storm windows are made by specialty contractors to fit the openings. Wood screens and storm windows are made in the shop and fitted by the carpenter on the job. If wood screens are to be made on the job the following procedure is followed:

Screens are usually made just half the full window size; full-sized screens are not very common.

The screening material may be copper, galvanized metal, aluminum, etc. The screening is purchased in standard widths and in rolls. For odd-sized windows there is usually some waste.

Screen frames are generally $1'' \times 2''$ stock. This is ordered in linear feet. The screen molding goes completely around the screen.

Screen moldings are joined in several ways, but for estimating the necessary amount of 1×2, the width and height are just doubled for regular half window screens.

Example 5. Two windows with identical dimensions need screens. What is the amount of framing needed if the windows are each 3'-0" long and require screens which are each 3'-6" high?

Solution: First find the perimeter of one screen, then double this figure to obtain the total length of material for the frame.

$$(2 \times 3.0') + (2 \times 3.5') =$$
13.0' for each screen

$$2 \times 13.0' = 27.0' \text{ needed for}$$
both screens

Although the required molding is slightly less than the framing (because it is nailed to the inside edge of the frame), most contractors assume that it is the same length as the framing.

Storm windows are generally mill made. No quantity take-off is necessary except for the windows them-selves—each counting as a unit.

Storm sash can either be bought as a stock item or made at the mill.

Sash Balances. These are used to hold up double-hung windows when opened. They are made part of the frame and need not be included as an item to be estimated. Operating devices for casement, awning and hopper windows are also part of the window and frame assembly.

When extensive remodeling work is done in old buildings, the estimator may have to figure on the work of replacing sash cords or modifying the windows to use modern sash balance devices.

Window Trim. Trim is usually bought in sets consisting of stools, aprons, casings, and stops. Linings are required as part of the set when the walls exceed 8″ thickness.

Stairs. Finished stairs are made in the mill or stair shops where all parts including stringers, risers, treads, railings, etc., are included. The stair is shipped to the construction site in parts. A main stairway is listed as "one flight, main stairs."

Basement and Attic Stairs. Basement stairs are not always shown in detail on the plans; the estimator must therefore do some preliminary calculations before doing a quantity take-off. Fig. 8-53 shows how a typical stairway is planned. Without the necessary details and elevations, the estimator (or contractor) must calculate or measure the *rise* and the *run* of the stairs. Once these dimensions are known, the length of the stringers is

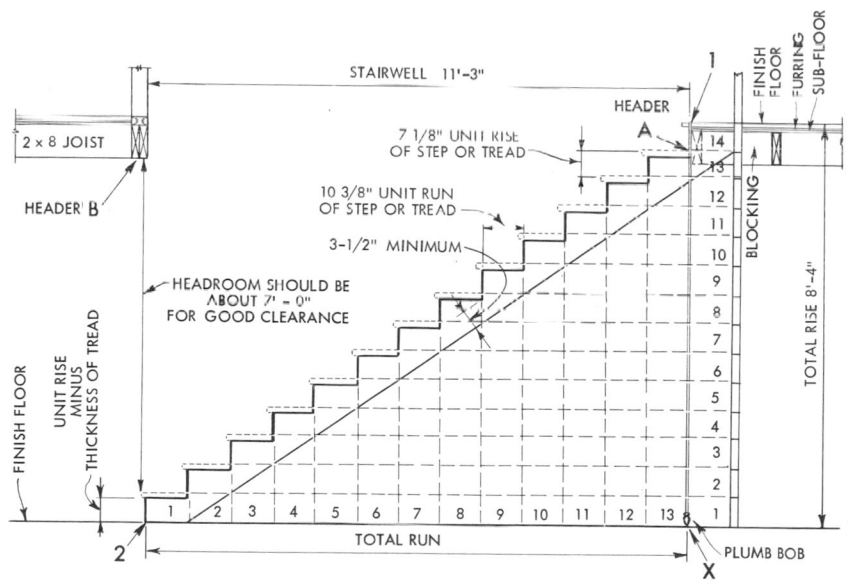

Fig. 8-53. How stairways are planned.

found by the Right Triangle Law: the total run of the stairs and the total rise would be the two right angle sides of the triangle with the hypotenuse the length of the stringer.

A 2 × 10 or 2 × 12 board is used for a stringer, keeping in mind that such lumber comes in lengths which are multiples of two. Except on very wide stairs, two or three stringers are normally required.

To find the tread and riser height and the number of risers, the following procedure is used:

If a 7½″ riser is required, divide the total floor to floor distance by 7½″; if the answer doesn't come out even, the height of each riser is increased or decreased by small amounts until it does come out even. The treads are always one less in number than the risers. The tread width is found by dividing the number of treads into the run of the stairway. (Sometimes there is only a limited distance available. Using that distance as a limit, the above process is used.)

When the total run of the stairs is not limited by the conditions of the building, the tread width is found by using a ratio established by the local building code. One such ratio is T + R = 17 to 18 inches.

Example: If the riser is 7¼ inches the tread is 9¾ or 10¾ or some measurement in between.

Attic stairs are calculated in the same manner as the above.

The material for treads is usually 2″ × 10″ stock but may be thinner. One-inch stock is ample for risers. Knowing the width of the stairs, the total length of tread and riser material is determined and lengths of 2″ × 10″ and 1″ stock are ordered, thus minimizing waste.

Stair Railings. Stock for railings is figured in linear feet.

Railings for carpenter-built stairs may be composed of 2 × 4 inch railing stock which is ploughed to receive the pickets or balusters. These are usually 1 × 2 or 1 × 3 members. All railing parts are ordered in long lengths.

Grounds and Backing. Both grounds and backing are measured in linear feet.

Grounds are guides for the plasterer to help him maintain plaster thickness and also to give the carpenter something to nail into when he applies trim.

Backing is used when structural members do not occur in places where needed to support lath or drywall.

Grounds are used around door openings as a plaster guide and nailing strip. The average door requires about 35′ of grounds.

The estimator or contractor must determine the places where backing is required or where grounds are necessary, and scale or calculate the linear footage. This is determined by a careful study of the plans and all the details. Door openings are easily calculated and ground locations can be scaled.

Special Details and Millwork. These details include case and cabinet work, shelf material, and small items which go into closets and storage areas.

Case and Cabinet Work. Ordinarily, cases and cabinets are mill made. The architect's designs are therefore sent to the mill. The cases or cabinets are delivered from the mill to the job ready to be installed. They are therefore listed as complete units in the quantity take-off.

If cases or cabinets are made by the contractor, the amounts and sizes of all materials must be calculated. This includes material for general enclosure, doors, drawers, glass, etc.

Shelf Material. Shelves are usually scaled to determine the required widths and lengths. The material is purchased in standard lengths and widths. Brackets for shelves are bought by the dozen. Wood strips, such as 1″ × 2″, are bought in standard lengths. Special mill made strips can be purchased for shelf supports.

Insulation Board for Storage Spaces. This item is found by determining the total area of the spaces or surfaces requiring insulation and then adding 6% for wastage.

Closets. For cedar closets the area must be found and then 20% added for wastage.

The actual lengths of clothes poles are determined from the drawings. They are usually the width of the closet. Two brackets are required for each pole. The material is ordered in linear feet.

Hook strips are estimated by taking the length of the closet plus two times the width or depth of the closet.

Example 7. What length of hook strips is needed for a closet whose dimensions are 4′-6″ long and 2′-0″ deep?

Solution:
$$4.5' + (2 \times 2.0') = 8.5',$$
or approximately 9′

(The answer is rounded to the nearest highest foot to allow for wastage.)

Shutters. These are ordinarily mill made. Mills stock standard sizes or will make odd sizes. In any case, the shutters are listed in the quantity take-off as units.

Trim. Trim is ordered in linear feet. The necessary amounts of each kind or size are determined by scaling the drawings or adding the given dimensions. When scaling drawings, care should be taken to scale only true lengths. For example, the trim for the cornice of the east and west ends of the main roof should be scaled using the east and west elevations.

Base Molding. This is figured as the perimeter of the room (in linear feet). Where many doors or a large opening exists, some deduction should be made. The perimeter of all rooms is added and the amount of base thus determined.

Quantity Take-Off—House Plan A. Figs. 8-54 through 8-58 are a quantity take-off for House Plan A, which may be found at the end of Chapter 2. The items contained in these tables have been discussed in this and the previous chapter. The door schedule (Fig. 8-57) and the window schedule (Fig. 8-58) are frequently found on the working drawings. (Chapter 13 discusses in more detail the estimating of glass for the windows and doors.)

Sheathing, roofing, insulation, etc., are either on the working drawings or in the specifications.

FRAMING LUMBER

LOCATION		SIZE	PIECES/LENGTH	LIN. FT.	B.F.M.
Sill	(2)	2x6	–	350	350
Floor Beams Front Left		2x10	21/14	294	490
" Front Middle		2x10	5/10	50	84
" Front Right		3x10	13/20	260	650
" Rear Left		2x10	18/16	288	480
" Rear Middle		2x10	3/18	54	90
" Rear Middle		2x10	12/14	168	280
" Rear Right		2x10	5/12	60	100
" Right Middle		2x10	16/12	192	320
Ceiling Joists Front Left		2x8	21/14	294	392
" Front Middle		2x8	5/10	50	67
" Front Right		2x6	15/16	240	240
" Rear Left		2x8	18/16	288	384
" Rear Middle		2x8	3/18	54	72
" Rear Middle		2x8	12/14	168	224
" Rear Right		2x8	5/12	60	80
" Right Middle		2x8	16/12	192	256
" Covered Porch		2x4	9/8	72	48
" Attic		2x6	15/16	240	240
Bow Window Header	(2)	2x8	2/12	24	32
Porch Header	(2)	2x8	2/12	24	32
Porch Header	(2)	2x8	2/8	16	22
Kitchen Header	(2)	2x12	2/18	36	72
Dinette Window Header	(2)	2x8	2/8	16	22
Living Room Arch Header	(2)	2x8	2/8	16	22
Living Room Arch Header	(2)	2x10	2/8	16	22
Box Header		2x10	–	176	294
Bridging		1x3	–	660	165
Rafters House		2x8	60/20	1200	1600
Rafters Shed Dormer		2x6	15/12	180	·180
Rafters Front Gable		2x6	26/14	364	364
Ridge at Gable		2x8	–	24	32
House Ridge		2x10	–	48	80
Collar Beams		2x4	16/12	192	128
Rafter Brackets		2x4	–	170	114
Post		4x4	1/8	8	11
Studs		2x4	650/8	5200	3467
Plates		2x4	–	1650	1100

NOTE: Under "Pieces" the first number denotes the number of pieces and the second number length of pieces, 21/14 is interpeted as 21 pieces, 14 feet long.

Fig. 8-54. House Plan A—Lumber take-off.

EXTERIOR TRIM			INTERIOR TRIM		
Location	Description	Amount	Location	Description	Amount
Fascia	1x8	250 lin. ft.	Base	1/2 X 3 1/4	650 lin. ft.
Frieze	1x6	130 lin. ft.	Clothes pole	1 1/2" diam.	45 lin. ft.
Dove cote	Screened plywood......	1 unit	Pole sockets	Wood	10 pair
Gable Peak	Texture 1-11 plywood ..	1 unit	Cleats	3/4 X 1 1/2	115 lin. ft.
Shutters	Fixed wood	8 pieces	Hook strip	1x4	85 lin. ft.
Soffit	Plywood	200 sq. ft.	Shelving	1x12	200 lin. ft.

Fig. 8-55. House Plan A—Exterior and interior trim.

SHEATHING, FLOORING, SIDING, INSULATION, ETC.

LOCATION	DESCRIPTION	ACTUAL	ACTUAL + 10%
Sub floor	Plywood	2217	2439 sq. ft.
Finish floor	Oak + 33 1/3%	1932	2576 "
Roof Sheathing	1x6 or plywood	2216	2438 "
Roofing	235# asphalt	2216	2438 "
Side wall sheathing	Plywood	2253	2479 "
Siding	Flitch Pine siding	431	475 "
Wall paper	15# felt	2253	2479 "
Floor paper	15# felt	1932	2126 "
Roofing paper	15# felt	2216	2438 "
Wall insulation	2" batts	1200	1320 "
Ceiling insulation	4" batts	2100	2310 "
Kitchen underlayment	Plywood	174	192 "
Kitchen	Linoleum	174	192 "
Areaways	Cor. metal	5	5

NOTE: The waste factor is determined by the type of material used and the method of installation.

Fig. 8-56. House Plan A—Additional material.

EXTERIOR DOORS

Size	Description	Amount
3'0" x 6'8" x 1 3/4"	9 lite	
	1 panel	1
2'8" x 6'8" x 1 3/4"	" "	2
	TOTAL	3
8'0" x 7'0" x 1 3/4"	GARAGE	4

NOTE: All doors are to be ordered from the door schedule complete with doors, door frames, trim, casing, saddles, etc.

INTERIOR DOORS

Size	Description	Amount
2'6" x 6'8" x 1 3/8"	Flush H.C.	6
2'4" x 6'8" x 1 3/8"	Flush H.C.	1
2'0" x 6'8" x 1 3/8"	Flush H.C.	9
1'4" x 6'8" x 1 3/8"	Louvered	2
2'0" x 6'8"	Sliding	5
2'6" x 6'8"	Bi Fold	2 units
3'0" x 6'8"	Bi Fold	1 unit
	TOTAL	26

Fig. 8-57. House Plan A—Door schedule.

Fig. 8-58. House Plan A—Window Schedule.

SIZE	DESCRIPTION	AMOUNT
2-2'8" x 4'2"	Gang of 2 D.H.	1
10'0" x 6'8"	Bow	1
2-2'4" x 4'2"	Gang of 2 D.H.	1
2'8" x 4'2"	D.H.	3
3'0" x 3'2"	D.H.	2
3'0" x 4'2"	D.H.	1
2-3'0" x 5'2"	Gang of 2 D.H.	1
4'0" x 3'0"	Slide	1
2'0" x 3'2"	D.H.	1
2-3'0" x 3'6"	Gang of 2 D.H.	1
2'8" x 1'8"	T.H. Hopper	4
Total		17

Aprons 1/2 X 3 1/4 80 L.F.
Stools 3/4 X 3 1/2 80 L.F.
All windows are to be estimated from the schedule in gangs as specified to include all frames, sash, mullion, trim, casing, etc.

ESTIMATING LABOR

Labor costs are calculated simply by multiplying the total man-hours by the hourly wages.

See Appendix D, pages 475 to 478 and 480 to 483, for Hourly Estimates for:

Framing, Sheathing, Decking: Commercial Store Fronts and Window Construction; Rafters; Stairways; Roof Trusses;

Roofing: Asphalt Roofing; Asbestos Shingles; Wood Shingles; Slate; Clay Tile; Metal Sheets; Corrugated Asbestos Sheets; Lightweight Concrete Slabs; Metal Work; Wood Gutters; Cant Strips;

Windows and Doors: Wood Windows; Metal Windows; Basement Sash; Commercial Projected; Interior Trim; Exterior Trim; Shutters, Storm Sash and Screens; Residence Doors; Glass Sliding Doors; Metal Door Frames; Outside Door Trim; Inside Door Trim; Thresholds; Hanging Doors; Weatherstripping; Louvers and Vents; Access Doors; Metal-Clad Doors;

Millwork and Trim: Wood Interior Trim; Metal Interior Trim; Wood Exterior Trim; Porch Rail; Stairs; Porch Columns; Mantels; Cabinets and Cupboards; Counter Tops; Misc. Factory-Built Cabinets.

Questions and Problems

1. What is the length of molding required for a window screen which is 3'-0" × 3'-6"?

    (Answer: 13.0')

2. Find the area of a ⅜ pitch roof whose attic floor area is 60'-0" × 20'-0". (Disregard area of cornice.)   (Answer: 1,500 sq. ft.)

3. Find the number of rafters needed for the triangular area **FWE** in Fig. 8-45. The spacing is 16" o.c.

    (Answer: 9)

4. A house has 75' of cornice; the plancier width is 12", the frieze is 8", and the fascia is 6". Find the board measure (to the nearest foot) necessary for the cornice.   (Answer: 163' BM)

5. A room is 10' × 13' with a 3'-0" wide door opening and a 6'-0" wide arched opening. Find the amount of base molding needed for the room.

    (Answer: 37')

6. How much ceiling molding is needed for the previous problem?

    (Answer: 46')

7. Calculate the amount of sheathing needed for a gable roof which is 40'-0" long and has common rafters 18'-0" long.

    (Answer: 1,440 sq. ft.)

8. What length of common rafter would you order for a building 20'-0" wide with a 1'-0" overhang? The slope is 5/12. (Remember to order in even lengths.)

(Answer: 14')

9. How many hours would be allowed for the installation, fitting and hanging by an unskilled craftsman of 4 double hung sash window frames 3'-0" × 3'-2"?

(Answer: 2 hours)

9

MASONRY

~~~~~~

The masonry section of the specifications includes brick, concrete block, tile, and other structural clay or concrete products, plus the necessary mortar and anchors. The estimator should keep the complete set of plans together when he is making his estimate of the materials and labor involved in this section of the specifications.

MATERIALS

Bricks and Blocks. The most common materials which the estimator will find in the specifications are brick, concrete block, and tile. Brick is usually one of four kinds: building brick, formerly referred to as *common brick,* firebrick, face brick, and SCR. The kind of concrete block most commonly encountered is the three-hole variety. Bricks and blocks, of course, are available in dozens of varieties and sizes.

Building brick is the brick usually used in construction work. It is made of ordinary clays or shales, has no special color or surface texture, and is not specially marked or scored.

Brick sizes are standardized just as lumber is standardized. The size of the standard American building brick is 8″ × 3¾″ × 2¼″. However, because of the somewhat uneven temperature at different places in the usual brick kiln, bricks will vary somewhat in size. In quantity, the differences in size will tend to even out, however, so that for fairly large surfaces this size may be used for a quantity take-off. See Fig. 9-1.

Modular brick is usually 7½″ × 2⅙″ × 3½″. The size is based upon the module (usually a 4″ unit). The two dimensions, which are factors in the width of a building, will work out to modular units when the width of the mortar is included.

Firebrick is used in fireplaces, furnaces, and other places subjected to high temperatures. It is made of a special kind of clay which resists crumbling and cracking at high temperature.

The usual dimensions for firebrick (with mortar) are 8″ × 3½″ × 2¼″

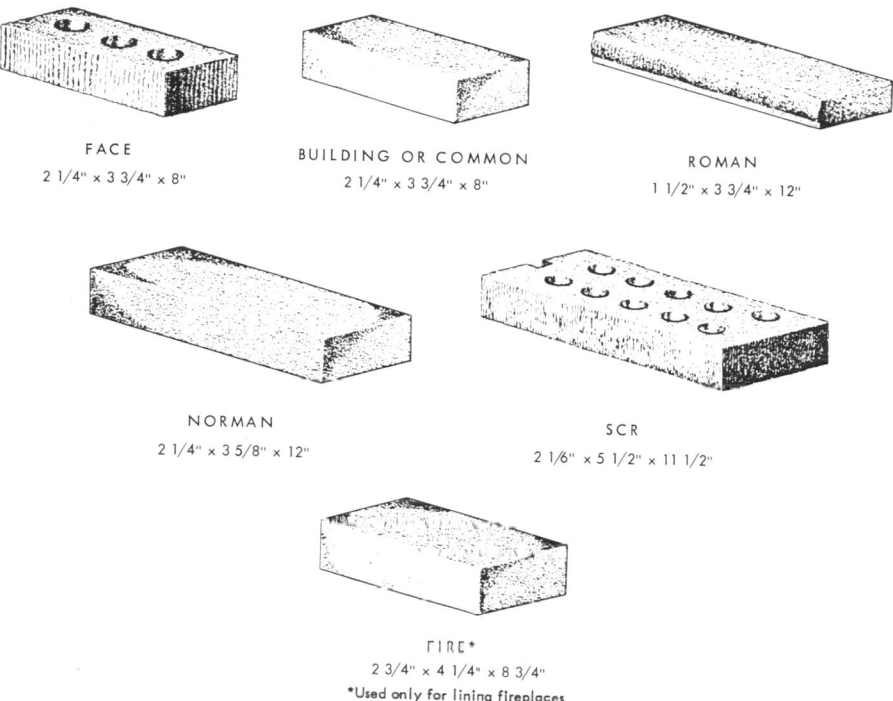

FACE
2 1/4" x 3 3/4" x 8"

BUILDING OR COMMON
2 1/4" x 3 3/4" x 8"

ROMAN
1 1/2" x 3 3/4" x 12"

NORMAN
2 1/4" x 3 5/8" x 12"

SCR
2 1/6" x 5 1/2" x 11 1/2"

FIRE*
2 3/4" x 4 1/4" x 8 3/4"
*Used only for lining fireplaces

Fig. 9-1. Various kinds of bricks. Sizes may vary in different localities.

or $9'' \times 4\frac{1}{2}'' \times 2\frac{1}{2}''$ but there are innumerable sizes and shapes available.

Face brick is manufactured under special conditions. The temperature and materials are closely controlled, so that the color and the texture of the brick can be controlled. Hardness, uniform size, and strength are all of a high quality. Face brick is available in the same sizes as building brick. See Fig. 9-1.

SCR brick is characterized by a greater width than other brick. It is $5\frac{1}{2}''$ wide. This width allows it to be used in single course construction where the added width gives good stability. See Fig. 9-1.

Glass brick, often called glass block, is a hollow block made of glass. It has the advantage of being translucent, that is, it admits light but is not transparent. There is also a clear glass block, but the objects seen through are slightly distorted. Glass blocks are made in different colors, sizes and patterns. Glass brick is never used as a load-bearing wall, but it is often used for decorative effects.

Concrete block, as its name suggests, is a type of block made of concrete. The block is used in much the same manner as brick and, when it is to be used in a building, will be specified in both the drawings and specifications. The standard wall block is

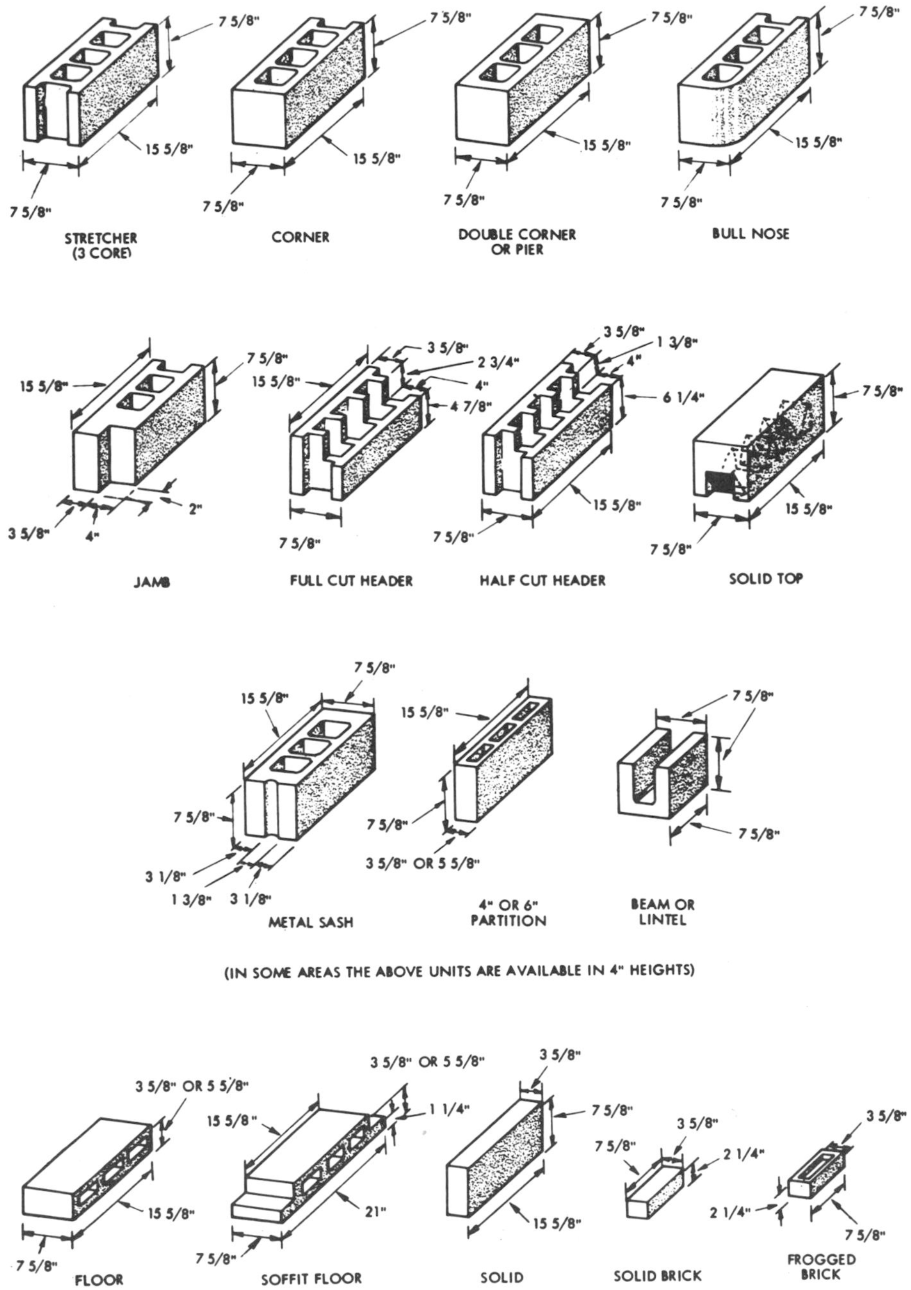

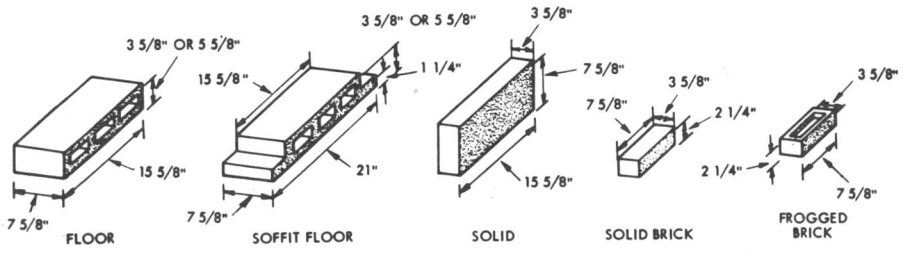

Fig. 9-2. The actual sizes are shown here. A 7⅝″ x 7⅝″ x 15⅝″ block is nominally 8″ x 8″ x 16″. (National Concrete Masonry Association)

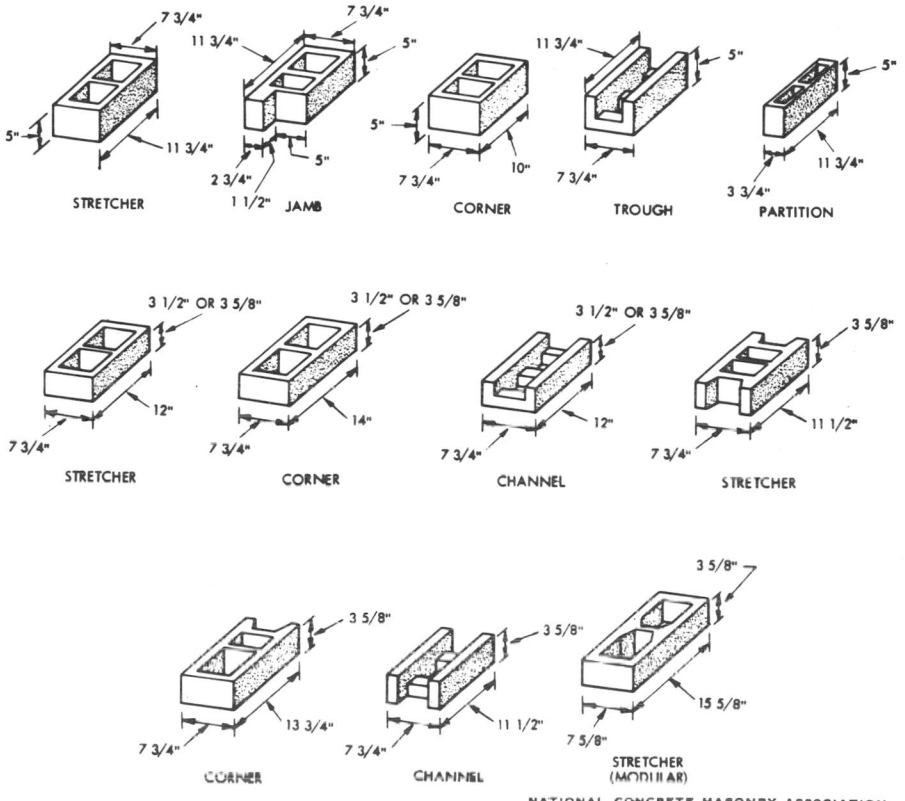

Fig. 9-2. Actual sizes of concrete block (cont.).

available with actual face dimensions of 15⅝″ × 7⅜″ and a variety of widths (7⅝″ is common). Nominal dimensions of 16″ and 8″ are used for estimating purposes. See Fig. 9-2.

Cinder block, a lightweight concrete block, is also in common use.

Terra Cotta. This material is a mixture of high quality clays with calcined clay. It is classified as architectural terra cotta and ceramic veneer. Terra cotta is used for interior and exterior trim and for store fronts. It is available glazed and in colors.

Brick Bonds. The strength of any masonry wall depends to a great extent upon the bond used in erecting the wall. A bond refers to the arrangement of brick or stone in the wall. The arrangements are designed to prevent the vertical joints between the masonry units from being directly above each other. There are many patterns for placing brick which will produce a structurally sound wall. The variation between bonds is brought about by the distribution of stretchers (the length of the brick laid parallel with the face of

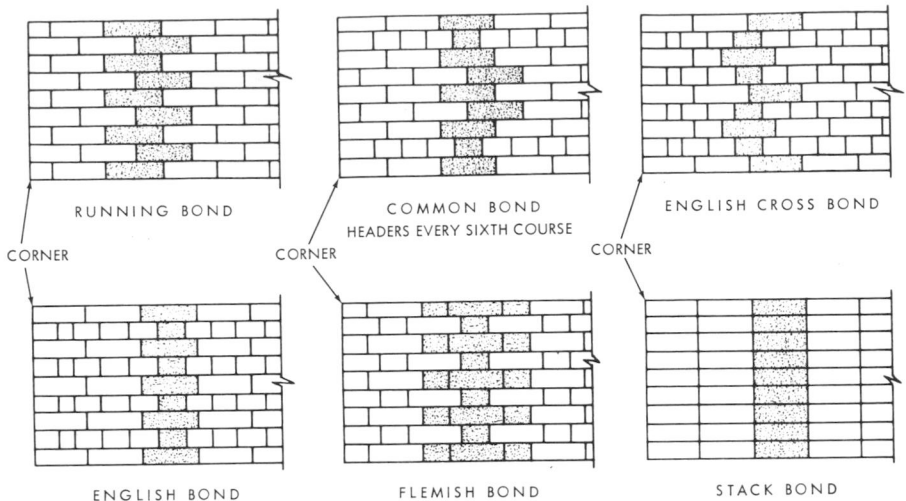

Fig. 9-3. Brick Courses.

the wall) and headers (laid with the length at right angles to the face of the wall) laid in various courses (rows). The different methods of bonding also affect the amount of material which will be used.

The following paragraphs list some of the basic bonds used in brick work.

Running or Stretcher Bond (Fig. 9-3 top, left). This bond uses stretcher courses with the joints breaking at the center of each brick immediately above and below. Face, building, Roman, or SCR brick is used for this bond.

Common Bond (Fig. 9-3, top, center). The common bond, or American bond as it is sometimes called, is a variation of the running bond, with a header course every 5th, 6th, or 7th course. This ties the wall to the backing masonry material. The header courses are centered on each other. Face or building brick is usually used in the common bond.

English Cross or Dutch Bond (Fig. 9-3 top, right). This bond uses alternate header and stretcher courses. Joints of the stretchers center on the stretchers two courses above and below; headers center on headers. This bond is usually building or face brick.

English Bond (Fig. 9-3 bottom, left). Alternate courses of headers and stretchers are laid so that the joints between stretchers are centered on the headers. Stretchers are centered on stretchers; headers on headers. Face or building brick is usually used for the English bond.

Flemish Bond (Fig. 9-3 bottom, center). Alternate headers and stretchers are in each course. Headers in one course are centered above and below stretchers in the other course. Face or building brick is used for this bond.

Stack Bond (Fig. 9-3 bottom, right). All courses are stretchers and all joints are in line. This is used primarily for aesthetic purposes — it has relatively little structural value. The most effective brick for this type of bond is Roman.

Many of the more ornamental bonds have been excluded from this discussion—they are seldom used because of the cost.

Masonry Walls. Masonry walls are porous and after a driving rain or period of severe cold, moisture may condense on the inside of the wall. To prevent this, the inner wall covering (covered with lath and plaster or other finish) is separated from the back-up masonry by furring or furring strips as in Fig. 9-4. This air space will stop any moisture transfer. Furring is adjusted to compensate for irregularities in the masonry wall and provides a nailing base for the wall covering. Furring may be either light steel channels or 1″ × 2″, 1″ × 3″, or 2″ × 2″ wood strips. These are positioned vertically and nailed to the inside face of the masonry unit. The spacing of the furring strips is determined by the type of interior wall covering. For example, if sheet rock is used, furring strips should be placed 16″ or 24″ o.c. to maintain the module of the standard size sheet (4′ × 8′). Placing the furring strips at a greater distance o.c. would cause too much flexibility when pressure was applied. The furring serves as a base for the interior finish: lath and plaster, drywall, gypsum board, paneling, plywood, etc.

Brick Walls. With mortar the standard brick width is roughly 4″. Therefore, brick walls are normally constructed in multiples of 4″: that is, in widths of 4″, 8″, 12″, and 16″. A residential building less than 35′ high normally uses an 8″ wall. A 12″ wall is recommended if there are high winds or earthquakes. Usually the outside layer of brick is backed up by an inside layer with a lesser grade of brick. Sometimes the outside brick wall is backed up by concrete blocks or hollow tiles. A brick veneer (one brick thick) may also be built over wood framing.

Concrete Block Walls. The most widely used size of concrete block is 7⅝″ × 7⅝″ × 15⅝″. If these are laid in a single wall thickness, they

Fig. 9-4. A brick veneer is often used over concrete block. Note the furring strips for the interior finish.

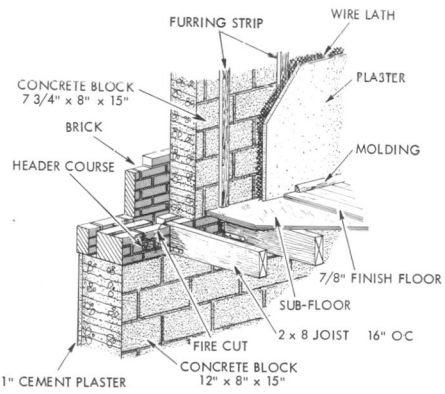

FURRING STRIP
WIRE LATH
CONCRETE BLOCK 7 3/4″ x 8″ x 15″
PLASTER
BRICK
HEADER COURSE
MOLDING
7/8″ FINISH FLOOR
SUB-FLOOR
2 x 8 JOIST   16″ OC
FIRE CUT
1″ CEMENT PLASTER
CONCRETE BLOCK 12″ x 8″ x 15″

Fig. 9-5. An attractive interior partition has been produced with a combination of stretchers and core blocks. (Portland Cement Association)

will produce roughly an 8″-thick wall. Blocks are also available in 4″, 10″, and 12″ widths, as well as other shapes, as shown in Fig. 9-2. The names given to each block are indicative of their use in construction.

The face of a concrete block wall will take on a new depth when laid with textured blocks (sometimes called shadow blocks). These may be randomly placed with plain blocks or used together to develop a pattern. Textured block may be used on both interior and exterior walls. A combination of blocks may produce an attractive interior wall as shown in Fig. 9-5.

SCR Brick Walls. In recent years the SCR (Structural Clay Products Research) brick wall has been introduced into the masonry construction field. This wall was designed primarily to compete with the frame wall. It has met with success in many areas. Fig. 9-6 shows a typical SCR brick wall. The wall is one unit or brick thick. A jamb slot or notched cut is made into one end of every brick to accommodate metal or wood windows. It is necessary with the SCR brick wall to use furring strips to provide a cavity for the installation of wiring and insulation. Attachment of the 2″ × 2″ furring strip is made by a patented clip which fits into the masonry joint.

Brick Veneer Walls. Brick veneer is commonly considered as a skin of brick over a frame house. See Fig. 9-7.

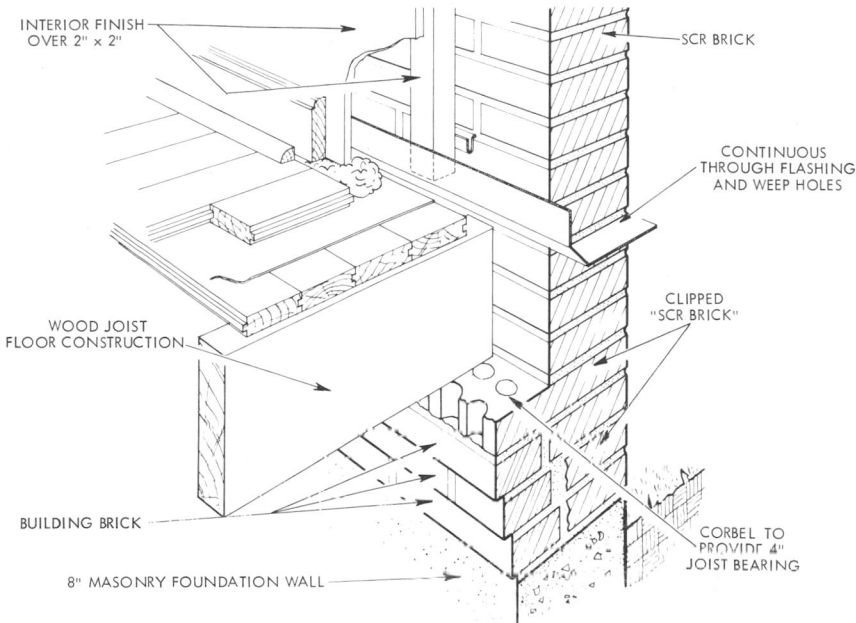

INTERIOR FINISH OVER 2" x 2"

SCR BRICK

CONTINUOUS THROUGH FLASHING AND WEEP HOLES

CLIPPED "SCR BRICK"

WOOD JOIST FLOOR CONSTRUCTION

BUILDING BRICK

CORBEL TO PROVIDE 4" JOIST BEARING

8" MASONRY FOUNDATION WALL

Fig. 9-6. SCR brick may be used to make a strong, durable wall with only one thickness of brick. Note the corbel used to provide joist bearing: the through wall flashing, and the furring strips fastened with clips to the wall.

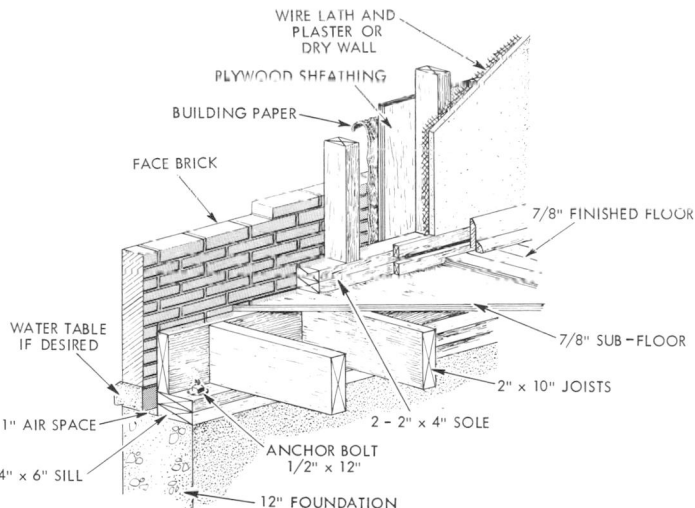

WIRE LATH AND PLASTER OR DRY WALL

PLYWOOD SHEATHING

BUILDING PAPER

FACE BRICK

7/8" FINISHED FLOOR

WATER TABLE IF DESIRED

7/8" SUB-FLOOR

2" x 10" JOISTS

1" AIR SPACE

2 - 2" x 4" SOLE

ANCHOR BOLT 1/2" x 12"

4" x 6" SILL

12" FOUNDATION

Fig. 9-7. Brick veneer construction is popular because it provides a brick exterior with the interior flexibility of frame construction.

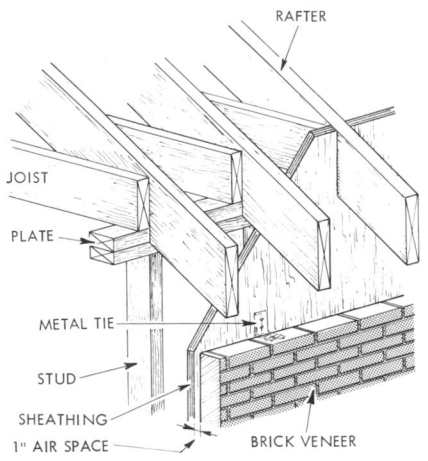

Fig. 9-8. Metal ties are used to hold the masonry wall to the frame super-structure.

Fig. 9-9. Split stone may be used to form an attractive wall. Stones must be laid so that no vertical cracks can develop.

The foundation wall must be wide enough to accommodate a course of brick and also the sill for the frame wall. Allowance must be made for ¾" sheathing on the frame wall and 1" air space. Balloon frame construction lends itself to this type of wall because there is a minimum of vertical shrinkage. Corrugated galvanized metal strips or wires are used to tie the brick wall to the frame construction. One end is nailed to the sheathing and the other end is embedded in the mortar joint. See Fig. 9-8. The ties are placed every fifth course and spaced 2'-0" o.c. horizontally. The masonry veneer part of the wall usually is not load bearing. The roof, floor, and ceiling joists are supported by the frame wall.

Stone Walls. The oldest examples of masonry houses in America are represented by the brick and stone buildings of colonial days. Stones in a wall must overlap so that joints are not directly above each other. The size and shape of the stone used, in addition to the color and texture, will determine how it will be laid in the wall. Today, almost all stone homes are veneered. The wall section will be essentially the same as brick veneer. The only difference is that stone replaces the brick, and the numerous types and cuts of stone produce many patterns. A few of the more common stonework combinations are displayed in Fig. 9-9.

Cavity walls are used to produce a watertight wall with good thermal and sound insulation. These are made up of two 4" walls normally separated by a 2" air space. These are tied together by ¼" metal ties placed in every 5th course and not more than 3' apart horizontally. See Fig. 9-10. Cavity walls should not exceed 25' in height.

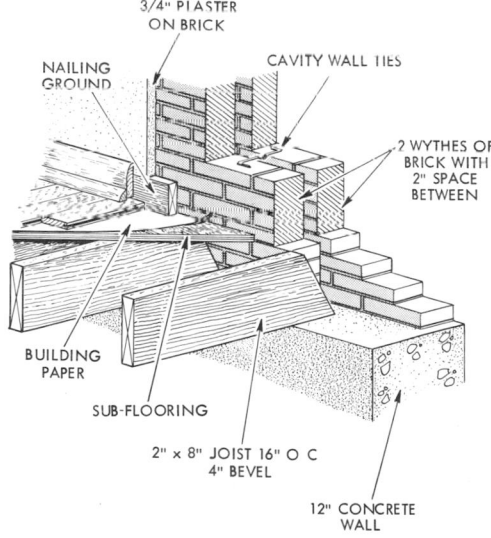

Fig. 9-10. A 10" cavity wall is formed of two 4" walls plus a 2" cavity. Note the metal ties used for support.

BRICK VENEER

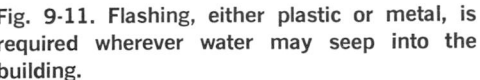

Fig. 9-11. Flashing, either plastic or metal, is required wherever water may seep into the building.

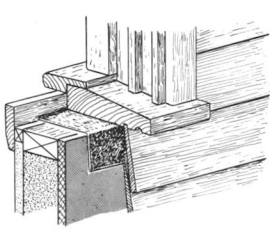

FRAME

Since the cavity wall prevents the penetration of heat and cold, it is important that the air space be kept free of mortar drippings or other obstructions.

A cavity wall with an interior plaster finish would need no furring strips, since the void between the outside brick and the back-up brick serves the same purpose as furring strips. In other words, a dead air space is provided. Plaster, therefore, may be applied directly to the brick work, or the masonry may remain exposed for its aesthetic value. Metal or wire lath is not necessary if plaster is to be used, since the rough surface will readily accept the scratch coat of plaster.

Flashing is placed wherever water may seep into a building, such as under windows and at the base of a wall. It is a continuous piece (usually metal) shaped to prevent moisture from entering the wall. See Fig. 9-11.

Chimneys and Fireplaces. Floor, roof, wall, or partition framing members must be kept free of the masonry chimney or fireplace. Figs. 9-12 and 9-13 show the headers and trimmers around a chimney and fireplace at a minimum distance of 2". This space between the masonry and structural members must be filled with non-combustible insulation. Mineral or glass wool is frequently used for this purpose.

The back hearth, the surface on which the fire is built, is lined with firebrick using fire clay joints. (Fire clay is a mortar-like material used to bond the fire brick.) Also note that the sides and back of the opening are laid up with firebrick and fire clay. Building brick and mortar are not used because they will not withstand heat. See Fig. 9-14 for the various parts of the fireplace.

Mortar is a necessary constituent of any masonry work. It is required for holding the bricks and blocks together. This material must be included in any quantity take-off of masonry work.

Anchors, like the *sleeper clip,* are

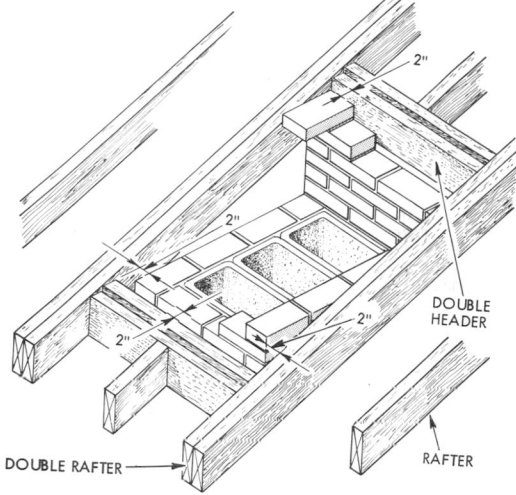

Fig. 9-12. A 2″ clearance must be provided between the chimney and the roof framing members. This space is filled with non-combustible material.

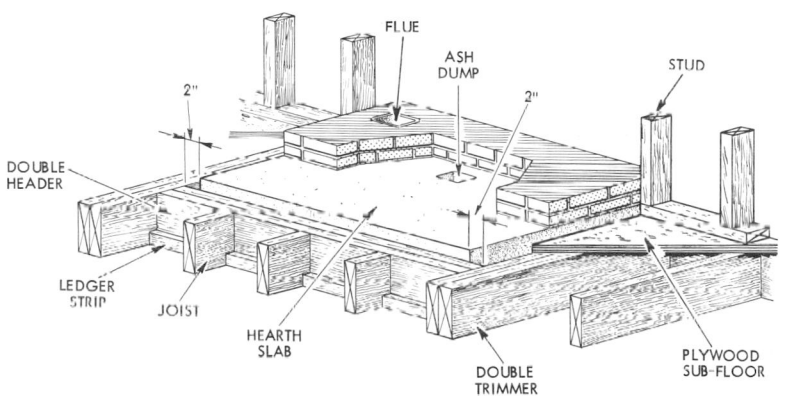

Fig. 9-13. A 2″ clearance must also be provided around the sides of the fireplace.

used to join masonry and concrete—particularly in those buildings which use a brick veneer. Fig. 9-15 shows a variety of anchors used in building construction. The cost of a single anchor of any variety is likely to be small and in the construction of a single residence the cost of anchors would be negligible. In estimating for large com-

mercial structures or for development tracts of hundreds of similar residences, however, this cost could cut severely into a contractor's profit—especially in the case of highly competitive bidding.

Lintels are required over basement doors, and in brick walls over all windows and doors. Lintels also are used

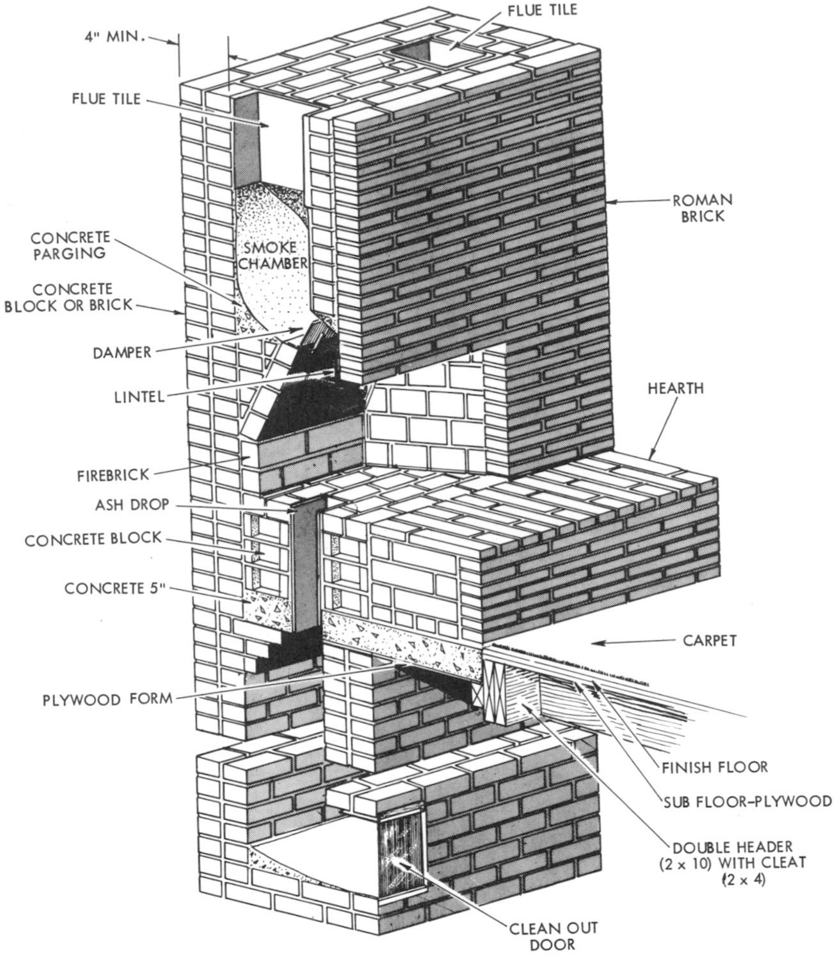

Fig. 9-14. Fireplace details and nomenclature.

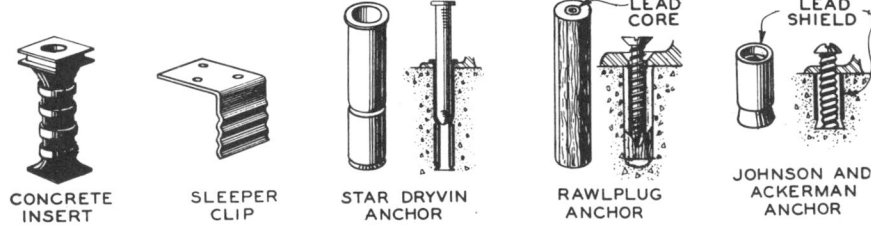

Fig. 9-15. Anchors used in building construction.

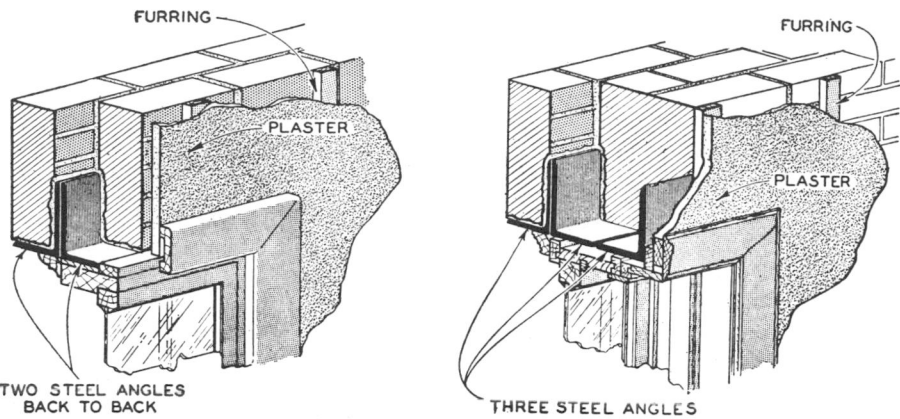

FURRING

PLASTER

FURRING

PLASTER

TWO STEEL ANGLES
BACK TO BACK

THREE STEEL ANGLES

Fig. 9-16. Lintels or steel angles used above windows.

in fireplace construction. See Fig. 9-16. These are made of steel ¼", ⅜" or ½" thick, the cross-section formed in the shape of a right angle and very often referred to as "angles." Their purpose is to support the weight of the masonry above the openings.

For example, a 3'-8" long angle is required above windows in a basement for any 3'-0" wide window opening.

Scaffolding. Many contractors use tubular scaffolding which can be readily assembled on the job. The cost of scaffolding materials is usually computed and spread out over the projected life span of the equipment. In cases where the contractor owns the scaffolding equipment, the estimator usually charges equipment rental to the job; this provides for replacement of the equipment when it becomes necessary to do so. However, scaffolding must sometimes be constructed on the job and, therefore, may be included with the cost of the masonry work.

Scaffolding may also be rented.

Schedules of the cost of equipment rental are compiled yearly by the Associated Equipment Distributors.*

ESTIMATING MATERIALS

When estimating the costs of materials, it is usual to work from the bottom upwards and from the outside in. In other words, the estimator begins by computing the foundation materials, then exterior walls above grade, and finally the interior masonry work. After computing exterior materials, he then computes such interior materials as the fireplace veneer and hearth.

Since brick must be cleaned after it is laid, it is wise to include the cost of cleaning materials and labor, especially for *tuckpointing*. The cost of cleaning materials is relatively small (compared to brick, block, and tile)

*For the cost of a copy of the current schedule, you should write to Associated Equipment Distributors, 615 W. 22nd, Oak Brook, Illinois 60521.

but it should be included in the quantity take-off. Since wood and metal are sometimes necessary for the completion of the masonry work, the cost of these must also be computed. Materials for lintels in the fireplace, dampers, and cleanout doors are usually computed last.

The preliminary steps in masonry work have been discussed in other sections of this book. Excavations, concrete work, footings, and foundations will not, therefore, be discussed here, although many estimators will consider them as part of the masonry cost of a building.

In general, when estimating masonry materials, the contractor should get the *delivered* cost of the materials. When taking a materials bid for a contract, he should assure himself that the quoted price includes delivery. In this way he can avoid the task of computing transportation costs, as well as the possibility of a sizeable error in the estimate which could result if he neglects transportation costs.

Bricks. Masonry materials are usually computed by the square foot. To estimate the number of masonry units in any building, the estimator first computes the area of the structure. Next, he subtracts the area which is not composed of masonry. (In a brick veneer residence, such as the one in HOUSE PLAN A, the areas of doors and windows are subtracted.) These

two steps yield the total area of masonry material. After finding the area, the estimator uses one or more of the appropriate tables to determine the number of masonry units per square foot. The material for the entire job is then computed by multiplying the area in square feet by the number of masonry units per square foot.

Example 1. Calculate the total area of brick required for a 2-car, brick veneer garage. The floor plan dimensions are 25'-0" × 25'-0". The garage has a flat roof; the elevation is 10'-0". There are two floor-to-ceiling doors, each 9'-0" wide, in the front of the garage.

Solution: First compute the area of the exterior walls.

$$25.0' \times 10.0' = 250 \text{ sq. ft.}$$

Since there are 4 such walls,

$$4 \times 250 = 1,000 \text{ sq. ft.}$$

Now, find the areas of the openings.

$$9.0' \times 10.0' = 90.0 \text{ sq. ft.}$$
$$2 \times 90 = 180 \text{ sq. ft.}$$

Deduct the area of the openings.

$$1,000 - 180 = 820 \text{ sq. ft.}$$

Example 2. The total area of brick work in the previous example was 820 sq. ft. Since the brick was unspecified, assume that building brick is to be used. Table I shows the number of building brick (or face brick) per sq. ft. (Since the mortar size is unspecified, assume a ½" joint.) Six and one-eighth bricks are needed for every

TABLE I.   STANDARD SIZE FACE BRICK PER SQUARE FOOT

JOINT	1/8"	1/4"	3/8"	1/2"	5/8"	3/4"	1"
NUMBER OF BRICKS	7 1/2	7	6 1/2	6 1/8	5 3/4	5 1/2	5

TABLE II. PERCENTAGES ADDED FOR VARIOUS BONDS

	PER CENT	
COMMON (header course every 5th course)	20	(1/5)
COMMON (header course every 6th course)	16 2/3	(1/6)
COMMON (header course every 7th course)	14 1/3	(1/7)
ENGLISH OR ENGLISH CROSS (full headers every 6th course)	16 2/3	(1/6)
FLEMISH (full headers every 5th course)	6 2/3	(1/15)
FLEMISH (full headers every 6th course)	5 2/3	(1/18)
TWO STRETCHER GARDEN WALL (full headers every 5th course)	4	(1/25)
THREE STRETCHER GARDEN WALL (full headers every 5th course)	2 7/8	(1/35)
DOUBLE HEADERS (alternating with stretcher every 5th course)	10	(1/10)
DOUBLE HEADERS (alternating with stretcher every 6th course)	8 1/3	(1/12)

square foot of masonry area. Calculate the number of bricks needed for the garage.

Solution: Multiply the area by the number of bricks per unit of area.

820 × 6⅛ = 5,022½ bricks

As Table I gives the quantities for standard size brick laid in running bond, additional allowances must be made for the number of brick when bonds using headers are used. The percentages in Table II are added to the number of required bricks (as calculated by the use of Table I) when the face brick are laid with the bonds indicated.

Table III can be used to estimate the number of face bricks required for laying the more common bonds. With this table the number of face brick for any wall or area may be figured easily. To make such an estimate for face brick, calculate the total area of wall surface and deduct the area of all openings except those containing less than 10 square feet. Then from Table III the number of face brick required can be obtained for half-inch joints. If other sized joints are to be used, corrections can be figured using the data at the foot of the table.

Example 3. Assume a total wall area of 2,546 square feet, with openings amounting to 276 square feet. How many face brick are required for a running bond?

Solution: Find the brick area and use Table III.

2,546 − 276 = 2,270 sq. ft.

2,000 sq. ft. take	12,320 face brick
200 sq. ft. take	1,232 face brick
70 sq. ft. take	432 face brick
2,270 total	13,984 total

If half-inch joints are used, 14,000 face brick will be required. If a ⅝″ joint is to be used, subtract 5% from 14,000 as shown at the foot of Table III.

If the bricks used in Example 3 are not of standard size then, as previously explained, the area of the brick, plus its mortar joint, must be determined and the bricks per square foot figured and multiplied by 2,270 square feet.

Soldier courses make no difference in the count, nor do *rowlock* courses, provided half brick only are used. If full brick are used in a rowlock course, however, once again as many face brick as the course requires must be added, while the same amount is sub-

TABLE III. NUMBER OF STANDARD SIZE FACE BRICK AND BUILDING BRICK IN MASONRY WALLS
Laid with 1/2" Joints in Various Bonds

Sq.Ft. of Wall	RUNNING			COMMON Header Course Every 7th Course			ENGLISH AND ENGLISH CROSS* Full Headers Every 6th Course			FLEMISH Full Headers Every 5th Course			DOUBLE HEADERS Alternating with Stretchers Every 5th Course			Sq.Ft. of Wall
	Face Brick	Building Brick in		Face Brick	Building Brick in		Face Brick	Building Brick in		Face Brick	Building Brick in		Face Brick	Building Brick in		
		8" Wall	12" Wall		8" Wall	12" Wall		8" Wall	12" Wall		8" Wall	12" Wall		8" Wall	12" Wall	
1	6.16	6.16	12.32	7.04	5.28	11.44	7.19	5.13	11.29	6.57	5.75	11.91	6.78	5.54	11.70	1
5	31	31	62	36	27	58	36	26	57	33	29	60	34	28	59	5
10	62	62	124	71	53	115	72	52	113	66	58	120	68	56	117	10
20	124	124	248	141	106	229	144	103	226	132	115	239	136	111	234	20
30	185	185	370	212	159	344	216	154	339	198	173	358	204	167	351	30
40	247	247	494	282	212	458	288	206	452	263	230	477	272	222	468	40
50	308	308	616	352	264	572	360	257	565	329	288	596	339	277	585	50
60	370	370	740	423	317	687	432	308	675	395	345	715	407	333	702	60
70	432	432	864	493	370	801	504	360	791	460	403	834	475	388	819	70
80	493	493	986	564	423	916	576	411	904	526	460	953	543	444	936	80
90	555	555	1110	634	476	1030	648	462	1017	592	518	1072	611	499	1053	90
100	616	616	1232	704	528	1144	719	513	1129	657	575	1191	678	554	1170	100
200	1232	1232	2464	1408	1056	2288	1438	1026	2258	1314	1150	2382	1356	1108	2340	200
300	1848	1848	3696	2110	1584	3432	2157	1539	3387	1971	1725	3573	2034	1662	3510	300
400	2464	2464	4928	2816	2112	4576	2876	2052	4516	2628	2300	4764	2712	2216	4680	400
500	3080	3080	6160	3520	2640	5720	3595	2565	5645	3285	2875	5955	3390	2770	5850	500
600	3696	3696	7392	4224	3168	6864	4314	3078	6774	3942	3450	7146	4068	3324	7020	600
700	4312	4312	8624	4928	3696	8010	5033	3591	7903	4599	4025	8337	4746	3878	8190	700
800	4928	4928	9856	5632	4224	9152	5752	4104	9032	5256	4600	9528	5424	4432	9360	800
900	5544	5544	11088	6336	4752	10296	6471	4617	10161	5913	5175	10719	6102	4986	10530	900
1000	6160	6160	12320	7040	5280	11440	7190	5130	11290	6570	5750	11910	6780	5540	11700	1000
2000	12320	12320	24640	14080	10560	22880	14380	10260	22580	13140	11500	23820	13560	11080	23400	2000

*The quantities in this column also apply to common bond with headers in every sixth course.
For other than 1/2" joints, the following percentages must be added to or subtracted from above results:
Add: for 1/8" joint, 21%; for 1/4" joint, 14%; for 3/8" joint, 7%.
Subtract: for 5/8" joint, 5%; for 3/4" joint, 10%; for 7/8" joint, 15%; for 1" joint, 20%.

tracted from the quantity of common brick. But for window sills, laid rowlock, no special provision need be made, as the usual allowances mentioned in the following paragraph will be sufficient.

If the workmen are careful to use *bats* (a piece of brick with one end whole, the other end broken off) for closures, instead of breaking whole bricks, no wastage need be figured. The area of small openings, not deducted in calculating quantities, and the doubling of brick at the corners, will allow a surplus of extra brick. Since it is customary to order the brick to the quarter-thousand next above the actual number calculated, ordinary small extras, as well as slight wastage, will be amply included.

For all practical purposes, common brick may be considered the same in size as the standard face brick. Any odd-sized building brick can be figured by calculating the area of the brick plus its mortar joint and dividing that area into 144 square inches. This gives the number of bricks per square foot for a 4″ wall. For an 8″ wall multiply by 2, and for a 12″ wall multiply by 3, etc.

Table III can be used to readily calculate the number of common brick, of standard size, required per foot. Because of the nature of the face brick bond used, it is only when there is a running bond that the number of backing brick in the 8″ wall equals the number of face brick. For additional thickness of the backing walls, however, this number should be added for each additional tier of backing in all bonds. This is due to the fact that the bonding face brick extending into the first tier of backing brick takes just so much more face brick and so much less backing brick; additional tiers of backing are not affected. The number for running bond applies to any building brick wall throughout, such as foundations, partitions, etc. For work of this kind, simply multiply this number by the number of brick tiers in the thickness of the wall.

While each tier of building brick takes about 20 less brick per vertical foot all around the wall, this saving may be disregarded in the case of the 8″ wall, as the extra brick will come in handy for fire stops and similar applications. If, however, two or more tiers of backing brick are used, an arithmetical progression by twenties per vertical foot should be deducted for each successive tier. Thus, beginning with the first tier at 20 brick, the series runs 40, 60, 80, etc.

Many estimators use the following simple rules. These rules apply only to building brick.

½″ Joints

1 sq. ft. of 4″ wall has 6 bricks
1 sq. ft. of 8″ wall has 12 bricks
1 sq. ft. of 12″ wall has 18 bricks

¼″ Joints

1 sq. ft. of 4″ wall has 7 bricks
1 sq. ft. of 8″ wall has 13 bricks
1 sq. ft. of 12″ wall has 20 bricks

Small chimneys. Table IV is used to estimate the amount of building brick for a 4″ wall around the flue lining, and at 13 courses to 3′ in height.

Chimneys that have more than one flue lining must have 4″ of brickwork between flues.

Fireplaces. In estimating for fireplaces, figure the portions projecting between the line of the wall, such as breast and ash pit, as solid areas; that is, the number of bricks for the surface multiplied by the number of tiers deep, deducting the number of bricks displaced by all flues and openings, face brick facing, and firebrick lining.

For chimneys above the roof, measure linear distance around the chimney and multiply by height above the roof. If there are 7 face brick per square foot, multiply the area by 7.

TABLE IV. BRICKS PER FOOT OF HEIGHT
FOR SMALL CHIMNEYS

NO. OF FLUES	SIZE OF FLUE INCHES	NO. OF BRICKS PER FT. OF HT.
1	8 x 8	26
2	8 x 8	44
3	8 x 8	63
1	8 x 12	31
2	8 x 12	52
3	8 x 12	74
1	12 x 12	38
2	12 x 12	60
..	......	..

In estimating the brick required for the chimney or fireplace, it is easier to estimate the cubic foot of chimney or fireplace and then deduct for the flue lining and the hearth. In addition to the ordinary brick we also must estimate the square foot of firebrick for the fire box and the fire box floor.

Note: to determine the amount of brick per cubic foot, refer to the Brick Table III and use the figure for a 12" wall. For the firebrick use the figure for a 4" wall.

Trim. In estimating the quantity of face brick used as trim, you measure the linear feet of trim and multiply by the number of brick per linear foot.

Example 4. Find the number of standard size brick, face and building, laid in a sixth course common bond with a ⅜" joint, for an 8" gable wall, 25' wide and 18' high from grade to eaves, and then 12' to ridge pole. The 12" cellar wall is 7½' high, 4½' being below grade level. There are four windows, each requiring an opening of 3'-6" × 5'-2" and one window requiring an opening for 2'-6" × 4'-2". The cellar windows, being less than 10 square feet, are disregarded.

Solution: First find the area in square feet for face brick.

Rectangle of wall 18' × 25' = 450

Gable triangle 12' × 25' = 300

$$300 \div 2 = \frac{450 \text{ sq. ft.}}{150 \text{ sq. ft.}}$$
$$\text{Total area} = \overline{600} \text{ sq. ft.}$$

Now, find and deduct the 5 window openings.

$4 \times 3.5' \times 5.17' = 72.3$ sq. ft.
$1 \times 2.5' \times 4.17' = 10.4$ sq. ft.

Total window openings 82.7 sq. ft.
(call it 83 sq. ft.)

$$\text{Face brick area} = 600 - 83$$
$$= 517 \text{ sq. ft.}$$

Using Table III, you can see that 3,-739 face brick are needed (for ½" joints). Adding 7% for a ⅜" joint,

$3,739 + 262 = 4,001$ face brick
(call it 4,000)

By using Table III for building brick, 2,668 brick are needed for backing.

Concrete Blocks. First find the total linear measurement of the wall in which concrete blocks are used. This includes all four sides of the basement (for most buildings), interior walls, etc. Then divide the total by the length of one block. (It is assumed here that blocks 16" long, 8" wide, and 8" high are used.)

Dividing the total length by 1.33' (16") gives the number of blocks in one course.

Next determine the height of the concrete block wall and then divide it by 0.67' (8"). This gives the number of courses. (Be sure to subtract openings.)

Mortar. Table V shows the estimated quantity of materials for various mortars based on laying 1,000 brick with ½" joints.

Brickwork below grade level usually is laid with ½" joints. For brick or tile above grade, in other than ½" joints, add to or subtract from the items under "4" wall," as a unit, the following quantities:

TABLE V. ESTIMATING QUANTITY OF MATERIALS FOR VARIOUS MORTARS

Quantities Based on Laying 1,000 Brick with ½" Joints

Proportions	4″ Wall			8″ Wall			12″ Wall		
	Cement Sacks	Lime Lump or Hydrated Bbls. Sacks	Sand Cu. Yds.	Cement Sacks	Lime Lump or Hydrated Bbls. Sacks	Sand Cu. Yds.	Cement Sacks	Lime Lump or Hydrated Bbls. Sacks	Sand Cu. Yds.
Cement Mortars									
Cement 1—Sand 2....	3.90	... .	.43	5.00	... .	.55	5.40	....	.60
Cement 1—Sand 2½..	3.30	... .	.43	4.20	... .	.55	4.40	....	.60
Cement 1—Sand 3....	2.90	... .	.43	3.70	... .	.55	4.00	....	.60
Lime Mortars									
Lime 1—Sand 2......	....	.75 3.00	.43	....	.96 3.80	.55	....	1.00 4.20	.60
Lime 1—Sand 2½....	....	.65 2.60	.43	....	.82 3.30	.55	....	.90 3.60	.60
Lime 1—Sand 3......	....	.54 2.20	.43	....	.69 2.70	.55	....	.70 3.00	.60
Cement-Lime Mortar									
Cement 1⎫—Sand 6... Lime 1⎭	1.75	.43 1.75	.43	2.20	.55 2.20	.55	2.40	.60 2.40	.60

Note: The quantities given under the heading 4″ wall are for all brickwork, whether backed with brick or tile, or as veneer, and for each tier of backing brick above grade and other work in which the joint between the tiers of brick is not filled. The quantities given under the heading 8″ wall and 12″ wall are for foundation work below grade or other places where the joints between the tiers of brick are filled with mortar.

For ⅛″ joints.... subtract ¾
For ¼″ joints.... subtract ½
For ⅜″ joints.... subtract ¼
For ⅝″ joints.... add ¼
For ¾″ joints.... add ½
For 1″ joints.... double

One cubic yard of sand, 6 sacks of cement, and 10% lime, well mixed, will lay about 1,000 concrete blocks of the 16″ × 8″ × 8″ size. The 10% lime is determined by first taking 10% of the cement volume.

ESTIMATING LABOR

Since there are many different kinds of bricks, blocks, and tiles, in addition to working conditions which vary greatly among regions, it is best to refer to tables and charts for your own area. The following is an estimate of the time required for the laying of 100 square feet of face brick. Remember that these figures can vary considerably in different regions.

Bricklayer	8 man-hours
Laborer	6 man-hours
Foreman	¾ man-hours
Misc. labor	¾ man-hours

Multiply each of these figures by the hourly wage rate (in your area) and then add the various labor costs. This will give you the total labor cost for 100 square feet of wall (about 7,000 brick).

For cinder blocks, one bricklayer with a laborer will lay almost 300 blocks (16″ × 8″ × 8″) in 8 hours. To this must be added about 1.5 hours of supervision and miscellaneous work.

ESTIMATING LABOR

See Appendix D, pages 472 to 474, for Hourly Estimates for:
Masonry: Masonry; Glazed Masonry Work;
Coping, Lintels, Beams, Columns: Wall Coping; Sills—Lintels;
Foundation Block; Concrete Masonry Beams and Lintels; Concrete
Masonry Pilasters;
Chimneys, Fireplaces: Chimneys; Fireplaces.

Questions and Problems

1. List the various kinds of brick used in the building trades.

2. How is scaffolding estimated?

3. How is brick trim estimated?

4. Estimate the wall area of the brickwork needed for a brick veneer house as shown in Fig. 9-17.
   (Answer: 1,051 sq. ft.)

5. How many bricks are needed for the walls of the house in Fig. 9-17?
   (Answer: 6,438)

6. What is the volume to be bricked in the chimney in Fig. 9-17?
   (Answer: 264 cu. ft.)

7. Estimate the area of face brick needed for a building 100'-0" × 100'-0", whose height is 20'-0". Window and door outs amount to 2,000 sq. ft.
   (Answer: 6,000 sq. ft.)

8. What is the area of backup brick needed for the building in the above problem? (The area of the spandrel beams is 1,000 sq. ft.)
   (Answer: 5,000 sq. ft.)

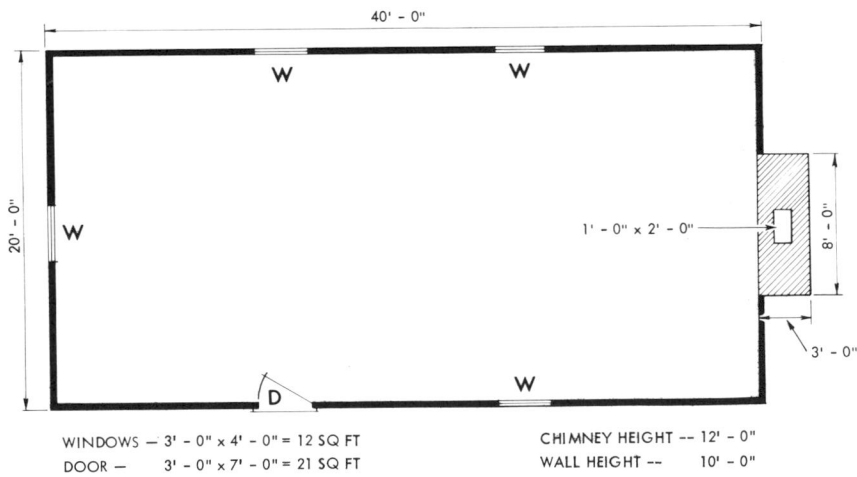

WINDOWS — 3' - 0" x 4' - 0" = 12 SQ FT
DOOR — 3' - 0" x 7' - 0" = 21 SQ FT
CHIMNEY HEIGHT -- 12' - 0"
WALL HEIGHT -- 10' - 0"

Fig. 9-17. House plan used for masonry calculations.

10

WET AND DRY WALLS

Wall coverings are divided into two major divisions—wet wall and drywall. Plaster and stucco are examples of wet wall coverings; plasterboard and wood paneling are types of drywall in common use.

It is impossible to fully explain all of the various wall coverings in a single chapter but this chapter will give you sufficient information to begin estimating this phase of construction.

WET WALL MATERIALS

Plaster. Plaster has two basic functions in a building. First, it provides a surface for finishing walls and partitions to make them acceptable from the standpoint of livability, and second, it serves as a means of sealing out dirt, cold, and heat.

Because of its smooth surface, plaster gives interior areas a pleasing appearance. Moreover, plaster offers a surface upon which paint, paper, canvas, and other decorative finishes may be applied.

Plastering is divided into two main areas of work: *exterior* and *interior*. The basic difference in these two areas is in the types of materials used. For exterior work the materials must be water resistant and structurally strong. The common materials used to bind together the aggregates are portland cement, lime and plastics.

For interior work almost any type of material can be used because the work will be protected from the elements. The most common binding materials are gypsum, lime, and portland cement. Gypsum is not water resistant and should not be used where high moisture conditions exist.

Plaster is applied in coats or layers to the desired thickness; the total thickness may vary from ⅛" for veneer plastering to 2" thick for solid plaster partitions. Various types of plastering materials are also used to provide sound control, insulation, fireproofing and ornamentation.

One of the prime advantages plaster has over most of the other interior and exterior finishes is its versatility. It can be formed into monolithic surfaces of any shape or size. There are almost unlimited supplies of the basic raw materials used in plastering. Gypsum, limestone, clay, sand, vermiculite and perlite are available worldwide with deposits located close to all markets.

The three main cementing materials used in plaster mortars are gypsum, lime and portland cement. Gypsum and lime have been used since ancient times. Portland cement is a relatively recent discovery. Gypsum and lime are made from natural rocks which are processed into several forms to produce the various kinds of cementing materials. Portland cement is a manufactured material that contains lime, gypsum, and other substances.

Gypsum Plasters. These plasters are often called plaster of Paris. Gypsum, a soft rock, is 20 percent water by weight or 50 percent by volume. Through pulverizing and calcining (heating or burning to make a powder) about 75 percent of the water is removed. Various materials are added and blended, and the mixture is then bagged. When water is added again, the reverse reaction occurs: the plaster

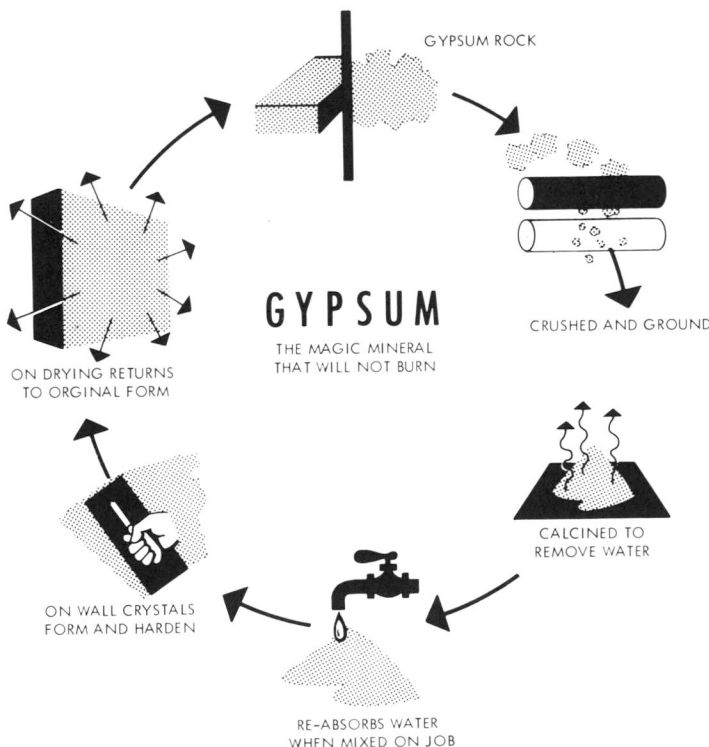

GYPSUM ROCK

CRUSHED AND GROUND

GYPSUM
THE MAGIC MINERAL
THAT WILL NOT BURN

CALCINED TO
REMOVE WATER

ON DRYING RETURNS
TO ORGINAL FORM

ON WALL CRYSTALS
FORM AND HARDEN

RE-ABSORBS WATER
WHEN MIXED ON JOB

Fig. 10-1. Cycle of gypsum. (United States Gypsum Co.)

takes up water and sets into the rock-like condition of its original state. See Fig. 10-1.

Lime. Lime for plastering is obtained by quarrying limestone. The lime is subjected to a process similar to that used for gypsum. The product of calcining lime is *quicklime,* a highly caustic substance that reacts violently with water. To avoid explosions when water is re-added, manufacturers slowly hydrate or *slake* the lime under controlled conditions. The lime is then pulverized and bagged. To use the hydrated lime, further soaking is required.

Portland cement. Since its invention, portland cement has replaced almost all other cements, both natural and artificial. This is due to low cost and superiority. Some form of lime is mixed with certain kinds of clay. This is ground and then calcined at temperatures near 2,700°F. The ingredients combine and form clinkers, which are then pulverized. A small amount of gypsum is added to control the setting properties.

Water starts a complex reaction, yielding crystalline substances in an amorphous gel which sets as a hard mass. While setting takes place in a short time, the cement requires several days for the hydration to become complete. As a result, cement continues to increase in hardness for about a month and in some cases for years.

Interior Plastering. When to apply two coats and when to apply three coats of plaster? This is a difficult question to answer, since the use of two or three coats of plaster depends upon many different considerations. The various surfaces upon which the plaster is to be applied sometimes dictate the number of coats required. For example, plaster applied over masonry seldom requires more than two coats. Metal lath generally requires three-coat treatment. Resilient lathing systems also require three coats of plaster.

Even upon wood laths, three coats make a better job of plastering than two. Extra strength and body are obtained by the addition of the extra coat, provided time is allowed for each of the coats to thoroughly dry out before the next coat is applied. On the less expensive residential work, however, two coats are customarily applied.

The best interior plaster work is worked to a final thickness of about ⅞". Of the three coatings, the first coat is the thickest, so that, when dry, it may be strong enough to resist the pressure of working the coat or coats to follow. A large part of the advantage of three-coat plastering is obtained by *thoroughly drying out each coat before applying the next,* thus securing the added density and strength made possible by forcing the next coating.

While it is generally the custom to add rough plaster finish on the second coat, in inexpensive work (such as summer residences), a very rustic effect is obtained by rough-working the surface of the first coat. Using but one coat, however, it is not possible to work the surface as true and as evenly as when two coats are applied.

Base coat plastering consists of the *scratch coat* and the *brown coat*. The third coat, which is applied over the other two is called the *finish coat*. The result is three coats of plaster called three-coat work. Two-coat work consists of one base coat and, of course, the finish coat. The finish coat is always the last coat.

The Scratch Coat. The reason for using a scratch coat in three-coat work is that it stiffens the lath. It forms either a mechanical key or an adhesive bond, depending upon the lathing used. In other words, the scratch coat (together with the lathing) is the foundation for the structure of the plaster wall. The scratch coat also provides uniform suction for the brown coat that follows.

After the scratch coat is applied, and before it sets, the surface is raked or scratched. The surface is usually scratched in two directions so that there is a cross-hatched effect of roughened surface. This allows for an adequate mechanical bond with the brown or second coat.

The Brown Coat. Browning is the term applied by the plasterer to a coat of mortar applied for the purpose of building up and straightening the surface. Its thickness is subject to existing conditions. Usually, however, the brown coat is from ½″ to ¾″ thick.

A brown coat is made poorer than a scratch coat. One reason for this is that a rich mortar is hard to rod because of its extreme stickiness. Another reason is that the brown coat is most often applied over surfaces that have fairly high suction power.

When the brown coat is applied as a first coat, it is applied over bases with some suction power in them. When the brown coat is applied as a second coat, it is applied over the scratch coat that has previously been applied and that provides the necessary suction.

Two-Coat Work. Most plaster work now consists of only two coats. The brown mortar used for the first coat should be made of fresh lime—used as soon as it is stiff enough to be worked. The first coat of mortar must always be put on with sufficient pressure to force the plaster through between the laths and thus ensure a good clinch. The face of this coat must be made as true and even as possible on angles and plumb on vertical surfaces. After the first coat is sufficiently set, it may be worked again, using a float that consists of a piece of hard pine about the size of the trowel. Sometimes the face of this float is covered with felt or other material to produce a rough texture on the plaster. The first coat should run a strong one-half inch in thickness, measured from the outer surface of the laths, and should be thoroughly dried out.

In general, it is inadvisable to trowel a two-coat job smooth. If the attempt is made to float the first coat when it is too thin or is insufficiently set, the tool is likely to leave marks on the wall and the plaster itself is liable to crack. If the plaster becomes too dry, it may be dampened by sprinkling water on it with the plasterer's broad brush. This is followed immediately with the float.

The finish second coat in two-coat

work is the same as the final skim coat in three-coat work.

The Finish Coat. The finish, skim, or white coat should never be applied until the earlier coats are thoroughly dry and hard. A simple putty coat should carry more sand than when the finish is hardened by the addition of plaster of Paris. If plaster of Paris is used, the mortar should always be *gaged* (that is, plaster of Paris should be mixed with the putty) *after* it is placed on the mortarboard. The usual method of gaging consists of making a hollow with the trowel in the lime putty lying on the mortarboard. This hollow is filled with water, the plaster of Paris is sprinkled upon it, and this mixed rapidly with the trowel and applied to the wall before the plaster has time to set. The proportion of lime and plaster of Paris, while variable, averages probably $\frac{1}{4}$ to $\frac{1}{5}$ plaster of Paris.

The finish is skimmed in a very thin coating that is generally less than $\frac{1}{8}''$ in thickness. It is immediately troweled several times, dampened with a wet brush, and again thoroughly troweled to smooth up the surface and prevent it from chipping or cracking. The water prevents the steel trowel from staining the surface, but the plaster should not be dampened too much as it will then blister or peel. The entire surface of the finish coat, whether of putty or hard finish, finally should be brushed over once or twice with a wet brush. If a polished (or buffed) surface is desired, it is obtained by brushing—without dipping the brush into the water—until a glossy surface is obtained.

Suction. This process is absolutely necessary in order to obtain the proper bond of plaster on masonry (or concrete). It is necessary in the first and second coats so that the final coat will bond properly.

Uniform suction helps to obtain uniform color. If one part of the wall draws more moisture from the plaster than another, the finish coat may be spotty.

Obtain uniform suction by dampening, but not soaking, the wall evenly before applying the plaster. A fog spray is recommended for this work.

Curing. Since portland cement plaster is generally exposed to severe use, it should be given every opportunity to develop its maximum strength and density through proper curing. The method of curing portland cement is simple: (1) keep brown and finish coats continuously damp for at least two days; begin moistening each coat as soon as the plaster has hardened sufficiently so that it won't be injured; the water should be applied in a fine fog spray and soaking should be avoided; (2) after the damp-curing period, allow the plaster coat to dry thoroughly before the next coat is applied.

Plaster on Masonry. If plaster is applied on a *stone* or *brick wall* or on tile, a scratch coat is seldom necessary as the scratch coat is used principally to insure a clinch back of the lathing and to provide a surface to receive the brown coat.

Fireproofing. Plaster is used with incombustible lathing for fireproofing walls, ceilings, beams, and columns. A $2''$ solid partition of gypsum lath and sanded plaster will qualify for a one-

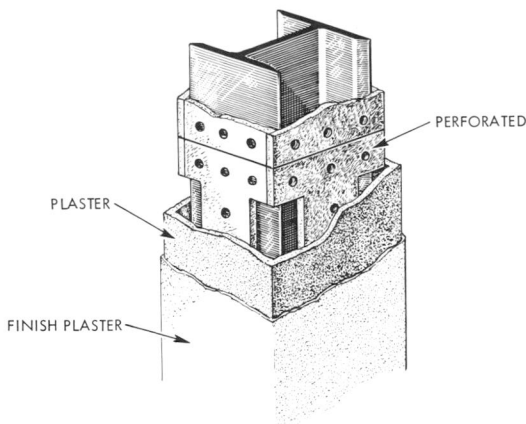

PERFORATED LATH

PLASTER

FINISH PLASTER

Fig. 10-2. Fireproofing a steel column with lath and plaster. (Bestwall Gypsum Company)

hour fire rating. Fireproof lath and plaster ceilings are often suspended from structural members. Fig. 10-2 shows a fireproof steel column.

Stucco. Another wet wall material, stucco, is commonly used on the exteriors of buildings. Stucco is composed largely of cement. It is applied to a wire mesh backup which is usually nailed to furring strips. On masonry buildings, no furring is required.

Lathing. Serving as a base for plaster, lathing also may provide definite resistance to the flow of heat or cold. The most common types of laths are:

1. wood
2. board
3. insulation
4. metal

Wood Laths. These laths are seldom used in modern construction. They have been largely replaced by board material such as gypsum board lath, and metal lathing. The estimator, however, may encounter wood laths in remodeling and repair work.

Wood laths are generally made of white pine, spruce, or hemlock. Other woods containing high percentages of pitch, or woods which are too hard, are not recommended. The common size for wood laths is ¼ " × 1½ " × 4'. This size may vary somewhat in width but, in general, it is the most common size. Laths are delivered to the job in bundles of 100 but are purchased by the thousand. Laths should be new, clean, and free from knots. Old or used laths do not provide a good bond, and dirt or knots may cause a stain to bleed through the plaster.

Wood laths are usually nailed with ¼ " spacing between them, although lath spacing up to ⅜ " is sometimes allowed.

Furring. As a means of preventing moisture from coming through the plaster, and also to create dead air spaces for insulating purposes, all masonry walls are furred before being plastered. Furring strips vary in size, but a 1 × 2 piece of wood is typical. These strips are nailed at intervals of

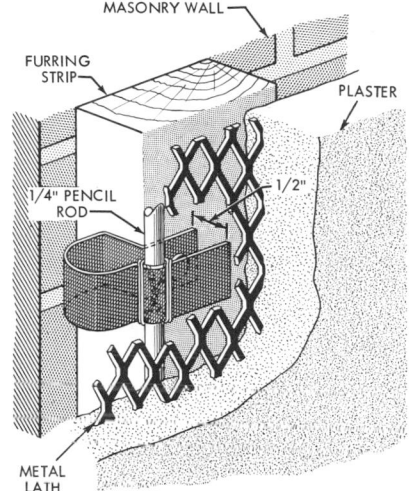

Fig. 10-3. Wood furring on masonry. Note the clip system.

16″ o.c. directly to the masonry wall, using a cut nail. They must be plumbed to insure a uniform plaster surface. The lathing is nailed to the studs and the plaster applied to the lathing. See Fig. 10-3.

The wood framing at corners often is made so that it will be easier to nail the lathing. Fig. 10-4 shows corner framing which provides nailing surfaces for lathing and helps in furring the walls.

Gypsum board (wet application) is generally ⅜″ or ½″ thick and available in sheets 16″ × 48″ or 2′ × 7′, 8′, 9′ or 10′. The paper surface is specially treated to give maximum suction to the plaster.

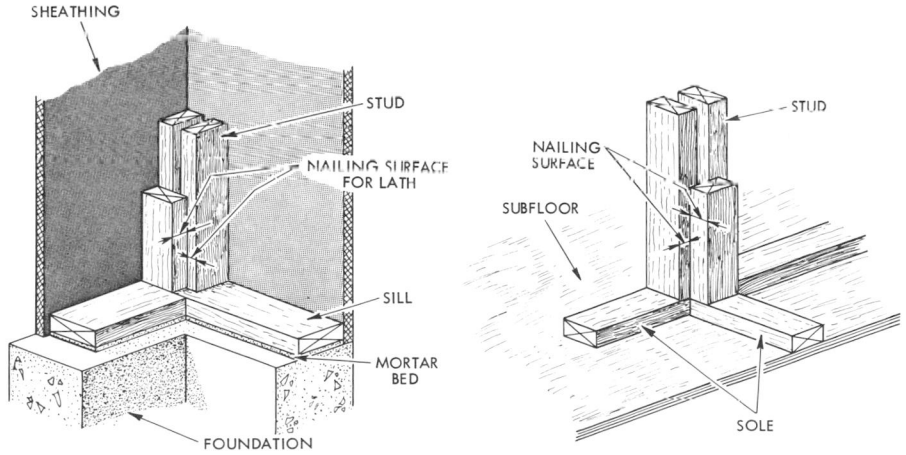

Fig. 10-4. Corner framing can be planned to create nailing surfaces for lathing.

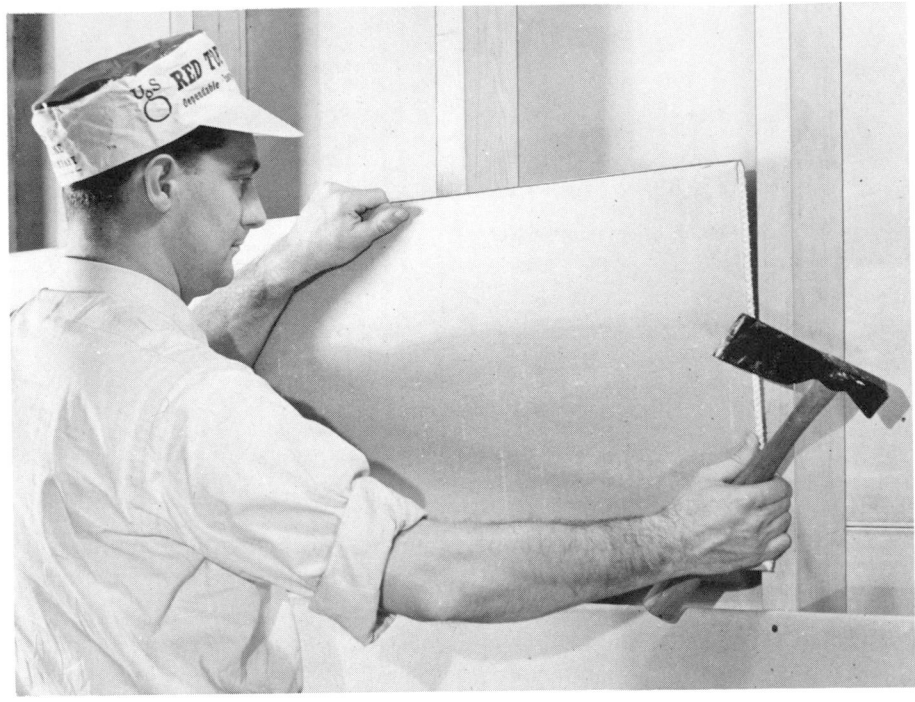

Fig. 10-5. Application of plain gypsum board.

It is nailed to the studding or furring strips in a brick pattern to avoid cracks in the plaster, Fig. 10-5. Gypsum board is the most widely used lath for interior plastering today.

The use of gypsum board lath saves time in lathing and can contribute some insulating value. Some contractors save one plaster coat when gypsum board laths are used, although such procedure should be indicated in the written specifications.

A *Satterly* partition is shown in Fig. 10-6. This is a method of erecting partitions without studs, using two layers of gypsum boards, one horizontal, the other vertical. The layers are held together by wire ties or clips.

Insulating board is being used to a great extent as plaster backing. The several types are made in variously sized sheets, with the usual 16″ o.c. spacing of studs considered. Like the board types, the insulating plaster backing is nailed to the studs. As with all other types of lathing, the plaster is applied directly to the insulating board. The material may be purchased in various thicknesses according to the insulating requirements.

Metal lath is perhaps the most versatile of all plaster bases. No other type of lath offers a better key for mortar.

Metal laths are manufactured in many varieties. Most laths are manu-

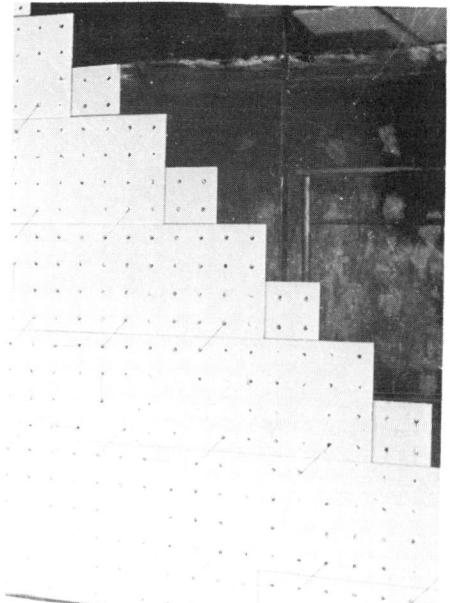

Fig. 10-6. A Satterly partition. Note the wire ties.

factured in sheet sizes of 24″ × 96″ and 27″ × 96″.

Metal lath is either nailed to wood studs or tied with wire to metal framing members. Staples, galvanized nails or blued nails may be used to attach the lath to studs. Sometimes a specialized nail such as that shown in Fig. 10-7 is used. Straight 6d or 8d nails may also be used and bent over to hold the lath. All metal lath should be nailed to studs at 6-inch intervals

Fig. 10-7. A lathing nail.

to prevent loosening or cracking of the plaster.

The lath should be well lapped at splice points and the overlap should be nailed securely.

Metal lath is especially suited for use around steel columns and beams where there is no woodwork. In such cases, it is wired on tightly. Corner beads are wired on at the same time. Metal lath is also employed in corners, on wood or steel framed arches, and where fireproof construction is required, as in garage ceilings. Where the lath is to be wired on, the estimator must include the cost of the wire and the labor in securing the laths.

Solid, non-load bearing partitions can be constructed by anchoring steel studs to the floor and ceiling. Metal lath is secured to one side of the stud and plastered so as to completely cover the studs on both sides. The size of the studs range from ¾″ to 1½″, depending on the height of the wall. They are spaced from 12″ to 24″ o.c., spacing depending on the type and weight of the lath. If the wall runs over 20 ft. in height, horizontal rod or channel stiffeners are secured to the studs and spaced not more than 6 ft. vertically on the channel side of the lath.

To estimate a job of this type, you must figure the total number of studs you need and the length of each stud. Studs are bought by the linear foot. The total area of wall to be covered must be computed to get the amount of lath you will need. The square foot area of the wall must be converted to square yards because metal lath is bought by the square yard. The esti-

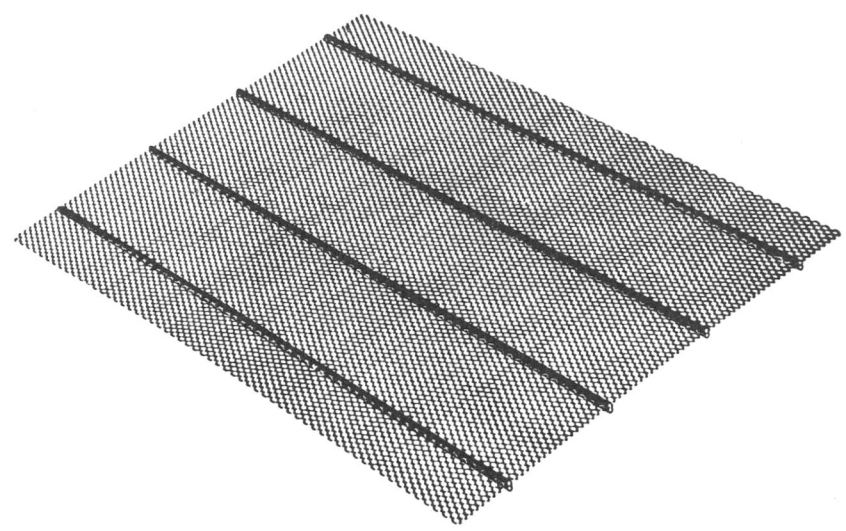

Fig. 10-8. Diamond mesh metal lath. (Metal Lath Manufacturers Association)

mator must also figure the total number of horizontal stiffeners for the job and the wire to secure them. The amount of plaster needed must also be estimated. The total area to be plastered and the number of coats and thickness must be known to get an approximate figure needed for the job.

Diamond mesh, Fig. 10-8, is used for all types of plastering.

Diamond mesh or expanded metal lath is produced by stamping or perforating a metal sheet and then expanding the sheet by pulling it apart until a meshlike material develops. The standard diamond mesh lath has a mesh size of $\frac{5}{16}'' \times \frac{9}{16}''$, the mesh is in a diamond pattern. The standardized weights for expanded diamond mesh are 2.5 and 3.4 lbs. per sq. yd. Lath is made in sheets of $27'' \times 96''$ and packed 10 sheets to the bundle (20 sq. yds.).

Diamond mesh lath is also made in large diamond mesh which is used for stucco work, reinforcement in concrete work, and as a support for rock wool and similar insulating materials. Sizes and weights are the same as for the small mesh. The small diamond mesh lath is also made into a self-furring lath by forming dimples into the surface which hold the lath approximately $\frac{1}{4}''$ away from the surface. This lath is used for fireproofing columns and beams or on flat wall and ceiling surfaces. Another form is the paper backed lath where the lath has a waterproofed or kraft paper glued to the back of the sheet. The paper acts as a plaster saver and moisture barrier.

Expanded Rib Lath. The expanded rib lath is much like the diamond mesh lath except that various size ribs are formed in the lath to stiffen it. Ribs

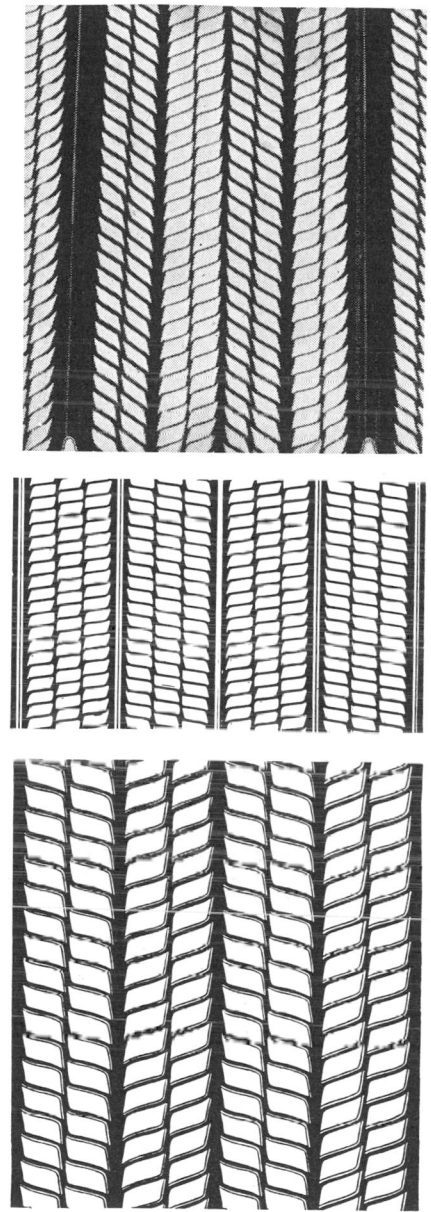

Fig. 10-9. Three types of rib lath. Top, with eight meshes between ribs. Center, with three meshes between ribs. Bottom, with two meshes between ribs. (Metal Lath Manufacturers Association)

run lengthwise of the lath and are made, for plastering use, in ⅛", ⅜" and ¾" rib height. The sheet sizes are 27" × 96" for the ⅛" and ⅜" rib lath and 29" width, for stucco work, reinforcement in concrete work, and as a support for rock wool and similar insulating materials. Sizes and weights are the same as for the small mesh. The small diamond mesh lath is also made into a self-furring lath by forming dimples into the surface which hold the lath approximately ¼" and 5, 10, and 12 ft. lengths for the ¾" rib lath. Weights run 2.75, 3.4, 4.0 lbs. per sq. yd., the ¾" rib lath runs .60 and .75 lbs. per sq. ft. Fig. 10-9 shows typical rib lath with varying mesh between ribs.

There is also available a diamond mesh lath with metal rods welded on both sides of the lath back to back. This forms an extremely stiff lath. All these laths are available in either painted steel or galvanized steel.

Wire Mesh Lath. This type of lath is made in two basic types, woven wire and welded wire. The woven wire lath is made of galvanized wire of various gages woven or twisted together to form either squares or hexagons. See Fig. 10-10. One popular type called "Keymesh," has 1" hexagonal openings and is woven of 20 gage wire. It is produced in rolls in various widths and is used over gypsum and fiber insulating lath as a reinforcement; it is also used for cornerite and stripite. It is very popular as a stucco mesh where it is placed over tar paper on open stud construction or over various sheathings.

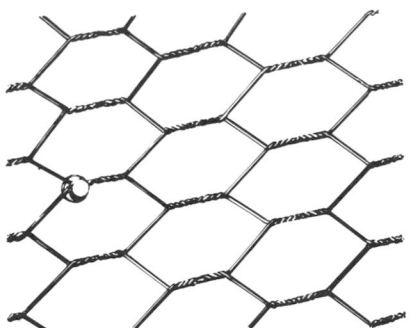

Fig. 10-10. Woven wire lath. (Keystone Steel & Wire Co.)

Another type is made with 2″ × 2″ mesh, 16 gage galvanized wire and the wire fabric is interwoven with a fibrous, absorbent paper backing. See Fig. 10-11. The backing is secured by a 17 gage wire that is corrugated every 4⅜″, which provides for an adequate plaster key to be formed. Sheet sizes are 30½″ × 49″. Each shipping carton contains 44 sheets or 50 sq. yds.

This type of lath is made also for stucco use and is popular on open stud construction because of its paper backing.

Welded wire lath has now become very popular because it is very adaptable for machine applied mortar.

Sheet Lath. This type of lath was very popular in the early days of metal lathing. It then seemed to fade in popularity, but now with the spread of gun applied plaster and cement mortars this type of lath is regaining favor. The lath is made from sheet steel and has slits punched into it plus ridges and cross bars. The design forms a stiff sheet while the slits provide the keys to hold the mortar. See Fig. 10-12. The design also provides a self-furring condition. Weight is 4.5 lbs. per sq. yd., sheet size 27″ × 96″ also in 24″ × 96″ sheets. Packed in 10 sheets to the bundle (20 sq. yds.) or 9 sheets (16 sq. yds.) per bundle on special order.

Fig. 10-11. Paper-backed lath. (Johns-Manville Co.)

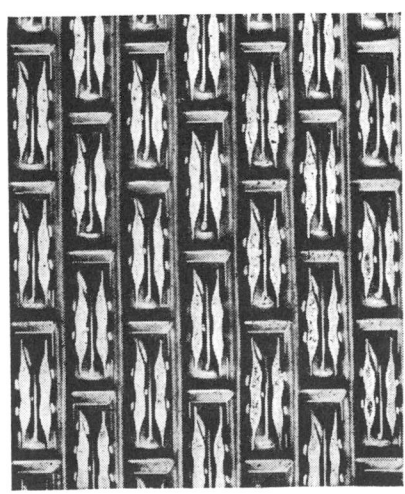

Fig. 10-12. An example of sheet lath. (Metal Manufacturers Association)

Lathing Accessories. Various metal accessories are usually required to complete a lathing job. These accessories serve several basic functions and in some cases serve one or more functions as individual units. See Fig. 10-13 showing some of the items in cross section and their names.

Plaster stops or *casing beads* define the limits of the plastered wall or ceiling and provide a thickness gage for the plaster at that point. See Fig. 10-13. They also protect the plaster from damage at this terminal edge.

Grounds or *screeds* (sometimes called *base beads*) are available in many types and sizes. See Fig. 10-13. They are installed primarily to control plaster thickness and alignment but sometimes do double duty in separating the plaster from various other materials, such as a cement base.

Corner Beads. Plaster will not stand severe bumps without chipping or cracking. This is especially true at corners where no casings are used. Also, it is difficult to make a perfectly uniform edge all around a plastered open-

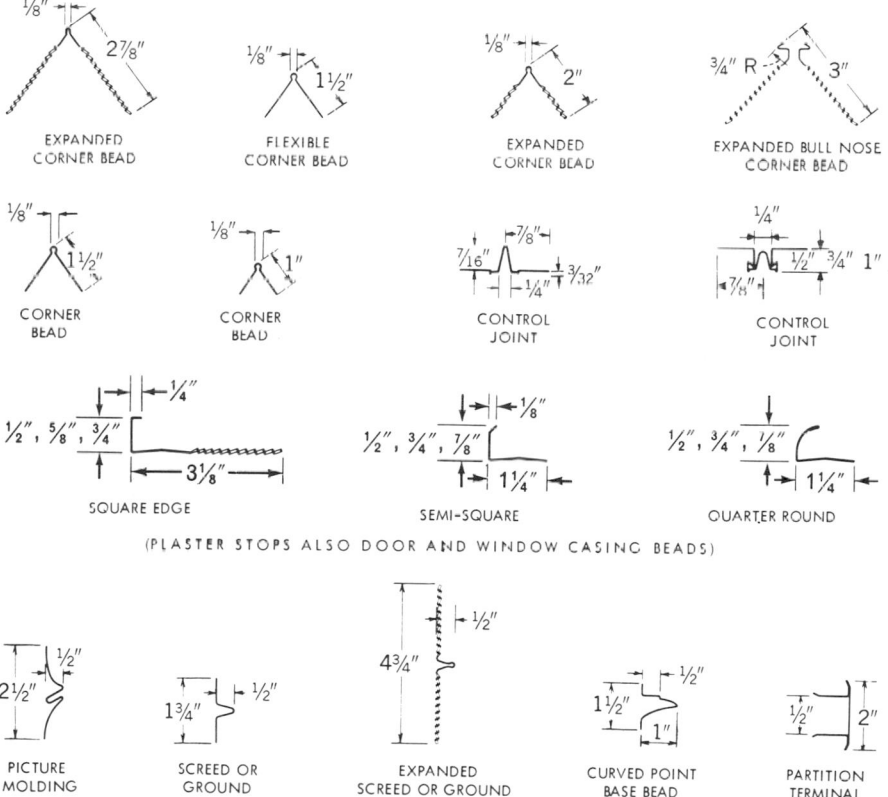

Fig. 10-13. Lathing accessories, showing some of the most commonly used items in cross-section.

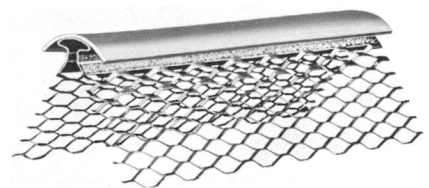

Fig. 10-14. Bull-nose corner bead. (Metal Lath Manufacturers Association)

ing or to keep a vertical edge perfectly straight. The time involved in this work is considerable. To overcome these problems, corner beads are used, Fig. 10-14. Corner beads are nailed to the corners and lathing is brought in close to the bead before plastering. After the plastering is completed, only the bead is visible and it can be painted, papered, or covered.

When corner beads are used with wired-on metal lath, binding wire is used to fasten the beads in place.

Metal trim consists of base, door and window casing, and various molded shapes providing terminal finishes for plastering. See Fig. 10-13.

Concealed picture molding provides a flush groove in the finished plaster into which picture hooks can be inserted anywhere along the wall.

Metal Arches. As previously explained, for an arched opening 2 × 4 framing is constructed over the arch. If the arch is to be plastered, metal arches (made of heavy sheet metal containing a surface of alternate holes and solid metal) are used for plaster keying. Such arches are purchased in many shapes and widths and are complete and ready to be nailed into place. Their use saves much labor time.

Control joints or expansion and contraction joints are available in many types and sizes for both interior and exterior work. See Fig. 10-11. These items help to provide room for expansion and contraction that takes place in any material due to temperature changes or structural movement.

All of these items are made in galvanized or zinc finish to prevent rusting. There are now on the market various accessories made from plastic and these are used for both exterior and interior work. Because they will not rust or stain they are used for some of the exposed aggregate and veneer coat plastering. This is only a partial listing as there are hundreds of variations available, each designed for a specific purpose or need.

Scaffolding. The plasterer generally scaffolds the room with boards at a sufficient height to enable him to easily reach the ceiling without stretching his arms too high to work the plaster coats evenly. The plaster for the upper part of the walls is also applied from the scaffolding, and the remainder of the work is completed from the floor. If too much time elapses in joining the coats at this point, the joint is likely to show. This is not serious unless the walls are to be left untreated. Occasionally two men working together, one on the scaffolding and one on the floor, finish their work at the same time.

DRYWALL

Wet wall construction is being replaced by drywall in much of today's modest-priced housing. Drywall is wall applied without the use of mortar or plaster and includes wood paneling, plasterboard, and a variety of acoustical and decorative types of paneling. (Chapter 11 discusses acoustical tile and special wall coverings.)

Wood paneling may be included in either the carpentry section of the estimate or under drywall. We have included wood paneling under wall framing in Chapter 7 and in Chapter 11 "Special Interior Coverings," but many estimators would include it with drywall.

Plasterboard, wallboard, and insulation board, however, are almost invariably included under lath and plaster in the specifications or description of materials. Plasterboard and insulation board have been discussed in the wet wall section of this chapter. These materials may, however, be used as drywall materials. The method of estimating them remains the same. When used as drywall materials, they will require additional treatment such as taping and spackling.

Gypsum Wallboard. One of the most common types of drywall used in residential construction is a gypsum

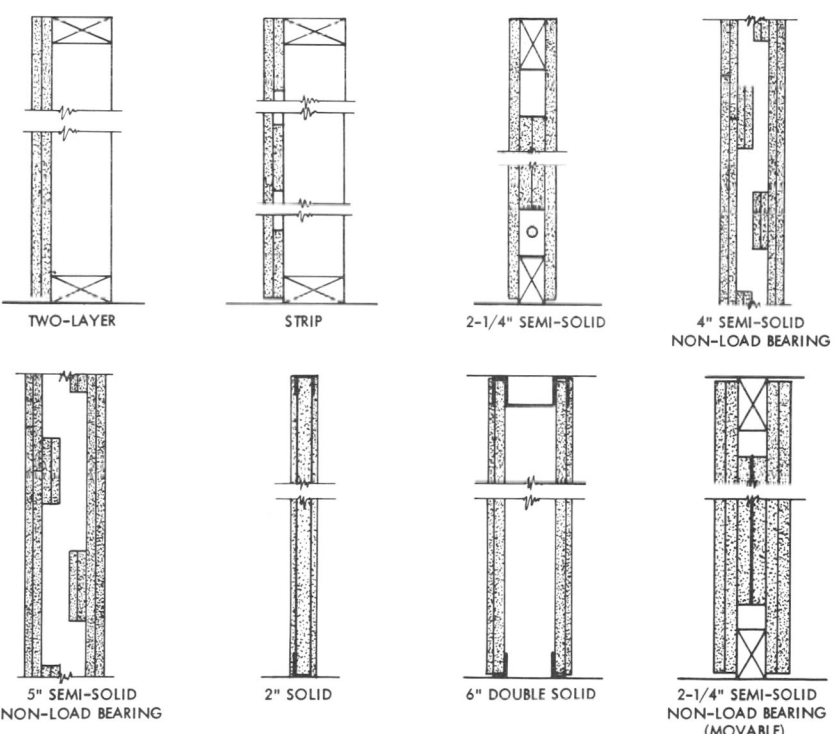

| TWO-LAYER | STRIP | 2-1/4" SEMI-SOLID | 4" SEMI-SOLID NON-LOAD BEARING |

| 5" SEMI-SOLID NON-LOAD BEARING | 2" SOLID | 6" DOUBLE SOLID | 2-1/4" SEMI-SOLID NON-LOAD BEARING (MOVABLE) |

Fig. 10-15. Laminated partitions. (Bestwall Gypsum Company)

wallboard. It comes in standard sizes of 4' × 8', 4' × 10', and 4' × 12'; other lengths up to 16 feet are also available, and widths of 2 feet may be ordered. The standard thickness is ½ inch although ⅜ inch and ⅝ inch are common.

Panels are composed of a gypsum rock base sandwiched between two layers of special paper. Insulating panels with an aluminum backing are available and standard fire resistant panels (with a base of gypsum rock mixed with glass fibers) are common.

Water resistant (W/R) gypsum wallboard has a core with a special asphalt composition. The face and back paper is chemically treated to combat the penetration of moisture. It provides an excellent base for ceramic, metal and plastic tile.

Panels may be either unfinished or finished. Vinyl finishes with various permanent colors and textures are available. Gypsumboard may be installed with nails, staples, drywall screws (Phillips) or clips. Adhesives may be used in combination with either nails, staples or screws.

Fig. 10-15 shows various kinds of *laminated partitions*. The material is glass fiber, reinforced gypsum wallboard. In the two-layer system, the first layer is nailed and the second is applied with adhesive and screws.

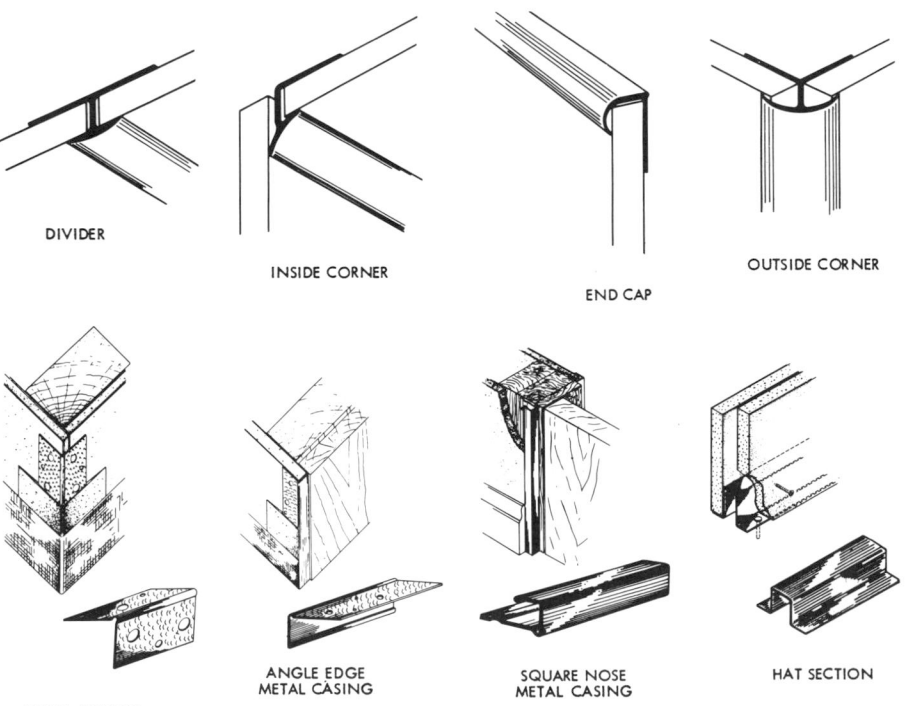

DIVIDER

INSIDE CORNER

END CAP

OUTSIDE CORNER

METAL CORNER

ANGLE EDGE METAL CASING

SQUARE NOSE METAL CASING

HAT SECTION

Fig. 10-16. Metal trim used in drywall construction. (Bestwall Gypsum Company)

Nailing. Nails used to hold plaster-board to studs or furring strips are driven in using a lather's hammer. This special hammer has a convex head which drives the nail beneath the surface of the plasterboard. The nail head is then filled, spackled, and painted.

Taping and Spackling. In many cases where plasterboard is used as wall covering, some method of con-cealing the seams is necessary. The basic material used for such conceal-ment is gauze-perforated tape applied with a gypsum cement which acts as a filler and adhesive. This tape is avail-able in 3″ and 4″ strips.

The tape is *spackled* after it has been applied to the seam. Spackling involves putting a very thin layer of finish plaster over the strip and blend-ing this with the texture of the board.

There are many other metal acces-sories for drywall aids and as trim. Fig. 10-16 shows some of the various metal accessories used in drywall construc-tion.

ESTIMATING MATERIAL

Plaster. Plaster is calculated by the square foot or by the square yard. This is true regardless of whether it is a one, two, or three coat job. With re-gard to the outs or deductions for openings, the method varies among contractors and estimators. Some will deduct all the openings and others will take them out completely and then fig-ure for the finishing of the opening in another part of the estimate.

Example 1. Find the plaster area of the inside walls in Fig. 10-17.

Solution: Find the total inside wall area and deduct the openings.

$$40' + 40' + 20' + 20' - 120'$$
$$120' \times 8' = 960 \text{ sq. ft.}$$

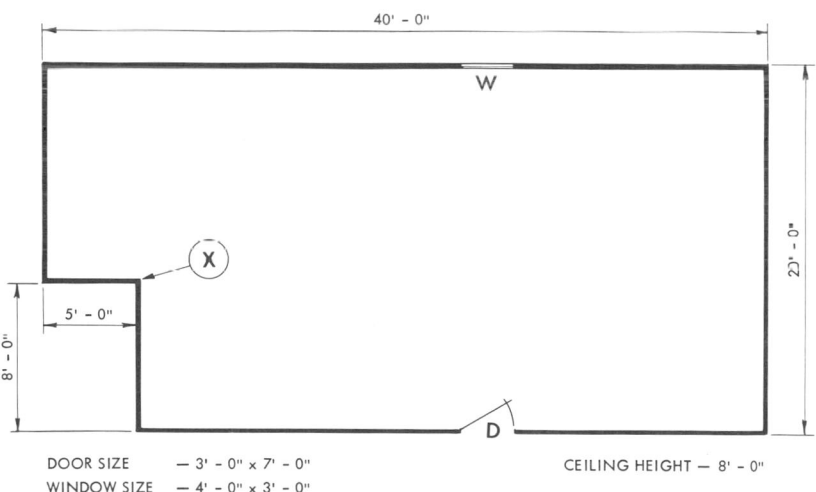

DOOR SIZE — 3' - 0" x 7' - 0"   CEILING HEIGHT — 8' - 0"
WINDOW SIZE — 4' - 0" x 3' - 0"

Fig. 10-17. An area requiring plastering of walls and ceiling. An outside (exterior) corner is shown at X.

Door area = 21 sq. ft.
Window area = 12 sq. ft.
Deductions = 21 + 12 = 33 sq. ft.
960 − 33 = 927 sq. ft. or 103 sq. yds.

Example 2. Determine the plaster area of the ceiling in Fig. 10-17.

Solution: Assume the room to be a rectangle and then deduct the area of the break.

$$40' \times 20' = 800 \text{ sq. ft.}$$
$$8' \times 5' = 40 \text{ sq. ft.}$$
$$800 - 40 = 760 \text{ sq. ft.}$$

Another method is to break the room into rectangles and then add together their areas.

$$35' \times 20' = 700 \text{ sq. ft.}$$
$$5' \times 12' = 60 \text{ sq. ft.}$$
$$700 + 60 = 760 \text{ sq. ft.}$$

Lathing. The lath or backup for wet walls is taken from the plaster calculations. The lath is calculated in square feet or in square yards. (Accessories are figured with the lath.)

Example 3. How many square feet of lath are required for the walls in Fig. 10-17?

Solution: In Example 1, there are 927 sq. ft. to be plastered. The same amount of lath is therefore needed for the walls.

Example 4. How much lath is needed for the ceiling in Fig. 10-17?

Solution: Since 760 sq. ft. of plaster are needed (see Example 2), the same amount of lathing is required.

Plasterboard and insulation board are calculated in the same manner. The sheet size used will depend on the height of the ceiling. The cost is found by multiplying the cost per sheet by the number of sheets.

Corner Beads. These are calculated by the linear foot. The corners requiring beading are determined and the total length of bead calculated for each corner. For example, around a plastered opening rectangular in shape, the floor-to-ceiling height is multiplied by four. Curved surfaces requiring beading are estimated as to length as closely as possible.

Example 5. How many linear feet of outside corner bead will be required in Fig. 10-17?

Solution: In Fig. 10-17 there is one outside corner (at *X*). Thus 8′ of corner bead will be sufficient.

Example 6. How many linear feet of inside corner bead will be required in Fig. 10-17?

Solution: There are 5 inside corners.

$$5 \times 8' = 40 \text{ lin. ft.}$$

Nails. A general rule for determining the number of nails necessary for metal lathing is to allow 9 pounds of 3*d* nails per 1,000 square feet of lathing. Nailing requirements, however, vary according to the type and weight of the lath. This item is often strictly controlled by local building codes; if not specified, you will want to consult the manufacturer's specifications.

For plasterboard and insulation board (and most other board backing), allow 10 pounds of 3*d* nails per 1,000 square feet of board. The number of nails per sheet varies, of course, with the sheet size and is controlled by the local building code. In either case, you should follow the building code or manufacturer's specifications.

For corner beads, allow 1½ pounds of 3*d* nails per linear feet.

Drywall. The square foot method is used for all drywall materials except acoustical tile, which may be calculated by the square foot or by the piece.

Plasterboard, insulation board, and wallboard are manufactured in a number of standard-sized sheets. If screws are used (especially for plasterboard), they are spaced not over 12″ apart on ceilings and a maximum of 16″ o.c. on walls. If wall studs are spaced 24″ o.c., the screw spacing must not exceed 12″.

Tape. Seam concealing materials are estimated by the linear foot. First determine the number of sheets of drywall material and the height of the seams. Then add the total number of feet needed for the tape to obtain the linear footage required for each room or job.

ESTIMATING LABOR

Wall construction is frequently given as a unit cost which includes materials, labor, and equipment such as scaffolding. Although labor time will vary among different regions, there are some general rules which can be used.

Wet Wall. Three coats of plaster on metal lath can be calculated at the rate of 1 man-hour for every 3 square yards of plaster. Two coats on masonry require 1 man-hour for 6 square yards of plaster.

On frame construction, stucco can be applied at the rate of 2½ square yards per man-hour. On masonry walls, stucco will be applied at the rate of 2 square yards per man-hour.

A good lather can install about 1,-000 square feet of metal lath in an 8-hour day.

Arches are normally installed at a unit cost per opening.

Drywall. For plasterboard or insulating board (which also may be used as lathing), one man can install over 1,050 square feet of sheets in 8 hours.

The estimator should check the labor times and costs for his own area. Remember that unit costs will include materials *and* labor.

See Appendix D, pages 479 to 480, for Hourly Estimates for:

Insulation: Rigid Board Types; Non-Rigid Types; Reflective Types; Pouring Types; Semi-Rigid Types; Vapor Seals, Moisture Barriers; **Walls and Ceilings:** Plaster Bases; Plaster; Metal Lath and Plaster Partitions; Gypsum Lath and Plaster Partitions; Suspended Ceilings; Plaster Bonding; Gypsum Board Walls and Ceiling; Plywood Panels; Patterned Panels; Beaded Wood; Insulation Board; Furring Plaster Grounds.

Questions and Problems

1. How many square yards of plaster are required for the walls in Fig. 10-18?

  (Answer: 83.6 sq. yds.)

2. How many square feet of plaster are required for the ceiling in Fig. 10-18?

  (Answer: 392 sq. ft.)

Each sheet is 8'-0" × 4'-0". How many sheets are needed?

  (Answer: 24)

7. Assume the same size plasterboard sheets are to be used for the ceiling. How many sheets are needed?

  (Answer: 13)

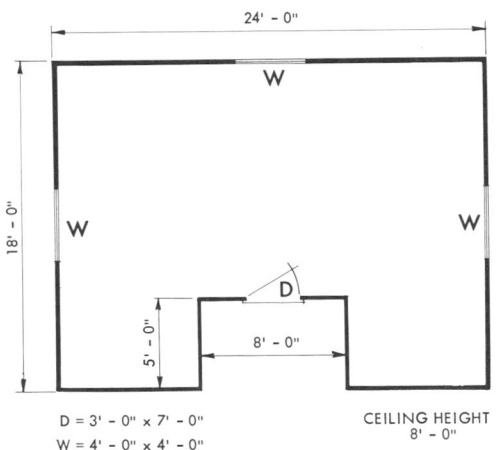

Fig. 10-18. A large room which is to be plastered.

D = 3' - 0" x 7' - 0"
W = 4' - 0" x 4' - 0"

CEILING HEIGHT
8' - 0"

3. How many linear feet of inside corner beads are needed in Fig. 10-18?

  (Answer: 48 ft.)

4. How many linear feet of outside corner beads are needed?

  (Answer: 16 ft.)

5. Assume that the ceiling will have tiles 18" × 18" installed. How many tiles are needed?

  (Answer: 175)

6. Assume that sheets of plasterboard will be used for the walls.

8. Based on the number of sheets used in Problem 6, above, how many linear feet of taping are required for the walls?

  (Answer: 192 ft.)

9. How many linear feet of ceiling corner tape are needed for this room? (Answer: 94 ft.)

10. Assume that the studs are spaced 24" o.c. How many metal studs are needed for the interior partitions of the room?

  (Answer: 47)

11

SPECIAL INTERIOR COVERINGS

~~~~~~

In addition to plain wood or concrete floors and painted or papered wall and ceiling surfaces, one or more of a number of special materials are frequently specified. This book cannot list all such special coverings, but enough are described to acquaint the inexperienced estimator with the various materials and their application.

The use of many of these special coverings requires special treatments for floor, wall, and ceiling subsurfaces. This definitely affects material and labor costs and therefore should be given careful attention by the estimator.

MATERIALS

Floors. Among the special floor coverings are plank floors, block floors, linoleum, terrazzo, and the various floor tiles.

Plank Floors. Fig. 11-1 illustrates a typical so-called colonial plank flooring. These flooring types may be obtained in such woods as oak, maple, and walnut. The planks are from 4″ to 10″ wide and come in random lengths. Sometimes nailing is done through holes in the ends of the boards, after which plugs of a contrasting color are inserted. Other types are nailed through the tongue. Some types, as shown in Fig. 11-1, have keys which add decoration to the floor and also

prevent spreading. Plank floors usually are finished natural and then waxed. They may require sanding.

Block Floors. Two types of block flooring are used. For installation over concrete floors, a mastic fill (hot or

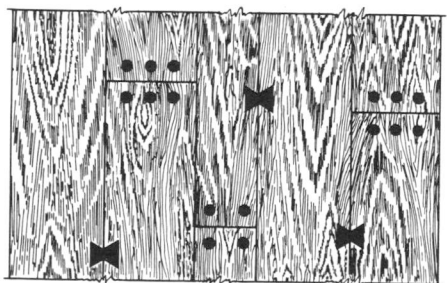

Fig. 11-1. Plank floors are available in various designs and dimensions.

Fig. 11-2. Block floors may be laid in a mastic fill either over concrete or over wood flooring. (E. L. Bruce Company)

cold) is spread and smoothed level. The blocks are arranged carefully in courses so as not to bleed the mastic (or cement-adhesive) over the block. Sanding may be required, but in many instances the wooden blocks (or squares) are pre-finished — in which case, the need for neat, clean installation is most important.

Fig. 11-2 illustrates a block floor being laid on a mastic fill. If the concrete is in contact with the ground or if it is below grade level, there could be a moisture problem, so polyethylene film or waterproof paper should be laid directly over the concrete. The mastic fill is applied directly over the damp-proofing.

For installation on a wood subfloor, tarred or asphalt felt is laid over the rough flooring and the blocks laid over the felt. Nailing is done through a special seat, or hole.

Resilient Flooring. This floor covering includes a wide variety of materials such as linoleum, vinyl, cork and rubber. It comes in sheets or tiles of various sizes and weights. See Table I. Resilient tile may be applied to concrete or wood subfloors.

Linoleum. Linoleum is sometimes specified for kitchen floors and is used in industrial and institutional installations. Fig. 11-3 shows the procedure for applying linoleum. For wood floors, felt is laid before the linoleum is applied; for concrete floors, the linoleum can be applied directly or with a felt lining. The contractor laying linoleum follows a definite procedure

TABLE I. TYPES, THICKNESS, WEIGHTS OF RESILIENT
FLOORING

MATERIAL	THICKNESS	WT./SQ. FT. LBS.
SHEETS		
LINOLEUM		
"Battleship"		
Heavy Duty	1/8	.83
Embossed,		
Inlaid, Standard	3/32	.60
Jaspe'		
Heavy	1/8	.83
Standard	3/32	.65
Marbelized,		
Heavy	1/8	.92
Standard	3/32	.65
Light	1/16	.45
Plain,		
Heavy	1/8	.83
Standard	3/32	.60
Straight Line Inlaid,		
Standard	3/32	.62
Light	1/16	.46
VINYL		
Embossed Inlaid,		
Standard	3/32	.71
Light	1/16	.43
Inlaid		
1/16	1/16	.43
TILES		
Asphalt, Plain	3/16	1.75
	1/8	1.16
Asphalt, Greaseproof	3/16	1.74
	1/8	1.17
Asphalt, Conductive	3/16	1.45
	1/8	.97
Asphalt, Industrial	3/16	1.35
	1/8	.90
Cork	5/16	1.70
	3/16	1.05
	1/8	.68
Linoleum	1/8	.83
Rubber	3/16	1.70
	1/8	1.24
Vinyl–Asbestos	3/16	1.90
	1/8	1.30
	3/32	1.00
	1/16	.67
Vinyl–Cork	1/8	1.30
Vinyl–(Homogenous)	3/16	1.06
	3/32	.93

Weights and thicknesses may vary slightly.

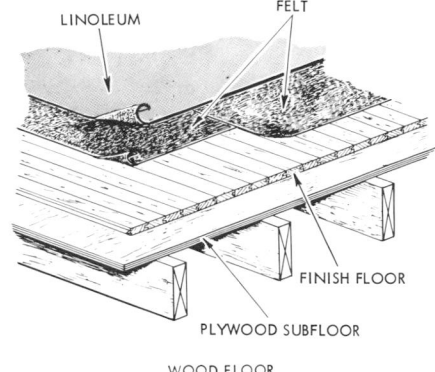

WOOD FLOOR

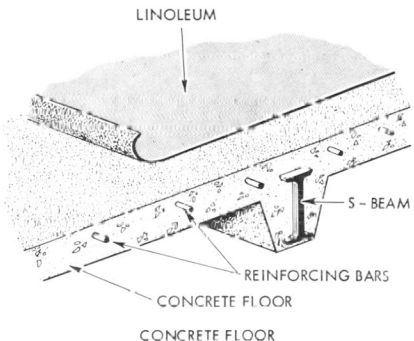

CONCRETE FLOOR

Fig. 11-3. A felt lining is commonly used between the linoleum and wood. It is not often used for concrete floors.

which the estimator should know in order to arrive at an adequate cost estimate.

Adhesives are necessary to cement the linoleum to the floor. Adhesives should have a spreading capacity of not less than approximately 100 square feet per gallon.

The felt used under the linoleum should be 1½ pound saturated or 1 pound unsaturated.

Concrete floors must be smoothed and filled where necessary. Plaster of Paris is used for filling holes or scratches. Cracks or holes in wood floors must be filled with strips or plugs of wood. All flooring boards must be nailed down tightly.

Most specifications require that when the felt is cemented to the floor a roller shall be used to smooth out any bubbles. The finished linoleum surface must be protected by heavy building paper. Ordinarily wax is applied. Some waxes require polishing, whereas others dry to the desired luster. Concrete floors must be tested for dampness and waterproofed if necessary. Waterproofing cement may be required at seams or around the edges.

A base is also run around the edge of the linoleum against the wall. See Fig. 11-4.

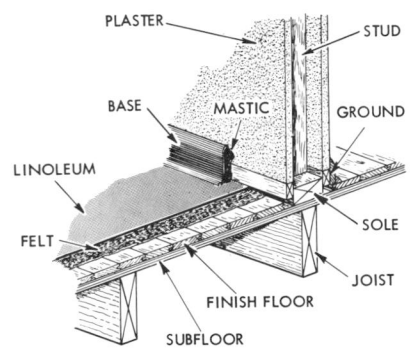

Fig. 11-4. A base is used around the walls, at the edge of the linoleum.

Vinyl Sheet Flooring. Vinyl sheet flooring is gaining in popularity. The application is similar to that of linoleum. Vinyl sheets are often cushioned and may be reinforced with fiberglass.

Asphalt Tile. This type of flooring is made in patterns of various sizes, and must be laid one piece at a time. When laid on wood floors all holes, rough spots, or cracks must be filled and smoothed. A felt lining should be used if the tile is laid over a wood subfloor. When applied to concrete, the surface (if new work) should be wood trowel finished. All holes or scratches

must be filled. The tiles are cemented to the base flooring. A prepared cement is spread evenly over the subfloor (or felt); it is allowed to set and the tile is applied by hand. Nailing is not required. Exposed edges must be protected by special metal strips made for the purpose. A base may be used against the wall. Generally, the contractor is required to wax or clean the finished surface.

Vinyl, Cork, and Rubber Tiles. These tiles are made in sizes similar to asphalt tile and are applied in the same manner.

Terrazzo. Two methods are used in laying terrazzo floors over concrete. One is to bond it to the concrete and the other to separate it from the structural slab.

The concrete fill can be made of cement, sand, and hard coal cinders in a mix of 1:1:6. Slag and broken stone, marble chips, or gravel are also good aggregates for concrete fill under terrazzo floors. Before the terrazzo contractor installs his underbed, he must see that his concrete fill is thoroughly cleaned of plaster droppings, wood chips, and other debris; it is then wetted to assure cohesion.

The second method is used in buildings where cracking is anticipated from settlement, or from expansion and contraction, or vibration. In this case the terrazzo contractor builds upon the structural floor slab. This method requires at least a total thickness of 3 inches. The concrete slab is covered with a thin bed of dry sand, over which a sheet of tar paper is laid. Over the paper the underbed is installed, as in

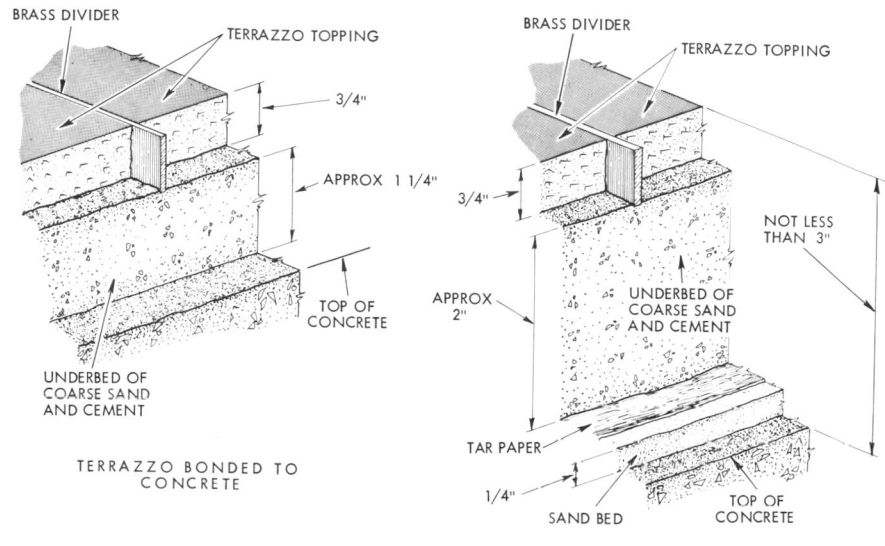

Fig. 11-5. Terrazzo may be either bonded to or separated from the concrete underbedding.

the first method, except that coarser aggregate can be used (such as cinders or fine gravel) where its thickness exceeds 2½ inches. See Fig. 11-5.

Where terrazzo is used over wood floors, a thickness of not less than 2 inches is required. The wood floor must be covered with tar paper and overlaid with galvanized wire netting. The underbed is laid on top of the paper and mesh. Brass dividers, see Fig. 11-5, may be required, depending on the pattern. Troweling is required to smooth the terrazzo surface even with the tops of the brass dividers. Gray or white portland cement is spread on the surface and surfacing is done with a special machine, using carborundum grit stones.

Monolithic Flooring. A relatively new type of flooring, *seamless monolithic flooring* is an aggregate of ce-ramic-coated quartz embedded in a polyester resin base. It is found to be extremely durable, retains its appearance and requires a minimum of care. It is non-porous and, therefore, waterproof, scuff-proof, acid-resistant, and unharmed by dirt. For these reasons it is becoming more popular in residential and industrial applications.

Ceramic Tile. This type of tile is used in bathrooms, powder rooms, and shower areas. A thin cement or mastic bedding is applied over the concrete base on which the tile are laid. Under very damp or otherwise severe conditions, portland cement mortar may be used as a base. This is much more expensive. The tiles must be thoroughly soaked in water before use.

The final step is to grout the joints with clean portland cement mixed with water.

If dampproofing is required, ceramic tile primer may be applied.

Mosaic Tiles. These are small tiles, of various patterns, mounted on sheets of paper of a convenient size. The sheets are laid on a bed of cement with the paper side up. The tiles are pressed into the bed and beaten down with a block until the mortar in the joints is visible through the paper. Finally, the paper is moistened and removed. White sand should be used on the surface so that as beating continues the cement will not adhere to the block.

out, commercially or residentially. Application may require cementing or taping.

Walls. There is a great variety of interior wall coverings ranging from tile and wallpaper to the many varieties of wall panels. This section, as in all previous sections, aims not to teach design or application, but to give the inexperienced estimator an appreciation of the amount of material and labor necessary for the various parts of a building.

Wallpaper and Fabric. There are

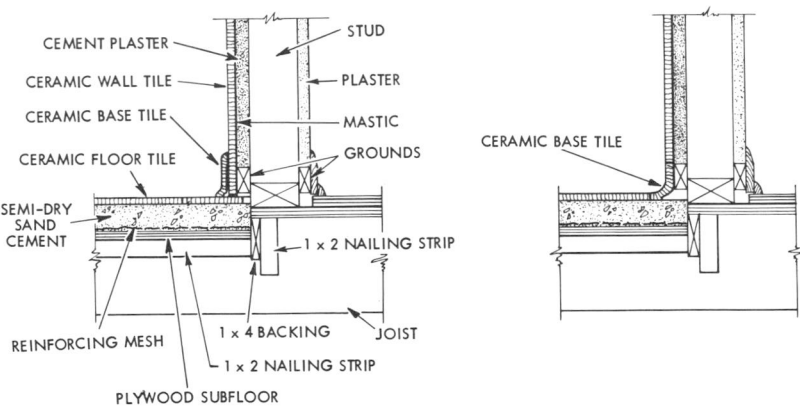

Fig. 11-6. Base details must also be included in the material and labor estimates.

The final steps are to level the entire area and to grout the joints in the same manner as for ceramic tile. Fig. 11-6 shows base details for a ceramic or mosaic tile floor. These details require additional material and labor must be included with the floor costs.

Carpeting. Carpeting comes in various natural and synthetic fibers and sizes range from wall-to-wall to 12″ x 12″ tiles. It may be applied indoors or

hundreds of wallpaper styles and many grades, from handmade papers of the finest quality to low cost conventional patterns. Wallpapers are made in narrow and wide widths in rolls of various footage. Fabric wall coverings (vinyl coated) are also applied to walls. This is almost identical to wallpaper in appearance and application.

Paste is supplied in powder form re-

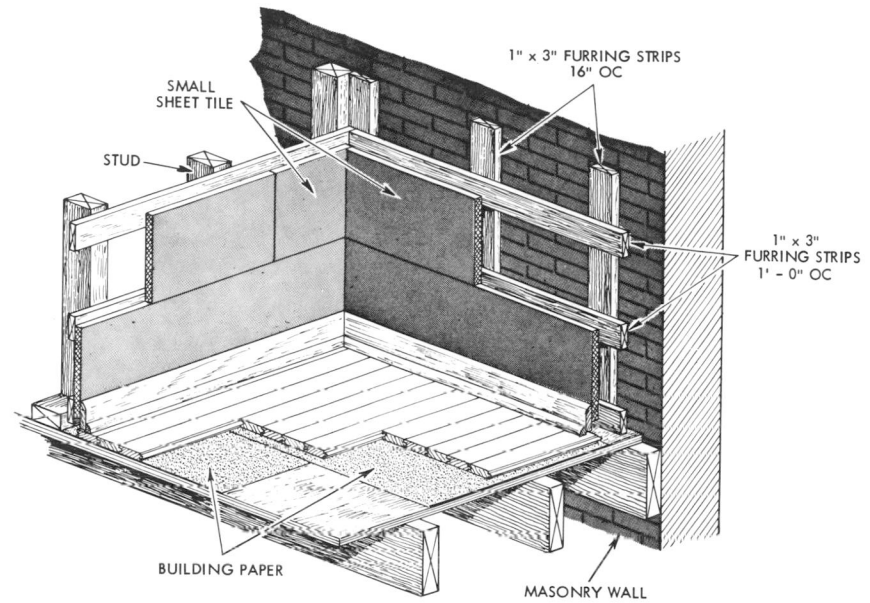

Fig. 11-7. Tile may be applied by using furring strips as a nailing base.

quiring only the addition of water and stirring. Plastered and dry walls are generally coated with sizing (a glue or paste) before papering or other treatment is applied.

Wall coverings also come with pre-pasted backings, to which only water is applied, or with adhesive backings.

Presdwood Tile. In bathrooms, kitchens, playrooms, etc., this style of tile may be used on walls. One such form of tile is manufactured of treated and formed wood pulp made up in large sheets or panels embossed in various designs. These tiles may be nailed directly to studs or furring strips or may be applied over old or new plaster.

Small Sheet Tiles. These tiles, manufactured in a variety of materials and shapes, are characteristically rigid.

Fig. 11-7 shows typical applications of small sheet tile to stud and masonry walls. In both instances furring strips (approximately 1″ × 3″ in section) are used as nailing strips. Fig. 11-8 illustrates a different method of application to a masonry wall.

Ceramic and Mosaic Tile. The preparation of wood stud walls for tiling is simple. It is necessary only to provide a firm and continuous sheet of plasterboard upon which the tiles may be bedded.

Wall tiles are set by two methods, called *floating* and *buttering*. In floating the tiles a portion of the bed is spread on the wall and the tiles are placed in position and tamped until firmly united. (Fig. 11-2 illustrates floating.) Buttering consists of spreading the mortar on the back of each tile,

310

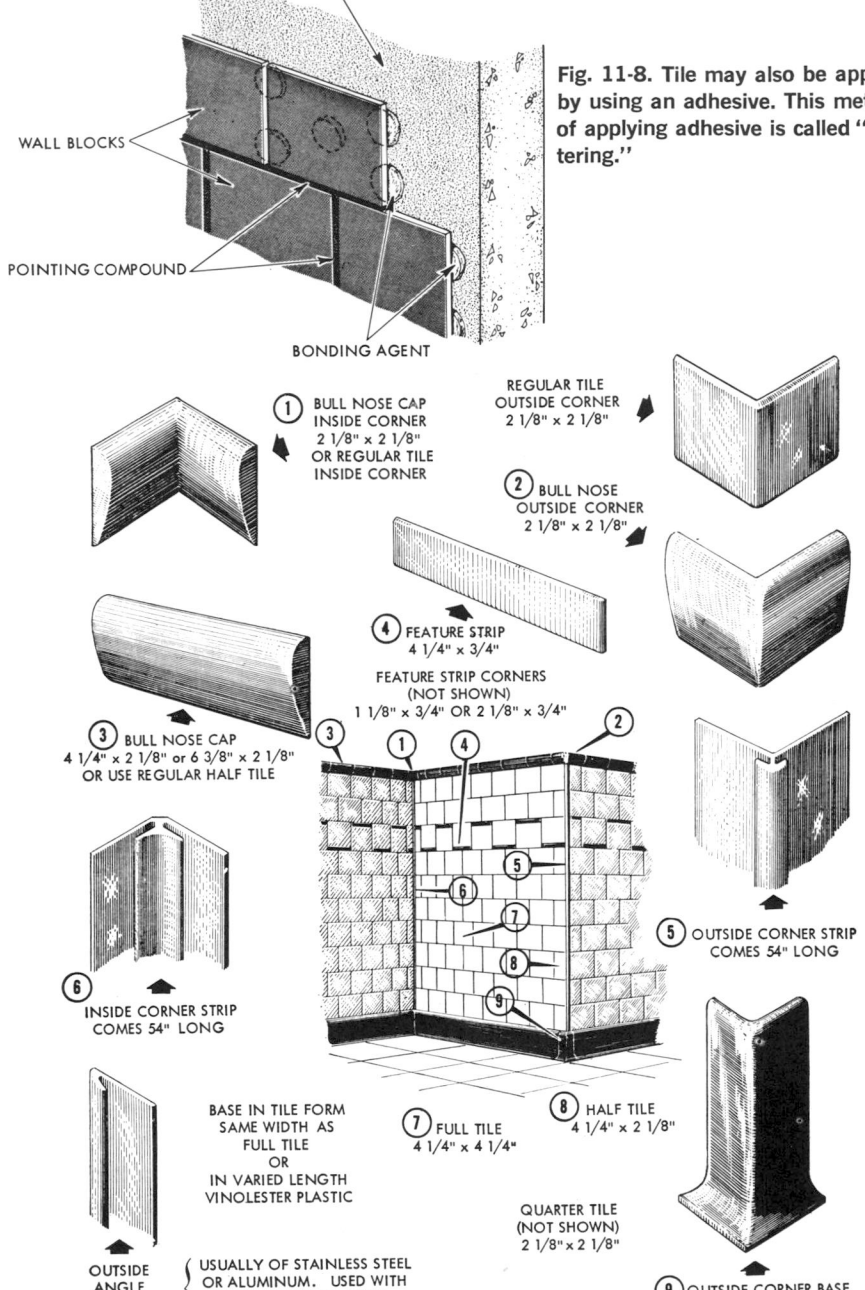

CONCRETE WALL

WALL BLOCKS

POINTING COMPOUND

BONDING AGENT

Fig. 11-8. Tile may also be applied by using an adhesive. This method of applying adhesive is called "buttering."

① BULL NOSE CAP INSIDE CORNER 2 1/8" x 2 1/8" OR REGULAR TILE INSIDE CORNER

REGULAR TILE OUTSIDE CORNER 2 1/8" x 2 1/8"

② BULL NOSE OUTSIDE CORNER 2 1/8" x 2 1/8"

④ FEATURE STRIP 4 1/4" x 3/4"

FEATURE STRIP CORNERS (NOT SHOWN) 1 1/8" x 3/4" OR 2 1/8" x 3/4"

③ BULL NOSE CAP 4 1/4" x 2 1/8" or 6 3/8" x 2 1/8" OR USE REGULAR HALF TILE

⑥ INSIDE CORNER STRIP COMES 54" LONG

⑤ OUTSIDE CORNER STRIP COMES 54" LONG

OUTSIDE ANGLE STRIP

BASE IN TILE FORM SAME WIDTH AS FULL TILE OR IN VARIED LENGTH VINOLESTER PLASTIC

⑦ FULL TILE 4 1/4" x 4 1/4"

⑧ HALF TILE 4 1/4" x 2 1/8"

QUARTER TILE (NOT SHOWN) 2 1/8" x 2 1/8"

USUALLY OF STAINLESS STEEL OR ALUMINUM. USED WITH METAL DOORS OR METAL WINDOWS

⑨ OUTSIDE CORNER BASE 2 1/8" x 4 1/4"

Fig. 11-9. Each different kind of tile must be considered in the estimate.

then placing the tiles against the prepared wall and tapping gently until they are united with the bed. (Fig. 11-8 illustrates buttering.) When the tiles are set, by either process, the joints are carefully washed out and filled with cement. When plumbing fixtures are to be installed, a wooden strip to which the fixtures can be secured is fastened to the wall flush with the rough bed and the tiles are laid over it. The setting of mosaic tile is exactly as explained for floors. Fig. 11-9 shows typical uses of tile.

Fig. 11-10. Wood paneling is used effectively for wall covering. It has beauty, durability, and requires a minimum of maintenance. (National Forest Products Assoc.)

Insulating Boards and Tile. Insulating boards and tile are manufactured from cornstalks, wood, and other substances and have definite insulating as well as decorative qualities. They are obtainable in many shapes, sizes, and thicknesses, and can be applied directly to studs or furring strips, or over old surfaces. They may also be used on ceilings. The surfaces have varied textures and finishes. Usually such insulation material is nailed.

In some instances large sheets are used; in others, workmen, using specially designed tools, cut or carve designs in boards, tile, or planks as they are installed.

Wood Board Paneling. Wood has become popular as a wall covering material, particularly in living, dining and recreational areas of the house. The number of different species of hardwood now available with a wide range of color and finish makes many choices possible. Some woods are light, others dark, some have imperfections such as burls, knots and markings which make them attractive. Some, such as pecky cypress or cedar, have deep fissures caused by insects or fungi. Others are sandblasted to show a swirl of hard and soft grain, or machined to give a striated effect. Much of the wood for wall paneling purposes comes prefinished, thus saving the builder much of the cost of decorating. See Fig. 11-10.

Paneling of pine, fir, and cedar comes in random widths of 4, 6, 8, 10 and 12 inches with a dressed thickness of approximately ¾ of an inch. The lengths are from 6 to 16 feet. The boards are tongue and groove and usually have a V joint or a more elaborate molded edge pattern.

When the paneling is applied horizontally, no furring strips are required because the paneling is applied directly to the studs.

Furring strips are required for masonry walls. On masonry walls, the furring prevents damage from moisture to the panels or planks; and on studs, when the planks are applied vertically, 2 × 4 headers (or cripples) are installed between the studs approximately 3 feet apart to provide a nailing surface for the planking. See Fig. 11-11.

Plywood Paneling. The plywood panels used for finished wall covering are generally 3 ply ¼ inch or 5 ply ⅜ inch. They come in several sheet sizes, thus providing the greatest economy of material. In general, panels are available in 48 inch widths and in lengths of 7 and 8 feet. Some are available in 10 foot lengths. Longer lengths may be ordered.

Oak, elm, walnut, cherry, birch and pine are some of the more common native woods used. Exotic and foreign veneers are also available.

Plywood is placed in a vertical position in most applications. The furring strips are placed vertically on 16-inch centers. The panels are then cut to reach from floor to ceiling and the backs of the panels marked approximately where the furring strips occur. The next operation is to apply two coats of contact cement to the furring or studs and to the backs of the panels

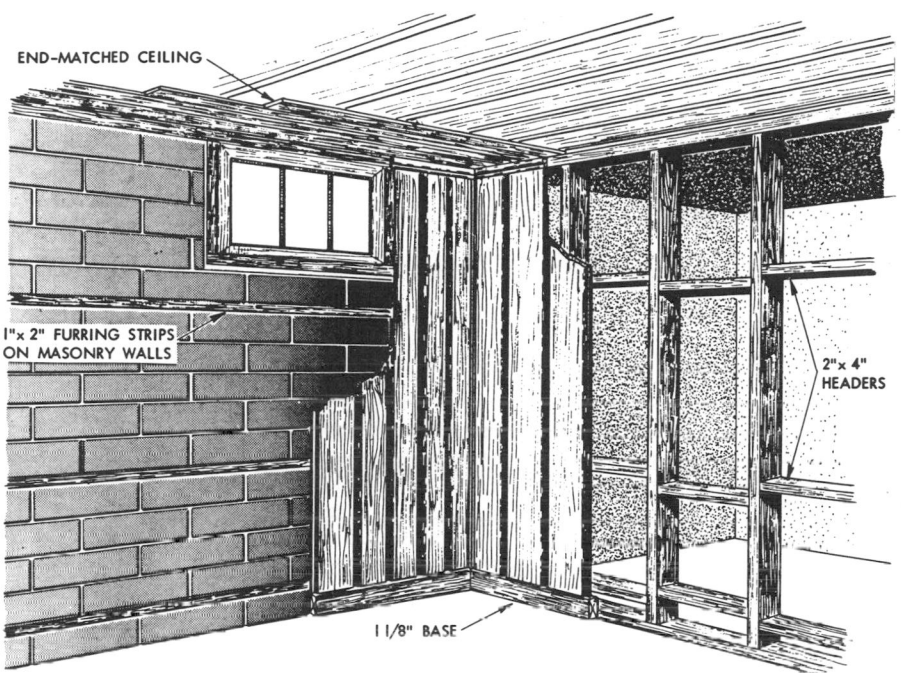

END-MATCHED CEILING

1"x 2" FURRING STRIPS
ON MASONRY WALLS

2"x 4"
HEADERS

1 1/8" BASE

Fig. 11-11. Boards may be nailed to headers placed between studs. Note furring strips on masonry walls.

where marked, allowing each coat to dry thoroughly. Then the plywood is put in place carefully and pressed against the furring strips to insure contact. The panels can be tacked at the top and bottom where nail holes will be covered by trim. Some carpenters use 4d finishing nails along studs to reinforce the cement bond. Nail holes are countersunk and filled with matching putty. A variety of moldings such as casing and trim, window stool, base, cove, inside and outside corners, etc. is available to match the factory finished panels.

Manufacturers have perfected clips which are completely hidden as each piece is put in place.

Hardboard. Hardboard is made by exploding wood chips into a fibrous state and then pressing them in heated hydraulic presses until they form a very dense, rigid board. It has a hard, smooth surface, great structural strength, and excellent wearing qualities; it takes finish readily, is easily worked with ordinary tools, and can be bent with comparative ease into curved shapes.

Hardboard is prepared for wall applications with a great variety of wood grain finishes simulating fine woods. Grooves are provided at random widths so that the finished wall has the appearance of a series of planks. One type resembles marble,

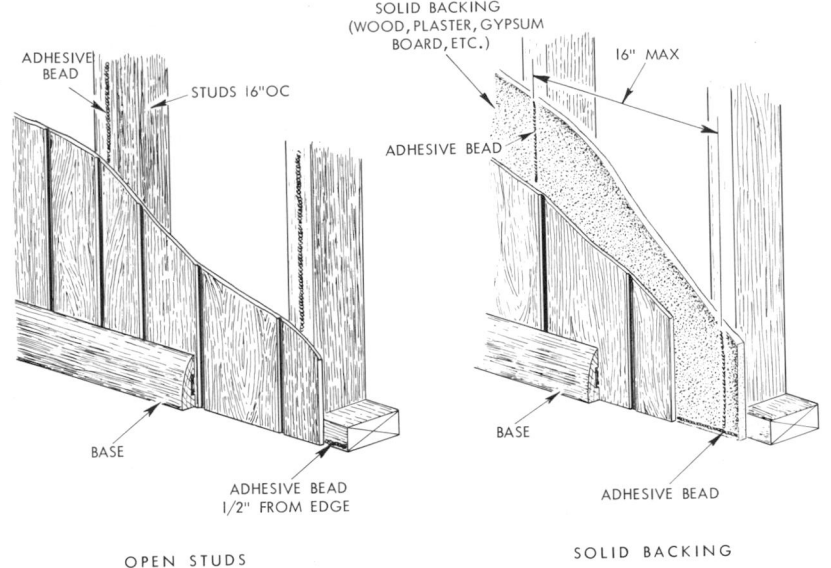

Fig. 11-12. Hardboard panels are fastened to the studs with adhesive. (Masonite Corp.)

other types have plain surfaces in many colors. A very durable plastic coating over the surface makes hardboard paneling highly wear resistant.

Most of the hardboard paneling is supplied in 4 × 7, 4 × 8 and 4 × 10 foot panels which are ¼ inch thick. It may also be obtained in 16 and 24 inch planks, 7, 8 and 10 feet long. Many builders prefer to apply the sheets by using adhesives. See Fig. 11-12.

Ceiling blocks 16 inches square and ¼-inch thick with tongue and groove edges are available for glueing to old ceilings or for fastening to furring strips with staples.

A series of matching metal or plastic moldings is provided to cover up inside corners, outside corners, and panel edges where panels end and where they join. Wood moldings are

provided for base, cove and also for inside and outside corners.

Ceilings. Many of the wall coverings are also used for ceilings. This is especially true of plasterboard and tile.

Acoustic Tile. Noise reduction through the use of sound-absorbing, soft or porous tile is a development of the last thirty years. Today, a great variety of tiles of this kind are available. Manufacturers offer many types and sizes of acoustic tile of rated noise-reduction ability.

Insulating Tile. Some mention should be made of insulating tile, which is very similar to acoustic tile. Insulating tile, of course, is produced to prevent the passage of heat or cold. Acoustic tile is produced to present a high degree of sound-absorbing capacity. Consequently, insulating tile and acoustical tile may often be similar in appearance. Insulating tile is, however, less expensive than acoustic tile.

Both acoustic and insulating tiles can be nailed or clipped on, or they can be stuck to the surface to which they are applied with mastic. Each of these methods of application is suited to given conditions. Often noise reduction is achieved as much by the use of the right suspension method as by the tile itself.

Tile Sizes. Certain types of tile, such as fire-resistant tile and grease-resistant tile, are made to suit specific applications. Sizes conform to architectural needs. Standard sizes are 12″ × 12″, 12″ × 24″, 24″ × 24″, 16″ × 16″, 16″ × 32″, 16″ × 48″, 24″ × 48″, and 48″ × 48″. Thicknesses vary according to the amount of noise reduction the job calls for. Some of the thicknesses are ½″, ⅝″, ¾″, ⅞″, 1″ and 1⅛″. Each thickness has a rated sound absorbing ability.

Fig. 11-13. Applying acoustic tile with mastic. (Insulation Board Institute)

The smaller sizes of tile are usually acoustic tile. Acoustic tile seldom exceeds 12″ × 24″ in size. The number of companies making acoustic tile for general use are few, but these manufacturers offer a total of more than twenty different types.

Methods of Attaching Tile. Most of the acoustic tile erected is stuck to the base with mastic. See Fig. 11-13. This method of erection is suited to bases such as concrete slab, gypsum board, and plastered metal or gypsum laths. Plaster and gypsum provide the best all-round base materials because they are straight, strong, fireproof and dustproof.

Tile may also be nailed or stapled to wood furring strips. See Fig. 11-14. This is done on some remodeling jobs and on new work when the ceiling is attached to the joists. This system is subject to seepage of dust through the joints. Moreover, it is not rated as fireproof, since the strips are of wood. Furring strips may also be nailed or tied to suspended metal channels.

Metal clips and special channels also are used for suspension of acoustic tile. These systems provide firesafe, positive-grip methods of erection, but they are not proof against filtration of dust through the joints.

Suspended Grid System. There are many installations incorporating the suspended grid system, which may fall under exposed grid (direct-hung suspension system) or concealed grid. These are for acoustical ceilings only. (The exposed grid system will show metal. Concealed zeebar is a system which shows no metal; only tile is visible from the floor.) In all cases the

Fig. 11-14. Applying acoustic tile by nailing.

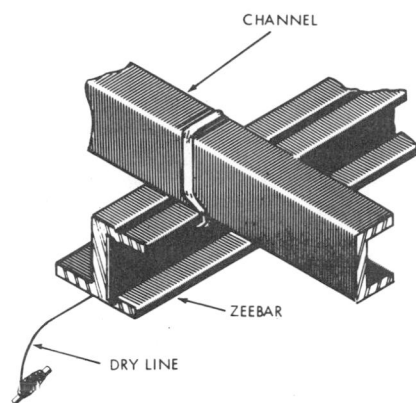

Fig. 11-15. Concealed zeebar. (1) Center zeebar over dry line; hold flush against channel. (2) Attach zeebar clip over channel; insert both ends under top flange of zeebar. (3) Fasten additional clips to zeebar if it intersects with channel.

recommended procedure for installation should be followed.

The concealed zeebars with splines provide an unbroken, monolithic appearance which is extremely popular today. This system uses 1½" channel which is suspended by the lather.

The zeebars are attached with clips at right angles to the 1½" carrying channels (see Fig. 11-15) at intervals specified, usually 12" on center, wherever the zeebars come in contact with the channel, generally four feet on center. Wall moldings are installed in the same manner as described for exposed grid.

ESTIMATING MATERIAL

The amount of material required to finish off a given floor, wall, or ceiling is determined by the total area to be covered—in square feet. Any large openings such as windows, doors, arches, stairs, fireplaces and hearths, heating units, etc., are treated as deductible items in the take-off.

Before the estimator begins his work he must study carefully the plans and specifications of the working area. He must know of what materials the existing or proposed subsurface consists. The subsurface material (which may be concrete, plaster, or wood) determines the kind of material and the extent of the work that is to be applied to the surface. To be sure of a correct estimate, the estimator should also have a knowledge of the various methods used to apply the special coverings.

Floors. In addition to the special flooring material, you must include any base molding and paste needed for proper installation.

Plank Floors. The amount of lumber required for special plank floors is estimated in square feet. This is calculated as the total area of floor to be covered.

Example 1. If a colonial plank floor is to be laid in a room which is 8'-0" × 12'-0", how much flooring is needed?

Solution: Simply find the floor area.
8.0' × 12.0' = 96.0 sq. ft.

Block Floors. The blocking can be computed by the square foot or by the piece (for which you must know the area to be covered). This is a costly item because of the millwork needed to achieve the desired course. The kind of subflooring used will also greatly affect the total cost. If the work is to be done on a concrete bed, a mastic fill must be used. Dampproofing may also be required. These items are calculated in relation to the amount of lumber which is to be laid down, or in relation to the room area.

Linoleum. Since many kitchen plans specify a floor of this type, it will be a common item in the estimator's work. The amount of linoleum needed is calculated and ordered by the square foot or by the square yard. The dimensions (usually wall-to-wall) are taken to the deepest part of the room. Cutting is usually necessary around closets and other obstructions; however, the excess is used to line the inside of the closets.

Example 2. How many square yards of linoleum are needed for the room in Example 1?

Solution: Since the area, in square feet, is already known, convert to square yards.

$$\frac{96}{9} = 10.6 \text{ or approximately}$$
$$11 \text{ sq. yds.}$$

If felt backing is necessary, it is ordered to the exact size of the linoleum. The adhesive paste is bought by the gallon. Allow one gallon of paste for each 100 square feet of linoleum.

If special metal moldings are necessary to protect the linoleum, they are ordered by the linear foot.

When waxing is specified, allow one gallon of wax for each 200 to 300 square feet.

Terrazzo. Since terrazzo may be applied in either of two ways, the estimator must know the type of subflooring.

If it is to be a bonded job, the amount of underbed fill must be calculated. This will be taken as if it were a poured slab. The amount of terrazzo topping is again calculated as a poured slab in relation to the area of the floor and the depth of the terrazzo finish.

If the terrazzo is to be separated from the concrete slab, a thin bed of dry sand is laid and tar paper is put over it. The underbed is put over this and the finished topping is installed. All fills are figured as if they were a slab. The tar paper is calculated as the area of the floor.

When terrazzo is used over wood floors, tar paper is again used, but an overlay of wire mesh is also necessary. This is figured as the area of the floor since the mesh will run under the entire surface. If any hand troweling is necessary, it is figured into the estimate. Brass dividers are necessary to separate the floor areas and to achieve the various desired patterns on the floor surface. These are figured in linear feet. The depths of the brass strips must also be considered when ordering or figuring the costs for this item.

Monolithic Flooring. Poured flooring depends on the thickness required and the number of square feet that a gallon of the material will cover. Roll or seamless flooring is bought by the square yard but the width of the material determines the waste factor.

Floor Tiles. The tiles, whether of vinyl, asphalt, cork, or rubber, are ordered and installed by the square foot or by the piece in relation to square feet of floor area. The area of a single tile is divided into the total area of the room to determine the number of tiles needed for the room. Some

TABLE II. ESTIMATING FLOOR TILE

SQUARE FEET	NUMBER OF TILES			
	9" x 9"	12" x 12"	6" x 6"	9" x 18"
1	2	1	4	1
2	4	2	8	2
3	6	3	12	3
4	8	4	16	4
5	9	5	20	5
6	11	6	24	6
7	13	7	28	7
8	15	8	32	8
9	16	9	36	8
10	18	10	40	9
20	36	20	80	18
30	54	30	120	27
40	72	40	160	36
50	89	50	200	45
60	107	60	240	54
70	125	70	280	63
80	143	80	320	72
90	160	90	360	80
100	178	100	400	90
200	356	200	800	178
300	534	300	1200	267
400	712	400	1600	356
500	890	500	2000	445
LABOR PER 100 SQ. FT.	2 HOURS	1.3 HOURS	3.3 HOURS	1.3 HOURS

TABLE III. TILE WASTE ALLOWANCES

```
  1 to   50 sq. ft. . . . . . . . . . . . . . . . . . . . 14%
 50 to  100 sq. ft. . . . . . . . . . . . . . . . . . . . 10%
100 to  200 sq. ft. . . . . . . . . . . . . . . . . . . .  8%
200 to  300 sq. ft. . . . . . . . . . . . . . . . . . . .  7%
300 to 1000 sq. ft. . . . . . . . . . . . . . . . . . . .  5%
Over 1000 sq. ft. . . . . . . . . . . . . . . . . . . . .  3%
```

To find the number of tile needed for an area not shown on the chart, such as the number of 9" x 9" tile needed for 850 sq. ft., add the number of tile for 50 sq. ft. to the number of tile needed for 800 sq. ft. The result will then be 1513 to which must be added 5% for waste (see table). Total 1589.

TABLE IV. FLOORING ADHESIVES

Type and Uses	Approximate Coverage in Square Feet
PRIMER--For treating on- or below-grade concrete subfloors before installing asphalt tile . .	250 to 350
ASPHALT CEMENT--For installing asphalt tile over primed concrete subfloors in direct contact with the ground	200
EMULSION ADHESIVE--Adhesive used for installing asphalt tile over lining felt	130 to 150
LINING PASTE--For cementing lining felt to wood subfloors	160
FLOOR AND WALL SIZE--Used to prime chalky or dusty suspended concrete subfloors before installing resilient tiles other than asphalt	
WATERPROOF CEMENT--Recommended for installing linoleum tile, rubber and cork tile over any type of suspended subfloor in areas where surface moisture is a problem	130 to 150

cutting of tile is almost always necessary, therefore a waste factor should be used. The waste factor is determined by the size and shape of the job.

Table II tells you how many tiles and their size for a given area. Table III gives you the waste factor for a given area. Table IV gives the approximate coverage for certain floor adhesives. Example, for 100 sq. ft. of floor to be covered with 12" × 12", look in Table II under the size of the tile and find the number of tiles (100); go to Table III and look under the area to be covered and find waste factor of 10%. You now have the total number of tiles needed to cover 100 sq. ft. of floor area. Table IV gives the amount of coverage based on one gallon of the adhesive.

Example 3. Determine the number of 9" × 9" asphalt floor tiles which would be needed for the room in Example 1.

Solution: Find the area of one tile and the room area (in square inches); then divide the tile area into the room area to determine the number of pieces.

Tile area $= 9'' \times 9'' = 81$ sq. in.

Room area $= 96 \times 144$
$$= 13,824 \text{ sq. in.}$$

No. of tiles $= \dfrac{13,824}{81} = 171$

The amount of adhesive is determined by the square feet of tile area needed to cover the floor. If a felt lining is used, it is the same size as the floor area. If any metal strippings are necessary, they are figured by the linear foot. The number of marble thresholds is taken from the floor plans and is dependent on the number of bathrooms. The size of the threshold is the same as the door width.

Ceramic Tiles. These tiles are calculated in the same manner as floor tiles. If any special molding is needed to finish the floor edges, it is calculated by the linear foot in relation to the perimeter of the base of the walls. Trim pieces such as (bullnoses, caps, base strips, etc.) are figured by the linear foot. If dampproofing is required, allow one gallon of ceramic tile primer per 125 square feet.

Carpeting. Wall-to-wall carpeting is estimated by the yard in relation to the length that is bought. In the final pricing, it is priced by the square yard.

If the room is 10'-0" × 11'-0", we must buy carpeting 12'-0" wide since the standard widths of room carpeting are 12'-0" and 15'-0". We would then buy 10 linear feet of carpeting. The amount of carpet we would buy and pay for would be 120 sq. ft. whereas the room requires only 110 sq. ft. As you can see, wall-to-wall carpeting usually has a built-in waste factor.

Carpet tiles may be estimated by methods used for any other type of tile.

Walls. Since the various wall coverings may be ordered in different ways, they must be calculated in relation to the contact area of the walls. Once again, it is to the estimator's benefit to familiarize himself with the methods of installing these materials.

Wallpaper. The amount of wallpaper is taken as the number of rolls needed to finish a room. The number of rolls is determined by the wall area

and the area of each roll (in square feet). If a border is required, it is ordered in linear yards. The paste is bought by the pound. The amount of paste depends on the weight of the paper. The percentage allowed for waste is affected by the pattern of the wallpaper.

Example 4. An 8'-0" × 12'-0" room has a ceiling height of 8'-0". The room has one window 3'-0" × 4'-0" and one door 3'-0" × 7'-0". How much wallpaper is needed?

Solution: Find the area of the walls and subtract the window and door openings.

$$\text{Wall area} = 8.0 \times (8.0' + 12.0' + 8.0' + 12.0')$$
$$= 320.0 \text{ sq. ft.}$$
$$\text{Window area} = 3.0' \times 4.0'$$
$$= 12.0 \text{ sq. ft.}$$
$$\text{Door area} = 3.0' \times 7.0'$$
$$= 21.0 \text{ sq. ft.}$$
$$320.0 - 12.0 - 21.0 = 287.0 \text{ sq. ft.}$$

Fabric material used as wall covering is treated in the same manner as wallpaper.

Tile. The amount of tile is determined by the size of the tile in relation to the wall area. If furring strips are used, they are estimated by the linear foot. Adhesive is figured by the pound or gallon, depending on the kind. If the walls are damp, primer is used. On the average, allow 130 square feet per gallon. Grout for filling the tile joints is figured at one pound per 15 square feet.

If the floating method is used, the take-off will be the same as for floor tile. The buttering method requires less adhesive. Trim pieces are figured the same as for floor tile.

Example 5. How many 6″ × 6″ tiles are needed for the room in Example 4? A 6″ cap and a 6″ base run all around the room.

Solution: Find the total area of the wall and then deduct the areas of the cap, base, and door; then find the area of each tile and divide into the area to be covered.

$$\text{Wall area} = 320.0 \text{ sq. ft.}$$

Cap area =
$$0.5' \times (40.0' - 3.0') = 18.5 \text{ sq. ft.}$$

Base area =
$$0.5' \times (40.0' - 3.0') = 18.5 \text{ sq. ft.}$$

$$\text{Door area} = 21.0 \text{ sq. ft.}$$

Area covered =
$$320.0 - 58.0 = 262.0 \text{ sq. ft.}$$
$$= 37,728 \text{ sq. in.}$$

$$\text{One tile} = 36 \text{ sq. in.}$$

$$\text{No. of tiles} = \frac{37,728}{36} = 1,048$$

Insulation Boards. These are figured in the same manner as tile.

Wood Panels. These are figured by the square foot of wall area. A more accurate method is to compute the square footage, taking into consideration the height of the walls in relation to the spacing of the studs. Furring strips are figured by linear feet.

Ceilings. Ceilings are also figured in square foot of surface area. Sometimes furring strips are required, especially for acoustical tile which is nailed or stapled. The amount of adhesive depends on the area to be covered and the absorption capacity of the backup wall.

House Plan A—Quantity Take-Off. Fig. 11-16 is a quantity take-off for the special floor and wall coverings for HOUSE PLAN A. The floor and wall coverings are listed by the total area in square feet of surface to be covered. The dimensions are obtained from the plans.

The caps, bases, and special moldings are listed in linear feet. Each corner, whether inside or outside, requires an angle cap to finish off the wall.

The waste factor used here is 10%. This is added to the amounts of materials after they are listed. This figure is used for *illustrative purposes only.* Actual waste factors would vary with the various materials.

After the estimator lists the materials and quantities, he calculates the total cost of the materials, the unit cost of labor, and the total labor cost.

ESTIMATING LABOR

In this type of work the cost of labor is quite high because of the special hand labor required. The estimator must visualize the type of job which is proposed and then know the various kinds of labor required to finish it.

The following labor estimates are furnished as a rough guide. They should only be used as an outline. The amount of work done in a given period of time depends on the working conditions and the skill of the workmen. The estimator familiarizes himself with the local working laws and wages. Since workmen's wages vary regionally, the hourly pay rates will differ among regions.

			SHEET /
QUANTITY TAKE-OFF			
SPECIAL, FLOOR AND WALL COVERINGS			*HOUSE PLAN A*

Location	Description	Unit	Amount	Amount + 10%	
Kitchen					
	Plywood base	sq. ft.	137 —	151 —	
	Vinyl floor	sq. ft.	137 —	151 —	
	Rubber base	lin. ft.	41 —	45 —	
	Paste	gal.	1.5	1.7	
Laundry					
	Plywood base	sq. ft.	51 —	57 —	
	Asphalt tile	sq. ft.	51 —	57 —	
	Rubber base	lin. ft.	17 —	19 —	
	Paste	sq. ft.	55 —	60 —	
Lavatory					
	Plywood base	sq. ft.	23 —	26 —	
	Ceramic tile	sq. ft	23 —	26 —	
	Ceramic base	lin. ft.	16 —	18 —	
	Wall tile	sq. ft.	75 —	83 —	
	Bullnose caps	lin. ft.	16 —	18 —	
	Angle caps	lin. ft.	27 —	30 —	
	Tile adhesive	sq. ft.	80 —	88 —	
	Marble threshold	lin. ft.	2 —		
Bathroom1					
	Ceramic tile floor	sq. ft.	40 —	44 —	
	" " base	lin. ft.	26 —	29 —	
	" " wall	sq. ft.	240 —	264 —	
	Bullnose caps	lin. ft.	26 —	29 —	
	Angle caps	lin. ft.	27 —	30 —	
	Adhesive	sq. ft.	280 —		
	"	lin. ft.	79 —		
	Marble theshold	lin. ft.	2.33		
Bathroom2					
	Ceramic tile floor	sq. ft.	30 —	33 —	
	" " base	lin. ft.	22 —	25 —	
	" " wall	sq. ft	128 —	141 —	
	Bullnose caps	lin. ft.	22 —	25 —	
	Angle caps	lin. ft.	18 —	20 —	
	Adhesive	lin. ft.	62 —		
	"	sq. ft.	128 —		
Note:	Paste and adhesive are ordered by the gallon.				

Fig. 11-16. House Plan A. Quantity take-off.

Floors. Labor for installation of special flooring is costly because of the special treatment of nailing and joining, and the necessary sanding and waxing of the finished surface.

One man can nail 1″ × 6″ tongued-and-grooved *planking* at the rate of 100 sq. ft. per 1½ man-hours. If special hand tooling is necessary, this time will be tripled.

A worker can lay 100 sq. ft. of *block flooring* over a pasted floor in 1 hour. If the flooring is nailed, the carpenter may take 3 hours per 100 sq. ft.

Labor for *linoleum* is not as costly for this item. If patchwork is necessary on the floor before laying, then this must be included. A man can put down approximately 4 to 5 square yards per hour.

Labor for installation of a *terrazzo* floor is costly. High quality and skill are necessary. Hand troweling is necessary to finish the surface; special surfacing is done with a grinder. Under average conditions, 100 sq. ft. of floor can be put down in about 10 man-hours. Allow 8 hours per 100 sq. ft. for grinding.

Floor tile is also costly. Work is tedious and slow; care must be taken to keep the tiles in a straight line. If troweling is necessary before the floor is laid, this must be calculated too. A tile setter can average 100 sq. ft. of tile per 2½ hours under average conditions. If a felt lining is used, then it can be placed at the rate of 200 to 300 sq. ft. per hour. Trim pieces are included in the general setting time. It takes approximately 15 minutes to set a threshold.

Ceramic tile can be set at the rate of 100 sq. ft. per 3 man-hours.

Walls. Labor rates in some areas are dependent upon such things as the kind and size of rolls of wallpaper.

A paper hanger can paste about one roll per hour. The rate of work is dependent on the pattern and the shape of the room. The mixing time for the paste is almost negligible and need not be included in the estimate. Preparing the wall for pasting, however, should be included.

Wall tiles can be calculated at the rate of 50 sq. ft. per 3 man-hours. This is affected by the size of the tiles and the amount of cutting.

Insulating boards are figured in the same manner as wall tiles.

Wall panels average approximately 50 sq. ft. per man-hour. The time varies according to the kind of panels and the shape of the room. The time for any roughing necessary to prepare the room must also be included.

Ceiling Coverings. This is figured at approximately the same rate as comparable work on the walls. *Acoustical tile,* for example, is figured at the same rate as wall tile.

Other Work. Preparing and finishing surfaces of floors, walls, and ceilings must also be included in the labor take-off.

Pouring and leveling a *cement underbed* for a floor is done at the rate of 500 sq. ft. per man-hour.

Furring (½″ wood strips) is done at the rate of 150 lin. ft. per man-hour.

Molding around the walls is installed at the rate of 100 lin. ft. per man-hour.

Hand troweling an average surface takes 1 man-hour per 100 sq. ft. Using an *electric sander* to finish the same surface requires 1 man-hour per 30 sq. ft.

See Appendix D, pages 476 to 477, 480, and 483 for Hourly Estimates for:

Subfloors, Finished Floors: Wood Finish Flooring; Resilient Type Flooring; Ceramic Tile on Concrete; Slate on Concrete; Marble on Concrete; Brick on Concrete; Patio Blocks on Concrete;

Walls and Ceilings: Insulating or Acoustical Tile; Wall Tile; Hardboard; Metal Sidewalls and Ceiling; Curtain Wall Panels;

Painting: Wall Paper.

Questions and Problems

1. How is the amount of floor tile estimated?
2. How are marble threshold sizes determined?
3. How is the labor figured for wall tiles?
4. How long does it take one man to finish 75 sq. ft. of floor with an electric sander?
   (Answer: 2.5 hrs.)
5. How many 9″ × 9″ tiles are needed for a floor area of 148 sq. ft.?   (Answer: 264)
6. A 10′-0″ × 10′-0″ room has one 2′-6″ door. How much base molding is needed? (Ans.: 37.5 lin. ft.)
7. How much ceramic tile is needed for the floor of a bathroom, 7′-0″ × 8′-0″, with a bath tub 2′-0″ × 5′-0″?   (Answer: 46.0 sq. ft.)
8. A room is 9′-0″ × 9′-0″; the height of the room is 10′-0″. There is a 3′-0″ × 4′-0″ window and a 3′-0″ × 7′-0″ door. How much wallpaper is needed?
   (Answer: 327.0 sq. ft.)

12

PAINT

∿∿∿∿∿∿∿∿∿∿∿∿∿∿∿∿∿∿∿∿∿∿∿∿∿∿∿∿∿∿∿∿∿∿∿∿

Painting is another one of the finishing touches in building construction. The estimator must study the specifications to determine just what is to be painted and how many coats of paint are needed for the job.

MATERIALS

Specifications. The specifications may designate all areas to be painted, varnished, enameled, or stained—or they may merely state that all exterior wood, for example, shall be painted. This means that the estimator must study the plans to determine the exact areas which require painting. A typical list of outside areas might be as follows:

siding or shingles
trim
cornice
sash
window frames
porches
fences
steps
floors
garage

If one or more outside items are omitted, it can easily reduce the contractor's profit.

In a similar manner, you must consider all interior painting, varnishing, enameling, waxing, etc.

Paint can be purchased in many grades. For this reason a specification should give definite instructions regarding the paint, either naming the brand or listing the various materials used in making the paint.

Often, due to climatic or other conditions, the specifications give exact materials and formulas for all areas to be painted.

For many years painting consisted of applying a mixture of white lead, oil, and color which the painter mixed on the job. Materials now used in paint have almost eliminated hand mixing.

The estimator should determine which type of paint is to be used and the method of application. Some paints are applied with rollers, others by spraying, and many are brushed on. Most paints must be applied to dry surfaces but there are some that can be applied to moist or damp wood, in fact they penetrate better.

Putty is included with paint. Formerly it was a whiting, lead and oil mixture, but new materials are being used that provide more lasting results and better adhesion.

The estimator should state clearly the material he used as a basis.

Miscellaneous Items. Some items belonging in a painting estimate, while seemingly unimportant, can reduce the contractor's profit should they be omitted from the estimate.

Cleaning. The painting contractor must remove all paint spots from finished work and leave the premises free of rubbish. Sometimes the specifications require the painters to clean the windows.

Protection of Work. The painting contractor has the responsibility of protecting his work and the work of all other contractors during the time that his work is in progress. He is responsible for any and all damage to the work or property of others caused by his employees or himself.

Sanding. On window sash, for example, sanding may be necessary to remove rough places left by machine sanders at the mill. Other items of trim, splices, joints, etc., often require sanding to give them the smoothness of recommended work.

Major sanding operations are frequently required on floors before finishing can proceed.

Puttying. All nail holes, cracks, etc., should be filled with putty to make a first-class finished job.

Scraping. Scraping is a time-con-suming item which should not be hurried. It is hard work and the workmen usually require frequent rests.

Preparation for Painting. The estimator must know of the preparations necessary for painting. First, there is the work of cleaning and preparing surfaces; second, the time spent in mixing paints; and third, the erecting of scaffolds, covering with canvas, etc.

Preparing Surfaces. New wood surfaces must be smooth. All rough places must be filled with putty. Dirt or other spots must be removed. Previously painted surfaces may be lightly sanded and dusted. Old blisters should be removed. Sometimes old paint must be removed with a blow torch. Old surfaces from which paint has been removed require a primer coat in addition to the usually specified second and finish coats.

Plaster surfaces, if new or not aged, require treating by a solution made of two pounds of zinc sulphate dissolved in one gallon of water. This is applied with a rag or a large brush. Aged or previously painted plaster does not require a priming coat.

Cracks in plaster should be filled; if this is to be done correctly, then considerable time must be allowed on the labor estimate.

Stucco, concrete, bricks, and stone also require special preparations. If stucco or concrete is not aged, a solution of zinc sulphate and water should be applied before any painting is done. New brick must be allowed to dry several days. On old brick surfaces, the joints frequently require tuckpointing before painting can be done. Small de-

fects in the brick surfaces are sometimes filled with putty.

Scaffolds. In many cases, ladders
and planks serve this purpose, but
in others special scaffolds must be
erected and taken down. The estimator must not forget to consider this
item in his painting estimate.

ESTIMATING MATERIAL

Areas to be painted are calculated in
square feet but the paint is bought by
the *gallon.* The estimator should know
how much area can be covered by a
gallon of paint, varnish, stain, etc.

Painting Materials. In addition to
paint, there are a number of products
which are included under the paint
estimate. These include stain, varnish,
shellac, enamel, putty and sizing.

TABLE I. COVERAGE FOR PAINTS
(Refer to manufacturers specifications and coverage figure)

MATERIAL	Sq. Ft. to Gallon		
	1 coat	2 coats	3 coats
Enamels	500	250	195
Flat wall paint: white/light colors on smooth finish	575	290	215
Flat wall paint: dark colors on rough sand finish	725	365	240
Inside floor paint	500	275	
Outside house paint: white/ light tints, porous woods	475	255	190
Outside house paint: white or light tints, close grained woods	525	275	190
Outside house paint: dark colors, greys, tans, etc., porous grained woods	525	280	215
Outside house paint: dark colors, greys, tans, etc., close grained woods	575	300	215
Stain, wood tints	500		

The estimator must find out whether
one or two coats will be required and
the approximate coverage for each
coat. See Table I.

The paint industry is undergoing
many changes in materials and methods at present and the latest information on the products used should be
obtained.

Floor varnish. Most contractors assume that 1 gallon of varnish will cover
700 square feet.

Stain covers about 500 square feet
per gallon.

Shellac can be estimated at 1 gallon for every 700 square feet of sur
face.

Trim varnish will cover about 700
square feet per gallon.

Doors. If exterior door frames are to
be the same color as the outside walls,
then no extra paint need be calculated. If the door (or window) frame
is to be a contrasting color, the exterior paint can probably be tinted in
sufficient amount for the frame color.
In such a case, about 3 quarts are figured for each 20 openings.

For interior door frames (and window openings) about 1 quart of paint
is required per coat for 4 openings.
The same allowance is made for bases,
picture moldings, etc.

Stain and filler for doors cover about
500 square feet per gallon.

On the average, allow 1 quart each
of stain, shellac, and varnish for every
5 doors. Although doors can be estimated individually, this method is sufficient for most jobs.

Interior paint covers about 700
square feet for priming coat and 800

square feet for second and third coats. The areas of doors are figured by multiplying length by width and then multiplying by two because of two sides. Sometimes an additional two or three square feet are added to take care of the bevels on panels, etc.

For *enamel* paint on doors, most estimators figure the coverage as the same as regular paint or up to 10% more.

For screen doors, one gallon of screen paint cut with ½ gallon of benzine generally will be ample to paint both sides of 25 screen doors with one coat.

Windows. The sash and frames are painted at the same time as the wall surfaces and the amount of paint figured for these surfaces is ample for the window sash and frames. No additional paint need be considered.

For brick walls, window frames and sash are considered as separate jobs because the exterior walls usually are not painted. About 3 quarts of paint per coat is required for 20 windows, or 20 openings. Ten openings require about ½ gallon of paint per coat.

About ⅓ the amount of paint required for wood sash usually is sufficient for steel sash. One quart of paint should be enough for one coat on 20 steel sash.

One half gallon of oil will prime the sash for about 25 windows.

When enamel is to be used for windows and other openings, allow 1 quart of undercoating for every four openings. The enamel provides about the same coverage as ordinary paint; sometimes it covers slightly less.

If window sills require spar varnish, allow 1 pint per 6 room residence. One pint stain and one pint shellac will also be needed.

For painting storm windows, allow 1 quart of paint for every 10 windows.

If the building has shutters, then the total area of the shutters, including both sides, must be considered when they are to be painted. The paint is assumed to cover 500 square feet per gallon.

Exterior Painting. Most painting contractors figure that 1 gallon of paint will cover about 700 square feet of *wood siding*. For *shingled walls,* allow about 600 square feet per gallon. Table II is a handy reference for estimating amounts of paint required. *Example 1.* A residence 30'-0" × 50'-0", 20'-0" high, has shingles on all walls. How much paint is needed for each coat?

Solution: First find the total wall area.

$$30.0' + 50.0' + 30.0' + 50.0'$$
$$= 160.0'$$
$$160.0' \times 20.0' = 3{,}200 \text{ sq. ft.}$$
$$\frac{3{,}200}{600} = \text{approximately 5.3 gallons}$$

5½ gallons should be sufficient.

If the *roof shingles* are to be stained, allow 1 gallon per 150 square feet (per coat).

Cornice paint is included in the paint for exterior wall surfaces on frame buildings. For brick buildings the cornice area is calculated and the paint bought on the basis of 700 square feet per gallon.

Two quarts of paint will usually

TABLE II. APPROXIMATE PAINT REQUIREMENTS FOR INTERIORS AND EXTERIORS

DISTANCE AROUND THE ROOM	CEILING HEIGHT 8 FEET	CEILING HEIGHT 8-1/2 FT.	CEILING HEIGHT 9 FEET	CEILING HEIGHT 9-1/2 FT.	PAINT FOR CEILING	FINISH FOR FLOORS	FOR EACH DOOR OR WINDOW
30 FEET	5/8 Gal.	5/8 Gal.	3/4 Gal.	3/4 Gal.	1 Pt.	1 Pt.	Each Window
35 FEET	3/4 Gal.	3/4 Gal.	3/4 Gal.	7/8 Gal.	1 Qt.	1 Pt.	and Frame
40 FEET	7/8 Gal.	7/8 Gal.	7/8 Gal.	1 Gal.	1 Qt.	1 Qt.	Requires 1/4 Pint
45 FEET	7/8 Gal.	1 Gal.	1 Gal.	1-1/8 Gals.	3 Pts.	1 Qt.	
50 FEET	1 Gal.	1-1/8 Gals.	1-1/8 Gals.	1-1/4 Gals.	3 Pts.	1 Qt.	Each Door
55 FEET	1-1/8 Gals.	1-1/8 Gals.	1-1/4 Gals.	1-1/4 Gals.	2 Qts.	3 Pts.	and Frame
60 FEET	1-1/4 Gals.	1-1/4 Gals.	1-3/8 Gals.	1-3/8 Gals.	2 Qts.	3 Pts.	Requires 1/2 Pint
70 FEET	1-3/8 Gals.	1-1/2 Gals.	1-1/2 Gals.	1-5/8 Gals.	3 Qts.	2 Qts.	
80 FEET	1-1/2 Gals.	1-5/8 Gals.	1-3/4 Gals.	1-7/8 Gals.	1 Gal.	5 Pts.	

DISTANCE AROUND THE HOUSE	AVERAGE HEIGHT 12 FEET	AVERAGE HEIGHT 15 FEET	AVERAGE HEIGHT 18 FEET	AVERAGE HEIGHT 21 FEET	AVERAGE HEIGHT 24 FEET
60 FEET	1 Gal.	1-1/4 Gals.	1-1/2 Gals.	1-3/4 Gals.	2 Gals.
76 FEET	1-1/4 Gals.	1-1/2 Gals.	2 Gals.	2-1/4 Gals.	2-1/2 Gals.
92 FEET	1-1/2 Gals.	2 Gals.	2-1/2 Gals.	2-3/4 Gals.	3 Gals.
100 FEET	1-3/4 Gals.	2-1/4 Gals.	2 3/4 Gals.	3 1/4 Gals.	3-3/4 Gals.
124 FEET	2 Gals.	2-1/2 Gals.	3-1/4 Gals.	3-3/4 Gals.	4-1/4 Gals.
140 FEET	2-1/2 Gals.	3 Gals.	3-1/2 Gals.	4 Gals.	4-1/2 Gals.
156 FEET	2-3/4 Gals.	3-1/4 Gals.	4 Gals.	4-1/2 Gals.	5-1/4 Gals.
172 FEET	3 Gals.	3-3/4 Gals.	4-1/2 Gals.	5 Gals.	5-3/4 Gals.

On interior work, for rough, sand-finished walls or unpainted wallboard, add 50% to quantities; for each door or window deduct 1/2 pint of materials for walls. For trim, add 1/8 to 1/5 of the amount required for the body. For exterior blinds, 1/2 gallon will cover 12 to 14 blinds, one coat. Requirements will vary due to condition of surfaces, type of paint and thickness of application.

cover 250 linear feet of *gutters* and *downspouts* when new if the metal has been primed with a red lead coating. Old gutters and downspouts require about 3 quarts per 250 linear feet.

Exterior *porch* floors are covered with at least three coats of paint in most instances. The priming coat covers about 600 square feet per gallon, and second and third coats cover about 700 square feet per gallon. The areas are calculated by multiplying the length by the width. If the floors are irregular in shape they can be broken into squares, rectangles, triangles, etc., for easier calculation.

For porch railings, columns, etc., the estimator must learn to judge by sight or else go through the tedious work of calculating the various items. For a picket railing, the pickets are assumed to be a solid surface and the areas of both sides are totaled. A hand railing is assumed to be cylindrical and the area is figured as the lateral area of a cylinder. Most paint used for such purposes will cover about 700 square feet per gallon.

For *brick, stucco,* and *concrete* walls, allow about 600 square feet per gallon of paint. Do not deduct the areas of openings, unless there are several, each as large as ordinary windows.

Example 2. How many gallons of paint are needed for each coat of a brick building which has 2,400 square feet of surface area?

Solution: As in Example 1, divide the total area by the coverage of one gallon of paint.

$$\frac{2,400}{600} = 4 \text{ gallons}$$

Allow about 400 square feet per gallon for concrete floors.

| | QUANTITY TAKE-OFF | | SHEET 1 | |
| | INTERIOR PAINT | | HOUSE PLAN A | |
Location	sq. ft.	lin. ft.	gallons	remarks
NOTE: 3 coats for all surfaces listed				
Dining–Living Room				
walls	500		2	
ceiling	325		1 1/3	
floor	325		1 1/3	shellac
base molding		64	1/4	
window trim		14	1/10	
bow window trim				not specified
Foyer				
walls	200		1	
ceiling	80		1/3	
floor treatment	80		1/3	
base molding		30	1/8	
door trim		35	1/8	
Family Room				
walls	200		1	
ceiling	170		1	
exposed beams, 6"x4"	320		1 1/3	shellac
floor treatment	170		1	
base molding		50	1/8	
window molding		30	1/8	
door molding		17	1/10	
Kitchen				
walls	280		1 1/3	
ceiling	135		1/2	
exposed beams	150		3/4	shellac
window trim		15	1/10	
door trim		35	1/8	
Laundry				
walls	170		1	
ceiling	50		1/4	
door trim		70	1/4	
Hall				
walls	325		1 1/3	
ceiling	63		1/4	
floor treatment	63		1/4	
base molding		20	1/10	
door trim		85	1/4	

Fig. 12-1. Quantity take-off. House Plan A.

	QUANTITY TAKE-OFF		SHEET 2	
	INTERIOR PAINT		HOUSE PLAN A	
Location	sq. ft.	lin. ft.	gallons	remarks
Lavatory				
walls	150		3/4	
ceiling	23		1/8	
door molding		16	1/10	
window trim		11	1/10	
Stairwell				
walls	160		1	
ceiling	40		1/5	
stair treatment	52		1/4	
Bedroom #1				
walls	500		2	
ceiling	180		1	
floor treatment	180		1	
base molding		70	1/4	
window trim		27	1/8	
door trim		42	1/8	
Bedroom #2				
walls	380		1 1/2	
ceiling	121		1/2	
floor treatment	121		1/2	
base molding		50	1/8	
window trim		28	1/8	
door trim		33	1/8	
Bedroom #3				
walls	360		1 1/2	
ceiling	100		1/2	
floor treatment	100		1/2	
base molding		60	1/4	
door trim		33	1/8	
window trim		15	1/10	
Bath #1				
walls	152		3/4	
ceiling	57		1/4	
door trim		17	1/10	
window trim		12	1/10	
Bath #2				
walls	220		1	
ceiling	30		1/8	
door trim		16	1/10	
window trim		12	1/10	

Interior Painting. When calculating the areas of interior walls (and ceilings), deduct the areas of windows and doors if they are not to be painted. One gallon of primer covers 800 square feet of plaster area. One gallon of second or third coat paint also covers 800 square feet.

The type of paint generally used for wood *basement partitions* usually covers about 700 square feet per gallon. If both sides of the partition are to be painted, multiply the area by two.

Paint for *interior stairs* will cover 700 square feet per gallon of primer, and 800 square feet per gallon for second and third coats. The areas of treads, risers, and stringers are calcu-

lated by multiplying the length by the width.

Thin enamel can cover 800 square feet per gallon, but if it is thick it may cover as little as 450 square feet per gallon, depending on the type of wood used for the stairway.

For an average flight of around 14 risers 3'-0" wide, some contractors allow the following:

½ gallon of paint
1 quart of enamel
1 pint of stain
1 pint of varnish
1 pint of filler (if open grain wood)

Undercoating for enamel walls is assumed to cover 500 square feet per

TABLE III. PAINTING TIME SCHEDULE

Item	Amount	Time (man-hours)	Remarks
Exterior siding	100 sq. ft.	1	
" windows	20	6	Included with siding unless brick walls
" doors	—	—	Included with windows
Exterior trim— contrasting color	—	—	no additional labor
Gutters—on ground	500 ft.	1	
" —hung	"	2	
Basement partitions	200 sq. ft.	1	
Stairs—interior	1 flight	5	1 coat
Interior walls and ceilings	200 sq. ft.	1	Includes trim
Interior windows and doors	4	1	1 coat
Steel sash	6	1	allow 25% more for old sash
Glusizing	200 sq. ft.	1/2	
Varnishing	1,500 sq. ft.	6	
" window sills	6 rooms	1	
Floors	200 sq. ft.	1	1 coat
Doors	4	1	same for stain, shellac, varnish
Pipes	2,500 ft.	6	3-inch diam.
Shingles—walls	150 sq. ft.	1	longer for priming coat staining
" —roof	200 sq. ft.	1	
Shutters	each	1/2	1 coat

gallon. The enamel for plaster walls will cover about 700 square feet per gallon.

Kitchen cabinets are usually estimated roughly in regard to area. The paint or enamel is then calculated, depending on the covering capacity. When cabinets are to be stained and varnished, some estimators will figure the cost at a fixed price per square foot.

If *pipes* are to be painted, allow one gallon of bronze for every 3,000 square feet of 3-inch pipe—other sizes in proportion.

House Plan A — Quantity Take-Off. The quantity take-off for the interior of HOUSE PLAN A is shown in Fig. 12-1. Each room has been estimated separately, but the areas and quantities of paint are rough figures. This esti-

mate is accurate enough, however, for most contractors.

ESTIMATING LABOR

Table III gives average times for various painting jobs. These times will vary with the working conditions, the skill of the workmen, and the set-up times, that is, the time necessary for erecting scaffolds, mixing paints, cleaning brushes, etc.

Example 3. How long will it take to paint 250 feet of gutters if the gutters have not been installed before painting?

Solution: Table III shows the average time as 500 linear feet per hour.

$$\frac{250}{500} = 0.5 \text{ hours}$$

See Appendix D, pages 483 to 484, for Hourly Estimates for:
Painting: Outside Walls; Inside Walls and Ceilings; Cabinets, Cupboards; Floors; Roofs; Trim; Doors; Windows; Shutters; Stairways; Wood Mantels.

Questions and Problems

1. How many gallons of inside floor paint would it take to cover 1275 sq. ft. with one coat?

    (Answer: 3 gallons)

2. Find how many sq. ft. could be covered by 2 gallons of trim varnish.

    (Answer: 1400 sq. ft.)

3. How long would it take to paint 10 medium sized windows in a brick veneer home?

    (Answer 3 to 3½ hours)

4. Find the amount of paint needed to paint the exterior of a house which is 30 ft. by 24 ft. by 15 ft. high.

    (Answer: 2¼ gallons)

13

GLASS
AND ACRYLICS

Glass estimating is one of the easiest items for the estimator. There are, however, numerous kinds of glass, thus prices for materials and labor can vary greatly.

MATERIALS

There are five kinds of glass of general interest to the contractor and estimator. These are cylinder glass, drawn glass, plate glass, vision-proof glass, and special glass.

Cylinder Glass. Cylinder glass obtains its name from its manufacturing process wherein molten glass is blown into a cylinder during its formulation. The United States Government has laid down certain standards for the classification of cylinder glass according to thickness and defects. Glass sheets are graded as *Single Thick* when they are approximately $\frac{1}{12}''$ thick and *Double Thick* when they are approximately $\frac{1}{8}''$ thick. Glass weighing 26 to 38 ounces per square foot and ranging from $\frac{1}{8}''$ to about $\frac{3}{8}''$ thick is called *Crystal Sheets.*

The government standards grade glass according to its defects, such as bubbles, curvatures, bows, streaks, waves, etc. Glass is graded ordinarily as *AA, A,* and *B*. Grade *AA* glass is the best of the blown glass while grade *A* is the next best and the type most generally used in residences. Grade *B* glass is used generally in factories and cheaper residences. The quality of glass is all the same and the classification given above refers only to waves, etc., as previously mentioned.

Drawn Glass. Another manufacturing process becoming more common is where the glass is drawn out into sheets instead of being blown. This process produces more grade *AA* and *A* glass than the blowing method.

Plate Glass. Plate glass is made by a process whereby the molten glass is poured on a hot flat surface and rolled until it is the required thickness. There are two kinds of plate glass, namely, *rough* and *polished*. The rough plate is made from plates which have been rolled and annealed and the polished plate is made by grinding and polishing the rough plate. Rough plates can be used as skylights whereas the polished plate is used for stores, residences, etc. Polished plate is divided into two classes depending on defects. These classes are *silvering* and *glazing*. The silvering class is generally used for mirrors and the glazing class for stores, etc., as mentioned above.

Plate glass varies between ⅛" and 1½" in thickness. Glazing thickness is generally about ¼" thick.

Thermal (insulation) glass is commonly used in residential construction. This glass is generally made of two layers of plate glass separated by an air space.

Vision-Proof Glass: For use in bath room windows and other places where light is required but vision obscured, there are many kinds of glass made. These are made of the same material but are cut, shaped, sandblasted, hammered, etc., to prevent or obscure vision but admit light. Glass blocks (see Chapter 9) are frequently used for this purpose.

Special Glass. This includes wire glass, one-way glass, tinted glass, prism glass, safety glass, and bulletproof glass. These, however, are not used generally in residential construction.

Glazing. Glazing is the setting of the glass in a frame. Most wood sash glazing has been done prior to delivery on the job. In *face glazing,* the glass is pressed into a putty bed, set firmly with stops and puttied. Channel glazing calls for a continuous metal stop or wood bead to form a channel that holds the glass in place.

Metal Sash. Putty or some kind of metal beading can be used for metal sash. Glass may also be placed in rubber neoprene gaskets or liners. Where heavy plate glass is used in doors and casement windows, the glass is bedded in putty and back-puttied and then held in place by wood beads. In metal settings, the glass should be somewhat smaller than the sash opening to allow for expansion.

Acrylic Plastics. The building code recommendations for glass are often interpreted or actually changed to include approved plastic glazing. Tough, resilient thermoplastic sheets are becoming a popular replacement for glass.

Acrylic sheets weigh half as much as glass and 43% as much as aluminum. On the other hand, they are less rigid and so require special care that channel or rabbet depths are sufficient to compensate for deflection. If the acrylics are bent into dome or ridged shapes, deflection decreases.

Acrylic sheets are available in a wide variety of thicknesses, colors, densities and textures, and they can be sawed, drilled, cemented, etc., much as wood can. In addition these acrylics can be heated and molded. Care is required, however, to avoid scratching surfaces. For this reason acrylic installations should be one of the last in the building.

Installation must allow for expansion since thermoplastics (acrylics) expand and contract eight times as much as glass. Acrylic panes may be face glazed or channel glazed. A special elastic glazing compound is required.

Acrylic sheets are used also for storm doors, shower doors, awnings, skylights, light diffusors, room dividers and decorative purposes.

ESTIMATING MATERIAL

In some localities, and especially for small jobs, windows, doors, etc., are glazed at the mill and shipped to the job ready to install. In such cases it is not necessary to estimate glass material and labor, because the carpenters install the windows and doors, and the cost of glass is included in the cost of the windows and doors. For large buildings such as high-rises, however, the windows are not glazed until the building is nearly finished. This is to avoid breakage.

Although most windows for high-rises can be installed from the inside, sometimes they cannot, and scaffolding is necessary. This item must then be included in the estimator's costs.

The first thing a contractor or estimator does is to study the working drawings and specifications. From these he obtains information about the grade and thickness of the glass to be used, kind of windows, what doors require glass, where plate glass is to be used, sizes of all windows and doors, number and size of panes, etc.

Lights. With a general idea of the job and specifications, the sizes and numbers of lights (individual panes of glass) is next determined. In glass take-offs (counting the amount of each size and kind), full inches are used to de-

scribe the sizes of the various lights. For example, if one light scaled 7½″ × 10¾″, the estimator or contractor calls it 8″ × 11″. In other words, all lights are listed in the take-off to the next full inch in excess of the actual size required. When the glazier is putting in the lights, he cuts them to the exact size.

Glass Sizes. There are many standard light sizes, all of which are noted in standard or official published lists as for example, that published by the National Glass Distributors Association. Standard sizes start with a minimum of 6″ × 8″.

Ordinary double-strength grade *A* *(DSA)* glass as used for window lights is generally priced per piece of glass. That is, there is a list price for the variously-sized lights. There is generally some reduction allowed from this list price. Such reductions vary widely, but whatever the reduction is it can be subtracted from the given list price. Any contractor or estimator can easily determine what the list price and reduction percentage is by consulting his local glass dealer.

Plate glass and mirrors are estimated by the *square foot*. The allowance for breakage is usually between 3 and 6 percent.

Acrylics. While acrylic sheets are initially more expensive than glass, their high resilience and other qualities often make their installation an economic saving. There is virtually no breakage allowance needed.

Acrylics are available in sizes 3′ × 4′ to 10′ × 14′ and in thicknesses .030″ to 4″. Manufacturers' recommendations should be followed in the selection, fabrication, installation (particularly glazing methods) and care of acrylic plastics.

Putty. The amount of putty used per light or per window is difficult to determine. It is not profitable to do such a take-off so the estimator uses an average figure based on his experience. Some estimators allow one pound of putty for every two square feet of glass.

House Plan A—Quantity Take-Off. Using HOUSE PLAN A as an example,

typical take-offs for glass quantities are shown in Fig. 13-1. This list by rooms is a more detailed list than most estimators or contractors would make, but it is given so that you will better understand the procedure used in estimating glass.

The next step would be to determine the actual prices for the glass.

ESTIMATING LABOR

Most contractors estimate their labor by assigning certain costs to each of

QUANTITY TAKE-OFF					SHEET 1	
GLASS					HOUSE PLAN A	
Location & Description	No.	Type	Window Size (In.)	No. of Lights	Kind of Glass	Light Size (in.)
Dining Room window	2	DH	32X24	24	DSA	10X12
Living Room window	1	BOW	10'-10"X6'-8"	15	DSA	
Bedroom #1 window	2	DH	28X24	24	DSA	8X12
window	1	DH	32X24	12	DSA	10X12
Baths window	2	DH	36X18	12	DSA	12X9
Bedroom #2 window	2	DH	32X24	24	DSA	10X12
Bedroom #3 window	1	DH	36X24	12	DSA	12X12
Family Room window	2	DH	36X30	24	DSA	12X15
Kitchen window	1	SLIDING	24X36		DSA	
Laundry Room door	1	REAR		9	PG	8X12
Lavatory window	1	DH	24X36	8	DSA	12X8
Garage window	2	DH	36X20	24	DSA	12X10

Note: All window are to be ordered from the window schedule complete in gangs to include all frames, sash, trim, mullions, etc.

Fig. 13-1. House Plan A—Quantity take-off.

the various-sized lights of glass. This unit cost system is fast and accurate.

In some localities, the glass contractors have an association which has fixed the costs for all of the various-sized lights. These costs are based on actual cost analysis data, obtained over a long period, and upon the prevailing wage scale for glaziers.

In localities where no such cost studies have been made, the contractor or estimator should keep his own cost records and after a time be able to make up such standard costs, based, of course, on the wage scale prevailing in his locality.

A glass installation in a wooden window would differ in labor cost from a glass installation in a steel or aluminum window. These costs also vary with the size of the job and climatic conditions. Outside work in the winter may add as much as 25% to the labor cost.

See Appendix D, page 483, for Hourly Estimates for:
Glazing: Glazing Wood Sash and Doors; Glazing Metal Sash; Store or Commercial Building Fronts.

Questions and Problems

See **Hourly Estimates** in Appendix D to answer the following:

1. Find how many hours to allow for a skilled craftsman to install 25 8″ × 10″ lights in wood sashes.
   (Answer: 2½ hours)

2. Find the amount of time it would take to glaze 10 lights, each 900 sq. in., in metal sashes with plastic stops.
   (Answer: 40 hours)

14

WINDOW COVERINGS

~~~~~~

Window coverings include shades, venetian blinds, and curtains. These are among the last items installed in a new building. Although these items are a relatively minor portion of the total construction cost, they are included in the estimate when they must be installed by the contractor.

MATERIALS

Window Shades. There are many kinds and qualities of shade cloth. All of them cannot be explained here but a few of the more common types are given in the following:

(1) *Vinyl cambric*—good quality cambric cloth coated with a vinyl finish that is washable and flame resistant.

(2) *Room darkening*—similar to the vinyl cambric but with an opaque finish that excludes light.

(3) *Hand-painted cambric*—a cambric base coated with translucent or opaque coating applied by hand.

(4) *Vinyl duck*—heavy duty shades with a cotton duck base coated with vinyl or other finish. Many of these shades can be obained in standard colors or with duplex colors in which one side is a different color than the other.

Shade cloth is manufactured in various standards widths as shown in Table I.

TABLE I. STANDARD WIDTHS OF WINDOW SHADES

SHADE CLOTH	WIDTH–INCHES							
VINYL CAMBRIC	36	42	45	48	54	63	72	77
ROOM DARKENING	36	42	45	48	54	63	72	77
HAND PAINTED CAMBRIC	36	43	45	48	54	63	72	77
VINYL DUCK	36	42	45	48	54	63	72	77

Normal lengths are 5, 6, and 7 feet. All manufacturers do not make all sizes.

339

Rollers. Rollers for shades are furnished in two types as follows: *Type A* — Wooden Rollers and *Type B* — Metal Rollers. The construction of roller ferrules and ends are of a grade known as standard in any of the following: malleable iron, coppered finish, stamped steel, nickeled finish, or aluminum.

The wooden roller, Type A, is made of a seasoned, straight-grained wood, and finished to a smooth, straight roller.

The metal roller, Type B, is made of tinned metal of sufficient thickness and strength to minimize bending.

Rollers should be selected and used according to the various widths of shade cloth required. Typical recommendations are shown in Table II.

distance between the ends of the round and rectangular shaped rods which enter the brackets. This length should be just enough less than the distance between stop-beads, for example, so that there is little play in the roller after it has been put in the brackets. This allows the roller to revolve freely. The brackets do not affect the length of the roller or end rods as the rods protrude through the brackets.

Fig. 14-1 shows the top portion of a typical window. Line *A* shows the distance to be measured if the roller brackets are to be put on the casing. Line *B* shows the distance to be measured if the brackets are to be put on the stops. The allowance for free rolling need not be over $1/16''$.

To determine the shade length,

TABLE II. SELECTING ROLLERS FOR SHADES

LENGTH OF SHADE—FEET	WIDTH OF SHADE—INCHES				
	28 TO 41	41 TO 46	46 TO 61	61 TO 73	73 AND OVER
4' TO 7' (INCL.)..........	1" W	1 1/8" W	1 1/4" W	1 1/2" M	1 3/4" M
8' TO 10' (INCL.).........	1 1/4" W	1 1/4" W	1 1/4" M	1 1/2" M	1 3/4" M
11' TO 12' (INCL.).........	1 1/2" M	1 1/2" M	1 1/2" M	1 1/2" M	1 3/4" M

W—indicates mounted on wood roller.
M—indicates mounted on metal roller.

Measuring for Shades. When measuring for window shades, make sure where the brackets are to be placed and measure accordingly—from jamb-to-jamb, stop-bead to stop-bead, or on the face of the casing. A steel or wood measuring tool is usually used for this measurement. Measurements should be made with care so that the shades will fit correctly when they are installed. The length of a shade is the

measure the distance from the top of the sash (allow room for roller to operate when it contains the full amount of shade) to the window sill. Then add at least 10″ so that the shade will not be strained and possibly jerked off the roller when the shade is pulled all the way down.

For casement windows, French doors, etc., it is usual practice to measure 1″ beyond the glass area on each

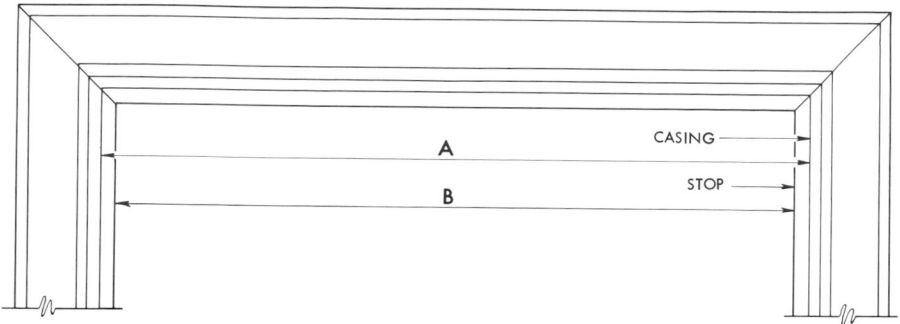

Fig. 14-1. How to measure lengths of shade rollers.

side so that the shade width will be sufficient to cover the entire glass area without any uncovered spaces on either side.

Drapery and Rods. Drapery is available in a variety of natural and synthetic fabrics, choice determined by use. Fireproofing and lining are but two specialties. Patterns are more expensive than plain drapes since some waste must be allowed in matching the pattern. Fullness of pleats also affects the cost.

Drapery rods have carriers, often nylon wheels, free to move along a channel track. The rods may be operated by hand, cords, or motor.

Venetian Blinds. Venetian blinds can be made to operate horizontally or vertically. They are made of either wood, metal (baked enamel), or vinyl plastic. The strips of the blinds are usually connected by cloth and operated by cords. The blinds may be made to fit either inside or outside the window casing.

Curtain Rods. Curtain rods require brackets for support somewhat the same as shades do. The rods are gen-

erally of brass about $\frac{5}{16}''$ diameter. Sometimes two rods are supplied at each window (one for curtains and one for drapes). Where two or more windows occur in a group, a single length rod is used.

ESTIMATING MATERIAL

Window shades are available in standard sizes. The cost of the shades is determined by the quality of the roller and the shade material. (The cost of making a shade, except material, is the same for all sizes. This includes cutting, accessories, brackets, rollers, slats, and cords.)

Venetian Blinds. These are usually estimated by the square foot of window area which they will cover. They are usually ordered in standard sizes but custom-made blinds may be ordered at extra cost.

ESTIMATING LABOR

Once they are assembled, window shades, venetian blinds, drapery and curtain rods can be quickly installed. Window shades can be installed at the rate of 12 to 15 per hour depending on the type of window or door. On ordinary windows and doors, the rate of 15 per hour is usually assumed.

One man can generally install 10 complete curtain rods in one hour.

Sample Specifications

A set of window shade specifications can be as detailed as these sample specifications. Note that these specifications also give details of materials and assembly for the shades.

GENERAL. (A) Each window is to be equipped with a window shade hung at the top of the opening and must operate smoothly and efficiently. Each shade is to hang perfectly level and have the spring tension of roller properly adjusted.

(B) All necessary measurements are to be taken after casings and jamb linings are in place.

(C) Shades are to be made in a thoroughly workmanlike manner, cut perfectly square and true and mounted on rollers in the same manner, using suitable substantial fasteners. All materials used in the manufacture of the shades are to be new.

CLOTH. All shades are to be made of *X* shade cloth.

COLOR. As selected by the owner, architect, or superintendent.

LENGTH. Finished length of each shade is to be 10 inches longer than the actual height of the window opening.

ROLLERS. All shades are to be mounted on guaranteed wood or metal rollers to carry shades of the respective widths and lengths properly.

HEMS. Hems are to be of ample width to include free inclosure of the slat; shall be turned, sewed, and the end backstitched to reinforce hems.

SLATS. Smooth white pine slats of the proper size are to be inserted in hems of all shades.

EYELETS. Aluminum rust proof eyelets for shade pulls are to be affixed to each shade, being inserted through the center of the slat.

RING PULLS. Crocheted ring pulls for operating shades are to be attached to each shade.

15

HARDWARE

~~~~~~

A hardware schedule for a building must be compiled from both specifications and drawings. To do this requires the ability to read plans easily and thoroughly.

Sometimes a quantity take-off is given to the estimator; the hardware is listed room by room and it is a simple matter to price the items and then estimate the labor time. If such a take-off is not available, the estimator should make one for himself.

MATERIALS

The hardware estimate includes *finish* hardware and not the rough hardware such as nails, which are usually estimated under the carpentry and millwork section.

Finish hardware is made up of a somewhat special group of wood fastenings in that, although widely used in carpentry, they are not considered *structural fastenings*. In other words, they are not typically used to give support to structural members, but are used mostly to permit the movement of parts, or to secure or close moving parts. Thus, most finish hardware consists of hinges, locks, catches, and pulls. Other finish hardware consists of miscellaneous items such as door stops and coat hooks.

A whole book could be devoted to the discussion of the various uses of finish hardware, their selection, and their method of installation. We can here only mention the large variety of hinges, locks, etc.

Obviously, the choice of finish hardware, whether it be a hinge, lock, or pull, depends primarily on its intended use. You would not, for instance, use a light strap hinge for a heavy hardwood door. A butt hinge would be more appropriate for such a door.

In other words, finish hardware should not only be chosen for *function,* but also for the *amount and length of service that should* be expected from it. In addition to function and service, finish hardware can also be selected for its *decorative effect*. There are many types of hardware that are similar in function and service requirements, differing only in their design or decorative value.

Hinges. A hinge is a movable joint upon which a door, gate, etc., turns. It consists primarily of a pin, and two plates which may be attached to a door and the door frame to permit the opening and closing of the door. Fig. 15-1 illustrates the parts and basic design of a common hinge.

There are four basic hinge classification of hinges (Fig. 15-2):

Full mortise

Half mortise

Full surface

Half surface

Hinges are further classified on the basis of whether a *loose pin* or a *tight pin* is employed. A loose pin may be removed; a tight pin is secured to the hinge.

Many different types of hinge are manufactured to meet various design requirements; however, the type of hinge which requires the most careful selection is the door hinge. The width, length, and weight of the door determine the weight of the hinge, and whether plain or ball bearing hinges are to be used. The width and thickness of the door determine the size of the hinge. A rule which can be applied to the selection of door hinges is as follows:

The width of the hinge for doors up to 2¼ inches is equal to twice the thickness of the door, plus the trim projection, minus ½ inch. For doors from 2½ inches to 3 inches thick, the same rule applies, but ¾ of an inch should be subtracted instead of ½ inch. If the result of this calculation falls between regular sizes, the next larger size should be selected.

Ball bearing hinges (Fig. 15-3) are

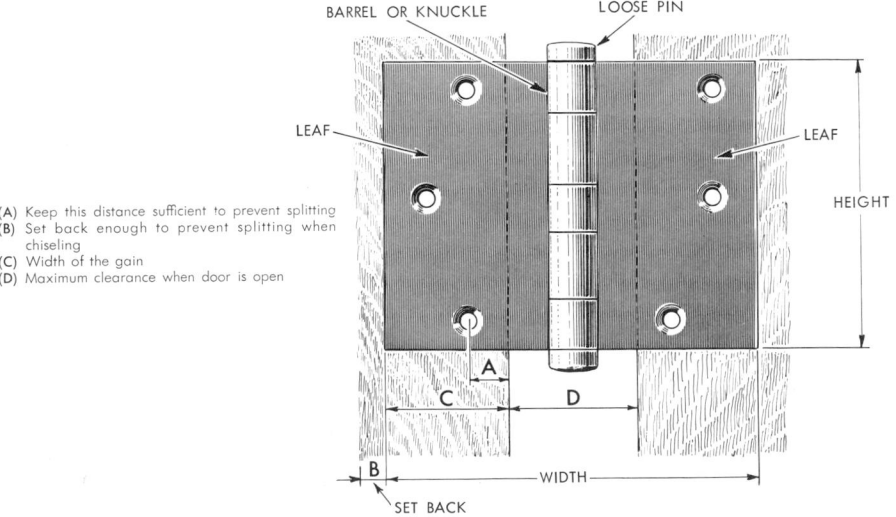

(A) Keep this distance sufficient to prevent splitting
(B) Set back enough to prevent splitting when chiseling
(C) Width of the gain
(D) Maximum clearance when door is open

Fig. 15-1. Butt hinges.

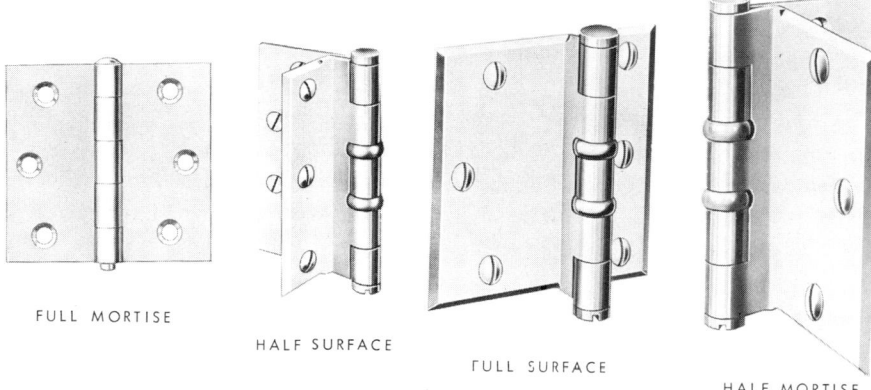

FULL MORTISE

HALF SURFACE

FULL SURFACE

HALF MORTISE

Fig. 15-2. A selection of door hinges. (Stanley Works)

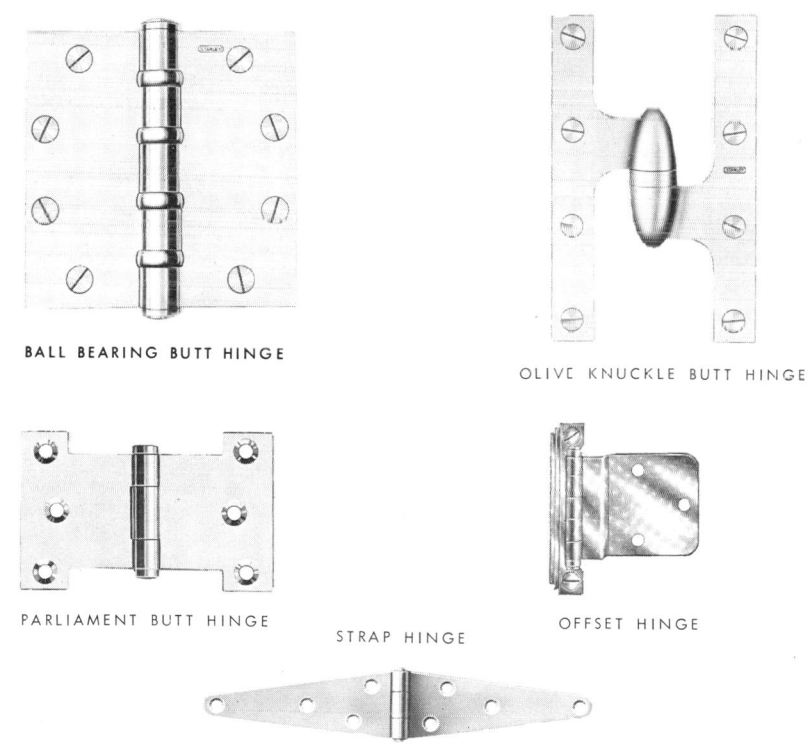

BALL BEARING BUTT HINGE

OLIVE KNUCKLE BUTT HINGE

PARLIAMENT BUTT HINGE

STRAP HINGE

OFFSET HINGE

Fig. 15-3. A selection of special hinges. (Stanley Works)

used for heavy doors, or on doors which are subjected to heavy use, such as doors in schools, office buildings and department stores.

Template hinges conform to government specifications. They will exactly fit the sinkage and screw hole locations in metal doors and jambs. They are also available with non rising pins.

Olive knuckle butt hinges (Fig. 15-3) have fixed pins with a loose leaf, and must be selected according to the way in which the door is to open. These hinges are longer than they are wide, and are used for cupboard and intercommunicating doors.

Parliament butt hinges (Fig. 15-3) have either a fixed or loose pin. They have a greater width than length and are used in churches on communion rails.

Offset hinges (Fig. 15-3) are used on lip cupboard doors.

Strap hinges and *T hinges* (Fig. 15-3) in most cases have a fixed pin. They are obtainable in light, heavy or extra heavy metal, according to their application. Strap hinges and T hinges are commonly used on carpenter-built doors and gates.

Double action spring floor hinge (Fig. 15-4), as its name implies, has a spring return action, which is effective in both directions. The spring action is generally concealed in the door

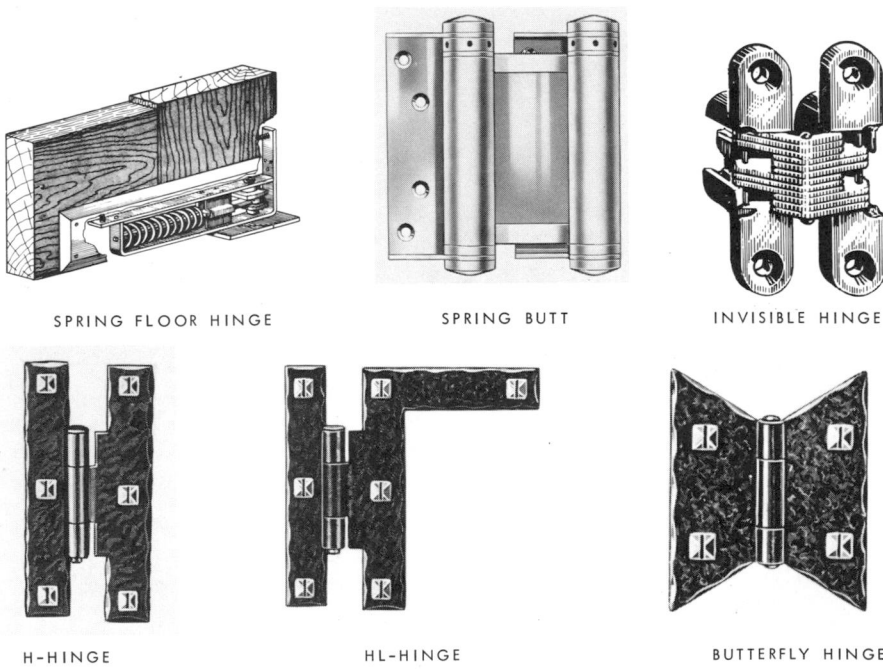

SPRING FLOOR HINGE   SPRING BUTT   INVISIBLE HINGE

H-HINGE   HL-HINGE   BUTTERFLY HINGE

Fig. 15-4. Special hinges and door hardware. (Ornamental hinges: Stanley Works)

in residential installations, and in the floor below the door in commercial installations. It is designed for doors which have a thickness between 1⅛ inches and 1¾ inches.

The *double action spring butt* (Fig. 15-4) is used in commercial installations. Its design requires a hinge strip on the jamb. It is designed for use on doors that can be pushed open from either side and will close under spring action.

Invisible hinges (Fig. 15-4) are made so that no portion of the hinge is visible when the door is closed. They fit snugly into a mortise cut into the door and jamb, and because of this, much of the weight of the door is taken off the screws. Invisible hinges are made to open 180 degrees and are reversible.

Many kinds of *ornamental hinges* are manufactured, and find their main use in the construction of cupboards. Illustrated in Fig 15-4 are an "H" hinge, an "H-L" hinge, and a butterfly hinge.

Door Trim. *Surface bolts* (Fig. 15-5) are generally used vertically on doors, cupboards and casement windows. They are available in various weights and lengths, and it is important to select the proper strike plate.

Flush bolts (Fig. 15-5) are used vertically on top or bottom of doors, or both. They are made in various sizes and for many purposes. Some have a flush type lever, while others have a knob.

Extension flush bolts (Fig. 15-5) are used vertically, set in the edge of the inactive door of a pair of doors. They

are available in various widths and their length ranges from about 5 inches to 48 inches.

Cremone bolts (Fig. 15-5) are used vertically and are designed for use with large French windows and doors. They are operated by means of a knob or lever handle and open from the inside.

Barrel bolts (Fig. 15-5) are a surface type of bolt used horizontally. They are a less expensive variety of surface bolts and are considered rough hardware.

Chain door fasteners (Fig. 15-5) are made in many styles. They permit exterior doors to be opened sufficiently wide for communication and yet prevent forcible entrance.

Foot bolts (Fig. 15-5) are used vertically on the bottom inside surface of garage and swing doors. As their name implies, they are designed to be foot operated.

Chain bolts (Fig. 15-5) are used vertically on the top inside surface of doors and are considered to be the companion bolt to foot bolts. They hold the top of the door closed.

Exit fixtures (Fig. 15-5). Automatic exit fixtures are used on doors opening outward, and are often called "panic bars." Public buildings are required by law to use this fixture on certain exterior doors. Because human life may depend upon their proper operation, it is of the utmost importance that they be properly fitted. They are specified according to the hand of the door, for they cannot be reversed, and are obtainable in rim or mortise type.

Pneumatic door closers (Fig. 15-5) are used for the same purpose as hy-

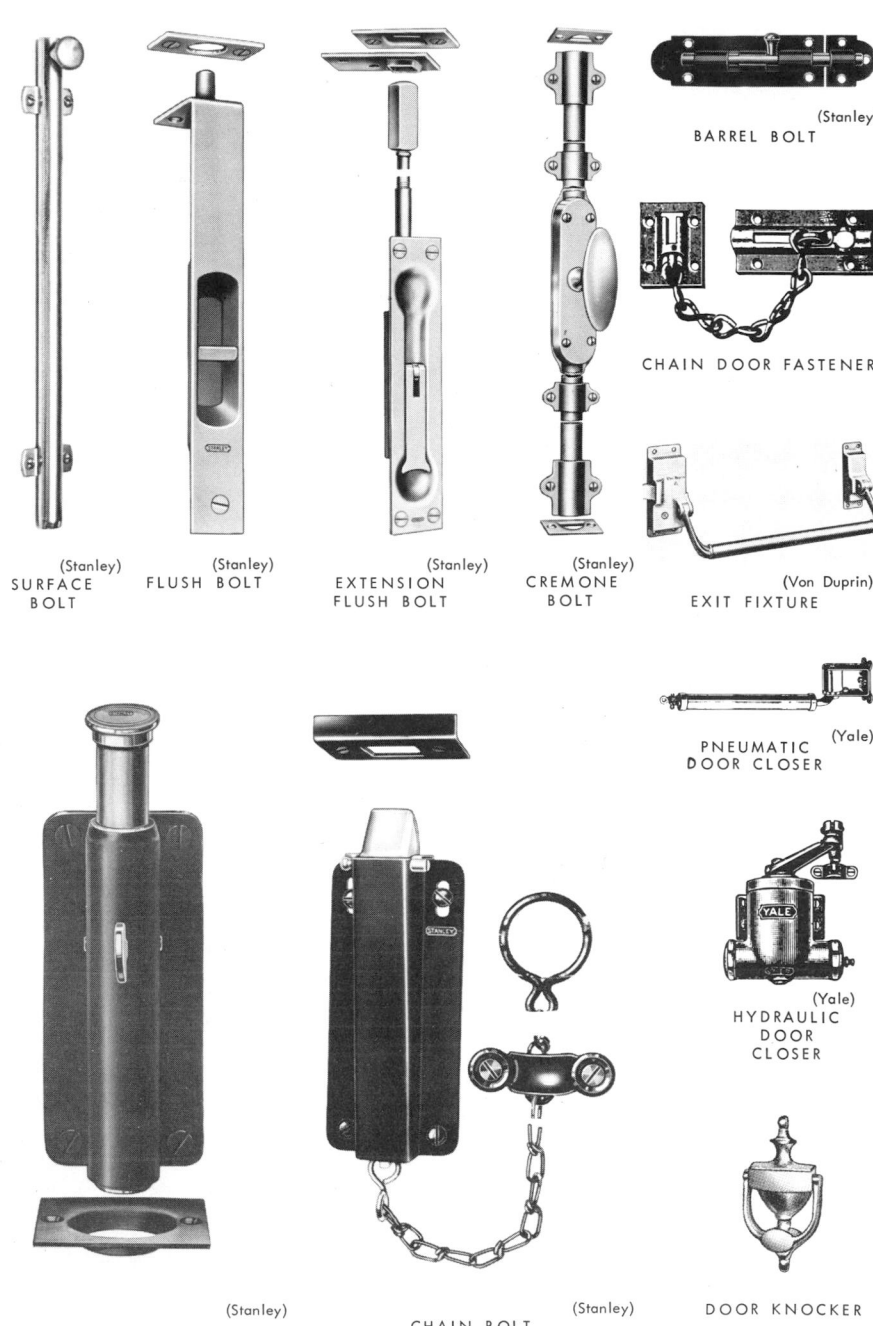

(Stanley)
BARREL BOLT

CHAIN DOOR FASTENER

(Stanley)
SURFACE
BOLT

(Stanley)
FLUSH BOLT

(Stanley)
EXTENSION
FLUSH BOLT

(Stanley)
CREMONE
BOLT

(Von Duprin)
EXIT FIXTURE

(Yale)
PNEUMATIC
DOOR CLOSER

(Yale)
HYDRAULIC
DOOR
CLOSER

(Stanley)

(Stanley)
CHAIN BOLT

DOOR KNOCKER

Fig. 15-5. Selection of bolt and door fixtures.

draulic door closers, but on lightweight doors. Their checking medium is air, and a set screw adjusts the time of closing. They are used on screen and combination doors.

Hydraulic door closers (Fig. 15-5) are mounted on the surface of the jamb, casing, or door bracket. The helical spring, either torsion or compression type, closes the door while the fluid checks it as it nears the door jamb and causes it to close slowly. The arm can be set to hold the door open.

Door knockers (Fig. 15-5) add to the appearance of a door, apart from their function. They are available in

DOOR STOP

BULLET CATCH

DOOR HOLDER

DOOR HOLDER

ELBOW CATCH

FRICTION CATCH

MAGNETIC CATCH

MORTISED

BAR

HOOK

SASH LIFTS

Fig. 15-6. Miscellaneous small building hardware. (Door stop and holder: Stanley Works)

Fig. 15-7. Pulls and knobs for doors, drawers, and panels. (Left: National Lock Hardware; Right: Stanley Hardware)

a great number of designs and sizes, for exterior doors as well as in smaller sizes called guest room knockers, which are becoming increasingly popular.

Other Building Hardware Used on Doors and Windows. *Door stops* (Fig. 15-6) vary in style. Some are used on the baseboard, while others are used on the floor. They protect the door locks and the wall from being marred or damaged.

Door holders (Fig. 15-6) vary in design; their purpose is to hold the door open at a given point.

Catches (Fig. 15-6) are used on cabinets and cupboards. The bullet catch is mortised in the edge of the door top, bottom side, or even the frame, whereas friction and elbow catches are applied to the inside surface.

Magnetic catches have a magnet mounted on the door or on the frame and a strike plate that comes in contact with the magnet when the door is closed.

Sash lifts (Fig. 15-6) are used on the bottom rail of double-hung windows. The flush sash lift is mortised into the rail while the bar and the hook are screwed to the rail.

Pulls and Knobs (Fig. 15-7) besides being functional add a variety of decorative effects. They are usually mounted from the inside of the drawer, door, or panel.

Locks. Three different types of locks are commonly used in residential building construction: the tubular, the cylindrical, and the mortise lock. *Tubular lock* sets are used mainly for interior doors, for bedrooms, bath-

rooms, passages, and closets. They can be obtained with pin tumbler locks in the knob on the outside of the door and turn button or push button locks on the inside. There are several variations to this arrangement. Fig. 15-8 illustrates a tubular lock set.

Cylindrical lock sets are sturdy, heavy duty locks, designed for maximum security. They are installed in exterior doors. See Fig. 15-9.

An ordinary *mortise lock* is illustrated in Fig. 15-10. More elaborate mortise locks are made with cylinder locks, with a handle on one side and a knob on the other side, or with handles on both sides. This type of lock is used principally on front or outside doors. The present trend is away from using mortise locks.

Ventilation Equipment. Ventilators are also classed as finish hardware. Fig. 15-11 illustrates six different types of venting devices. Frequently two of these are used in combination, such as the ridge ventilator and the continuous under-the-eave (soffit) ventilator. Venting devices come in a wide variety of materials and finishes. Sizes vary, depending on the structure and the venting needs.

Miscellaneous Finish Hardware. Items not found on many hardware schedules, but which are costs, include the following: mailboxes, street numbers, foot scrapers, door knockers, etc.

Bathroom accessories such as towel racks, bars, rings and ladders; soap holders; tumbler and toothbrush holders; medicine cabinets; paper holders; shower curtain rods; and grab bars are

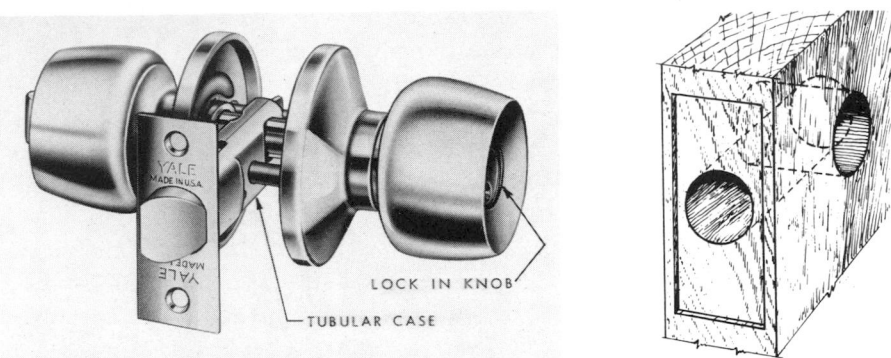

Fig. 15-8. Tubular lock set is installed by drilling 2 holes and mortising lock face. Locks of this type are supplied for several different applications. (Yale & Towne Mfg. Co.)

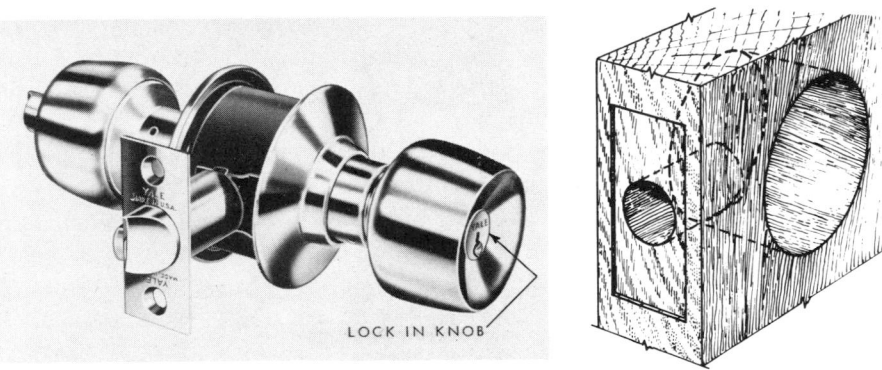

Fig. 15-9. Cylindrical lock sets for heavy duty exterior use. (Yale & Towne Mfg. Co.)

Fig. 15-10. Mortise lock. (Yale & Towne Mfg. Co.)

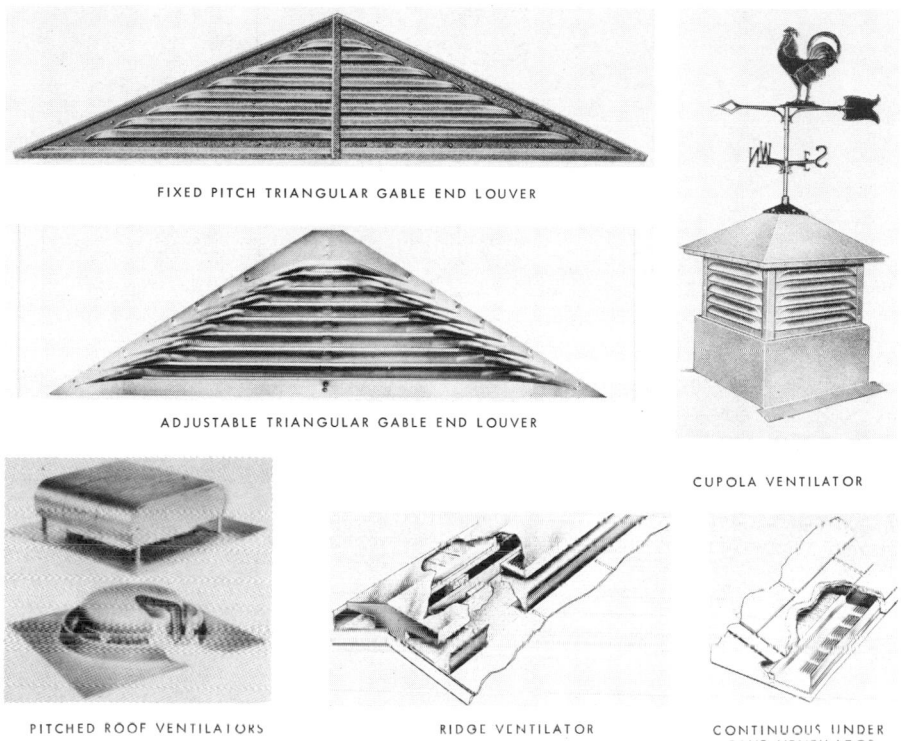

FIXED PITCH TRIANGULAR GABLE END LOUVER

ADJUSTABLE TRIANGULAR GABLE END LOUVER

CUPOLA VENTILATOR

PITCHED ROOF VENTILATORS

RIDGE VENTILATOR

CONTINUOUS UNDER EAVE VENTILATOR

Fig. 15-11. Ventilating devices. (H. C. Products Co.)

also classed as finish hardware. The variety in design is nearly unlimited.

The pantry might include hooks for utensils and dishes, knife sharpeners, special knobs or fasteners for cabinets, brackets to support shelving, etc. First-floor windows may require special locks which allow windows to be raised only a few inches. The front and rear doors may have special bolts or chains. Closets may include special devices for hanging up ties, suits, dresses, etc. Storm windows may require special locks and devices allowing them to be only partially opened. Such items are usually listed under hardware.

ESTIMATING HARDWARE

The first step in estimating hardware is to study the hardware specifications along with the carpentry and millwork specifications to see if the latter contains anything which should properly be included under hardware. The window and door schedules should also be used. A detailed hardware schedule is needed for a complete and accurate estimate.

In many construction projects, there is an allowance for hardware in the specifications. This permits the builder or owner to choose his materials after

	QUANTITY TAKE-OFF	SHEET 1	
	HARDWARE	*HOUSE PLAN A*	

Hardware	Location Description	Size	Amount
EXTERIOR DOORS			
BUTTS 4½"x4½" STL DULL BRASS	DUTCH DOOR	3'-0"x7'-0"x1¾	2 pair
BUTTS 4½"x4½" STL DULL BRASS	DUTCH DOOR	2'-6"x6'-8"x1¾"	2 pair
HINGES, 1 SET O H DOOR (RAIL sets)	OH DOOR (4 PANEL)	8'-0"x6'-8"x1¾"	4 pair
BUTTS 4½"x4½" STL	9 PANE, 1 PANEL	2'-8"x6'-8"x1¾"	1½ pair
BUTTS 4½"x4½" STL	SCREEN DOOR	2'-6"x6'-8"	1 pair
INTERIOR DOORS			
BUTTS 4½"x4½" STL	FLUSH HC	2'-6"x6'-8"x1⅜"	4 pair
AUTOMATIC DOOR CLOSER	FPSC	2'-6"x6'-8"x1⅜"	1 set
BUTTS 4½"x4½" STL	FLUSH HC	2'-4"x6'-8"x1⅜"	1 pair
BUTTS 4½"x4½" STL	FLUSH HC	2'-0"x6'-8"x1⅜"	5 pair
SLIDING HARDWARE	SLIDING	2'-0"x6'-8"x1⅜"	4 sets
BUTTS 4½"x4½" STL	FLUSH HC	1'-6"x6'-8"x1⅜"	1 pair
BUTTS 4½"x4½" STL	LOUVERED	1'-6"x6'-8"x1⅜"	2 pair
NOTE: 1 door bumper for each door—total 23. Armored pick-proof			
cylinder door locks for front and rear door—total 2.			
WINDOWS			
SASH LOCKS	GANG OF 2 DH	2(2'-8"x4'-2")	2
" LIFTS	" " " "	"	4
(FIXED)	BOW WINDOW	10'-0"x6'-8"	1
SASH LOCKS	GANG OF 2 DH	2(2'-8"x4'-2")	2
" LIFTS	" " " "	"	4
SASH LOCKS.	DH	2'-8"x 4'-2"	3
" LIFTS	"	"	6
SASH LOCKS	DH	3'-0"x3'-2"	2
" LIFTS	"	"	4
SASH LOCKS	DH	3'-0"x4'-2"	1
" LIFTS	"	"	2
SASH LOCKS	GANG OF 2 DH	2(3'-0"x5'-2")	2
" LIFTS	" " " "	"	4
(ALREADY EQUIPPED)	SLIDING	4'-0"x3'-0"	1
SASH LOCKS	DH	2'-0"x3'-2"	1
" LIFTS	"	"	2
SASH LOCKS	GANG OF 2 DH	2(3'-0"x3'-6")	2
" LIFTS	" " " "	"	4
LATCH	TH	2'-8"x1'-8"	5
TOP HINGES	"	"	5 pair

Fig. 15-12. House Plan A. Hardware take-off.

the construction contract has been awarded. For example, if there is a $10,000 hardware allowance for a construction project and the builder chooses materials for $12,000, the contractor is entitled to an extra $2,000 for the materials plus a percentage for installation, overhead, and profit.

House Plan A—Quantity Take-Off. A hardware schedule for HOUSE PLAN A is shown in Fig. 15-12. This schedule contains the finish hardware needed for the doors and windows. It is based on the door and window schedules for the house plans.

As previously explained, a complete hardware schedule includes more than door and window hardware. This schedule is sufficient, however, to give the inexperienced estimator an idea of how an actual hardware take-off is compiled.

ESTIMATING LABOR

The installation of hardware is usually done by carpenters and the cost is therefore included under the carpentry contract. The cost of the hardware materials, however, is established under a separate hardware contract.

Labor rates vary in different regions, but the estimator can determine the prevailing rates by contacting the local union.

Installation time for the various pieces of hardware can vary from one minute to several hours. You should therefore keep accurate cost records as you gain experience in this kind of estimating.

See Appendix D, page 484, for Hourly Estimates for:
Hardware: Lock Sets; Door Accessories; Sash Hardware; Closet Rods; Garage or Heavy Swinging Door Accessories.

Questions and Problems

See **Hourly Estimates** in Appendix D to answer the following:

1. Find how many lock sets could be installed on 4 plain outside doors in 2 hours.

    (Answer: 4 sets)

2. How long would it take a skilled craftsman to install locks and lifts on 12 double-hung windows?

    (Answer: 1-1/5 hours)

16

ELECTRICAL WIRING

The architect does not show details of the wiring on the building plans. However, he does show where the ceiling lights, wall or bracket lights, convenience outlets, wall switches, electric motors, heating appliances, etc., are to be located. Generally there is a set of typewritten specifications giving more details in regard to what is desired. The details of the particular make or kind of apparatus are usually left to the judgment of the electrical contractor who obtains the architect's approval. The specifications usually require that all electrical wiring and materials shall comply with the National Electrical Code and applicable portions of codes and ordinances prevailing in the area of construction.

WIRING METHODS

Estimating the cost of electrical wiring is one of the most complicated areas of the estimating field. In most cases it is left to a specialist and the building estimator is never called upon to give more than a rough idea of the cost.

Although many different wiring methods are recognized by the National Electrical Code (NFPA No. 70) as suitable for use in a residential occupancy, only a few of the more flexible methods are actually used because of the problems of installation in residential frame construction.

The four most common wiring methods used for the internal wiring of a residence are:

1. *Flexible Metal Conduit,* generally known as *Greenfield* or *flex.*
2. *Metal Clad Cable, type AC.* Generally known as *armored cable* or *BX.*
3. *Nonmetallic-Sheathed Cable, types NM and NMC.* Generally known as *romex.*
4. *Electrical Metallic Tubing.* Generally known as *EMT* or *thin-wall conduit.*

The wiring method used for the electrical service would usually be:

1. Electrical Metallic Tubing.
2. Service Entrance Cable, Types USE or SE.
3. Rigid Metal Conduit.

The underground wiring to a garage or other structure would usually consist of underground feeder and branch circuit cable, type UF or rigid metal conduit containing conductors that have a moisture resistant insulation.

The particular wiring method used on any job depends upon the requirements of the situation and the local ordinances. Concealed knob and tube work is virtually obsolete. *It is specifically prohibited by the building codes of many communities* and for this reason, will not be discussed here. Non-metallic sheathed cable and armored cable are widely used for residential wiring in smaller cities, towns and rural areas. In large cities and for commercial and industrial applications, however, wiring is usually required to be rigid conduit or electrical metallic tubing for exposed work and flexible metal conduit or armored cable for wiring which will be concealed when the building is completed.

Electrical Terms

AMPERE (OR AMP) is the measure of the rate of flow of electricity. Branch circuits, fuses, and circuit breakers are rated in "amperes" to designate the electricity they will carry safely.

APPLIANCE OUTLET, or heavy-duty outlet, is a 3-wire outlet for equipment such as a range, dryer or water heater which operates on 240-volt current, with rating of 30 to 50 amperes. Only one major appliance should be used on a circuit.

BRANCH CIRCUITS are the wires which carry electricity from the main service panel, or from load center panelboards, to lights, appliances and electrical equipment.

CIRCUIT BREAKERS like fuses are automatic protective devices that cut off the flow of electricity in excess of their ampere rating when the demand due to overloads or short circuits is greater than the main or branch circuits are designed to carry. (Unlike fuses, they need not be replaced—only reset—after they do their job.)

DIMMER CONTROL is a switch which lets you vary the amount of light.

ELECTRIC SERVICE ENTRANCE is a combination of intake wires and equipment, including the service entrance wires, electric meter, main switch or circuit breaker and main distribution or service panel, through which the supply of electric power enters the house.

ELECTRIC CIRCUIT is the path of electricity from the electric service entrance to the point of use.

FUSES are safety devices which shut off the flow of electricity in main or branch circuits in case of overloads or short circuits. Fuses are rated in "amperes."

KILOWATTHOUR is 1000 watthours, abbreviated "kwhr." This term on electric bills shows the amount of electrical energy used. (Example: a 100 watt lamp may burn for 10 hours, consuming one kwhr of electrical energy. 100 watts × 10 hours = 1000 watthours or one kwhr.)

LIGHTING OUTLET is an electrical outlet for a lighting fixture (luminaire) or other fixed lighting (valance, cornice light, cove lighting or recess units).

MAIN SERVICE PANEL includes the main electrical switch or circuit breaker and the circuit panel box which houses the circuit breakers or fuses for the branch circuits. These may be separate or combined within one panel box.

RECEPTACLE OUTLET is an electrical outlet for the plug-in connection of portable electrical appliances or lamps. In a residence the NEC (National Electrical Code) requires that all of the receptacle outlets in the kitchen shall be connected to at least two heavy-duty 20-ampere appliance circuits. Many of the larger electrical appliances used in the kitchen such as self-defrost combination freezer-refrigerator, freezer or portable dishwasher, should have their own individual 20-ampere circuit. The receptacle outlets installed in the pantry, family room, dining room and breakfast room are required to be connected to heavy-duty 20-ampere circuits. These receptacles may be connected to the heavy-duty kitchen circuits but usually additional heavy duty circuits are installed to feed these receptacles. The NEC requires that at least one receptacle outlet connected to an individual heavy-duty 20-ampere circuit must be installed in the laundry area. Generally the receptacle outlets installed in the other areas of the house such as living room and bedrooms are connected to 15-ampere branch circuits. All receptacle outlets shall be of the grounding type. (See Fig. 16-1.)

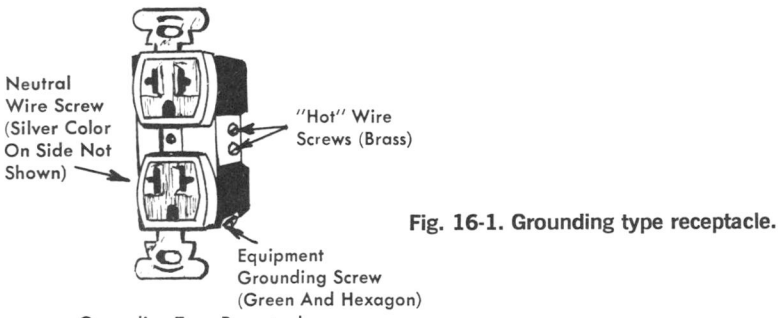

Neutral Wire Screw (Silver Color On Side Not Shown)

"Hot" Wire Screws (Brass)

Equipment Grounding Screw (Green And Hexagon)

Grounding Type Receptacle

Fig. 16-1. Grounding type receptacle.

VOLT is a measure of electric force. It is the force behind the current (AMPs) flowing through a wire.

WALL SWITCH, sometimes called a snap-switch or toggle switch, turns the current "on" and "off" for a lighting outlet, or for convenience outlets like those for portable floor or table lamps in a room not equipped with a ceiling lighting unit.

WATT is a unit of power that does work electrically. It is the product of amperes and volts. Most electrical appliances and equipment are rated in both "watts" and "volts" on the nameplate.

WATTHOUR is the measurement of electrical energy used. It represents "one watt used for one hour."

Electrical Conductors. The wires which carry electricity are called conductors and are made of a copper or aluminum wire which is covered with an insulating material such as thermoplastic or heat resistant rubber, Fig. 16-2. The minimum size conductor recognized by the National Electrical Code (NEC) for general wiring in a residence is No. 14 American Wire Gage (AWG). A No. 14 copper conductor is listed as being capable of carrying 15 amperes and shall be protected by a fuse or circuit breaker not exceeding 15 amperes. Aluminum is not as good a conductor of electricity as copper, therefore, if an aluminum conductor is used the size must be increased to No. 12 for a 15 ampere circuit. Table I indicates capacities of various sizes of copper conductors.

TABLE I. CARRYING CAPACITY OF COPPER CONDUCTORS

Size of Conductor AWG	Circular Mills	In Raceway or Cable	
		Type RUW Thermoplastic Types T, TW	Rubber Types RH, THW, THWN, XH, HW
14	4,110	15	15
12	6,530	20	20
10	10,380	30	30
8	16,510	40	45
6	26,240	55	65
4	41,740	70	85
3	52,620	80	100
2	66,630	95	115
1	83,690	110	130
0	105,600	125	150

Not more than 3 conductors in raceway or cable or direct burial (based on ambient temp. of 30%C, 86%F).

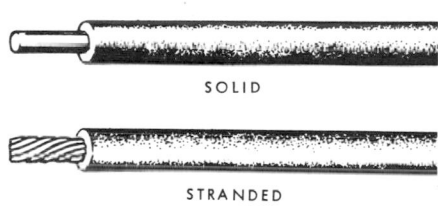

SOLID

STRANDED

TYPE THW MOISTURE RESISTANT
PLASTIC INSULATED WIRE
FOR USE IN WET OR DRY LOCATIONS
MAX. OPERATING TEMP. 75° C
COLORS: 14, 12, 10—BLACK, WHITE, RED, GREEN,
ORANGE, BLUE, BROWN, YELLOW, PURPLE

Fig. 16.2. Building wire. The moisture-resistant, flame-retardant thermoplastic insulation is without braid or wrap, strips easily and has a smooth, glossy surface which will not support combustion and fishes easily.

Metal Clad Cable (Armored Cable or BX). Metal clad cable is a factory fabricated assembly consisting of rubber or thermoplastic insulated conductors protected by a flexible galvanized steel armor. It is manufactured with or without a lead sheath under the armor for protection against excessive dampness, Fig. 16-3. Armored

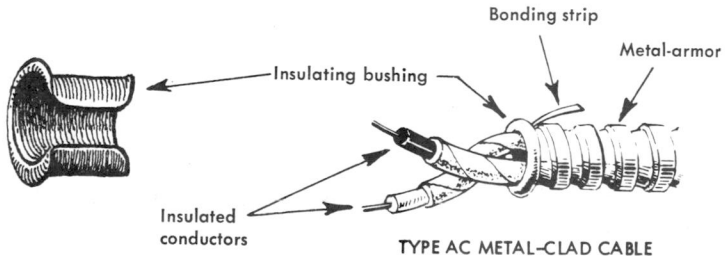

Bonding strip

Metal-armor

Insulating bushing

Insulated conductors

TYPE AC METAL–CLAD CABLE

Fig. 16-3. Type AC metal-clad cable without lead sheath and with insulating bushing.

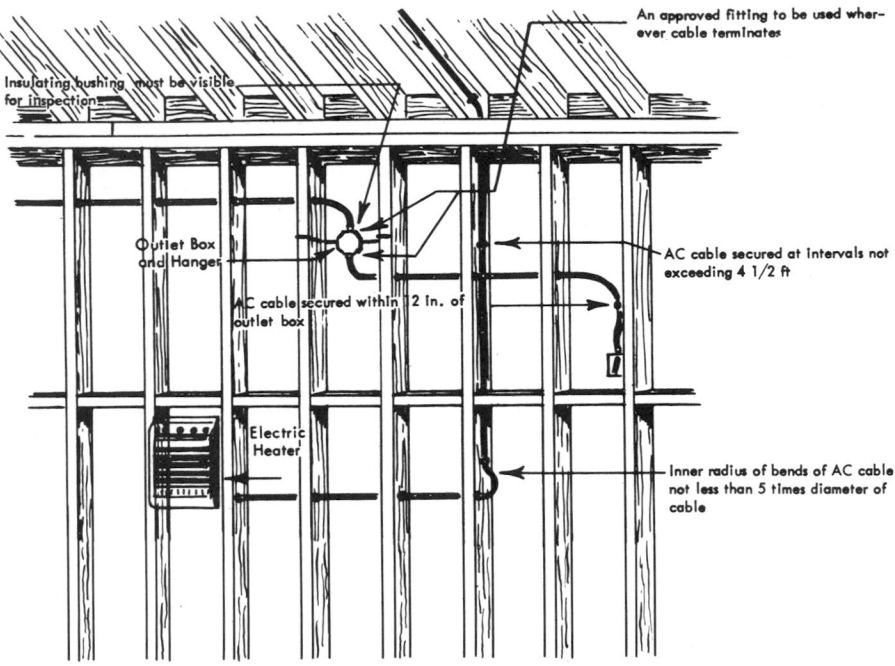

An approved fitting to be used wher-ever cable terminates

Insulating bushing must be visible for inspection

Outlet Box and Hanger

AC cable secured within 12 in. of outlet box

Electric Heater

AC cable secured at intervals not exceeding 4 1/2 ft

Inner radius of bends of AC cable not less than 5 times diameter of cable

Fig. 16-4. Installation of metal-clad cable (Type AC) in frame construction.

cable (Type AC) is the type used for residential wiring and is regularly available in sizes No. 14 to No. 2 AWG, Fig. 16-4. The NEC requires that type AC armored cable shall have an internal bonding strip of copper or aluminum in intimate contact with the armor for its entire length. The wires

Fig. 16-5. BX connector.

at the end of the cable are required to be protected by a fiber bushing called an antishort insulating bushing, Fig. 16-3. This bushing protects the insulation of the wires from the rough edges of the cut ends of the armor and must be visible for inspection after the cable connector has been installed, Fig. 16-5.

Metal Conduit. This system, used for general purpose wiring is made of either steel or aluminum. The three common types recognized by the NEC are rigid metal conduit, electrical metallic tubing (EMT or thinwall conduit) and flexible metal conduit (Greenfield). Metallic conduit wiring systems provide mechanical protection and electrical safety to both persons and property as well as convenient and accessible raceways for the conductors. A well designed electrical raceway system is one which has adequate capacity for future expansions and is readily adapted to changing conditions.

When using metal conduit as a wiring method, the conduit, fittings, and boxes are installed complete without the wires as soon as the floor joists, rough floor and studding are in position and before any insulating or lathing is done. The wires are not installed until the walls and ceiling are completed. The switches, receptacles and fixtures are not usually installed until all of the painting is completed.

Rigid Metal Conduit. This system is constructed from ferrous or nonferrous metals such as wrought iron, steel or aluminum. It has approximately the same dimensions as ordinary water and gas pipes and is recognized by the NEC in standard lengths of ten feet. Each length shall be seamed and threaded on each end and shall include one coupling, Fig. 16-6. Where a conduit enters a box the wires must be protected from abrasion by an approved bushing. Rigid metal conduit is seldom used for the internal wiring of a residence because of the high cost of material and labor. This type of

wiring may be required if underground conduit or conduit encased in a concrete slab in contact with the earth is necessary.

Electrical Metallic Tubing (EMT or Thinwall). This tubing is similar to rigid conduit, but it has a thin wall (about 40 percent that of rigid conduit) and is not threaded, Fig. 16-7. Clamp

Fig. 16-7. Electrical metallic tubing (EMT or thinwall conduit). Comes in 10-ft. lengths—no couplings.

or set screw types of fittings are used to join one piece to the next, Fig. 16-8, or to outlet boxes, Fig. 16-9. It is lighter and is easier to cut, bend and install than rigid conduit. Thus the labor and material cost is less than with rigid conduit, yet the advantages of having all the wires inside a metal raceway, where they can be removed or replaced when necessary, is retained. EMT is not normally used in wood frame construction, but is often used when installed exposed on the bottom of the floor joists in an unfinished basement or where run down a concrete block wall in the basement

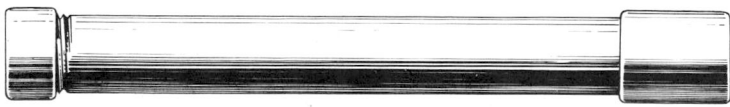

Fig. 16-6. Rigid metal conduit (10 ft. lengths—one coupling and one thread protector included).

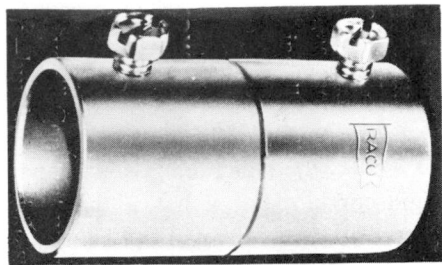

Fig. 16-8. Fittings to join lengths of EMT.

Fig. 16-10. Typical installation of EMT to a switch box and receptacle on the block wall in the basement of a residence. The wireman will install straps, pull wire and install device after building is completed. Owner will install paneling and complete room at a future date.

Fig. 16-9. Fittings used to join EMT to outlet boxes.

for receptacles in a recreation room, Fig. 16-10.

Flexible Metal Conduit (Greenfield). This is constructed from a single strip of aluminum or galvanized steel, spirally wound and interlocked so as to provide a round cross section of raceway with a high mechanical strength and great flexibility, Fig. 16-11. Flexible metal conduit is available

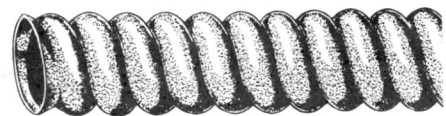

Fig. 16-11. Flexible metal conduit (Greenfield).

Fig. 16-12. Set-screw flexible conduit (Greenfield) connectors.

in lengths of 25 to 100 feet depending upon the size of the conduit. The NEC requirements governing the number of conductors in the conduit and the type of insulation on the conductors are the same as for rigid conduit or EMT. Outlet or switch boxes must be installed for all outlets or switches and the conduit must be securely fastened to the boxes, Fig. 16-12. The conduit must be continuous from outlet to outlet and when used in frame construc-

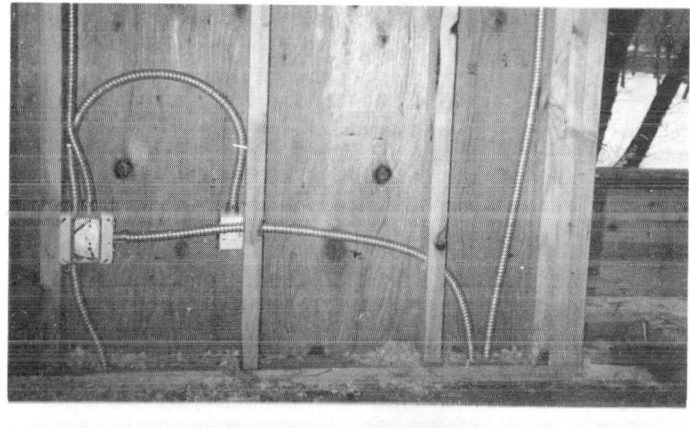

Fig. 16-13. Greenfield used to (top) tie boxes together (inside and outside receptacles) and (bottom) feed switch and receptacles.

tion must be securely fastened to the studs and joists by straps or other approved means, Fig. 16-13. This wiring system is permitted by the NEC to be used for many kinds of wiring and in some cases is preferable to rigid conduit or EMT. Because it is flexible and comes in continuous lengths it is easy to install in frame construction where rigid conduit or EMT in 10 foot lengths is difficult and economically impractical to install, Fig. 16-14. Flex-

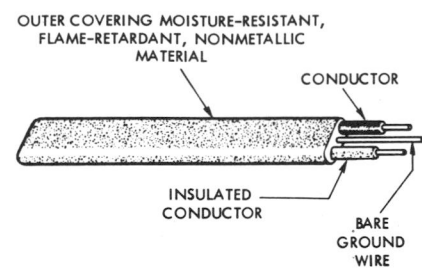

Fig. 16-15. Nonmetallic sheathed cable (romex) with ground wire.

tardant, nonmetallic material. The outer sheath in most of the cable now being used is made from a thermoplastic material. This cable is available with and without a ground wire. A ground wire is installed in the cable by laying a bare conductor between the insulated circuit conductors which are then encased by the outer covering. See Fig. 16-15.

This cable is usually installed by drilling holes in the studding or joists and pulling the cable through the holes. When run in the same direction as the studs or joists, it is attached to them with straps or staples driven in the wood, Fig. 16-16. The cable may be run in notches cut in the wooden studs or joists if protected by installing a steel plate at least $\frac{1}{16}$ inch in thickness

Fig. 16-14. Greenfield in frame construction.

ible metal conduit is widely used for residential wiring where local codes and ordinances require a metallic raceway system.

Nonmetallic Sheathed Cable (Romex). This cable is an assembly of two or more rubber or thermoplastic insulated conductors having an outer sheath of moisture-resistant, flame-re-

Fig. 16-16. Method of fastening nonmetallic sheathed cable to floor joists and ceiling with straps.

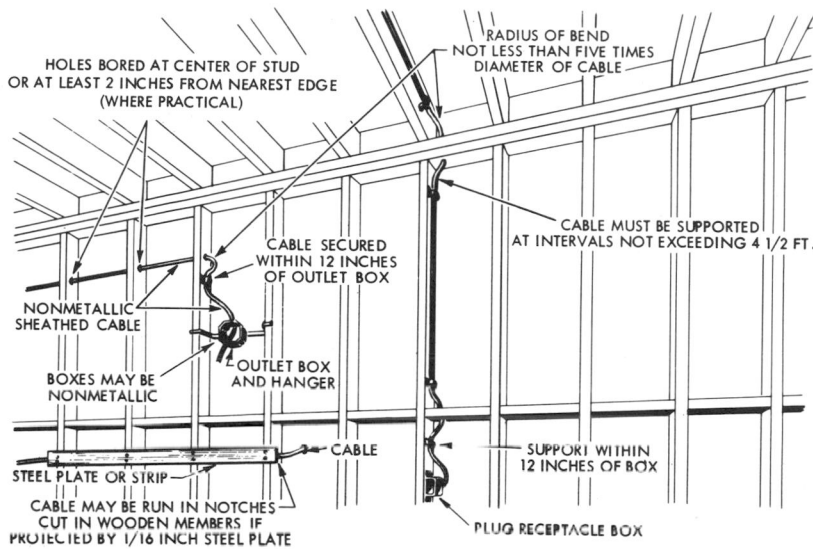

HOLES BORED AT CENTER OF STUD OR AT LEAST 2 INCHES FROM NEAREST EDGE (WHERE PRACTICAL)

RADIUS OF BEND NOT LESS THAN FIVE TIMES DIAMETER OF CABLE

CABLE SECURED WITHIN 12 INCHES OF OUTLET BOX

CABLE MUST BE SUPPORTED AT INTERVALS NOT EXCEEDING 4 1/2 FT.

NONMETALLIC SHEATHED CABLE

BOXES MAY BE NONMETALLIC

OUTLET BOX AND HANGER

CABLE

STEEL PLATE OR STRIP

CABLE MAY BE RUN IN NOTCHES CUT IN WOODEN MEMBERS IF PROTECTED BY 1/16 INCH STEEL PLATE

SUPPORT WITHIN 12 INCHES OF BOX

PLUG RECEPTACLE BOX

Fig. 16-17. Installation of nonmetallic sheathed cable (romex) in studding and joists.

before building finish is applied. See Fig. 16-17. Either metallic or nonmetallic outlet boxes may be used with this wiring system, Fig. 16-18.

A cable connector is used to connect the cable to the box, Fig. 16-19.

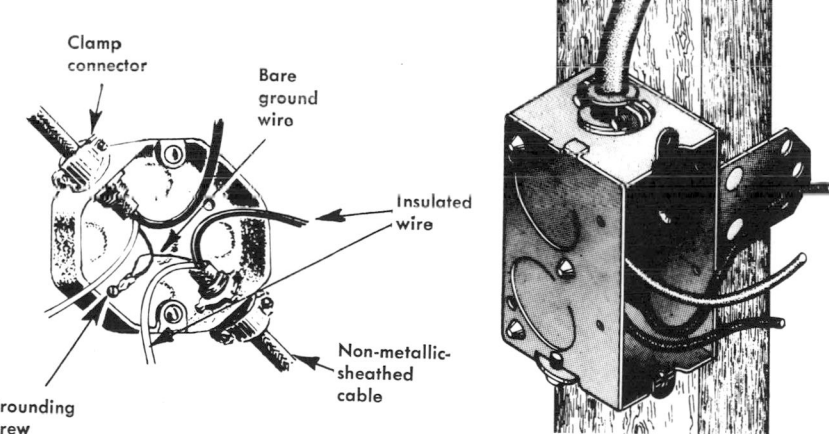

Clamp connector

Bare ground wire

Insulated wire

Non-metallic-sheathed cable

Grounding screw

Fig. 16-18. Method of installing a lighting outlet (left) and a switch or receptacle outlet (right) using romex and a metal outlet box.

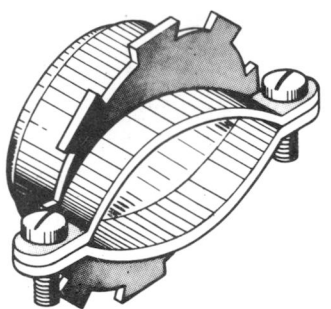

Fig. 16-19. Romex cable connector (used to connect romex to box).

ELECTRIC SERVICE

All electric energy supplied to electrical equipment within a building must enter through the electric service where it is metered, grounded, protected and distributed throughout the building by feeders or branch circuits. The size of the service is determined by the amount of electrical energy that will be needed to supply the electrical equipment within the building. Normally the electric service to a single family residence will consist of three conductors, each capable of carrying 100 amperes, with a voltage of 115/230 volts. The National Electrical Code requires that if the electrical load is 10kw or more or if there are over five 2-wire branch circuits, a minimum service of 100 amperes, 115/230 volt must be installed. With the increased use of electrical appliances, air conditioning and electric heat, it is often necessary to install a service larger than 100 amperes.

Service Entrance Wiring. The electric service to a residence will consist

of either overhead or underground conductors depending upon the location of the power company distribution lines.

Overhead Service. The electric light and power company usually will install the overhead service drop conductors from the pole to the secondary rack on the house at no cost to the customer, Fig. 16-20. The electric me-

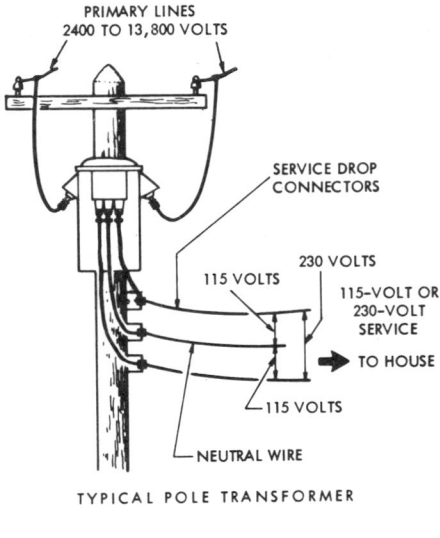

TYPICAL POLE TRANSFORMER

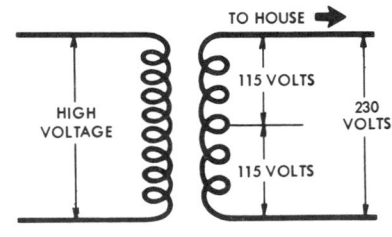

INTERNAL WIRING DIAGRAM OF TRANSFORMER

Fig. 16-20. A transformer steps down the power line voltage to 115/230 volts before it reaches the home.

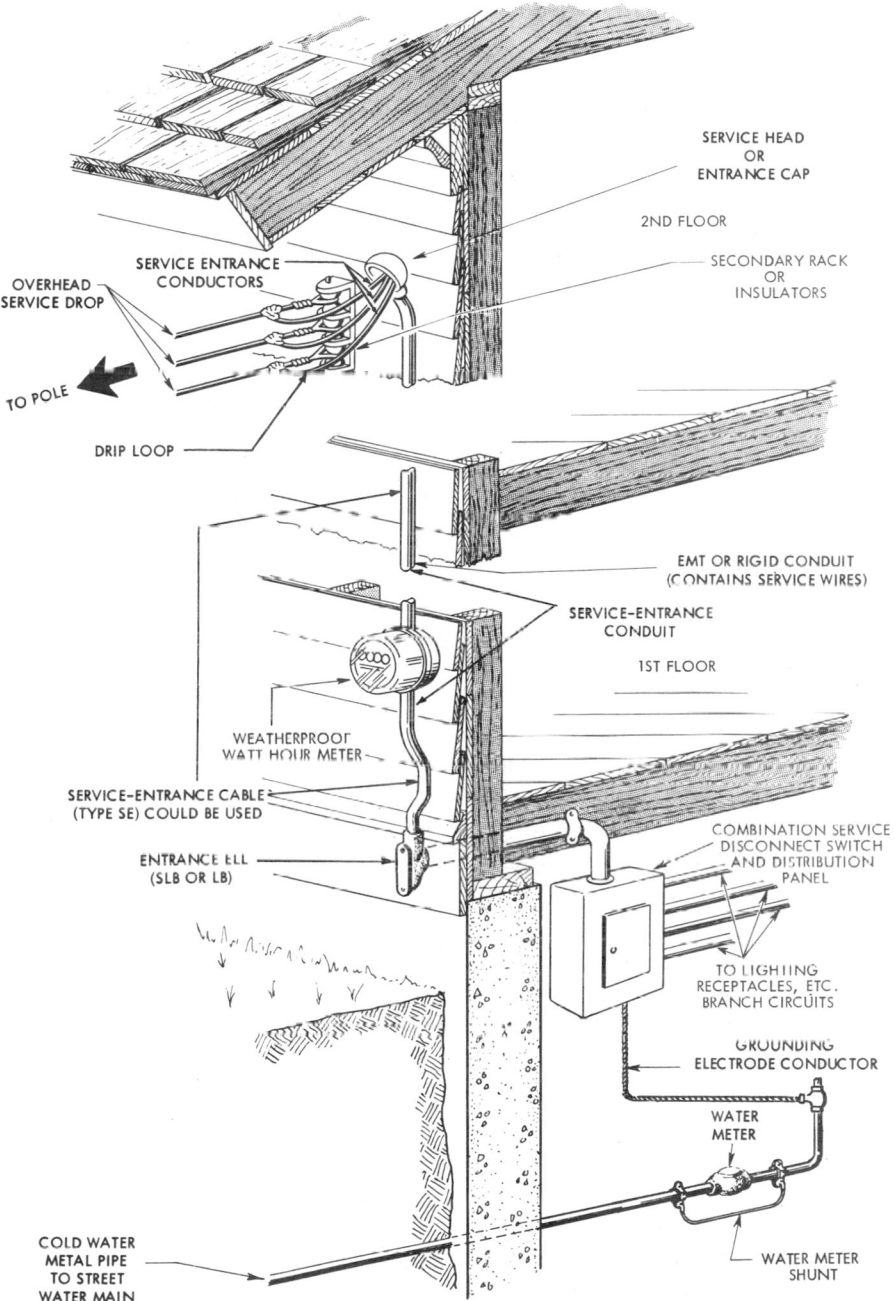

Fig. 16-21. Principal parts of an overhead service.

Fig. 16-22. Typical overhead service to a single family rambler. Rigid 2-inch conduit is used as a mast to give required distance above ground. Triplex cable is used as service drop from the pole to the house.

ter is installed by the power company after the building is completed. The electrical contractor will install the wiring from the secondary rack on the outside of the building to the meter socket and the service switch or distribution panel, Fig. 16-21. The wiring method used will often be determined by the local electrical code or by the power company but normally will be electrical metallic tubing, rigid metal conduit or service entrance cable, Figs. 16-21 and 16-22.

Underground Service. In areas where the power company has an underground distribution system it is usually the builder's responsibility to install the underground service conductors from the transformer to the meter which is usually located on the building. It is the architect's or the electrical contractor's responsibility to make the necessary arrangements with the power company for the installation of the underground service conductors, Fig. 16-23.

Most power companies have specific requirements governing the location, depth, size of conductors and wiring methods to be used. The electrical contractor usually will install the complete underground service but the power company workman will make the final connection to the power distribution lines. The wiring method usually will be rigid conduit or underground service entrance cable (USE), Fig. 16-24.

Service-Entrance Cable (Type SE). Type SE cable is approved by the National Electrical Code as a wiring method to bring electrical power from the power company overhead connection on the building to the meter and to the service switch or distribution panel in the building, Fig. 16-21.

The three-conductor cable shown in Fig. 16-25 is an assembly consisting of two insulated conductors with one concentric-stranded uninsulated conductor (commonly known as the grounded neutral conductor) wound around the two insulated conductors. This oval shaped assembly has an outer flame-retardant, moisture-resistant covering. The strands of the uninsulated conductor are twisted together to permit the electrician to make terminal connections at each end. This

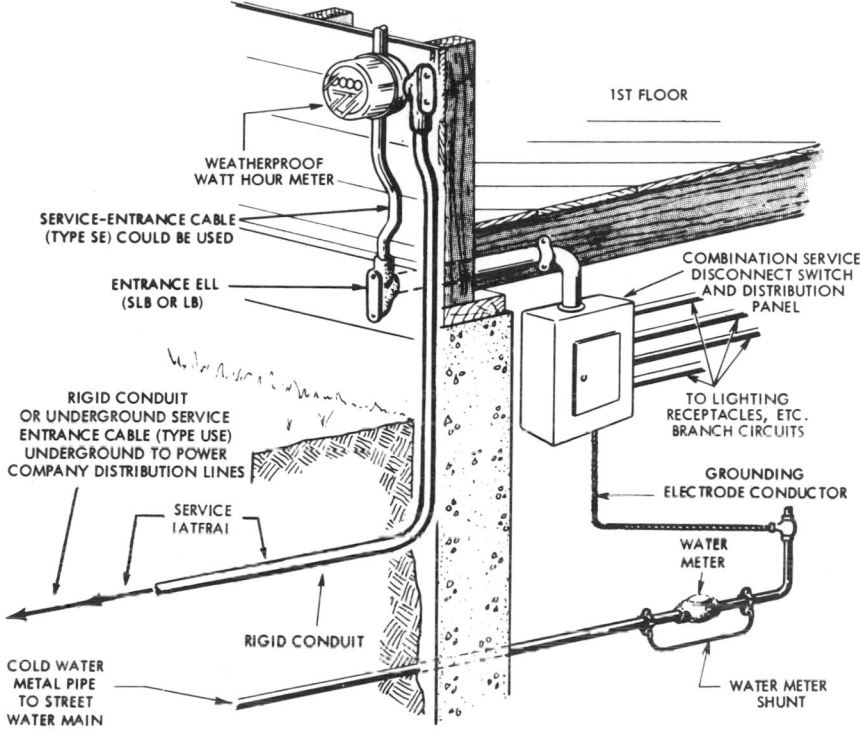

1ST FLOOR

WEATHERPROOF
WATT HOUR METER

SERVICE-ENTRANCE CABLE
(TYPE SE) COULD BE USED

ENTRANCE ELL
(SLB OR LB)

COMBINATION SERVICE
DISCONNECT SWITCH
AND DISTRIBUTION
PANEL

RIGID CONDUIT
OR UNDERGROUND SERVICE
ENTRANCE CABLE (TYPE USE)
UNDERGROUND TO POWER
COMPANY DISTRIBUTION LINES

TO LIGHTING
RECEPTACLES, ETC.
BRANCH CIRCUITS

GROUNDING
ELECTRODE CONDUCTOR

SERVICE
LATERAL

WATER
METER

RIGID CONDUIT

COLD WATER
METAL PIPE
TO STREET
WATER MAIN

WATER METER
SHUNT

Fig. 16-23. Principal parts of an underground service.

4 0 600 V TYPE USE OR RHW OR RHH

Fig. 16-24. Service-entrance cable (Type USE).

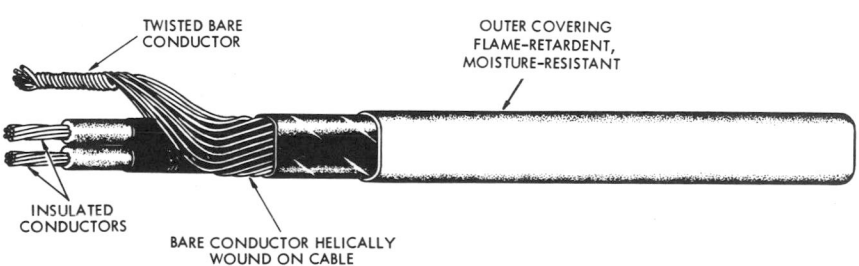

TWISTED BARE
CONDUCTOR

OUTER COVERING
FLAME-RETARDENT,
MOISTURE-RESISTANT

INSULATED
CONDUCTORS

BARE CONDUCTOR HELICALLY
WOUND ON CABLE

Fig. 16-25. Service-entrance cable (Type SE).

type of cable is not required to have inherent protection against mechanical abuse. Type SE cable is often not permitted by local codes and ordinances in the larger cities but is sometimes used in smaller towns and rural areas.

Service-Entrance Cable (Type USE). Type USE cable is recognized by the NEC as an underground wiring method to bring electric energy from the power company underground distribution system to the electric meter which is normally located on the building, Fig. 16-23. This cable is available as a multi-conductor cable or as a single conductor. When in a cable assembly, the insulated conductors are required by the NEC to have moisture-resistant covering, whereas a single conductor having rubber insulation specifically approved for the purpose does not require an outer covering, Fig. 16-24.

The distribution of electric energy

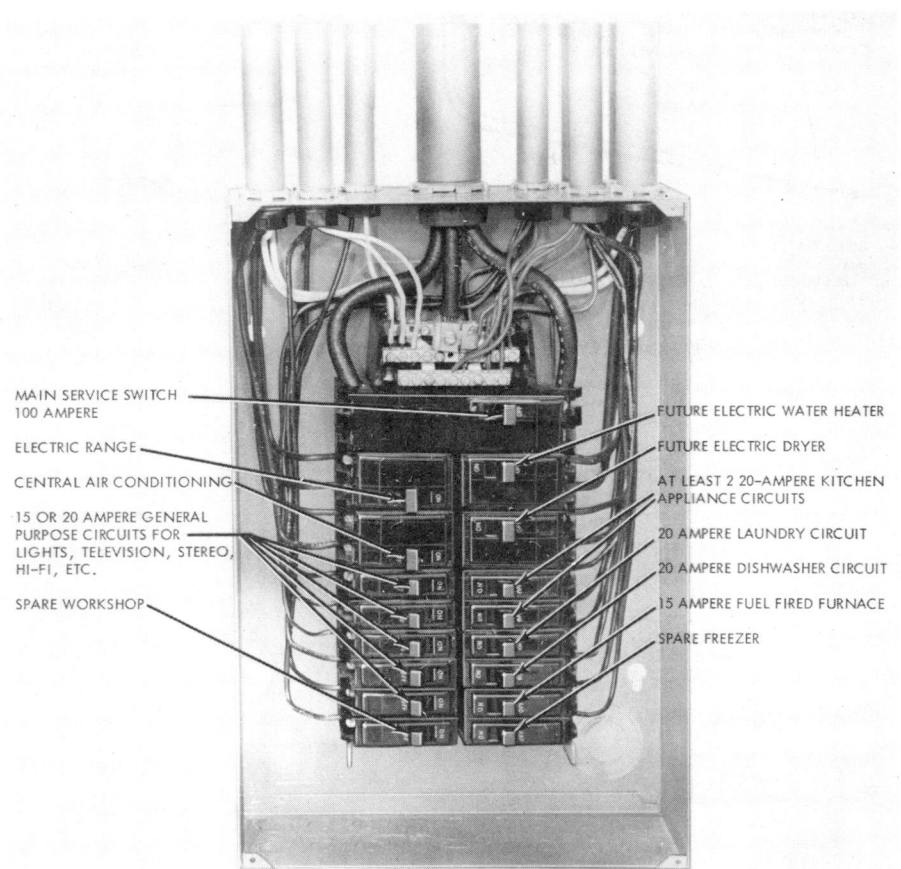

MAIN SERVICE SWITCH 100 AMPERE

ELECTRIC RANGE

CENTRAL AIR CONDITIONING

15 OR 20 AMPERE GENERAL PURPOSE CIRCUITS FOR LIGHTS, TELEVISION, STEREO, HI-FI, ETC.

SPARE WORKSHOP

FUTURE ELECTRIC WATER HEATER

FUTURE ELECTRIC DRYER

AT LEAST 2 20-AMPERE KITCHEN APPLIANCE CIRCUITS

20 AMPERE LAUNDRY CIRCUIT

20 AMPERE DISHWASHER CIRCUIT

15 AMPERE FUEL FIRED FURNACE

SPARE FREEZER

Fig. 16-26. Lightning and appliance panelboard (circuit breaker type). (Courtesy Square D Company)

under the exclusive control of the electric utilities is exempt from the requirements of the NEC therefore other types of underground service entrance cables which are not covered in the NEC could be required by the electric utility in specific areas. A careful check should be made with each electric utility as to local requirements.

Service Switch and Distribution Panel. After passing through the meter, the electricity goes to the service disconnecting means (service switch). The service switch shall provide for disconnecting all conductors in the building from the service-entrance conductors thereby cutting off all electricity in the house. The service switch may be a circuit breaker or a fused switch and in a single family residence it is normally incorporated in the same enclosure as the branch-circuit distribution panelboard, Figs. 16-26 and 16-27. Each undergrounded service-entrance conductor shall have overcurrent protection which is normally provided by fuses in the service switch or by a circuit breaker. The distribution panelboard is supplied by the service switch and it distributes electrical energy through many branch circuits throughout the house, Fig. 16-26. The branch circuits are protected in the panelboard from short circuits, grounds and overloads by fuses, Fig. 16-28, or circuit breakers, Fig. 16-29.

Fuses and Circuit Breakers. Fuses may be of the plug, cartridge or knife blade type. Plug fuses are rated at 0 to 30 amperes and are limited to branch circuits with voltages not exceeding 150 volts to ground. The cartridge or ferrule type is rated from 0 to 60 amperes and the knife blade type is rated from 70 to 600 amperes, Fig. 16-28. When plug fuses are used on circuits of 0 to 30 amperes they shall be of the Type S fustat type with a Type S adaptor. Fuses are used in panelboards but modern practice favors the use of circuit breakers because when the cause of overload is cor-

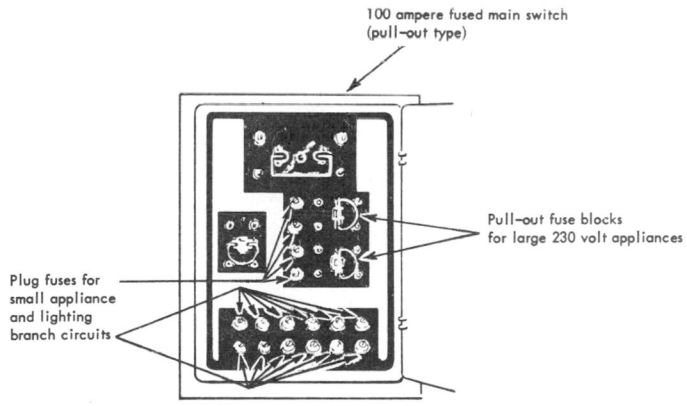

100 ampere fused main switch (pull-out type)

Pull-out fuse blocks for large 230 volt appliances

Plug fuses for small appliance and lighting branch circuits

Fig. 16-27. Lightning and appliance panelboard (fuse type).

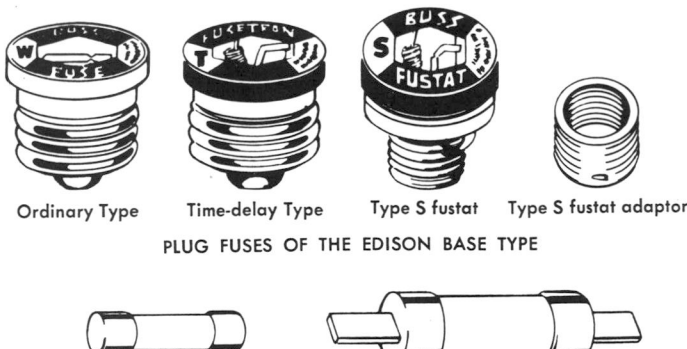

BUSS
FUSTAT
Ordinary Type
Time-delay Type
Type S fustat
Type S fustat adaptor
PLUG FUSES OF THE EDISON BASE TYPE
0 to 60 amp, Ferrule type
70 to 600 amp. Knife-blade type

Fig. 16-28. Various types of fuses.

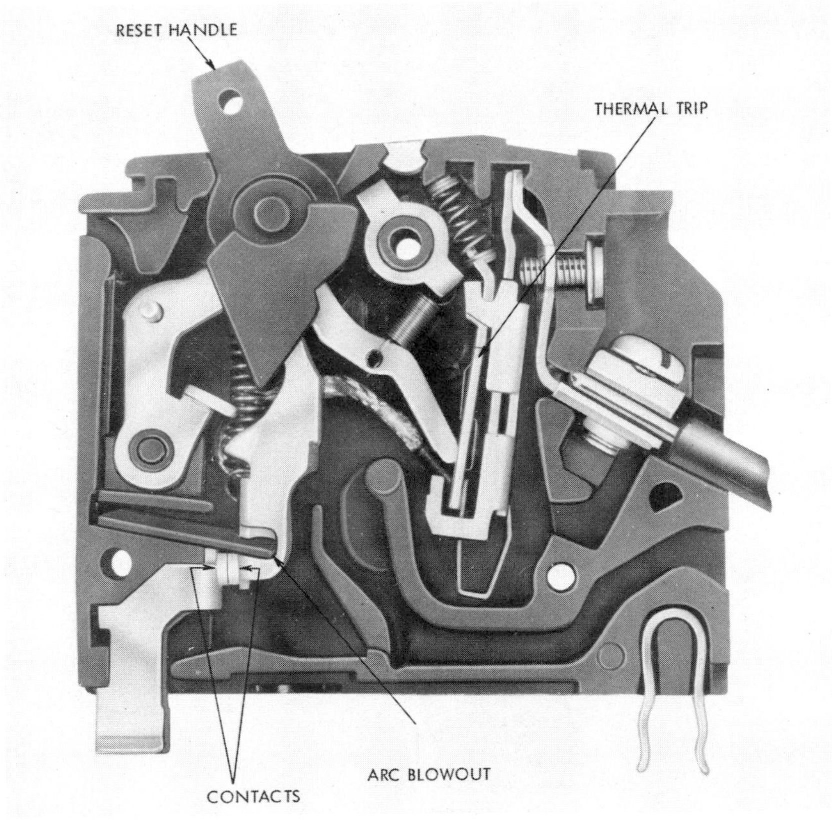

RESET HANDLE
THERMAL TRIP
ARC BLOWOUT
CONTACTS

Fig. 16-29. A cutaway view of a circuit breaker. (Courtesy Square D Company)

rected they may be switched back on and used over and over again, Fig. 16-29.

ESTIMATING WIRING

Two methods of estimating are described below. The first is a precise method that is used for industrial and commercial work and wherever the conditions of the job differ markedly from the average. Its successful application, however, requires a great deal of experience in this sort of work. For this reason it is impractical for the average residential estimator and will be discussed only briefly. Though he will not use this system himself, the general estimator should know something about it. This will help him to check the estimate received from the electrical subcontractor for obvious errors due to carelessness, omissions, etc. In specialized work, involving high voltage or high-power wiring, the estimate is done by an experienced electrical estimator or electrical engineer.

The second method is simple and useful for a non-specialist. So long as it is applied only to average jobs involving residential dwellings, it is sufficiently accurate for most practical purposes.

Method 1. The most accurate method of estimating wiring cost is undoubtedly one which makes a detailed analysis of all required electrical work. An analysis of this type is made from plans and specifications drawn up by an electrical engineer or the architect. The materials needed for a job are determined from these plans. If such detailed plans or specifications are not available, however, it is necessary for the electrical estimator to prepare his own plans and have these layouts approved by the architect before beginning the estimate. These layouts will show all of the necessary wiring and equipment. They will give the size and type of electrical service and the size and location of all feeders and distribution panelboards. They will also show the size and location of all motors and their controllers as well as the type, number and location of all fixtures, switches and receptacles. The estimator makes a detailed count of all of the needed material and equipment and from his scaled plans measures the length of each circuit to get total length of cable needed for the wiring. The total cost of material is obtained by consulting the appropriate catalog.

The amount of time necessary to install the wiring and equipment is determined by using appropriate labor units. A labor unit is the average time needed by an electrician to install a specific type of material, such as an outlet box on a wood stud or a receptable box on a concrete wall. When the total amount of electrical material and equipment is known the estimator converts them into labor units and calculates the total time needed to make the complete installation. There are many things which must be considered before preparing a final estimate. Will there be a temporary service or temporary wiring for power tools during construction, will the elements such as rain or snow, hot air or cold weather affect your labor hours, is there an un-

HOME WIRING ESTIMATOR

JOB NO. *19-*
DATE *May 1, 1972*

JOB LOCATION *164 Oak Street*
OWNER *Marvin Johnson* BUILDER *Home Acres Builders, Inc.*

BRANCH CIRCUIT REQUIREMENTS

Ref. to 1971 NEC		
220-2(a)	**A. GENERAL LIGHTING LOAD** *2000* Sq. Ft. x 3 Watts . . . . = - *6000* Watts	
220-3(a)	**B. MINIMUM NUMBER OF BRANCH CIRCUITS** ✓ General Purpose *6000* Watts ÷ 115 Volts = *52* Amperes	
	which means (*4-* 15 amp. 2 wire or) (*3-* 20 amp. 2 wire circuits) · · · · · · · · · · · · ·	*3-* 1 pole
220-3(b)	✓ Small Appliance* *2 -* 20 amp. 2 wire circuits	*2-* 1 pole
220-4 220-4(j) 220-3(b)	✓ Individual Equipment	

✓ Individual Equipment

- ☒ Range (12 kw) . . . . *50* amp. 3 wire 115/230 volts . . . . *1-* 2 pole
- ☒ Water Heater . . . . *20* amp. 2 wire 230 volts . . . . *1-* 2 pole
- ☒ Clothes Washer . . . . 20 amp. 2 wire 115 volts . . . . *1-* 1 pole
- ☒ Clothes Dryer . . . . . . (☒ 30 amp. 3 wire 115/230 volts, or) . . . . . *1 - 2* pole
- ☐ Dishwasher . . . . (☐ 20 amp. 2 wire 115 volts) . . . . ____ 1 pole
- ☐ Food Waste Disposer. . . . . . . 20 amp. 2 wire 115 volts . . . . ____ 1 pole
- ☐ Attic Fan . . . . ____ amp. 2 wire ____ volts . . . . ____ 1 pole
- ☒ Bathroom Heater No. 1 . . . . 20 amp. 2 wire 115 volts . . . . *1-* 1 pole
- ☐ Bathroom Heater No. 2 . . . . 20 amp. 2 wire 115 volts . . . . ____ 1 pole
- ☒ Heating Plant--Air Cleaner . . 20 amp. 2 wire 115 volts . . . . *1-* 1 pole
- ☐ Electric Home Heating . . . . (____ amp. ____ wire ____ volts . . . . ____ pole
- (____ amp. ____ wire ____ volts . . . . ____ pole
- ☒ *Power Saw* *20* amp. *2* wire *115* volts . . . . *1 - 1* pole
- ☐ _____ ____ amp. ____ wire ____ volts . . . . ____ pole
- ☐ _____ ____ amp. ____ wire ____ volts . . . . ____ pole
- ☐ _____ ____ amp. ____ wire ____ volts . . . . ____ pole

Total Circuits *12*

Total Poles *15*

LAYOUT OF LOADCENTERS AND FEEDERS

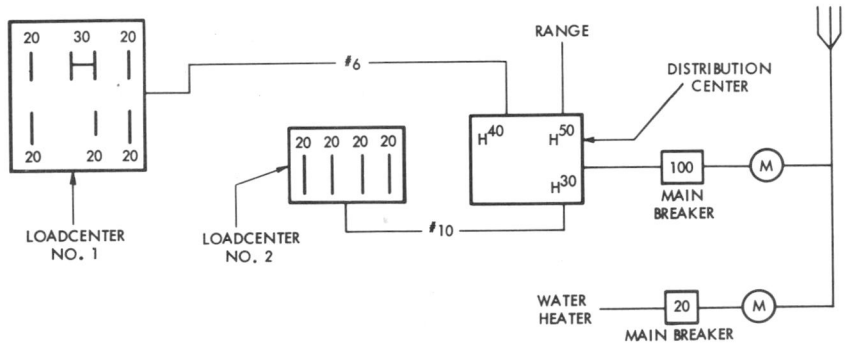

*NEC minimum (220-3(b), 1971 NEC) . . . . 2 Circuits
For more adequate service . . . . 3 Circuits
Recommended for "Electrical Living" . . . . 4 Circuits

Fig. 16-30. Home wiring estimator.

derground service and will the installation be your responsibility? When all of these factors have been evaluated, an experienced estimator can prepare an accurate bid for the complete electrical installation.

Method 2. The second method of estimating residential electrical wiring costs is to do the figuring on an average cost per outlet basis. The estimator determines the cost of the materials and labor which go into installing what he believes to be a typical outlet. He does this for each different kind of outlet as well as for the service entrance conduit, meter, service distribution panelboard, etc. The total cost of the job is then determined from these figures. Variables which must be considered are the wiring methods used, local codes and ordinances, material costs, labor regulations and labor costs. Because of economic and local variations, it is not possible to quote exact cost figures for these items. Outlet, receptacle and switch costs are usually figured on the basis of a 15-foot run of cable (or conduit).

The costs of materials can be found by consulting manufacturers' catalogs. No. 14 and No. 12 armored cable and nonmetallic sheathed cable (romex) are normally purchased in 250 foot lengths, ½- and ¾-inch flexible metal conduit (Greenfield) are normally purchased in 100-foot lengths and rigid conduit and thinwall conduit comes in 10-foot lengths. (The prices do not include fittings.) Prices of other materials, switches, outlet boxes, fittings etc. are also obtained from such catalogs.

The "Home Wiring Estimator," Fig. 16-30, will assist the student in computing the type of wiring needed for his set of plans. It will also give him a list of the equipment and the amount of each in a convenient and organized manner.

After calculating the load requirements (no. of square feet of residence $\times$ 3 watts), indicate the various branch circuits. Finally calculate the total load.

Quantity Take-Off—House Plan A. Fig. 16-31 is a branch circuit schedule for HOUSE PLAN A. It is a tabulation of the various outlets and fixtures shown on the plans for the house.

One electrical lighting outlet in the basement is turned on by switches at the top and bottom of the stairs. This is so that the light can be turned on before descending the basement stairs. These are called 3-way switches and the electrical symbol on the print is S_3, Fig. 16-32.

There are no ceiling or wall lighting outlets in the living room. The receptacle outlet by the foyer entrance is controlled by a switch located at the foyer entrance. The passageway from the foyer through the living room and dining room into the kitchen does not have any doors so switches could be located on either side of the openings. The switch for the foyer light is shown at the living room entrance where it is conveniently located to switch the lights on or off when entering or leaving the house.

The kitchen light can be turned on from the family room or kitchen. A branch-circuit panelboard is located

BRANCH CIRCUIT SCHEDULE

LOCATION	LIGHT OUTLET		SWITCH OUTLETS			RECEPTACLE OUTLETS		SPECIAL
	CEILING	WALL	S1	S3	S4	GENERAL USE	HEAVY DUTY	
BASEMENT	6			1				FURNACE
LIVING ROOM			1			4		
DINING ROOM	1		1			2		
FOYER	1		1					
HALL	1			2				
FAMILY ROOM				2			3	
KITCHEN	1			2			3	ELECTRIC OVEN FAN
LAUNDRY ROOM	1			2			1	
GARAGE	2	1	1			1		
PORCH	1		1			2		2 W.P. COVERS
PANTRY	1							
REAR ENTRY		1	1					
STAIRWAY				1				
BEDROOM #1			1			5		
BEDROOM #1 ENTRY	1		1					
BEDROOM #2			1			4		
BEDROOM #3			1			4		
REAR PORCH ENTRY		1	1					
LAVATORY		1	1			1		
FRONT ENTRY	1		1					
BATHROOM #1	1		1			1		
BATHROOM #2	1		1			1		
CLOSETS (5)	5							
	24	4	15	10	0	25	7	

Fig. 16-31. Branch circuit schedule for House Plan A.

on the wall between the kitchen and the porch. It is located where it is easily accessible and can feed all circuits of the house. (Another distribution panel is located in the basement near the stairs.)

Special circuits are normally installed for special loads such as the dishwasher and the furnace. These circuits are listed under the special column in the branch circuit schedule and must be considered separately. The thermostat wiring for the furnace must also be considered unless this is done by the heating contractor.

A ventilating fan is installed over the kitchen cabinets to the right of the sink. This fan may not be installed un-

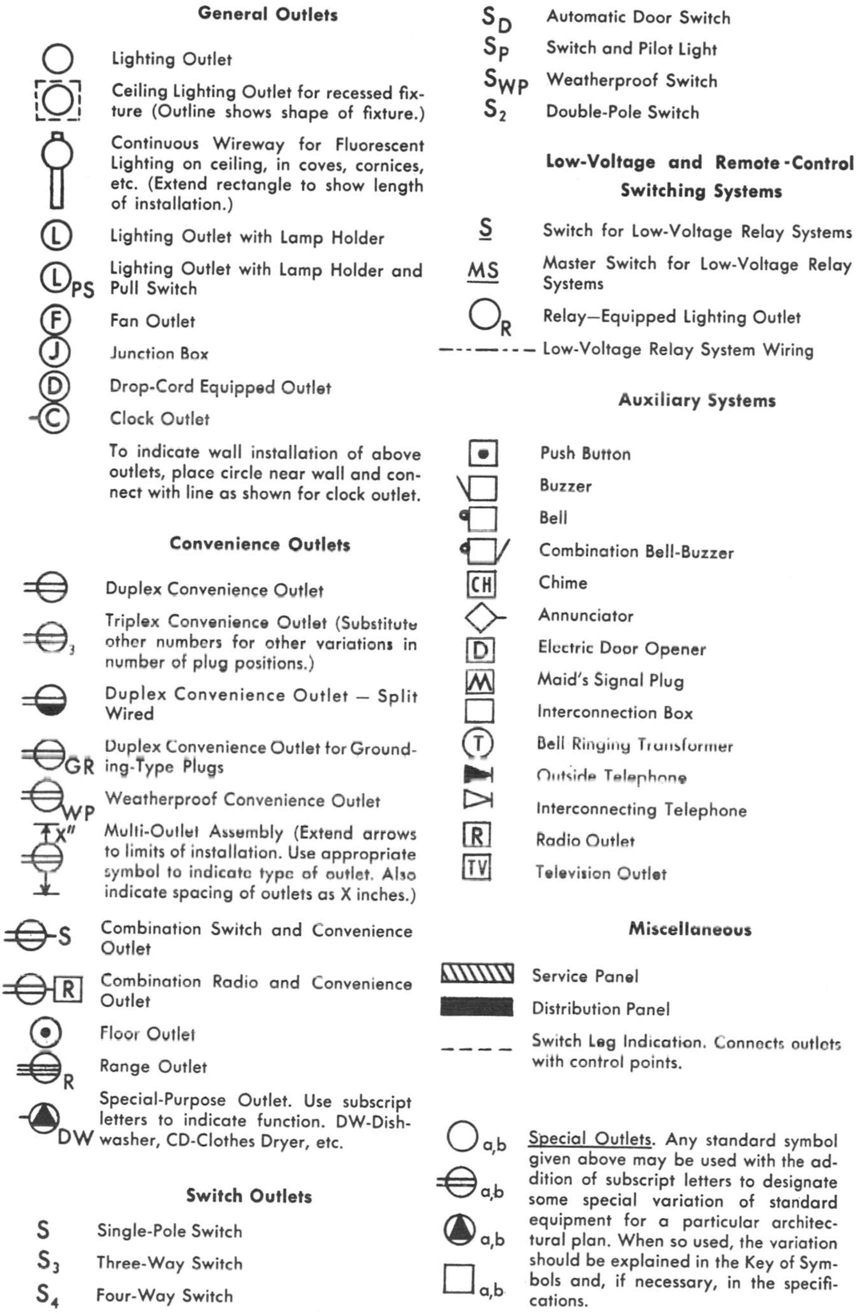

General Outlets

Lighting Outlet

Ceiling Lighting Outlet for recessed fixture (Outline shows shape of fixture.)

Continuous Wireway for Fluorescent Lighting on ceiling, in coves, cornices, etc. (Extend rectangle to show length of installation.)

Lighting Outlet with Lamp Holder

Lighting Outlet with Lamp Holder and Pull Switch

Fan Outlet

Junction Box

Drop-Cord Equipped Outlet

Clock Outlet

To indicate wall installation of above outlets, place circle near wall and connect with line as shown for clock outlet.

Convenience Outlets

Duplex Convenience Outlet

Triplex Convenience Outlet (Substitute other numbers for other variations in number of plug positions.)

Duplex Convenience Outlet — Split Wired

Duplex Convenience Outlet for Grounding-Type Plugs

Weatherproof Convenience Outlet

Multi-Outlet Assembly (Extend arrows to limits of installation. Use appropriate symbol to indicate type of outlet. Also indicate spacing of outlets as X inches.)

Combination Switch and Convenience Outlet

Combination Radio and Convenience Outlet

Floor Outlet

Range Outlet

Special-Purpose Outlet. Use subscript letters to indicate function. DW-Dishwasher, CD-Clothes Dryer, etc.

Switch Outlets

S   Single-Pole Switch

S_3 Three-Way Switch

S_4 Four-Way Switch

S_D Automatic Door Switch

S_P Switch and Pilot Light

S_{WP} Weatherproof Switch

S_2 Double-Pole Switch

Low-Voltage and Remote-Control Switching Systems

$\underline{S}$ Switch for Low-Voltage Relay Systems

$\underline{MS}$ Master Switch for Low-Voltage Relay Systems

O_R Relay—Equipped Lighting Outlet

— · — · — · — Low-Voltage Relay System Wiring

Auxiliary Systems

Push Button

Buzzer

Bell

Combination Bell-Buzzer

Chime

Annunciator

Electric Door Opener

Maid's Signal Plug

Interconnection Box

Bell Ringing Transformer

Outside Telephone

Interconnecting Telephone

Radio Outlet

Television Outlet

Miscellaneous

Service Panel

Distribution Panel

Switch Leg Indication. Connects outlets with control points.

Special Outlets. Any standard symbol given above may be used with the addition of subscript letters to designate some special variation of standard equipment for a particular architectural plan. When so used, the variation should be explained in the Key of Symbols and, if necessary, in the specifications.

Fig. 16-32. Electrical symbols. (American National Standards)

til after the wiring is finished and will probably mean a special trip to the house by the electrician. Only about one hour is required to do the actual work, but the required traveling time should be taken into consideration.

ESTIMATING LABOR

Electrical work is usually divided into two parts, the rough work or wiring, and the finish work, or fixtures. A specific amount is normally allowed in the estimate for the electrical fixtures which are often selected by the owner. If he should select fixtures which exceed allowance, he would have to pay the difference.

The rough work is done by electricians who may be assisted by apprentices or helpers. On small jobs, one electrician and and one helper, or two electricians, are often sufficient. The amount of time needed for rough work depends on the size and kind of wiring as well as the special conditions of the particular job. (Labor is least with nonmetallic or armored cable, greatest with heavy, rigid conduit.) Electricians' wages vary in different parts of the country and the estimator must consider the pay scale used in his locality.

In order to compute the labor cost, the estimator once again looks at each circuit separately. He must be able to determine the time required to wire each circuit. Since this time depends on how much work (such as boring) the electrician must do for each circuit, the estimator must be able to visualize those features of each circuit (or the total job) which will tend to make the work go more slowly. This can be done with reliability only if the estimator has considerable experience in this type of work. Since it is unlikely that the average building contractor will possess this kind of experience, it is recommended that he avoid trying to make detailed estimates of this type and leave such work to the electrical subcontractor.

Questions and Problems

1. Briefly describe the two methods of electrical estimating.
2. What is the minimum service to be installed in a residence having more than five 2-wire branch circuits?
3. If 48 electrical outlets are to be installed at an average cost of $36.00 per outlet, what would be the total estimated cost?
   (Answer: $1728.00)
4. If it takes 0.4 of an hour to run 100 ft. of 12-3 cable, how much is the labor cost at $6.50 an hour to run 550 feet?
   (Answer: $14.30)

17

PLUMBING

To estimate plumbing jobs, the plumbing estimator must have a complete knowledge of the local building codes and have worked in the field as a plumber. He must know the proper sizes of pipes and fittings to use for each different type of installation. He must be able to visualize the plumbing requirements of a building because at times some items are omitted from a plan or specifications, but the estimator must include these items so the installation can be completed. He must know the different types of piping and their proper uses. He must know of all the different types of fixtures, faucets, shower controls, and valves.

MATERIALS

Piping. There are five basic kinds of pipe used today: steel and wrought iron pipe, cast iron pipe, seamless brass and copper pipe, copper tubing and plastic pipe. In addition, fiber pipe, vitrified clay pipe, and concrete tile pipe are used for the outdoor parts of drainage systems, particularly in connection with septic tank fields. The use of lead pipe has been practically discontinued.

Steel and Wrought Iron Pipe. Galvanized steel pipe is sometimes used in plumbing installations as a drain for small fixtures, but its chief use is for venting. The sink and lavatory stack is made of galvanized steel. This stack also goes through the roof but drains into the catch basin if one is provided.

Water service pipes are also sometimes made of galvanized steel so that rust does not contaminate the drinking water.

Galvanized wrought iron pipe is better fitted for plumbing installation than steel pipe. In addition, it resists acid wastes more favorably than steel. These pipes come in sizes ⅛" to 12" and may be standard or extra strong.

Black steel pipe is sometimes used for gas, but the gas companies are now using more plastic pipe underground.

Cast Iron Pipe. This type of pipe has been used for drainage installations for many years. It is best suited for drainage piping which is laid under ground or concrete floors. Cast iron pipe is also used for soil, waste and vent pipes. It may also be used to replace vitrified clay sewer pipe because it will not sag in unstable soil or become clogged as tile pipe will with tree roots seeking water.

Cast iron pipe is used for the soil stack which goes from the basement through the roof to drain the water closets and let the odor escape above the roof. This pipe drains directly into the septic tank, waste storage tank, or city sewer.

Cast iron pipe comes in 5-foot lengths, which are cast in two forms. Single-hub pipes have a hub on one end and a spigot on the other; double-hub pipes have hubs on both ends. Cast iron pipe comes in standard and extra heavy types.

Cast iron pipe is affected to some extent by corrosion, which may be the result of chemical action in the system impossible to control.

Seamless Brass and Copper Pipe. Brass pipe is used extensively for waste installations in buildings constructed under federal specifications. Brass is a superior metal for this purpose because of its smooth interior and its resistance against acids. Brass pipe is drawn into 20-foot lengths but must be supported at 8 to 10-foot intervals. It comes in regular and extra strong types, ranging from ⅛ to 8 inches.

Copper pipe is used primarily for water pipes. It may be used for waste and vent pipes, but because of the cost this use is rare. Copper oxide forms as a protective coating inside waste pipes. Ammonia in the wastes, how- ever, often dissolves the protective coating and again exposes the pipes to damaging acids.

Standard lengths are 12 and 20 feet. Copper pipe comes in sizes ⅛″ to 12″ in regular and extra strong types.

Copper Tubing. Copper tubing is most commonly used for water pipes. It may be hard-drawn (rigid) or soft-drawn (flexible). Copper tubing is easier to install than galvanized steel, has a smoother interior finish (decreasing friction and facilitating water pressure) and cannot rust. Plumbers find that the saving in labor more than offsets the additional cost of the material because they just bend the pipe at corners instead of cutting, threading and installing fittings.

Plastic Pipe. Plastic pipe is, in fact, becoming more and more popular in drainage, waste, and vent systems. Polyethylene (PE) pipe is used mainly for underground services. Because it comes in long coils, fewer fittings are required. Polyvinylchloride (PVC) is used in drainage, waste and vent systems (DWV) and may be joined either by threads or solvents. Acrilonitride-butadiene-styrene (ABS) pipe is also used in DWV systems.

Plastic pipe, in comparison with other pipe, is extremely lightweight and durable. Fittings are eliminated; internal friction and corrosion are minimal. Because of this, plastic pipe is less expensive in material, installation and maintenance costs.

Fittings. Pipes may be welded or joined by fittings which are made of the same material as the pipes that they join. However, where pipes of two different materials are joined (such as copper tubing and steel) corrosion may eventually result. Special fittings must be used for such joints.

Exterior Water Service. A water service pipe is required from the street main to the inside of the foundation wall. For the ordinary residence 1″ pipe is sufficient.

Generally, the trenches required for water service are about 5′-0″ deep to keep the pipe below the frost line (which is above 5′-0″ in most areas).

Permits. If parkways (areas between sidewalk and curb) are city controlled, a deposit for digging a trench is generally required. This also applies to sewer trenches. The street opening permits for water service are the same as for sewers. For connecting the pipes to the water main a charge of about one-third of this amount is generally made by the city. Charges for this item, however, vary among cities. In some cities the water main is so situated, or ready-to-use connections provided, that the street does not have to be torn up. Thus the street permit would apply only for the sewer. Such information can be obtained from the sewer department of the city prior to making your estimate.

Modular Plumbing. As one area of construction, plumbing is now available in a variety of module subsystems, Fig. 17-1. These subsystems

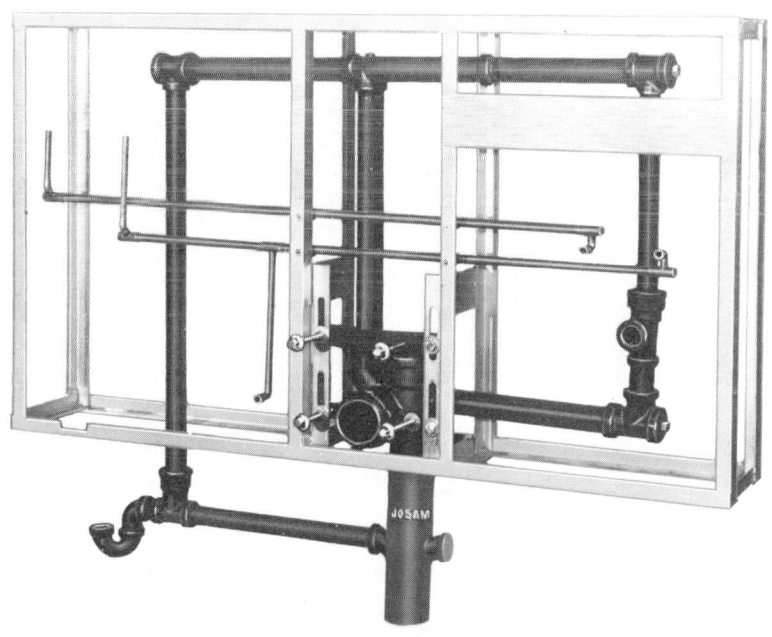

Fig. 17-1. Modular plumbing subsystem. (Josam Manufacturing Co.)

come ready to install in kitchen, bathroom, or a combination of rooms and are assembled for residential or commercial installation. A variety of piping is also available including copper, ABS or PVC plastic types, and galvanized steel.

Modular plumbing may easily reduce material and installation costs as well as simplify the task of the estimator.

Fixtures. See Fig. 17-2. The plumbing fixtures in the average house consist of the following:

Water Closets. There is the type with the flush box and the type with the flush valve which requires a pump in the basement to provide additional pressure for the valve. Wall-mounted types facilitate cleaning, Fig. 17-3.

Bathtubs. They are available in different sizes, styles, and colors and come in cast iron, steel, and plastic. Some installations are made with a shower head above and a cloth curtain, glass or plastic doors to prevent the water from splashing on the floor when taking a shower.

Kitchen Sinks. Previously the sinks were made of enameled iron with one

Fig. 17-2. Sound plumbing is **necessary** for the health of the occupants. Plumbing fixtures not only are functional but they may also be pleasing in appearance.

Fig. 17-3. Wall-hung toilet. (American Radiator and Standard Sanitary Corporation)

Fig. 17-4. Kitchen sinks are designed for their esthetic as well as their practical value. (Kohler Co.)

or two drainboards on enameled iron legs. The kitchen sinks nowadays are made of enameled steel, stainless steel, plastic, or cast iron without drainboards. They come in either single or double bowl and are installed directly into the plastic covered (Formica) top of the kitchen cabinet. See Fig. 17-4.

Lavatories. Previously the bathroom sinks were made similar to the old kitchen sink, and some are still used today. In the majority of the new homes, the lavanette (sometimes called vanity or vanitory) is used. This is a china bowl, installed in a plastic-covered or marble top and set on top of a wood or metal cabinet. See Fig. 17-5. The faucets are drilled through the top.

Shower Stall. Some people want their shower separate from the bath-

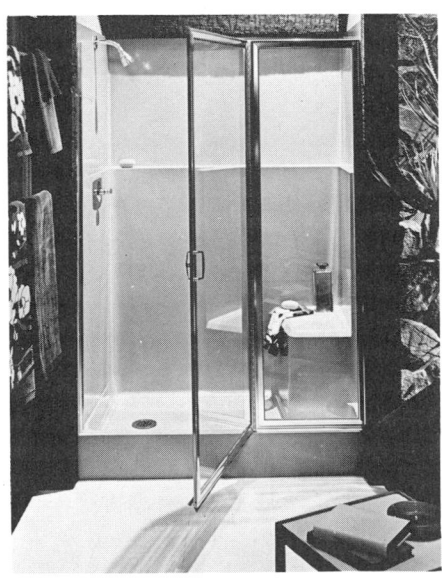

Fig. 17-5. Contemporary lavatories. *Top,* common wall hung; *Center,* set in plastic top, bowl flush with cabinet top; *Bottom,* bowl recessed in marble top. (American Standard Inc.)

Fig. 17-6. One-piece recessed shower. (American Standard Inc.)

tub. The shower stall, Fig. 17-6, can be had in a complete unit of enameled steel with a steel or ceramic floor and a curtain or glass door. Or, the stall can be framed out with 2 × 4's, covered with plasterboard and plastic or ceramic tile on the walls and a ceramic or tile floor.

Sump Pump. See Fig. 17-7. This is a pump which is installed in the base-

ment below the floor level in a sump pit. In case of seepage or flooding of the basement, the water flows into the sump pit and is pumped into the house sewer which is on a higher level. The sump pump has a float and, when the water level is raised, the float trips the pump switch and it starts pumping. When the pit has emptied, the float drops and shuts off the switch.

Water Softener. In the localities where the water is "hard," some people have a water softener, Fig. 17-8, installed in their homes. In most instances, this is not included by the original builder. The softened water is better (like rain water) for the laundry and washing. The softener is a small tank filled with sodium pellets through which the hard water enters and comes out softened. The cold wa-

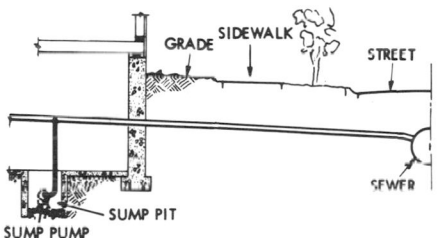

Fig. 17-7. A sump pump may be used to raise the waste water up to the house sewer.

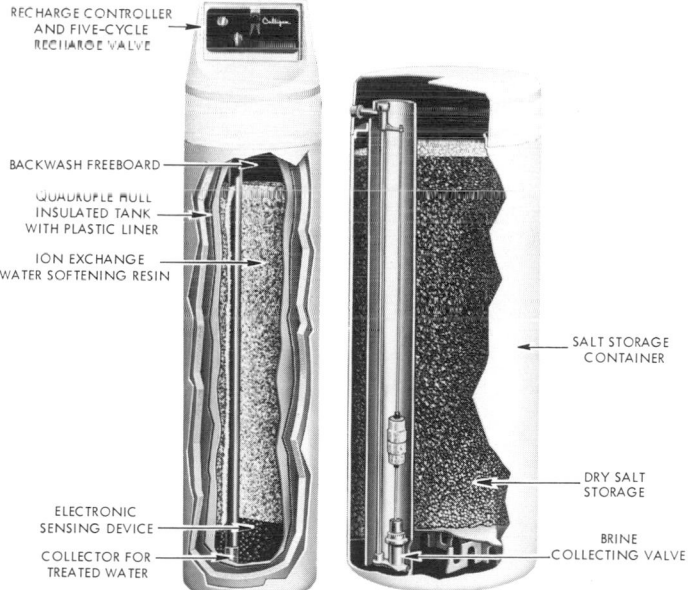

Fig. 17-8. Automatic water softener with electronic sensing device. (Culligan, Inc.)

ter (for drinking and cooking) should not run through the softener because the sodium is unhealthy.

SEWAGE DRAINING

Catch Basin. This is an underground tank used mainly in localities which have a city sewer system. The city does not allow the water that drains from the lavatory, kitchen sink, and bathtub to go directly into the sewer because the soap and grease coat the inside of the pipes and cause clogging. The grease can then be removed manually from the catch basin when it fills up.

Catch basins are generally constructed of concrete blocks and have concrete bottoms ranging from 6″ to 8″ in depth. Generally, the lid is of cast iron set in a prepared stone cap. The total depth of catch basins ranges from 6′ to 8′ and the inside diameter is generally 2′-6″ for a single family dwelling. At this diameter, about five blocks are required for a 6′-0″ depth. On the ordinary catch basins, the time for digging and backfilling as required is about six hours. The work can be done by a laborer. The basin itself requires about six hours for installation by a sewer builder.

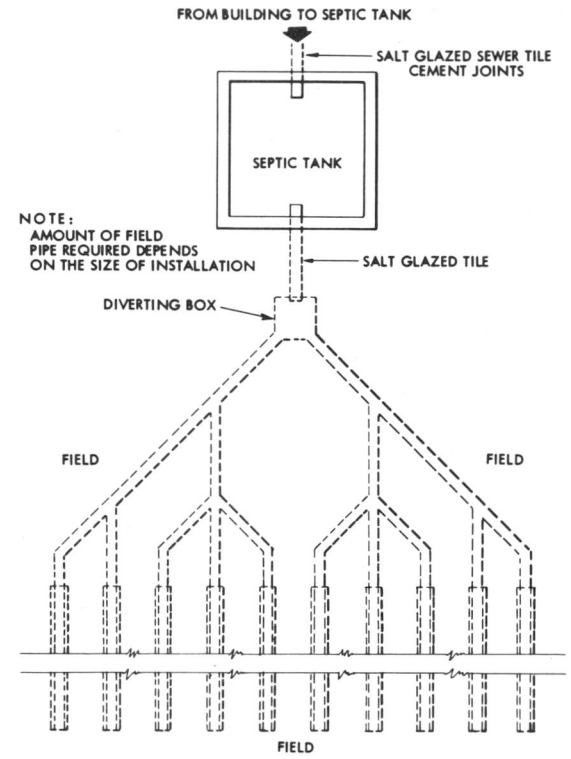

Fig. 17-9. Sewage disposal system.

Septic Tanks. In small towns, at summer resorts, in summer cottages, and throughout rural communities the septic tank is used extensively as a means of treating and safely disposing of sewage. There is considerable labor involved in the installation of a septic tank as will be seen once the complete system is understood.

Septic tanks are manufactured in various sizes and styles. They are made of metal, of reinforced concrete, and of glazed tile. Sometimes tanks are built on the job. The smaller tanks generally are made of metal and are simple to install. For private residences, usually the small manufactured tank is installed. Where facilities are needed for a large number of people, tanks made to specifications or any number of ready-built units are installed.

There are many styles and kinds, so the estimator must acquaint himself with the style indicated for the job.

Here's how the septic tank operates.

The waste from all the plumbing fixtures drains into the tank at one side. The solid material falls to the bottom, and the liquid flows out of the tank into a drain field consisting of a series of perforated tiles laid together, which allows the liquid to seep into the earth and become absorbed therein. See Fig. 17-9. While the waste is settling in the tank, a bacterial action takes place which helps to liquify some of the solid waste. When the tank finally fills up, a special tank truck removes the sludge through the manhole at the top. See Fig. 17-10.

Waste Storage Tanks. Many rural localities are now banning the use of septic tanks because they claim that the flow from the drain fields contaminates the water of the nearby wells.

Instead of the septic tank, the waste storage tank is used. This is a round, steel tank, installed horizontally in the ground with one or two manholes and two pipe ventilators projecting approximately one and one-half feet above the ground. These are for allow-

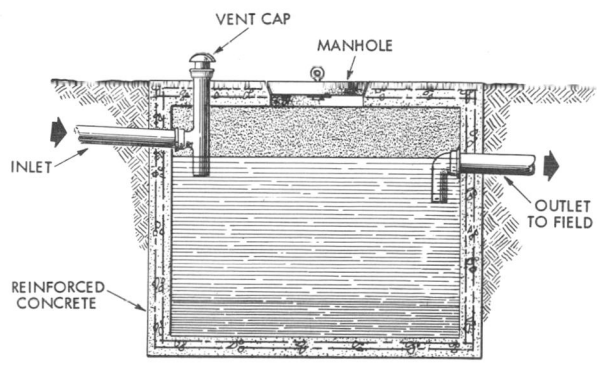

Fig. 17-10. Typical concrete septic tank.

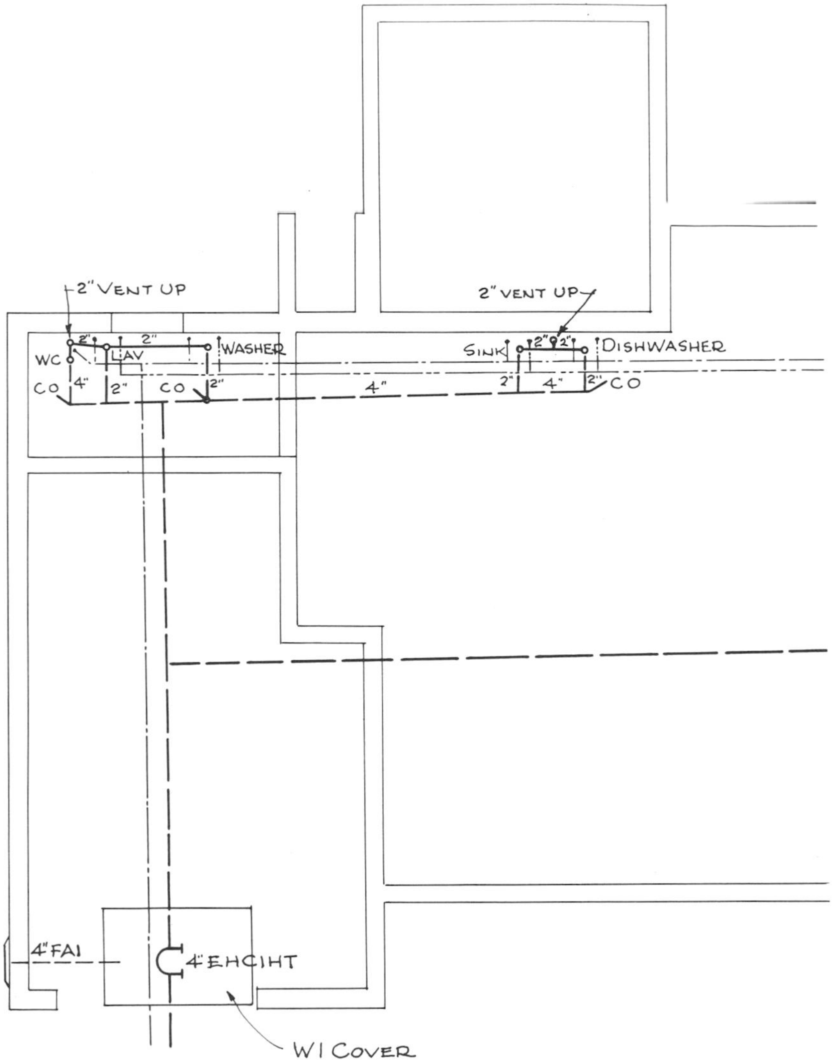

PLUMBING PLAN

Fig. 17-11. Plumbing diagram for House Plan A. The following abbreviations are commonly used: FAI—fresh air intake, CO—cleanout, WI—wrought iron, and EHCIHT—extra heavy cast iron house trap.

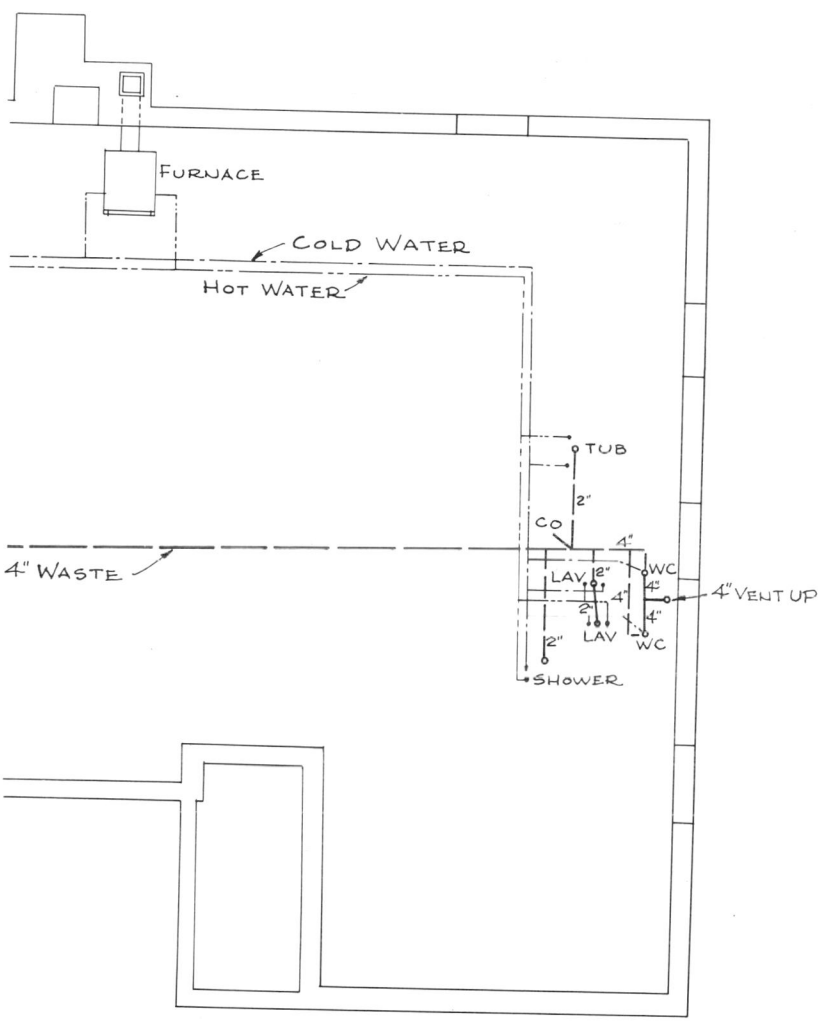

ing the sewer gas to escape and some of the liquid to evaporate. The minimum size is 3,000 gallons for a single family residence. When the tank gets full, a special tank truck is called to empty it. The labor for installing this tank is the same as for a septic tank without the drain field.

Rain Water Drainage. Every residence requires some means of disposing of rain water from the roofs. There are a great many ways of doing this, among which are (1) draining to sewer. (2) draining to cistern, and (3) draining to open ground.

The first method is used extensively in city residences where no use is made of soft (rain) water. In smaller towns and throughout rural districts, rain water is drained into a storage cistern for future use. Such cisterns may be a short distance away from the house, in the basement, or under the garage or other ordinarily unexcavated areas. The third method sometimes uses tile to drain the water a short distance from the house, where it empties on the ground.

The estimator obtains his information about such systems by studying the plans and written specifications.

Cisterns. Cisterns are sometimes built of poured concrete and lined with cement to make them watertight. Others are constructed of concrete blocks lined with cement.

Because of the materials used, the construction of tile systems, cisterns, sump systems, etc., may be done by the masonry contractor or the plumbing contractor, or both, according to the local regulations.

Piping Systems. The estimator must determine the amount and sizes of pipes required for water supply, sewage, etc. Occasionally an architect may supply a completely designed plumbing system as part of the plans. In such cases the estimator has but to refer to the drawings and calculate the lengths of the various pipes. Where no plumbing design is supplied, the estimator must, after noting all the fixtures, determine and plan the piping system himself. This will require a careful check of the plans and specifications to determine the size and length of each and every pipe indicated.

Most architects, when designing a building, will make it a point to specify the use of 2×6 studs in one partition near both the bathrooms and the kitchen. The reason for this is to provide a wall space ample for $4''$ soil pipes having bell connections almost $6''$ in diameter. If a plumbing design is not supplied, this one feature indicates the position, at least, of the soil and vent stacks.

Plumbing Diagrams. These types of drawings, Fig. 17-11, are required on every building plan by many cities and localities. It leaves no doubt as to how the plumbing installation is to be made so it conforms to the local plumbing code.

Water Diagram. This diagram, Fig. 17-12, shows the water entering the building from the city main through the valve, which is controlled through the buffalo box at the surface, then through the water meter. The cold water flows to water closets, lava-

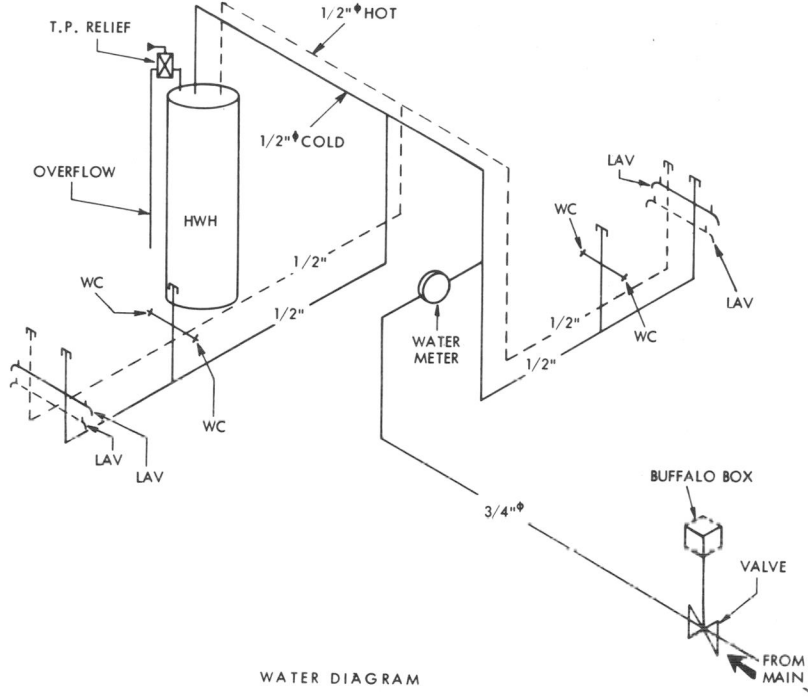

Fig. 17-12. Water diagram.

tories, sinks, and hot water heater. The water, after it is heated, flows from the water heater to lavatories and sinks. This diagram also shows the pipe sizes and the relief valve which allows hot water to escape in case the water becomes excessively hot on account of a thermostat malfunction. Without this, the hot water heater can explode with the build-up of steam.

Waste Diagram. This drawing, Fig. 17-13, shows the waste water from floor drains, lavatories, and water closets flowing into the main sewer. It also shows the proper sizes of pipe to use.

The dotted lines show the venting system. These pipes lead into the stack which extends through the roof. This system allows the odors and sewer gas to escape and also let air into the pipes to keep the water flowing freely. The small curved lines signify a trap under each fixture in which water lies, preventing any odors from coming into the room through the fixture drain. The clean-out is a stub wherein, in case the sewer pipe is clogged, the cover is removed and a clean-out rod can be inserted to break up the stoppage. The increaser is the piece of pipe above the roof, of a larger diameter than the stack, which gives the venting system a better updraft.

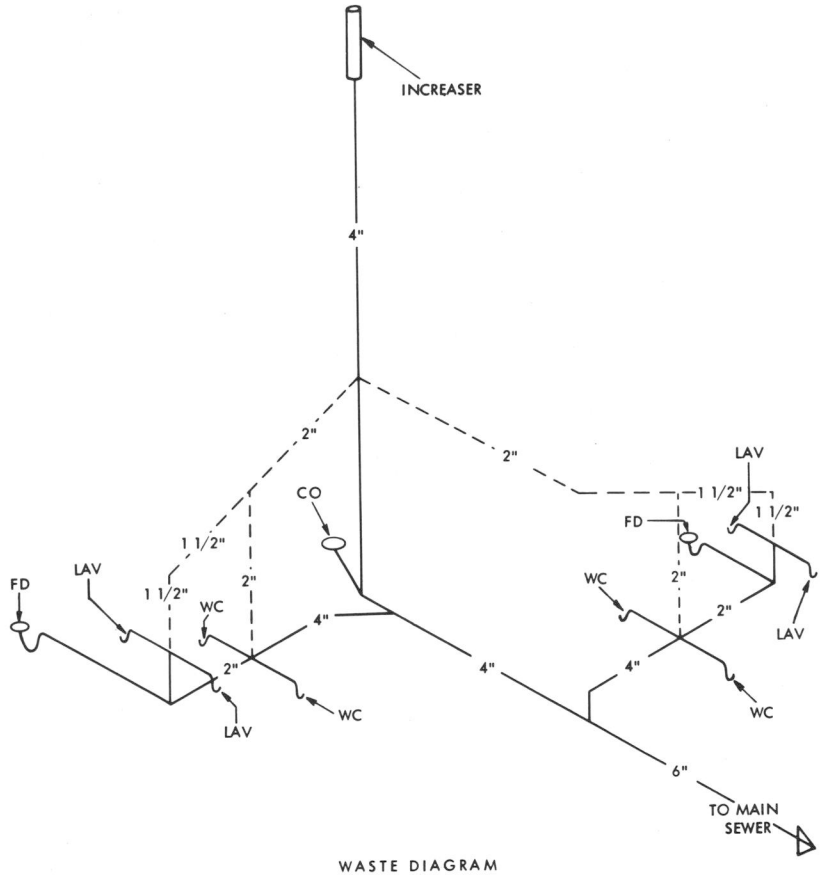

Fig. 17-13. Waste diagram.

ESTIMATING MATERIALS

The plumbing estimator lists all the fixtures, pipes, and pipe fittings. He then gets a price from the plumbing supply house, or lists the prices himself from a catalog and/or price lists he had previously received from the supply house. He must add the regular overhead expense to this material.

Unit Estimating. On the smaller re-modeling jobs, where the estimator-salesman tries to sell the job in one call, he uses a unit estimating system. Bear in mind it is a dangerous practice to use unit prices unless you have data or experience to back up your figures. After consulting with his plumbing contractor, the estimator makes up a price list for each item. The prices should include the cost of the pipes, fixtures, fittings, faucets, and labor for the average installation. He must be careful not to forget any items. The price list reads as follows:

Closet stack (4″)	1-story	$_____
	2-story	$_____
	3-story	$_____
Sink stack (2″)	1-story	$_____
	2-story	$_____
	3-story	$_____
Water closet		$_____
Bathtub		$_____
Kitchen sink		$
Lavatory		$_____
Shower Stall		$_____
Sump pump		$_____
Septic tank	small	$_____
	medium	$_____
	large	$_____
Drain field		$_____
Catch basin		$_____
Waste storage tank		$_____
Cistern		$_____

This system can also be used for arriving at an approximate cost for the plumbing on a new building.

The prices are for average installation and average price fixtures. If the installation is more complicated or better fixtures are wanted, additional charge must of course be made.

ESTIMATING LABOR

Estimating plumbing labor can be confusing because of the numerous ways in which it can be done. Contractors may either estimate their labor by a certain percentage of material costs plus detailed calculations for trenching and such other items as have little material, or they may estimate labor at so much per portion of an average job. Sometimes labor is based on the number of pipe joints plus a certain percentage of fixture values and detailed calculations for trenching.

Fixture Labor. Usually, the installation of fixtures for nominal cost residences is taken as 50% of total fixture cost. For higher quality residences, where the fixtures are more elaborate, only about 40% is assumed for labor.

For hot water heaters and for pumps for cistern water labor is figured at about 30% of the material cost.

Questions and Problems

1. If 2 bathroom fixtures were to be installed in each of 6 apartments at an estimated rate of $540 a fixture, what would the total estimated cost be?

   (Answer: $6480)

2. If 460 feet of 6-inch sewer pipe is laid at the average rate of 12 linear feet of pipe per hour, assuming an hourly rate of $9.00, what would the total cost be?

   (Answer: $345.00)

3. A 50-gallon gas hot water heater is priced at $98.00. Figuring 30% for labor what would be the total estimated cost?

   (Answer: $127.40)

18

SHEET METAL

~~~~~~

USES OF SHEET METAL

Because of the variety of sheet metal applications in building construction, a careful check of both specifications and drawings should be made. Some applications, such as termite shields, Fig. 18-1, and water table flashing, are near the foundation. Other applications, such as chimney flashing and ridging, are on the roof. Sheet metal is also used on and around doors and within the house.

The most common applications of sheet metal are as barriers and shields, flashing, stops, drops and chutes, ridg-

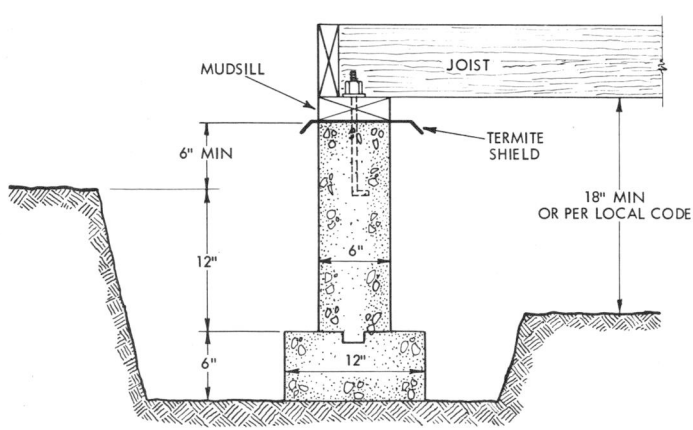

Fig. 18-1. A termite shield used between the concrete foundation and wood sill.

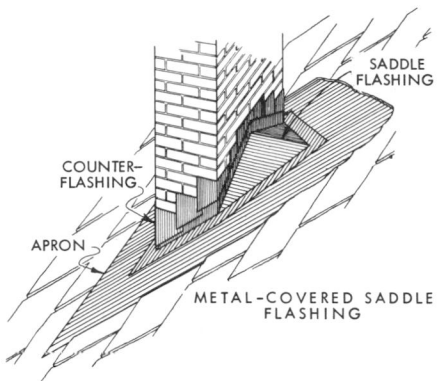

Fig. 18-2. Saddle flashing and sheet metal around a chimney.

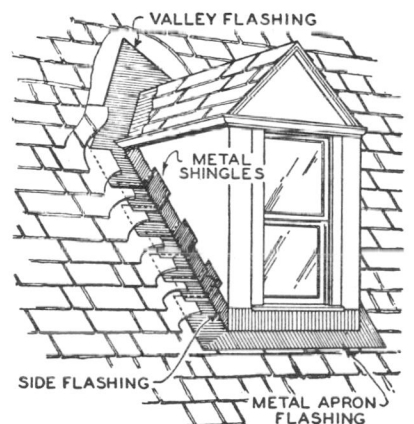

Fig. 18-3. Valley flashing and sheet metal around a dormer.

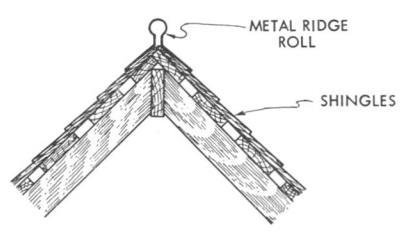

Fig. 18-4. Sheet metal is used along ridge boards to waterproof the top of the roof.

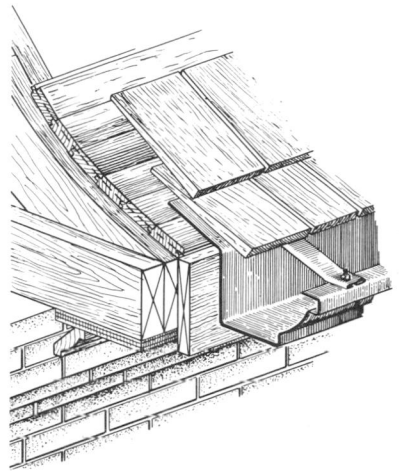

Fig. 18-5. A sheet metal gutter attached to the edge of a roof.

ing, gutters, leaders (downspouts), and roofing.

Fig. 18-2 shows sheet metal being used as saddle flashing to divert water around a chimney. Fig. 18-3 shows valley flashing at the intersection of a roof and dormer. It serves the same purpose as saddle flashing.

Fig. 18-4 illustrates how sheet metal is used along the lengths of the ridge boards.

Fig. 18-5 shows how a sheet metal gutter is attached to a wood roof. Note the metal strap bolted to the gutter. Fig. 18-6 is a cross section of a sheet metal gutter which hangs free of the roof. The metal straps are the only support for the gutter.

Some sheet metal work such as heating and air conditioning ducts and flashing for plumbing vents are usually figured as part of the heating and plumbing estimates.

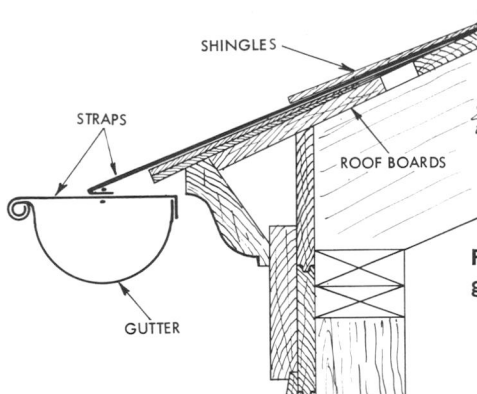

Fig. 18-6. Cross section of a sheet metal gutter.

MATERIALS

A wide variety of metals are used for the various sheet metal applications. Although the majority of jobs will specify galvanized sheet steel for most work, any of the following metals could be specified for a portion of a sheet metal job: aluminum, copper, lead, monel, tin, or zinc.

The various metals can differ greatly in cost and labor required for installation. It is therefore important to know not only where sheet metal is to be used, but the kind of metal which is to be used.

Along with sheet metal are various screws, bolts, nails, paints, and solder necessary for proper installation and finishing.

ESTIMATING MATERIALS

Table I shows the various thicknesses and weights of iron and steel sheets and plates. Contractors and estimators assume round numbers for weights in order to simplify calculations. For example, 26-gage sheet metal is shown as weighing .75 pounds per square foot. To make the calculations easier and to add a safety factor to the estimate, you can assume that the metal weighs exactly 1 pound per square foot.

Roofs. Most estimators scale the ridges of the roof to find the linear footage of sheet metal *ridging.* Sheet metal for ridging is commonly 13″ wide and comes in 8′-0″ lengths. Some overlapping of sheets is considered in the estimate.

Dormer pans using sheet metal are measured or scaled in linear feet. The material generally comes 8′-0″ lengths and in various standard widths.

Sheet metal for *dormer decks* is calculated in square feet. Allow for cutting and overlapping.

The amount of sheet metal needed for *valleys* is determined by scaling from the plans. Some allowance for overlapping should be made in determining the required length; 2″ for overlaps should be sufficient.

TABLE 1
UNITED STATES STANDARD GAGE
FOR SHEET AND PLATE IRON AND STEEL

Number of Gage	THICKNESS		WEIGHT		Number of Gage
	Approximate thickness in fractions of an inch	Approximate thickness in decimal parts of an inch	Weight per square foot in OUNCES avoirdupois	Weight per square foot in POUNDS avoirdupois	
0000000	1–2	.5	320	20.	0000000
000000	15–32	.46875	300	18.75	000000
00000	7–16	.4375	280	17.5	00000
0000	13–32	.40625	260	16.25	0000
000	3–8	.375	240	15.	000
00	11–32	.34375	220	13.75	00
0	5–16	.3125	200	12.5	0
1	9–32	.28125	180	11.25	1
2	17–64	.265625	170	10.625	2
3	1–4	.25	160	10.	3
4	15–64	.234375	150	9.375	4
5	7–32	.21875	140	8.75	5
6	13–64	.203125	130	8.125	6
7	3–16	.1875	120	7.5	7
8	11–64	.171875	110	6.875	8
9	5–32	.15625	100	6.25	9
10	9–64	.140625	90	5.625	10
11	1–8	.125	80	5.	11
12	7–64	.109375	70	4.375	12
13	3–32	.09375	60	3.75	13
14	5–64	.078125	50	3.125	14
15	9–128	.0703125	45	2.8125	15
16	1–16	.0625	40	2.5	16
17	9–160	.05625	36	2.25	17
18	1–20	.05	32	2.	18
19	7–160	.04375	28	1.75	19
20	3–80	.0375	24	1.5	20
21	11–320	.034375	22	1.375	21
22	1–32	.03125	20	1.25	22
23	9–320	.028125	18	1.125	23
24	1–40	.025	16	1.	24
25	7–320	.021875	14	.875	25
26	3–160	.01875	12	.75	26
27	11–640	.0171875	11	.6875	27
28	1–64	.015625	10	.625	28
29	9–640	.0140625	9	.5625	29
30	1–80	.0125	8	.5	30
31	7–640	.0109375	7	.4375	31
32	13–1280	.01015625	6½	.40625	32
33	3–320	.009375	6	.375	33
34	11–1280	.00859375	5½	.34375	34
35	5–640	.0078125	5	.3125	35
36	9–1280	.00703125	4½	.28125	36
37	17–2560	.0066406	4¼	.265625	37
38	1–160	.00625	4	.25	38

Flashing. For vertical members, flashing lengths are scaled from the plans. The flashing is generally in 8'-0" lengths and at least 12" wide. Allow 2" for overlap at each joint of the material. Little or no waste need be added to the quantity take-off.

Right-angle flashing is also measured in linear feet, allowing 2" for overlap. Usually the sheets are purchased in 8'-0" lengths and then bent into the shape of a right angle. No waste need be considered.

Chimney flashing, on a large chimney, generally is applied using pieces 10" × 10", or using stock 10" wide and 8'-0" long. Where the 10" × 10" pieces are used, allow 1" for overlap.

Roof openings, such as those made for access to flat roofs, usually have the appearance of an inverted but shallow rectangular or square pan.

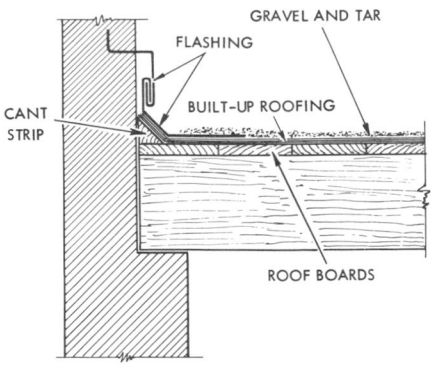

Fig. 18-7. Flashing used at the intersection of a brick wall and a flat roof.

The combined areas of sides and top are calculated, allowing a little for overlap, and the sheet metal estimated in square feet.

Flashing for flat roofs and side wall intersections is measured in linear feet. Fig. 18-7 shows how flashing is constructed at the intersection of a flat roof and brick wall.

Cap and base flashing can be used on flat roofs that intersect a brick wall. The base strip is laid over the roofing material and cant strip and run up to the reglet block. The cap is then fastened in the insert strip in the reglet block and run back over the base strip. This is estimated like all flashing, by the linear foot.

Gutters and Leaders (Downspouts). The working drawings should show where the gutter is to be placed—or the specifications should explain in detail where and how the gutter is to be installed. The drawings can be scaled to determine the linear footage. Porch and dormer gutters are measured in the same manner.

Leaders should be indicated on the drawings. If they are not indicated, then the contractor or estimator must decide for himself where such piping is required. On repair jobs it is simply a matter of locating the original pipes and following the layout. The vertical dimensions on the elevation drawings can be used, or actual distances can be scaled.

Drip caps over windows, doors, etc., are scaled in linear feet. Then about 25% more is added for wastage, because the material comes in 8'-0" lengths and is not usually spliced.

Doors. Sheet metal is used to frame doors and to cover wood doors. For door frames, the jamb and casings must be covered all the way around. Sheet metal 10" wide would accomplish this in most cases. The linear footage is the length of the 2 sides plus the top. The ordinary 2'-8" × 6'-8" door will require 16'-0" of sheet metal around the frame.

For covering a wood door, the amount is first calculated in square feet and then in pounds. The door area should include both sides and enough extra to lap the sheet metal of one side around the edge of the door and over the metal on the other side; an inch or two overlap is sufficient.

Accessories. Gutters often hang free of the roof, so they must be connected by *brackets* or *hangers*. Gutter brackets are generally spaced 3'-0" apart. Other gutter hangers have the same spacing, or less when the gutter is exceptionally large. To find the required number, divide the total length of gutter by the hanger spacing; add 4 to 6 to cover wastage and loss.

Elbows are often needed for leaders and should be specified and/or shown on the working drawings. These are counted by studying all four elevation views of the building. If the drawings do not show the leaders and elbows, the estimator must consult the architect or make his own layout.

Leader brackets and *clamps* are ordinarily spaced about every 10 feet. On a one story building, the leader is fastened at the top and bottom. On a two-story building, perhaps 3 brackets or clamps are used. All short lengths such as for low dormers, should have the brackets at the top and bottom of the leader.

Waterproof felt is sometimes installed on roofs, under the sheet metal for dormer decks. For one layer, it is necessary only to measure or scale the dimensions and find the area. No waste allowance is necessary as the felt can be cut and laid as required.

Sheet metal is often covered with *red lead paint*. Two quarts of paint will generally cover about 250 linear feet of gutter, ridge, valley, leaders, etc. This is for new work. For old work an extra quart is allowed for the same amount of work. Paint goes a long way on metal. (Chapter 12, "Paint," discusses paint materials and labor in detail.)

Screws used to fasten sheet metal to a wood door or door frame are difficult to estimate because the spacing between screws can vary with different doors and frames. The spacing is generally about 3″ for most doors. Allow about 75 screws for a door of 21 square feet, 150 screws for a 42 square foot door, etc. The cost is usually so small that it is often considered a part of the general overhead expenses. For the metal frame allow about 75 screws for a 21 square foot door.

Nails for applying sheet metal are not usually calculated in detail because all exposed metal is usually nailed every 2″. Allow 6 or 8 pounds of nails for an average job. This can also be added to the overhead.

Estimating *solder* is also difficult and not worth the time which it would take to estimate. For most jobs you can assume 1 pound of solder for the average residence, 2 pounds for large residences, etc.

House Plan A. The sheet metal used in HOUSE PLAN A is of several different materials. An aluminum shield is used for termite protection under all wood construction. Copper flashing is used on the roof. The gutters and leaders are also of aluminum, as is the window (and entrance door) head flashing. Zinc weatherstripping is used for the entrance door.

ESTIMATING LABOR

Labor is estimated by multiplying the workman's hourly salary by the number of hours it takes him to complete his work. Typical labor times are given in this section. Use these times only under average conditions.

Generally one man can put on about 20 linear feet of *ridging* per hour. When there is a great deal of cutting and fitting to be done, reduce the work rate to about 15 linear feet per hour.

For *dormer pans* averaging 10″ wide, where 6″ of metal is under the frame and 4½″ on the roof, the labor time is about 1 hour for every 10 linear feet.

One man can lay about 25 square feet of metal *dormer deck* per hour.

Valley flashing is installed at the approximate rate of 27 linear feet per hour.

Flashing for vertical members is done at the rate of about 12 linear feet per hour. The flashing for right angles is installed at the rate of about 25 linear feet per hour. For a chimney the size of that in HOUSE PLAN A, the flashing time is approximately 2 hours. For a small, one-flue chimney, the labor time can be taken as 1 hour.

The labor time for *gutter* work can vary widely depending on the lengths of each gutter run. For example, on a roof where a great many short gutter runs are installed, the time might be as much as 1 hour per 15 linear feet of gutter. This is because the ends of the gutters must be closed, and on short runs this requires considerably more time than when the runs are long. On a roof where there are few dormers, and where the gutter runs are long, one man can install about 40 linear feet per hour. On the average residence you can assume at least 17 linear feet per hour.

Gutter hangers are included in the gutter labor and require no special labor consideration.

Labor for *leaders* varies according to the lengths. A long run naturally requires less labor per linear foot than a short run. A leader pipe must be soldered to the gutter and sometimes fitted into a sewer pipe connection. The labor for fastening the leader to the wall or other place amounts to very little per linear foot. Thus a number of long leaders can be installed cheaper than a like number of short ones. When many short runs are used, a rate of 10 linear feet per hour is a good average; for long runs, up to 30 linear feet per hour can be used as an average figure.

Drip caps are installed at the rate of 17 linear feet per hour.

For an average-sized *door* of 2′-8″ or 3′-0″ width, it takes approximately 2 hours to cover the door with sheet metal. The sheet metal angles used for a metal door frame require about 1½ hours to install for an average-sized door opening up to 3′-0″ wide. Larger doors take proportionally longer.

Waterproof felt can be installed over dormer roofs at the rate of about 75 square feet per hour.

Where *clothes chutes* are used, the labor is figured at ½ hour per foot for a chute which is 12″ in diameter.

See Appendix D, pages 477 and 478, for Hourly Estimates for:
Roofing: Metal Sheets; Metal Work.

19

AIR CONDITIONING: HEATING AND COOLING

‹‹‹‹‹‹‹‹‹‹‹‹‹‹‹‹‹‹‹‹‹‹‹‹‹‹‹‹‹‹‹‹‹‹‹‹

No matter where a residence is located, some consideration should be given to air-conditioning systems. (Air-conditioning is understood to include both heating and cooling systems.) For utmost comfort, air-conditioning must perform four basic functions: (1) temperature control, (2) humidity control, (3) air circulation and ventilation control, and (4) air filtering.

This chapter is primarily concerned with estimating heating systems but it also includes basic information on central cooling systems. Modern heating and cooling are frequently combined in a year-round system; they use the same ducts, fans, control equipment, etc.

In southern areas of the United States, heating units for occasional heating are popular, and complete central heating plants are rare. Electric heaters are used extensively for occasional heating. If complete summer cooling is required, however, then a central plant for both cooling and heating is more practicable and economical.

In northern areas where severe winter conditions exist, central heating plants are absolutely necessary where the full comfort of modern heating is desired.

Central plants are located in basements or utility rooms of homes and the heating medium is distributed through ducts or pipes to the various rooms.

HEAT DISTRIBUTION

There are two basic methods by which heat may be distributed to the rooms: pipe and duct. Pipe requires the least amount of space to transmit the heat. Pipes are used to carry hot water or steam to a radiator, baseboard unit, or radiant panel located in the floor, ceiling, or wall. Ducts require more space than pipes. However, the ease with which ducts may be used for cooling and for humidification from one central location is a distinct advantage. (A pipe system may also be used for cooling if blower-equipped convector outlets are used.) Ducts are placed between floor or ceiling joists and between studs in walls.

Major types of heating systems include (1) warm air, (2) hot water and steam and (3) electrical systems.

Different methods are available to distribute heat: (1) baseboard, floor, and wall diffusers, used for warm air systems; (2) convectors, radiators, and baseboard (radiant and convector) type units, used for steam and hot water systems; and (3) radiant panel units, used for hot water and electrical systems.

Thermostat. All types of heating equipment installed in new construction may be regulated by a thermostat which can be set for a desired temperature. If the temperature falls below this indicated point, the space heater will automatically turn on. Most thermostats are designed to maintain a constant temperature. Some are designed to permit a low temperature for a specified length of time, and at a certain time the temperature will be raised to the desired degree.

Humidity. During the heating season it is desirable that moisture be used to condition the indoor air. Outside air during the cold season is normally much drier than inside air and when mixed they tend to lower the indoor relative humidity.

Storm windows, weatherstripping, and tighter house construction has decreased the amount of infiltrating outdoor air. Although moisture is added to the indoor air by cooking, bathing, dishwashing, and laundering, it is not sufficient to raise the humidity to a desirable, healthful level. New homes with effective vapor barriers need less humidifying than normally required to maintain the desired indoor relative humidity.

Generally a relative humidity of 50 percent is considered ideal. This amount provides the best protection against airborne infections and has a soothing effect upon the nose and throat. In addition, sufficient humidity permits woodwork and other home furnishings to retain their moisture; it also retards rapid deterioration.

When the outdoor temperature drops below 40° the relative humidity should be lowered enough to prevent excessive condensation on windows and structural components.

Forced Warm-Air Heating. Heating systems in many new homes being built use forced warm air. This system is an advancement over the gravity warm-air system. The gravity system employs large round pipes that extend octopus-like from the furnace to wall registers and cold-air returns. The forced-air or blower system has replaced the old outlets with smaller ducts and neat, functional wall-type or base-type grilles and it insures a more direct and even circulation.

The forced warm-air system should provide approximately four to six air changes per hour for the average size five or six room house. The warm air leaves the air chamber at approximately 155°F and arrives at the room at approximately 140°F. Some of the heat is lost in the transmission from the furnace to the rooms.

Forced-air furnaces do not depend upon the different weights of hot and cold air to provide circulation within a room. The forced-air furnace, either gas or oil fired or heated by electric coils, has a blower that draws cool air into the air chamber and expels the heated air through the ducts to the various room outlets.

Warm-Air Perimeter-Loop System. Perimeter-loop heating, used almost exclusively in basementless slab houses, has a duct system which encircles the slab. Fig. 19-1 shows the general principle of the perimeter-loop system with four feeder ducts supplying the perimeter duct. The ducting may be sheet metal, concrete pipe, vitrified tile, or other precast materials. Because the perimeter-loop system is used essentially in slab homes, a downdraft furnace is necessary. (Be sure to note the differentiation in furnaces when selecting from a catalog.) Warm air is forced through the ducts and is discharged into the room through floor or baseboard diffusers. The air is returned to the furnace via air intakes either on an inside wall or in a hallway ceiling close to the furnace.

This type of system is economical to install and needs little floor space.

Because the perimeter duct is imbedded in the floor and is connected to the furnace by feeder ducts, cold floors are eliminated. A humidier and filter may be installed to eliminate dryness and dirt.

Forced Warm-Air Extended-Plenum System. Air warmed by the heating unit may also be distributed by a blower through the *plenum* and ducts to the baseboard diffusers. (A plenum system functions by placing the air under a higher *pressure*.) The blower draws the room air back to the furnace through the cold-air return and return ducts to be filtered, reheated, and humidified. After it has been heated, the air is then distributed again. Fig. 19-2 illustrates a portion of an extended-plenum installation. Note the baseboard diffuser is placed below a window and the return air intake is located along an inside wall.

In comparison to hot water and steam heating systems, the extended-plenum system has the advantage of being more economical to install. This system is well adapted to houses with

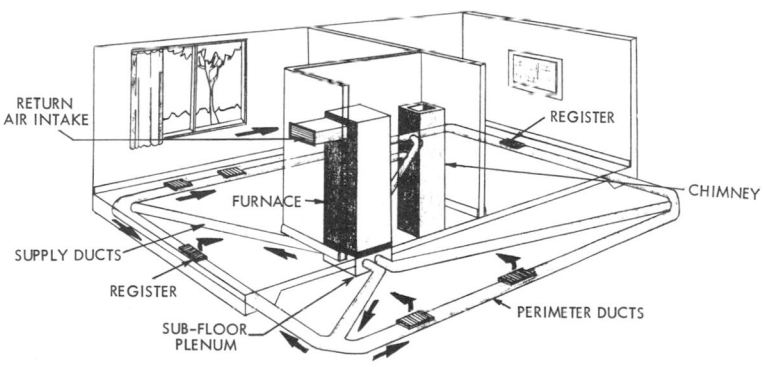

Fig. 19-1. Warm-air perimeter-loop system.

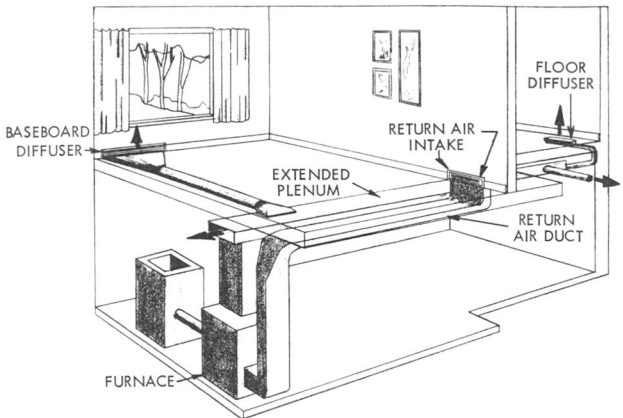

Fig. 19-2. Forced warm-air extended-plenum system.

and without basements because the blower maintains air circulation.

A warm air system should be designed and selected by a reputable heating engineer who understands the calculations of heat losses and gains, etc. This service may be an additional expense but is most likely an architectural service charge rather than a builder's expense.

Hot-Water and Steam Systems. Hot water systems require a boiler, piping, radiators, valves, vents, and miscellaneous boiler and control equipment. The boilers can be automatically fired, using gas or oil burners. In many cases the pipes are insulated with a specially shaped, more or less rigid, insulation.

The one-pipe, forced hot water system, shown in Fig. 19-3, is probably

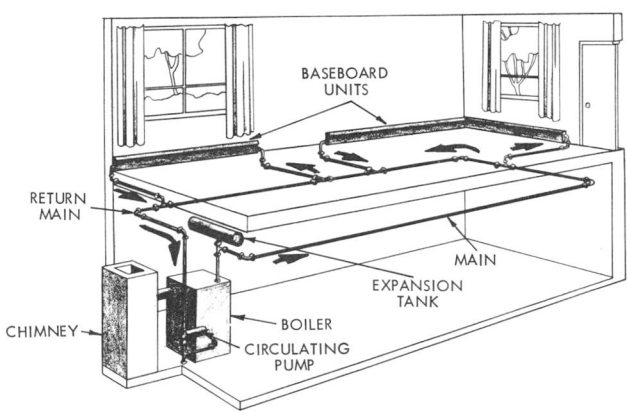

Fig. 19-3. Forced hot water system.

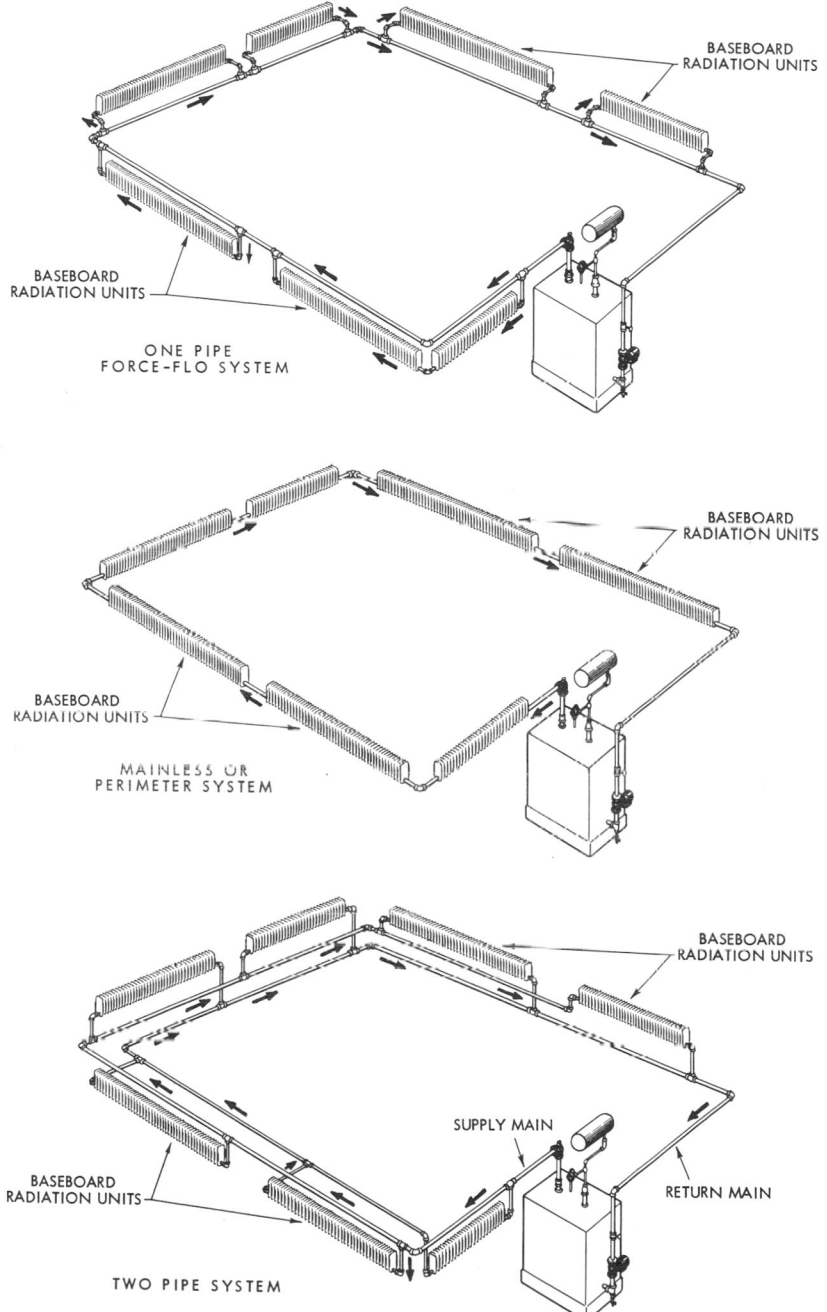

BASEBOARD RADIATION UNITS

BASEBOARD RADIATION UNITS

ONE PIPE FORCE-FLO SYSTEM

BASEBOARD RADIATION UNITS

BASEBOARD RADIATION UNITS

MAINLESS OR PERIMETER SYSTEM

BASEBOARD RADIATION UNITS

SUPPLY MAIN

RETURN MAIN

BASEBOARD RADIATION UNITS

TWO PIPE SYSTEM

Fig. 19-4. Various types of piping systems. The two-pipe system is the most complex.

the most widely used of all *hydronic* (heating or cooling with water) systems in residential construction. Heat for this system is generated by an automatically controlled gas or oil burner that heats the water in the boiler. With the one-pipe system the boiler may be located in the basement or in the utility room of a basementless house. A single *main* pipe usually follows the perimeter of the building, and *branches* or *risers* connect to the radiators, convectors, or baseboard units. Special flow fittings are placed at the return of each radiator or convector so that a single pipe may be used for supply and return of the water. The flow fittings separate the cooled water from the hot water leading to the next unit. As the *main* leaves the boiler, a flow valve is placed at the first elbow. This valve opens under the flowing water pressure when the circulating pump is operating and closes when the pump is not operating. If this flow valve were not placed in the main, the water would continue to circulate, causing overheated radiators and convectors. An expansion tank permits the water to expand and contract with the changes in temperature, thereby keeping the boiler and radiator filled with water.

Fig. 19-4 shows various types of piping systems, some of which require two connections to each radiator. The two-pipe system requires considerably more piping and consequently a greater amount of labor.

A steam system is essentially the same as a hot-water system except that steam instead of hot water is the heating medium.

Radiant Heating. The radiant heating systems of the Romans and Koreans were of the floor panel, hot-air variety, and similar systems are in use today. However, by far the greatest majority of installations in use and under construction at the present time are of the hot-water variety.

Panels. Panels for a hot-water radiant system may be located in the walls, the ceiling, or the floor. While the actual location of the panels depends to some extent upon the architectural design of the building, the differences in the percentages of heat transferred by convection and radiation for each panel, as shown in Table I, also should be a factor in the selection of the panel location.

Baseboard Radiation. This is a system of heating in which the heating units occupy space ordinarily taken by the wooden baseboard, projecting into the room only a little more than the conventional baseboard and blending

TABLE I
PERCENTAGE OF TYPE OF HEAT TRANSFER BY PANEL LOCATION

Panel Location	Convection Transfer	Radiant Transfer
Wall	48%	52%
Ceiling	35%	65%
Floor	52%	48%

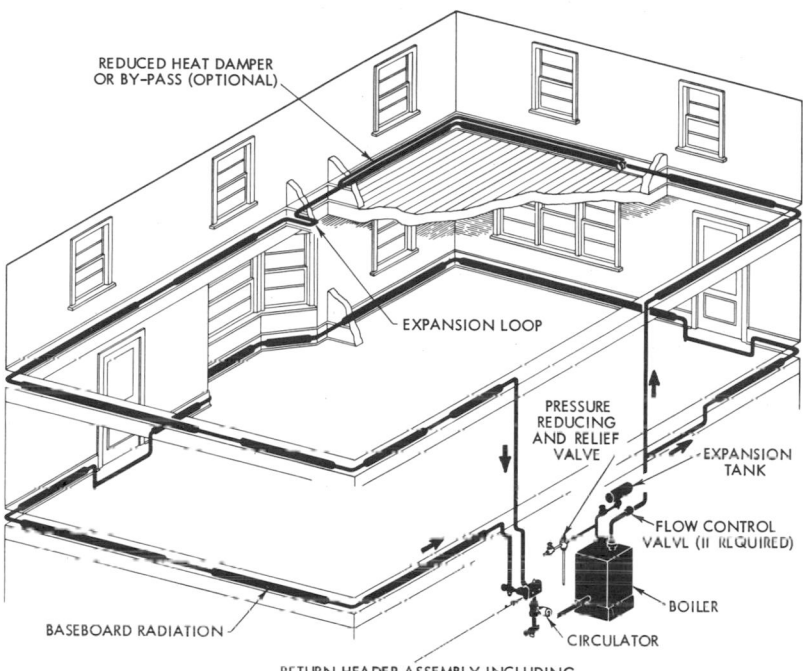

REDUCED HEAT DAMPER
OR BY-PASS (OPTIONAL)

EXPANSION LOOP

PRESSURE
REDUCING
AND RELIEF
VALVE

EXPANSION
TANK

FLOW CONTROL
VALVE (IF REQUIRED)

BOILER

BASEBOARD RADIATION

CIRCULATOR

RETURN HEADER ASSEMBLY INCLUDING
PURGE AND BALANCE VALVES AND AIR VENT

Fig. 19-5. The perimeter system uses no mains and requires a minimum of piping. (Warren Webster and Company)

into the wall. Baseboard radiation units consist either of hollow, cast-iron panels, with or without fins on the back, or of ferrous or nonferrous finned tubing behind metal enclosures.

Heat is radiated from the baseboard units to the walls and furniture from which it is reradiated. Due to its location at floor level, drafts, cold floors, and hard-to-heat areas are positively eliminated. Furniture can be placed wherever desired. Since heat is distributed near the floor, the floor-to-ceiling variation in temperature is reduced to about 2°F. Thus, there is a tendency for more uniform temperatures throughout the room. Finally, base-board radiation units are clean in operation.

Baseboard radiators can be installed as part of a two-pipe system, a conventional one-pipe or monoflow system with basement mains, or as a perimeter (mainless or series) system in which the baseboards themselves are used as mains. These systems are illustrated in Fig. 19-4. An example of the perimeter or series system is shown in Fig. 19-5.

Electrical Heating Methods. Heating the home by electricity offers a number of decided advantages over the "conventional" methods previously discussed. Electric heating provides an even, draft-free warmth without periodic cooling or overheating. Temperatures in each room may be maintained at the desired level since individual thermostats or automatic controlling devices are installed with each unit. To economically heat the home by electricity, consideration must be given to proper insulation to reduce the heat loss. This added insulation not only serves as a sound deadener, both from the outside and from floor to floor, but also aids in reducing the inside temperature during the summer months.

With many electric heating devices, maintenance is cut or practically eliminated since the individual room control thermostat is usually the only moving part. To many, the silence of the electric heating system is a great asset. In new construction, often an electrical heating system can be installed for the same amount as the regular gas- or oil-fired systems. Since the chimney and space for fuel storage is eliminated, and in most electric heat-ing systems duct work is nonexistent, lower building costs can be achieved.

There are, however, disadvantages. In some areas, in comparison with other fuels, operating costs are higher. Some power companies, however, give very favorable rates to consumers having electric heat. Electric power failure is a disadvantage. However, any system using a blower will not function without electricity. More insulation is required to hold the heat than with conventional heating systems. Humidity control can be a problem with resistance type radiant heaters because these do not employ moving air. To counteract the absence of humidity, a humidistat-operated ventilating fan may be necessary.

Various means are available for the builder to equip the home for electric heating. Principles of operation vary.

Baseboard Electric Heating Units. Baseboard units give a uniform degree of heat over a wide area. Some of these units emit heat by both radiation and

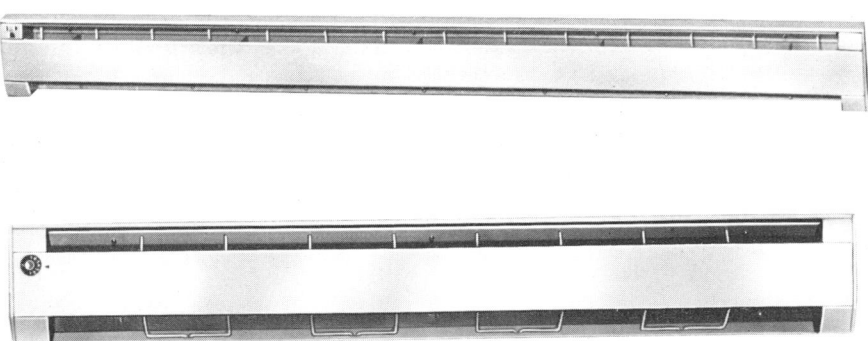

Fig. 19-6. Electric baseboards heat by radiation and convection. Some convection baseboard units are equipped with an individual thermostat. (Berko Electric Mfg. Corp.)

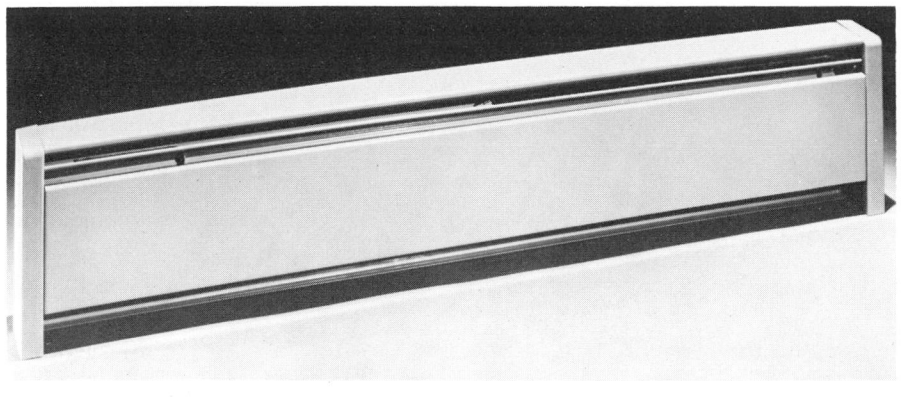

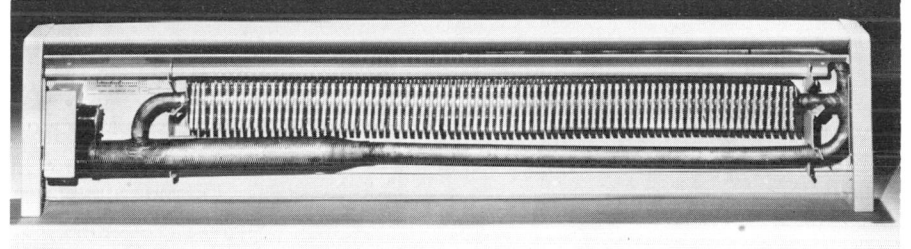

Fig. 19-7. Hydronic baseboard heaters are a relatively new innovation in electric heating. Each unit is self contained. (International Oil Burner Co.)

convection and are designed to replace the standard base and shoe moldings. See Fig. 19-6. These units have an opening at the bottom that allows the air to enter and pass over the resistance element. Another opening at the top allows the air to escape. Some baseboard units have attached thermostats (Fig. 19-6, bottom) but frequently they are controlled by a wall mounted device. A recent development in baseboard heating is the *hydronic* unit (Fig. 19-7). Each hydronic baseboard unit is filled with water and is heated by an immersion element, thus producing more evenly distributed heat since the water-filled tubes will not cool as rapidly as a resistance element.

Both the convector and hydronic units are used to heat entire residences. They are also used to heat individual rooms that have been added to an existing dwelling.

Ceiling and Wall Electric Heating Units. Electric ceiling units or electric wall units warm the occupants of a room with radiant heat rays. The most common type of ceiling or wall unit is the resistance heater that radiates heat through a protective metal grille or screen. Fig. 19-8 shows a ceiling-type resistance radiant heater. Radiant glass panels (Fig. 19-9) are similar to the resistance heater with the exception that the resistance element is covered or imbedded in glass. Since the

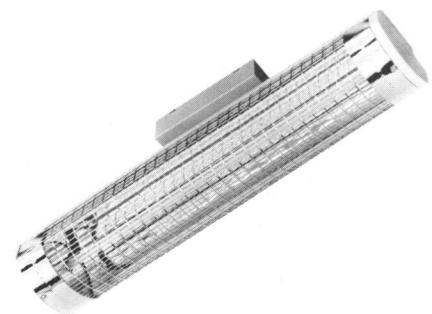

Fig. 19-10. Quartz tube heaters provide infrared rays for heating. (Electromode, Division Friden, Inc.)

Fig. 19-8. Electric ceiling heaters, such as this resistance type, are used in bathrooms to provide warmth after leaving the bath. Some units are equipped with a quiet fan. (Markel Electric Products, Inc.)

Fig. 19-9. Radiant glass panels are used in the ceiling or wall to provide spot heating. In some instances a series of panels are used to heat rooms. (Sun-Heat Inc., Electric Heat Division of Insto-Gas Corp.)

glass becomes hot, the panel is covered with a grille. Some wall and ceiling heaters may be equipped with a thermostat and/or fan to promote heat circulation. Frequently, rather than installing a resistance heater, an infrared unit may be desired for the wall or ceiling (Fig. 19-10).

Wall and ceiling heaters are used in bathrooms where additional heat is required for short periods of time. Occasionally these are placed in kitchens, recreation areas, family rooms, or enclosed porches. Series of radiant glass panels may be used to heat an entire house. Infrared units are used mostly in bathrooms.

Electric Cable Units. Electric wire used as a radiant cable converts electrical energy to heat energy. This works in the same way as the common resistance unit. Cable units (see Fig. 19-11) placed in plastered ceilings or walls are commonly covered with plastic or non-flammable vinyl plastic material. The electric cable is installed before the drywall, plaster ceiling, or

Fig. 19-11. Various types of radiant cable are available for heating ceilings, walls, and exterior house areas. (Ceil Heat, Inc.)

wall surface is applied. The whole wall or ceiling then radiates heat to the room. This form of equipment is inexpensive, but it does not respond as well as other types of heating devices.

Cable may be used to heat one room or the entire home. It may also be placed in concrete floors, driveways, sidewalks, steps, downspouts, gutters, or roof edges to melt snow and ice. Cable is a temporary measure that may be used to eliminate ice dams that cause damage to gutters and shingles.

Duct and Floor Insert Heaters. Electric duct heaters (Fig. 19-12) are placed in forced-air ducts. The duct heater, in essence a resistance heater, is controlled by an integral or wall-located thermostat. The floor insert heater, shown in Fig. 19-13, is similar to the duct unit and it is placed in the floor. This unit is thermostatically controlled. These heaters are used to give additional heat or to provide complete heating requirements. The floor heater is ideally suited for installation below

Fig. 19-12. The electric duct heater is inserted directly into the ductwork. Thermador Division of Norris-Thermador Corp.)

Fig. 19-13. Floor insert heaters are particularly suited for installation below floor to ceiling windows or sliding glass doors. (Electromode, Division Friden, Inc.)

ceiling to floor windows and sliding glass patio doors.

Electric Furnaces. Electric furnaces heat *only* by convection. The resistance heating elements (Fig. 19-14) are arranged in a series and operate only as heat is needed. The motor driven blower propels the air over the heating elements, through the ducts, and into the rooms. Electric furnaces are not limited to forced-air heating; they may also be used with a boiler to supply hot water for radiators, convectors, and radiant panel heaters.

Some electric furnaces for residential heating are small enough to be placed in a closet or suspended from the basement ceiling These are always used to heat an entire house.

Electric Heat Pumps. One of the most recent developments in home heating has been the heat pump (Fig. 19-15). The natural heat from the out-

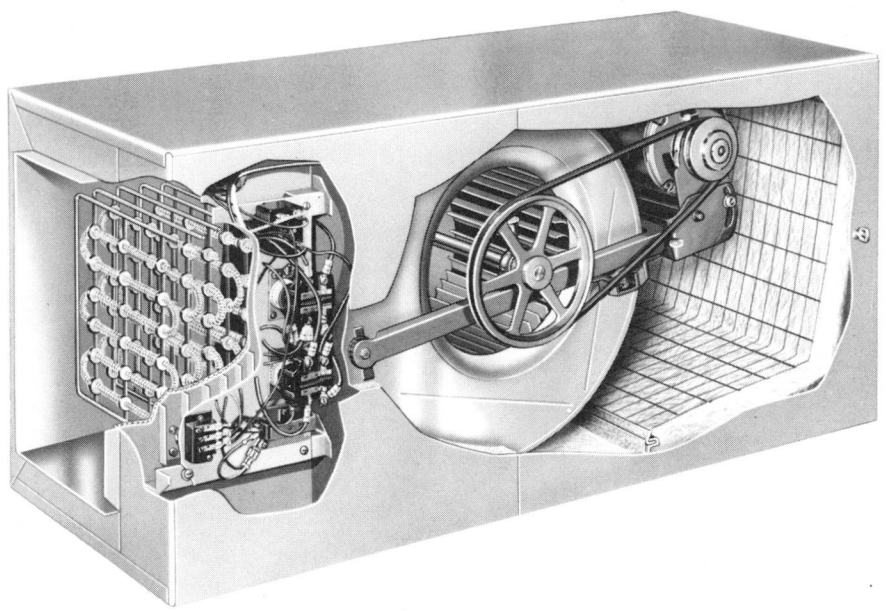

Fig. 19-14. This view shows the interior of an electrical furnace. Note the resistance elements, blower, and filter. (Many electric furnaces are small enough to be concealed in a closet.) (Lennox Industries, Inc.)

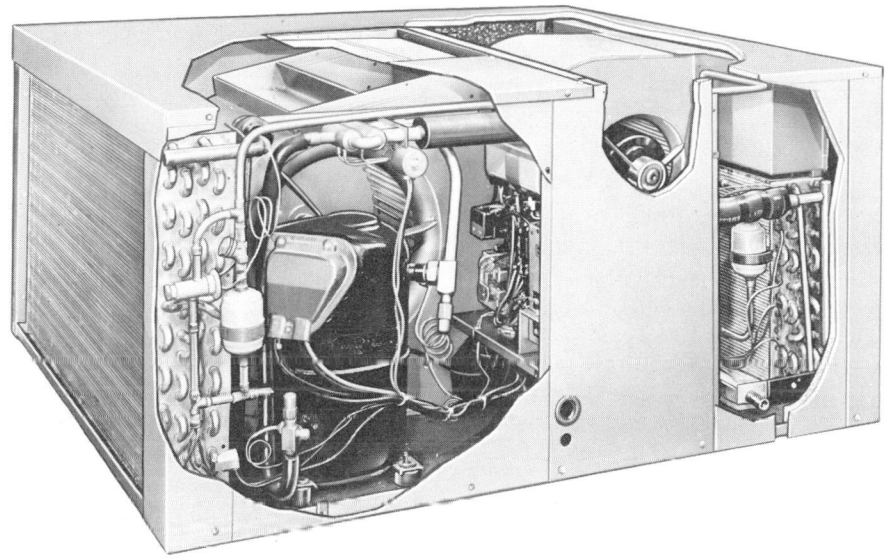

Fig. 19-15. Heat pumps may draw heat from air, earth, or water. This is an air-cooled heat pump. Ductwork is required. (Worthington Air Conditioning Co., Climatrol Div.)

side air may be utilized with either a hot water or a forced warm-air system. Earth and water may also be used as a source of natural heat. The basic principle of the heat pump is similar to reverse refrigeration. In the summer months the heat pump acts as a cooling unit. The processes of heating and cooling are controlled by one or more thermostats located in the house.

For example, if water is the source of heat, the water will be pumped through a pipe coil buried below the frost line. As the water circulates through the coils of the heat pump, the heat is removed, thereby lowering the water temperature The chilled water is then returned to the underground coil. The heat that has been removed is concentrated and distributed. When outside temperatures are low, 20°F or lower, heat pumps using the outside air or earth as a source of heat operate at a materially reduced level of efficiency. This requires a supplemental heat source. Some heat pump manufacturers employ an outdoor temperature control that disconnects the unit when the temperature drops below 20°F; the entire heating load is shunted to supplemental heaters. In areas where temperatures dip low, the size of the heat pump is determined by the summer cooling load. The additional heat that is required during the winter months is obtained from resistance heaters located in the duct work.

Insulation. The conveniences of

electric heat can pay dividends *only if the house is properly insulated.* Proper insulation will mean greater comfort, heating efficiency, and economy. Common insulating materials used in electrically heated homes are: (1) loose fill (blown or poured) cellulose fiber, mineral wool, perlite, or expanded vermiculite; (2) blankets and batts of mineral wool or cellulose fiber; (3) roof deck panels, blocks, and slabs; (4) exterior fiber sheathing; (5) perimeter slabs, and block for concrete floor perimeters; (6) reflective insulation; and (7) interior panels, blocks, and tiles.

The insulation manufacturers' association has adopted a uniform method of rating the effectiveness of all types of insulation when installed according to instructions. The purpose of any insulative material is to *resist* heat flow; the "resistance" of a specific thickness and type of insulation is indicated by an "R" value. This number designates the amount of resistance the material has to heat passage. For example, a piece of insulative material of a given thickness may have an R value of 12 (this is written as R-12). The insulation just specified, having a value of R-12, would offer only ¾ as much resistance as one having a value of R-16. *The higher the R value, the higher the resistance.* The following quantities of insulation are recommended for electrically heated residences:

Ceilings: 8″ or equivalent to R-28
Side walls: 4″ or equivalent to R-13
Floors: 4″ or equivalent to R-14

ESTIMATING MATERIALS

Hot Air Systems. All warm-air heating systems (forced-air or mechanical furnaces) include: (1) a *firebox* in which the fuel (gas or oil) is burned, (2) an *air chamber* in which the air is heated, (3) a *cold air box or chamber* in which the air is returned for circulation, and (4) the *ducts* in which the heated air is carried to various rooms of the house. *Humidifiers,* either evaporating plate or atomizer, are installed to add moisture (50 percent relative humidity) to the air. This makes the air more comfortable to breathe. Air filters may be installed in the cold-air return of the furnace to remove dust and dirt particles prior to their entry into the heating chamber. The filter may be an easily cleaned permanent type, or an electrostatically charged type. Any of the three methods of filtering the air is satisfactory.

Fig. 19-16 is a cut-away drawing of a residence, showing the supply and return ducts, grilles, various fittings, etc. Note that the supply ducts are carried to the upper parts of the various rooms; that the return ducts are at the baseboard level; that all supply ducts are in interior partitions; and that all return ducts are in exterior walls.

Ducts. In forced air systems the hot and cold air pipes are known as ducts. They are rectangular or circular in cross section.

Duct Lengths. The layout drawings generally are traced from the regular working drawings. If they are not

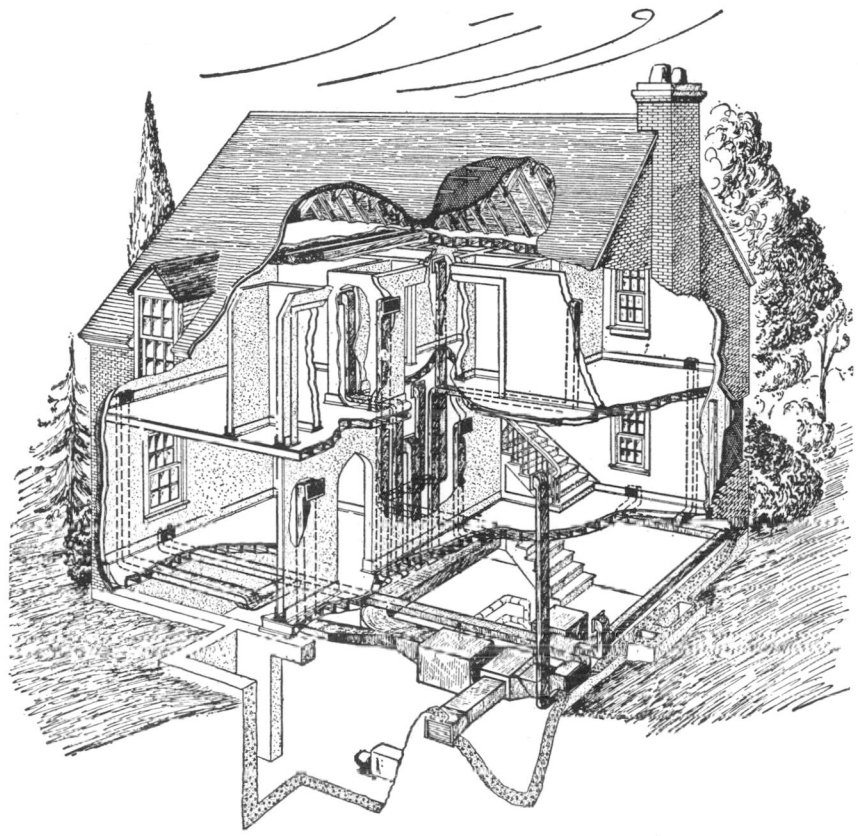

Fig. 19-16. A cut-away view of the duct systems in a typical residence. The supply ducts are in the interior partitions and the return ducts are in the outside walls.

traced, the layout drawings should be drawn to scale. In either case the layouts are scaled to determine the lengths of various ducts in the basement or wherever the central heating plant is located. The stack lengths of both warm-air and cold-air ducts can be determined easily from vertical dimensions shown on the drawings plus the known distance above the floor.

The number and kinds of various fittings required are counted by tracing the various ducts throughout their lengths. If you learn to visualize a duct system in the manner shown in Fig. 19-17, you will not be likely to make omissions.

Fig. 19-18 shows the floor plan for HOUSE PLAN A, and section views of typical duct work which can be installed in this house. Try to visualize all the ducts in the walls of the house and compute their lengths.

Ducts can be either fabricated as

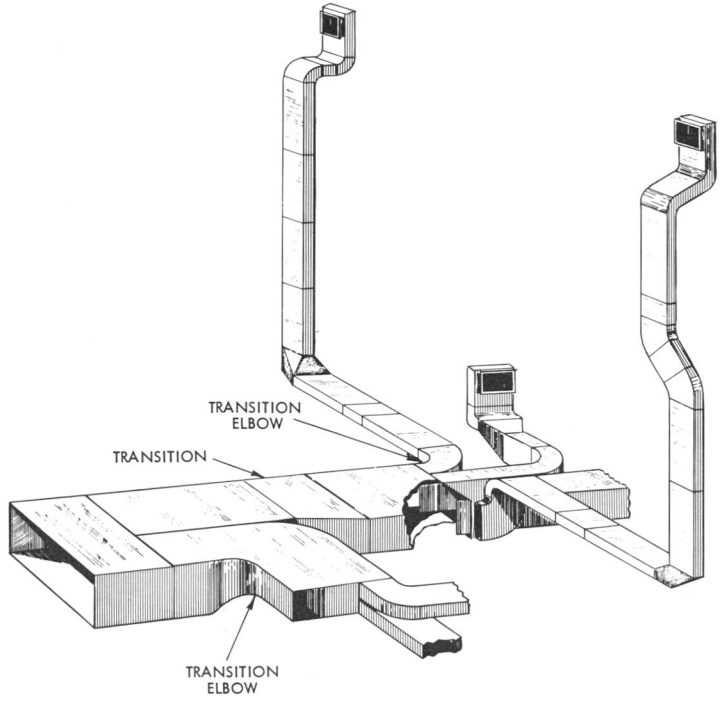

Fig. 19-17. A diagram of a typical duct system used for a mechanical heating system. Learn to visualize duct systems in this manner. (Henry Furnace Co.)

required on the job, using the gages designated in the manufacturer's specifications, or prefabricated using standard parts and fittings. The use of prefabricated ducts and fittings is naturally considerably cheaper because they are made on a production basis, whereas job-fabricated ducts and fittings require considerably more labor.

Job-Fabricated Ducts. If the duct work is to be made on the job, the total weight of sheet metal required is figured as explained in Chapter 18, "Sheet Metal." The total weight is usually multiplied by a figure which includes both material and labor costs.

This figure varies in different regions. The estimator's cost analysis records, if they have been carefully compiled and kept up to date, can be readily used to create such a figure.

Prefabricated Ducts. The manufacturers of prefabricated ducts have catalogs in which all parts and fittings are shown together with figure or part numbers. As each new size or part is found on the layout it is written down in the quantity take-off. Each time this part is repeated on the layout it is noted on the take-off (usually by a check mark after the first entry).

Duct work is being installed in an

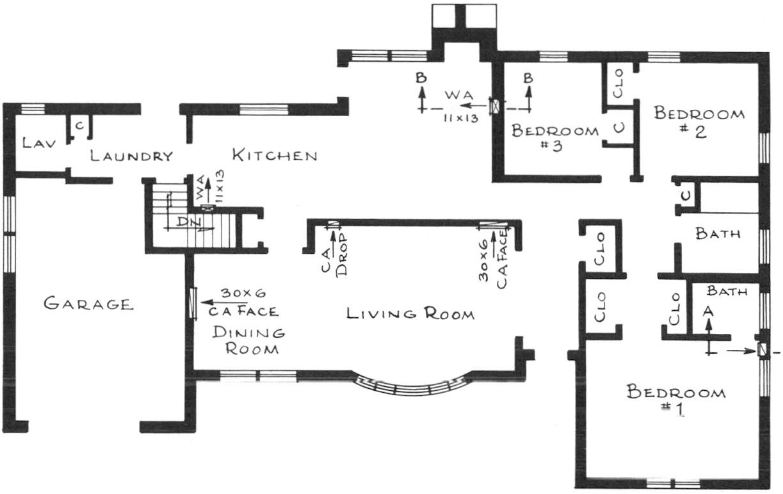

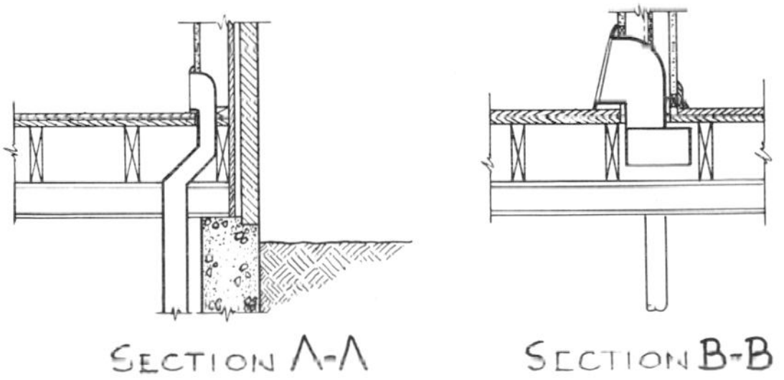

Fig. 19-18. This is a plan view of House Plan A with section views of air ducts in interior and exterior walls.

SECTION A-A

SECTION B-B

increasing number of cases, using pre-fabricated ducts and fittings. This shortens the installation time and naturally lessens the cost by a considerable amount. With prefabricated ducts and fittings there is little cutting and fitting of sheet metal other than around joists when the joists are used as ducts for cold-air returns See Fig. 19-19.

The use of prefabricated ducts makes estimating much more simpli-

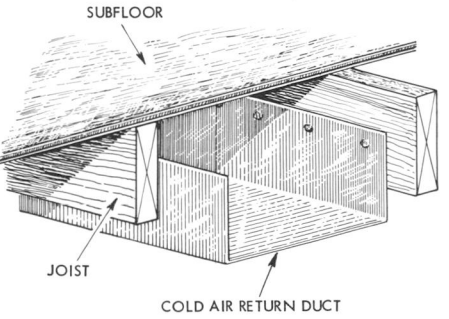

SUBFLOOR

JOIST

COLD AIR RETURN DUCT

Fig. 19-19. Joists can be used as part of the duct for cold-air returns.

fied and the keeping of cost records easier. If a contractor has kept a good time record for installing ducts on previous jobs, he can easily determine the cost of installing ducts per foot; he can make up an average cost for ducts based on the number of hot-air registers, or make up an average cost of installing complete duct systems, de-

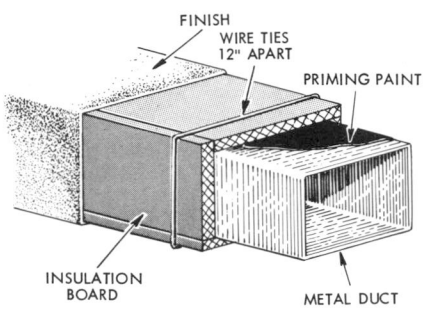

Fig. 19-20. Sometimes ducts must be insulated, especially when they are in outside walls.

pending on the number of cold and hot air ducts or registers.

You must be careful to note anything irregular in any of the duct runs. Sometimes one or more ducts may require insulation or, because of structural conditions, other expensive operations may be necessary. You should know whether prefabricated or a combination of prefabricated and job-made ducts are to be installed. The use of prefabricated ducts has brought about a practically standard method of estimating.

Insulation. Sometimes one or more ducts run through outside walls or other places of excessive heat or cold. For such ducts some form of insulation is necessary as illustrated in Fig. 19-20. Prefabricated ducts can be purchased already insulated. If the contractor is to do the insulating, he must calculate the amount of insulation re-

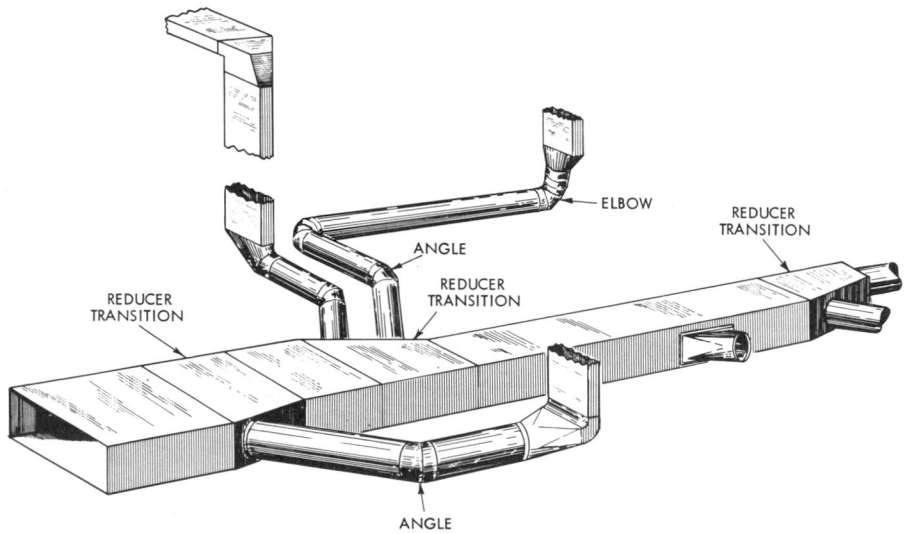

Fig. 19-21. Typical fittings used in a duct system. (Henry Furnace Co.)

quired and add this to the material take-off.

Fittings. Figs. 19-17 and 19-21 give some idea of the various fittings used in mechanical or hot-air systems. You should obtain catalogs showing all types of fittings and their separate parts. In the system illustrated, the basement trunk duct is suspended from the ceiling. Where individual duct systems are specified the ducts are often run between joists, where they are out of sight. Various types of air-conditioning equipment require widely differing connections for the ducts to the equipment. Cold-air ducts are generally laid out so as to enter the equipment in one or two trunk ducts. However, these items will be shown on the layouts. Study Figs. 19-16, 19-17, and 19-21 until you are able to visualize the entire system. The cost of fittings is determined by the complexity of the fitting. For example, a T-connection is more costly than a 90° elbow.

Grilles. Fig. 19-22 shows a box for

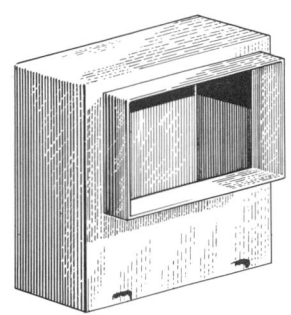

Fig. 19-23. The top of a stack for a wall grille is mounted near the ceiling of a room.

a baseboard grille. A sheet metal cover may be used during construction to keep out dirt. This cover is later replaced by a grille. Fig. 19-23 illustrates the fitting for a wall grille. The details of these boxes vary according to stack and grille sizes.

Baseboard and wall grilles are similar to those shown in Fig. 19-24. They are delivered ready to install. Grilles are estimated by their size, by whether they are a movable or fixed grille system, and by the material from which

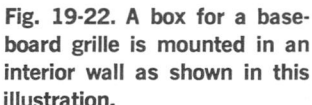

Fig. 19-22. A box for a baseboard grille is mounted in an interior wall as shown in this illustration.

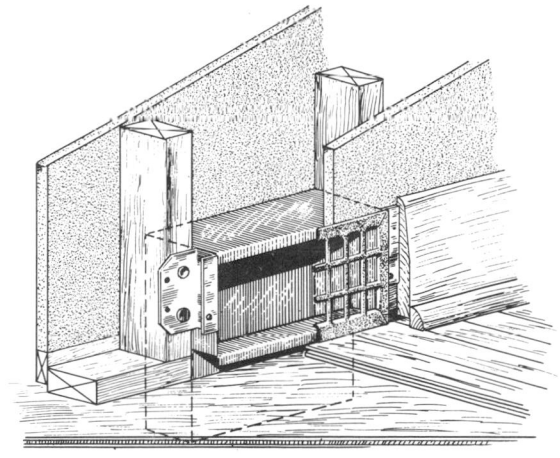

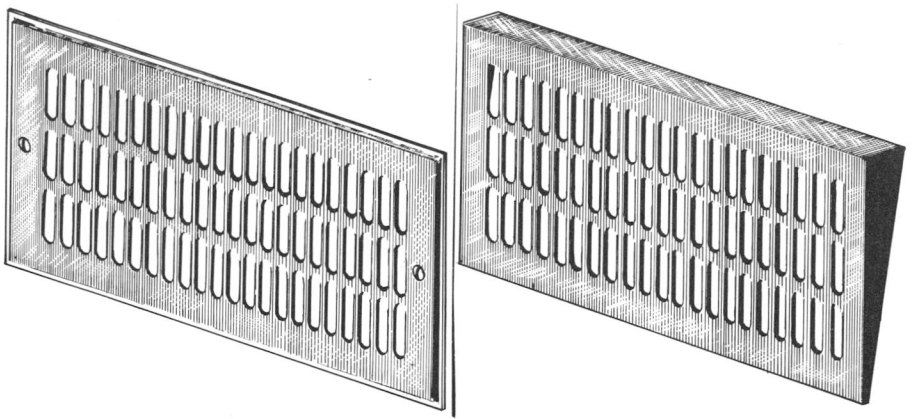

Fig. 19-24. Grilles are easily attached to the openings. At the left is a baseboard grille; at the right is a wall grille.

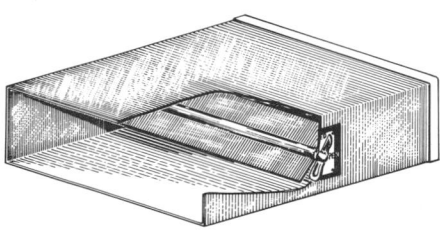

Fig. 19-25. This is a typical damper which is mounted in a rectangular duct.

the grille is made, such as cast iron, sheet metal, brass, etc.

Dampers. The typical damper in Fig. 19-25 can be used in any rectangular duct. Dampers are built to fit only one size of duct, but many damper sizes are available. Dampers are sometimes used behind grilles as a means of controlling the air flow in a room. In mechanical furnace systems, dampers (in the ducts) balance the flow of air through the system. Dampers must be estimated by their size and the type of material from which they are made.

Controls. Fig. 19-26 illustrates a typical control system for a year-round air-conditioning plant having a mechanical furnace as the primary heat source.

Where heating, cooling, and air conditioning are specified, the matter of automatic controls becomes complicated and requires a great deal of additional costly equipment and labor. One must first estimate the cost of the equipment and, based on the complexity of the equipment, the labor may be determined.

Pipe Lengths. Layouts for steam heating are seldom made. Generally the positions and sizes of radiators or convectors are indicated on the working drawings and the estimator makes his own calculations for the amounts of piping and various fittings, such as elbows, nipples, tees, etc. If full architectural service is engaged for a new residence, it is probable that a heating layout will be furnished. If a layout is not furnished, the contractor must design the pipe sizes before the estimator

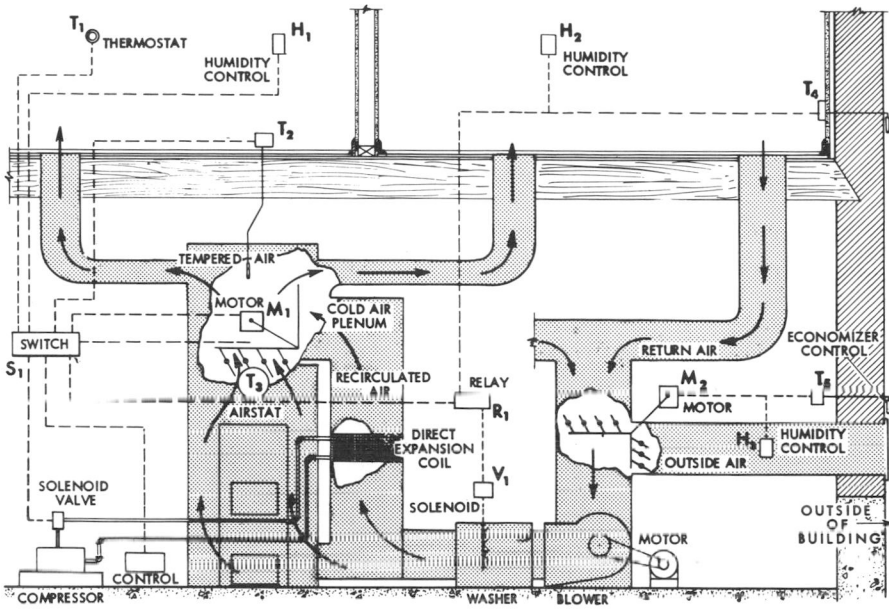

Fig. 19-26. A typical control system for a year-round air-conditioning system. A mechanical furnace is the basic heat source.

can do a quantity take-off. The pipe lengths are determined from dimensions on the drawings and by scaling. A heating layout for a hot-water system will usually indicate the following: correct radiator and convector sizes and styles, pipe lengths and sizes, slopes of pipes, fittings, radiators, valves, expansion tank, etc. With such a layout and the working drawings, all dimensions can be found.

Radiant Heating. Radiant panels may be considered a type of water system or a type of electrical system. Since radiant heating is one of the most popular methods of heating by water, it will be dealt with in detail here. Most of the component parts are the same for any hot-water system and

so estimating in this section is applicable to all water and steam systems.

A typical radiant heating job for an average residence is either a floor panel or a ceiling panel installation. While the costs of the completed systems are approximately the same regardless of panel location, some difference exists because the nature of the installations is different. For this reason, the estimator must be familiar with every step in either system.

An experienced estimator bases his calculations on his previous experience. As a result, it is generally unnecessary for him to itemize each factor going into a job as a means of attempting to approximate the cost in advance. Instead, a job estimate is more likely to be made on the basis of previous installations which were similar. It is obvious, however, that there must be a starting point somewhere for the individual who has no experience in estimating. The following material, therefore, may differ somewhat from the methods used by experienced estimators, the main difference being the detail into which the estimating is carried.

Gravel Fill. In order to achieve the maximum efficiency from a radiant floor panel installation, it is necessary to combine several building materials and construction techniques which are peculiar to radiant heating alone. While a gravel fill; moisture barrier, asphalt topping, and insulation do not constitute a part of the radiant heating system and, therefore, should not be figured in an estimate for such a heating system, nevertheless, the work and materials are necessary and must be paid for.

Good installation practice requires at least 6″ of gravel be placed over undisturbed earth. If the earth has been disturbed or if any fill has been made, the soil should be rolled so as to prevent any settlement after the installation is complete.

The amount of gravel necessary for the fill is found by multiplying the length of the floor panel area by the width, then multiplying by 6″ or 0.5′. Since gravel is usually estimated in cubic yards, it is necessary to divide the number of cubic feet by 27. The cost of gravel is found by multiplying the number of cubic yards by the prevailing cost per cubic yard.

Cinders are not used as fill material. Sulphur-bearing materials such as cinders will react with moisture to form sulphuric or sulphurous acids. Such acids, of course, are highly corrosive and would cause deterioration of the coil piping.

Moisture Barrier. A moisture barrier is placed over the gravel fill, particularly if there is any reason to believe that there will be inadequate drainage for the building site. The moisture barrier should be lapped at the edges up to 12″ and should be carried around the edge of the slab so that there is a barrier between the slab edge and the foundation, Fig. 19-25.

The amount of material required for the moisture barrier is found by determining the number of square feet to be covered. To this is added an additional 36½% for the material which will be "lost" due to the 12″ overlap which is recommended for each joint, and for spoilage or waste. Once the total amount of moisture barrier is known, the number of square feet is multiplied by the current price per square foot.

In order that each overlapped joint will be as resistant to water as the moisture barrier itself, each overlapped section must be covered with asphalt topping. In addition, the surfaces of the foundation over which the moisture barrier is laid (see Fig. 19-27) also must be covered with asphalt. The area of the overlapping sections is the same as that "lost" due to overlapping. Therefore, 36½% of the total area covered by the moisture barrier must

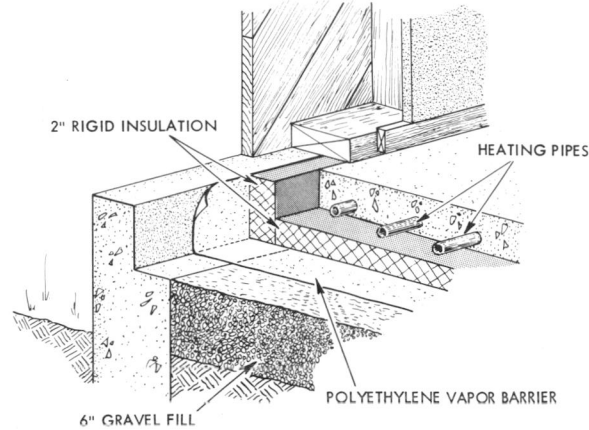

Fig. 19-27. The proper use of gravel fill, moisture barrier, and insulation for a concrete floor slab used as a radiant panel will minimize heat losses.

be asphalted. Added to this is the area of the vapor barrier where it overlaps on the foundation. Assuming the foundation in Fig. 19-27 to be 8″ thick, the area covered by the moisture barrier is 10/12 or 5/6 feet multiplied by the length of the foundation perimeter. For ease in calculation, convert the 5/6 to a decimal (0.83′).

When the total area to be asphalted is determined, allow 2 gallons for every 100 square feet.

Insulation. In the past, it was considered necessary to insulate the slab perimeter only. Tests have indicated, however, that the flow of heat into the ground directly under the slab is great enough to merit the cost of the insulating material necessary to prevent this flow. The types of insulation which seem best adapted to this purpose include cellular glass, fiberglass, and insulating board with supplementary integral preservative treatment for

protection against dry rot and decay. Whichever type of insulation is selected, it should be thoroughly coated on all sides and edges with asphalt or pitch before being applied.

The area of the floor panel has been previously determined in connection with the gravel fill and the moisture barrier. Therefore, the cost of the insulation is determined simply by multiplying the area in square feet by the cost of the insulation per square foot. The necessary asphalt is determined on the coverage basis previously established of 2 gallons per 100 square feet.

Insulating Concrete. The use of an insulating concrete in place of 2″ of insulation has proved advantageous in many instances. Insulating concretes are composed of lightweight cellular aggregates (vermiculite, for example) instead of the customary sand and gravel. In the event the insulating concrete is considered more desirable than the 2″ insulation, the quantity required can be determined by finding the area of the panel in square feet as was done previously, then dividing by 3 (⅓′ or 4″, the recommended thickness of the insulating concrete). This figure is then divided by 27 to convert it into cubic yards.

The installation of a radiant heating system in the ceiling requires no elaborate preparation as was necessary for the floor panel. Once the ceiling joists are in place, the coils for the radiant panel can be formed and fixed in position. While some installations have been made in which an expanded metal lath was installed first, the ceiling coils then being placed in position, the better method is to place the metal lath *below* the coils. When the coils are placed below the metal lath, the plaster must cover the tubing with sufficient thickness to prevent cracking. Even with the tubing of small diameter, the plaster would have to be exceptionally thick. When the coils are placed above the expanded metal lath, the plaster can be carried over the whole of the ceiling area at an even thickness. This means that there is less weight for the expanded metal lath to support, less risk of sagging and cracked plaster, and less chance of darkened ceilings. Another factor to be considered is that the tensile strengths of some plasters are considerably reduced by the constant application of heat.

Coil Piping. Regardless of whether the coil installation is to go into the floor or ceiling, the architect and/or a heating engineer will have determined the quantity, size, and material for the pipe or tubing necessary to handle the heating load. The quantity and size of pipe or tubing required for each room is expressed in terms of inside diameter and linear feet. In addition, a detailed plan will have been prepared indicating the exact placement of the coils, proper spacing, etc.

The total amount of pipe or tubing needed for both ceiling and floor installations is roughly estimated from the findings of the architect or engineed when the system was designed. (Each room will have been allowed a certain number of feet of piping, based on the known heat loss of the room.) A more accurate method would be to scale the drawing. In this way, an exact figure is possible.

When the total amount of piping is known (including piping around the boiler or heating unit and the supply and return mains, all of which will be indicated on the drawings), the number of linear feet is multiplied by the prevailing cost per foot and the material cost thus determined.

There is one other item of material cost which must be considered in connection with the coil piping for floor and/or ceiling installations. This is the quantity of balancing valves, shut-off valves, tees, elbows, and air vents which must be incorporated in the system at the time the coils are fabricated and installed. These fittings are taken from the drawings. The number of each type of fitting is multiplied by the unit cost per fitting in order to determine their cost.

To some extent, the location of the coil is a determining factor in the method by which the coil is formed. It has been found economical, for example, to fabricate coils of copper tubing, intended for ceiling installation, on forms built on top of benches. The entire completed coil is then raised into position at one time. Bench-formed coils are made either at the job or in the shop, whichever is more economical and practical. It is also possible to form ceiling coils of soft copper tubing in place on the ceiling joists without the use of forms or a bender. If steel or wrought-iron tubing is specified for a ceiling panel, the coil is formed on the floor and raised into position, or the coil could be prefabricated at the shop. Copper, steel, or wrought iron coils, intended for use in a floor installation, will usually be formed in place, although some contractors prefer to prefabricate the coil in the shop.

If copper tubing is used instead of wrought iron or steel, the time which must be allowed for bending will be approximately the same since a bender is necessary for bending hard copper tubing, and desirable for soft copper tubing. While soft copper tubing can be bent by hand, a mechanical bender will produce even, uniform bends.

If steel or wrought iron is to be used, the approximate number of welds which must be made can be determined rapidly and easily by dividing the total number of linear feet of piping by 21. Steel and wrought-iron pipe come in uniform lengths of 21'-0", Therefore, every 21 linear feet of coil piping will require a weld.

In addition to these welds, there are those which must be made for the installation of shut-off valves, air vents, balancing valves, etc. These items must be counted, using the plan prepared by the heating engineer and/or architect which gives the location and quantity of every valve and fitting to be used in the system. For each such fitting, two welds will be required.

It is possible to obtain copper tubing in lengths of 40, 60, and 100 feet. Therefore, if copper tubing is to be used, fewer joints are necessary because of the greater length in which the tubing is supplied. However, the same number of soldered joints must be made at balancing valves, air vents, shut-off valves, etc., as welds were made for the installation using steel or wrought-iron pipe.

Welding. Another item of material cost which must be considered is the amount of welding rod, oxygen, and acetylene gas necessary to make the welds that will be required. If copper tubing is to be used, the cost of the solder, flux, and gas used for making the soldered joints must be calculated.

Table II gives the materials *and* time required for making pipe welds. Time and materials for welding can be presented in tabular form (likewise for cutting) because their nature permits them to be reduced to an approximate average. A fairly definite quantity of gases is consumed by a blowpipe in operation with respect to the thickness and length of the steel welded or cut. Likewise, the amount of welding rod depends on the length and thickness of the joint; and actual time for weld-

TABLE II
WELDING RATES AND MATERIAL CONSUMPTION BASED ON PIPE WALL THICKNESS
SINGLE FLAME WELDING*

Pipe Wall Thickness in Inches	Backhand Technique 70 Degree Vee, Rotation Welding			Forehand Technique † 90 Degree Vee, Rotation Welding		
	Linear Rate of Welding In. per Minute	Rod per In. of Weld in Lbs.	Oxygen per In. of Weld in Cu. Feet	Linear Rate of Welding In. per Minute	Rod per In. of Weld in Lbs.	Oxygen per In. of Weld in Cu. Feet
$3/16$	3.0	0.020	0.27	1.3	0.025	0.33
$7/32$	2.9	0.020	0.30	1.3	0.027	0.37
$1/4$	2.8	0.022	0.35	1.2	0.029	0.44
$9/32$	2.6	0.028	0.44	1.1	0.039	0.58
$5/16$	2.4	0.030	0.48	1.0	0.041	0.75
$3/8$	1.8	0.041	0.71	0.85	0.059	1.05
$7/16$	1.2	0.051	0.87	0.65	0.078	1.40
$1/2$	0.95	0.070	1.25	0.53	0.100	1.85

* The welding rates and material consumption in this table are based on continuous welding on or near the top of the pipe; work being done by a properly qualified pipe-welding operator; and actual welding only without allowance for tack-welding, starting, or finishing. Allowance for starting and finishing is approximately one-half to one minute per weld for pipe up to 10′ in diameter.
     These figures may vary 10 to 15% either way, depending on the size of the vee, the flame, the rod, and the diameter of the pipe.
     † These figures will also apply for forehand welding with the neutral flame providing welds to be made are of minimum size.
           Courtesy Linde Air Products Company

ing and cutting is also approximately proportional to the amount of work to be done.

Estimates for the remaining time-consuming operations—laying out and marking of cuts, joint alignment, tack-welding, etc.—as well as the unproductive time of the welder and his assistants can be determined only by experience, since these vary so much from one job to another.

While Table II is used to determine precisely what each welded joint should cost in terms of time and material, it is seldom necessary to estimate costs so closely. Rather, an estimate is made on the basis of previous experience. Moreover, the difficulty in arriving at an accurate estimate, due to uncontrollable and unforseeable events, makes such close figuring largely wasted effort.

If it is necessary or desirable to estimate welding costs as closely as possible, the first step is to determine the total number of welds. The length of each weld (pipe circumference) also must be found. The total amount of welding necessary (in terms of linear inches) is then multiplied by the prevailing cost, thus arriving at the materials cost for the welding.

The amount of material — solder, flux, and gas—necessary for a radiant panel job in which copper tubing is used is figured in the same manner as when steel or wrought-iron pipe and welding are used.

Boiler. The estimate for the boiler must be made on the basis of the schedule of heat losses and the specifications. The schedule of heat losses will have been prepared by the architect, a heating engineer, or some other competent person who knows every detail of the construction and insulation of the building. An allowance also must be made for heating domestic hot water, piping tax, and heating-up or starting-up.

The specifications will call for either a gas- or an oil-fired boiler, coal seldom being used for radiant heating installations. If a gas boiler is specified, stainless-steel flue liners, tees, elbows, and condensation drains are sometimes considered necessary for the chimney. Vitreous sewer tile with bell-mouth joints also make an acceptable lining. See Fig. 19-28. These special flue linings and other fittings are necessary because of the corrosive nature of the condensate from the combustion gases.

If the stainless-steel liner is used, its cost (which will be greater than any other type of flue lining) must be determined from manufacturers' or suppliers' literature and added to the estimate.

Whatever the type of flue lining used, it is customary to include a stainless-steel tee where the smoke pipe or breeching enters the chimney, and a drain fitting of stainless steel to which is attached a piece of plastic tubing. See Fig. 19-28. The condensation from the chimney is led off through this fitting and tubing to the nearest floor drain. Special fittings and boilers are cataloged and their costs listed in manufacturers' literature.

A certain amount of piping is necessary in bringing the gas to the boiler in installations where gas is specified. The amount of piping is determined by the location of the meter and the boiler. This distance is scaled from the drawings. The various elbows, valves, and other fittings also can be taken from the drawings, although the quantity needed, in terms of cost, can be approximated by allowing 100% of the cost of the necessary pipe.

If an oil-fired boiler is specified, the estimate would include a tank or tanks for the storage of oil, sized and placed according to the architect's specifications. The estimate is based on costs supplied by manufacturers or suppliers. The labor, as well as the material cost, is slightly higher for an oil-fired job than for a gas-fired installation since the storage tanks must be placed in addition to running the lines for the oil supply, setting and connecting the boiler, etc.

If the boiler is purchased as a packaged unit, all boiler controls are included in the price and will be an integral part of the unit. If the boiler is purchased in any other form, an electrical contractor, in most cases, is asked to bid on the job of installing all electrical materials. This bid should be added to the cost of the job and will include hooking up all the controls for the boiler, the controls for the panels, and the circulator.

The efficiency and the effectiveness of the radiant heating system is dependent to a large extent on automatic controls. There are a number of manufacturers producing automatic con-

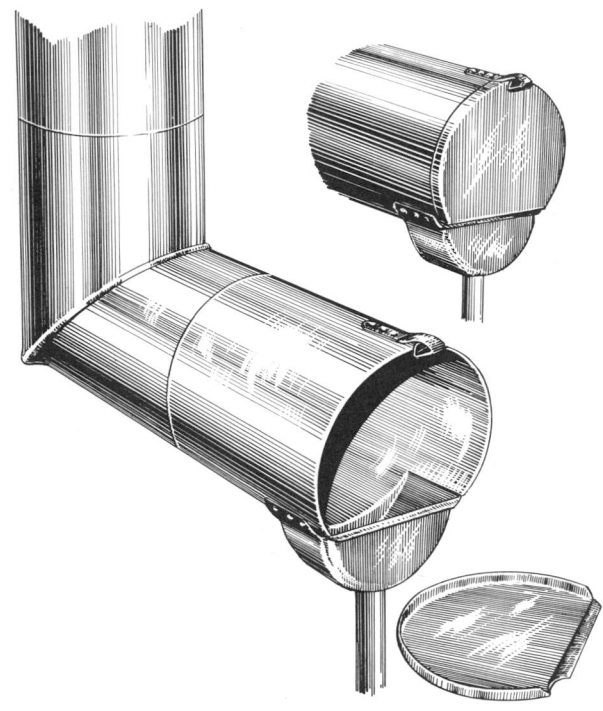

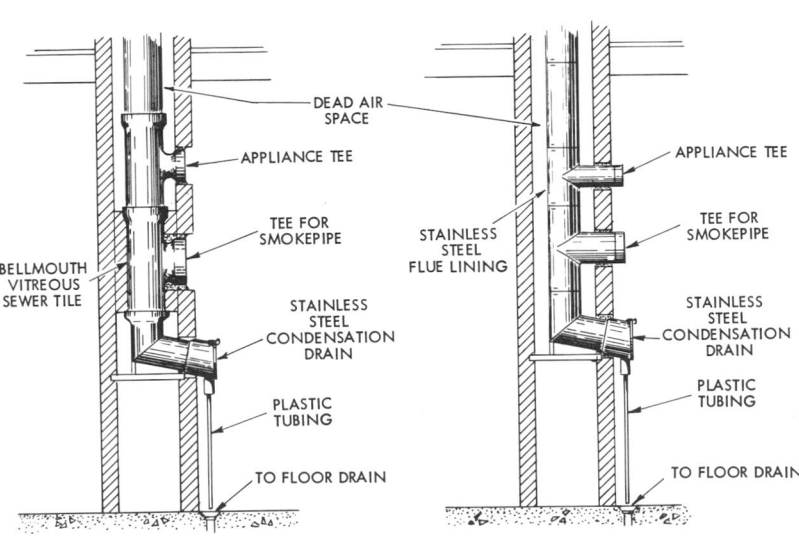

Fig. 19-28. Gas-fired boilers require stainless-steel condensation drains (top) and vitreous clay tile or stainless-steel flue liners (bottom).

trols for radiant panel systems which are entirely satisfactory. The estimate of the cost of any particular control system is based on the cost for the material and the labor required to make the installation. The manufacturer of the equipment or one of his local distributors should quote prices on request.

When a complete cost list of the materials and labor needed for a job has been compiled, most contractors add 15% to the total cost estimate for overhead or operating expenses. To this new total is added about 10% for profit. Other additions could be as follows: 7% for employee insurance, 3% for employee welfare, 6% for testing, 2½% for cleaning up the job,

and 10% for the handling of materials. The final or *grand total* is presented in the estimate.

Snow Melting. Where the annual snowfall presents a removal problem the radiant panel coil or grid has been successfully adapted and is becoming more popular.

A snow melting system consists simply of a coil or grid embedded in the concrete of the walk or driveway which is to be kept free of ice and snow. Through these coils or grids is circulated water to which an antifreeze has been added. The water may be heated in any of several ways, but the most common and convenient method is by means of a heat exchanger or heater which is attached to the boiler.

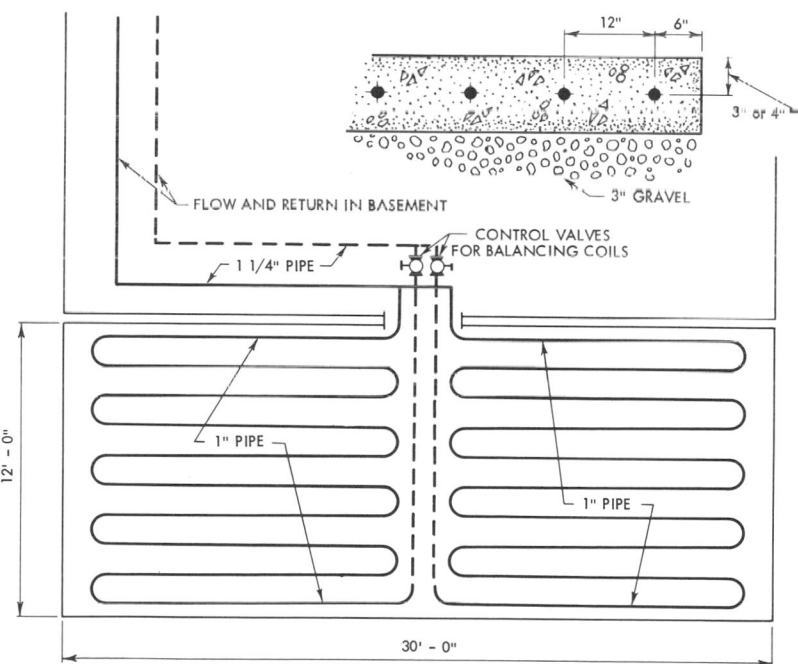

Fig. 19-29. Radiant heating coils in concrete are used for melting snow. (Taco, Inc.)

A typical snow removal installation is shown in Fig. 19-29. The similarity between a snow removal system and a radiant panel is such that a more detailed description is unnecessary.

Electrical Systems. Installation costs of a heat pump are naturally higher than a resistance installation. However, the costs are comparable to separate heating and cooling systems. Reports indicate that operating costs of heat pumps are lower than the operating costs of resistance type units.

Additional Parts. These include such items as controls, electrical work, and plumbing equipment.

Controls. These often come as part of the regular equipment. Where a more elaborate control system is required, all additional instruments must be calculated as extras and their costs added to the estimate.

Electrical Work. The controls used on furnaces and especially air-conditioning equipment are electrically operated. In most instances an electrical contractor is asked to bid on this part of the work (installing all electrical equipment) and his bid should be added to the cost of the job.

Plumbing Work. Where furnaces have automatic humidifiers or air washers as part of the general equipment, a plumbing contractor may be asked to bid on that part of the work. His bid is added to the cost of the job.

Permits. In most cities it is necessary to purchase a permit before installing a furnace. Permit costs vary among cities so each estimator must determine the costs in his locality.

Zoned Heat Control. Zone control of the heating system is becoming more important because of the changing habits of the home owner, as well as the contemporary methods of construction and architectural styles. One thermostat located in a central position cannot satisfy the heating requirements for an entire house. This is particularly true in large homes and in split-level or rambling ranch-type homes. For some dwellings several control devices may be necessary to provide the desired heat ranges throughout the house.

The architectural trends toward large windows, rambling and less compact buildings, and multi-level dwellings have complicated the problem of temperature control. Frequently, one section of the house may be sheltered, while another section may be exposed to a hard driving winter wind. One section may be exposed to the warm afternoon sun, while another may be shaded. Each section of the home has its own requirements according to its orientation and the desires of the family.

To accommodate these differences in heating requirements, the distribution of heat may be *zoned.* A zone is an area in which the temperature is controlled separately from another area of the building. A dwelling may be zoned by rooms, groups of rooms, or by levels. The size, orientation, protection surrounding the house, and the living habits of the family will determine the number of zones required.

Zoned heat is based on the premise that the piping, air ducts, or radiant panels are arranged so that heat to each area can be controlled with automatic valves, circulating pumps, or dampers. Hydronic systems probably offer more advantages than other systems because of their obvious adaptability to zone control. Overheating the entire house to increase the heat in one area is eliminated with multiple controls. The cost of zoning a house, of course, is greater because additional equipment is necessary. Over the years, however, the extra investment will be returned in added comfort and lower fuel costs.

EXPENSE OF ESTIMATING

Some sheet metal and furnace contractors and estimators break up their estimates into small sections and besides considering the material cost in a detailed manner go even further and base their labor cost on a per foot of duct, per special fitting, per control instrument, etc. Other contractors and estimators think it is just as accurate to base their estimates, so far as labor is concerned, on the number of runs plus certain other charges which will be explained later. If a labor estimating method could be worked out whereby the labor could be accurately determined without the necessity of scaling all runs, etc., not only could much time be saved, but the resulting estimates would be lower.

A person contemplating the installation of a new furnace might easily consult as many as five different contractors in an effort to obtain what he felt was the most reasonable figure. This practice requires all contractors

to make many estimates that are not accepted. There is a considerable cost involved in making an estimate, especially if a detailed method of estimating labor is used. Thus if a contractor makes five or six estimates in one week and is fortunate enough to get one job out of the five or six estimates, the cost for making the other estimates is lost. This has caused many contractors to do less actual work on estimates by making the figure high enough to insure its being more than sufficient to cover the cost of the estimate. This is poor practice, however, and it penalizes the home owner.

The specifications made by the architect must be read and understood so that every direction in them will be followed exactly. In the specifications there often is a section directing that the heating contractor is to make a full analysis of heat losses and prepare a complete layout. Sometimes the contractor can obtain this service from his supplier or manufacturer. The manufacturers usually maintain an engineering department which is at the service of contractors. The engineering department will use the drawings and specifications of the job to provide the contractor with a complete design of the job including layouts, heat loss calculations, size of furnace, and in some cases even a complete list of materials. Then the contractor has only to estimate his labor and add this to the quantity take-off.

In other cases where the contractor has no opportunity of obtaining such information from a manufacturer, he must do the work himself or have it done by a heating engineer. This will increase the total costs for the job and must be considered in the estimate.

Whether the work is done by a manufacturer or a contractor the process is the same and the following description, schedule, and layouts are applicable. Not many contractors have the facilities for doing all the engineering work themselves. They can figure a great deal of the work easily enough but the design of air-conditioning systems requires complete engineering preparation, which some of the smaller contractors are not in a position to have.

Sometimes the architects maintain a mechanical engineering department as part of their service. In such a case they can design the complete system and call for bids from contractors to supply material and labor. For our purpose, it is assumed that the manufacturer makes the layout.

Summary. Other heating systems are so similar to those explained herein that special explanations are unnecessary, because the estimator is interested only in the amounts of materials and labor involved.

Unless detailed layouts are a part of the regular working drawings, the inexperienced estimator should either have a complete knowledge of all heating principles or seek aid from a more experienced person in estimating materials and labor.

This chapter does not include every material or labor item encountered in heating installation. Each job presents its own special features which the experienced estimator knows how to anticipate and estimate. Some estimators specialize in air conditioning estimating just as others specialize in electrical work.

In heating work, as in any other trade, a checklist is of great value if fully detailed. The estimator can make such a list easily, and he may add to it as much detail as he deems necessary or helpful.

ESTIMATING LABOR

Labor estimating is done in any one of a number of acceptable forms. However, in many large cities the sheet metal contractors have formed associations which carry out studies of labor costs. The purpose of such studies is to find easy but accurate methods of estimating labor.

The following labor times are all average figures and will vary greatly among different regions, differing climatic conditions, etc.

For *radiant heating,* one laborer can place approximately 100 square feet of *moisture barrier* an hour. The total number of square feet of barrier, divided by 100, will give the number of man-hours required for its placing.

The labor required for mopping asphalt is determined by dividing the total area to be mopped by the rate at which a laborer works — approximately 60 square feet per hour. The total number of hours required for placing and mopping the barrier is then multiplied by the prevailing wage scale.

The cost of labor for *welding* is found by multiplying the length of each weld (pipe circumference) by the time necessary to complete the weld. To this must be added the time taken by the welder for starting and finishing. The total time required for making all welds is converted into hours, then multiplied by the prevailing hourly wage rate for welders.

The amount of labor necessary to make the connections in a *copper tubing radiant panel* installation is considerably less than for a wrought-iron or steel installation. An inexperienced journeyman and helper can jig, raise, and wire into place approximately 700 to 900 feet of copper tubing per 7-hour day. This includes all couplings bench soldered as part of the work.

For the floor panel installation, some time must be allowed for blocking up the coils to the level specified in the drawings. This operation will vary widely with each job, and experience alone will provide the best measure for estimating amount of time that will be necessary for this part of the job. However, approximately four hours' work by a journeyman fitter should be sufficient for the average six-room residence.

When the complete floor- or ceiling-panel system is in place and all soldered or welded connections have been made, a hydrostatic test must be made on the system. Where welding has been used, each joint should be rapped smartly with a hammer while the system is under pressure. A flow test also should be made in order to make certain there are no obstructions in any of the coils or other piping.

Usually 6% of the total labor cost is figured for testing.

Supervisory time will be required during the pouring of the concrete for the floor slab. Since the heating contractor is responsible for the satisfactory operation of the completed system, he will want to make certain that the coils remain in place and that the piping is not dented or otherwise injured during the placement of the concrete. This cost is usually included in the 15% charged for overhead or operating expenses for the entire job.

Installation time for *boilers* varies greatly. Some sample times indicated that, on the average, two journeymen fitters will take one day (about 16 man-hours) to set and connect a boiler. Two journeymen fitters will take one-half day (about 8 man-hours) to install the boiler breeching. To run the gas piping and connect the burner will require two journeymen fitters about one day (16 man-hours).

For an *oil installation*, two journeymen fitters will carry in, set up, and pipe two 275 gallon storage tanks in 1½ days (about 24 man-hours).

The labor involved in the electrical work is usually included in the electrical contractor's bid.

SAMPLE HEATING ESTIMATE

To demonstrate the itemizing methods and procedures described in this chapter, a typical radiant floor panel installation is used for a sample quantity take-off and labor estimate.

In estimating any heating system, the first thing that should be done is

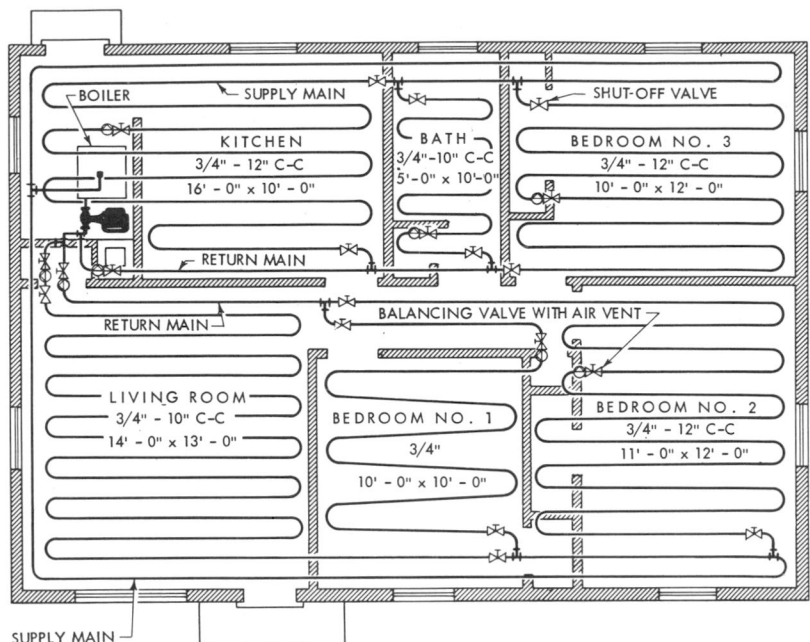

Fig. 19-30. A heating layout for a five-room residence. The layout of the radiant coils and fittings is accurate enough for estimating purposes.

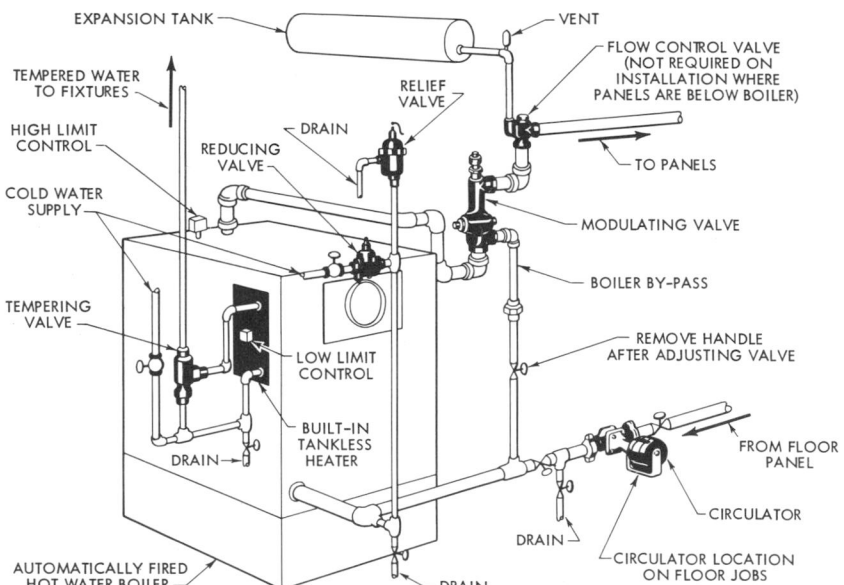

Fig. 19-31. Typical boiler details for the heating layout shown in Fig. 19-30.

to study the plans and specifications, making a note of all equipment on which it is necessary to obtain manufacturer's or distributor's quotations. If this is done at once, you will not have to wait long for prices after you have finished the quantity take-off.

A set of drawings prepared by the architect or a heating engineer is necessary for determining just what is needed for the job in the way of materials. The floor plan and boiler detail shown in Figs. 19-30 and 19-31, respectively, illustrate every feature which will be required in a typical radiant heating system with the exception of the automatic controls. These two illustrations are more helpful than the conventional working drawings.

In addition to the drawings submitted to the heating contractor for his bid, a set of specifications is provided, describing in detail the various components which make up the completed heating system. Use the following specifications for this estimate.

TYPICAL SPECIFICATIONS FOR A RADIANT FLOOR PANEL HEATING SYSTEM

AIR VENTS: Automatic air vents shall be provided where indicated and shall be of the float-operated type. Small, manually operated air vents shall also be placed where indicated.

INSULATION: All exposed heating piping in the boiler room shall be covered with three-ply asbestos air cell pipe covering provided with a pasted canvas jacket and three bands per section. All fittings shall be covered with asbestos cement and enclosed in a pasted canvas jacket.

TESTS: The heating work shall include the testing of all piping and apparatus in the system for leaks, faulty joints, improper operation, etc.

When all heating piping is installed, with the exception of the boiler and its equipment, the system shall be tested to a hydrostatic pressure of 125 lbs. This pressure shall be maintained for at least four hours. This test shall be performed before the floors are poured into place.

Prior to installation of the boiler and its equipment, circulator, and controls, the contractor shall thoroughly flush out the system.

After installation of the boiler and its equipment, the entire system is to be filled with water, and a hydrostatic test pressure built up to equal the maximum operating pressure for which the boiler was designed as a means of checking the complete system for leaks.

Any defective joints or piping shall be immediately replaced and the tests repeated until found to be tight. All tests are to be performed in the presence of the owner or the architect.

Before final acceptance of the work, the contractor shall put the entire system into operation and demonstrate that all parts are in proper working order.

GUARANTEE: The contractor shall guarantee all material and workmanship for a period of one year.

MATERIAL

1 – Gas boiler complete	$ _____
20' 0" – Stainless steel flue liner	
1 – Stainless steel drain fitting, tees, and plastic drip tubing	
1 – Boiler breeching	
15' 0" – 1" gas piping with necessary fittings . .	
1 – $\frac{3}{4}$" circulating pump	
1 – $\frac{1}{4}$" combination reducing and relief valve .	
1 – $\frac{3}{4}$" modulating valve (including cost of automatic control system complete) . . .	
1 – Ten-gallon expansion tank	
20 – $\frac{3}{4}$" gate valves	
6 – $\frac{3}{4}$" balancing valves	
10 – $\frac{3}{4}$" tees	
11 – Elbows	
900' 0" – $\frac{3}{4}$" black iron pipe	
	Total $ _____

LABOR

Set and connect boiler	$ _____
Boiler breeching	
Gas piping	
Circulator	
Supply and return piping	
Forming and installing coils	
	Total $ _____

RECAPITULATION

Material	$ _____
Labor	$ _____
Overhead (15% of total)	$ _____
Profit (10% of new total)	$ _____
Labor insurance (7%) and welfare (3%) (figured on labor cost)	
Labor for testing (6% of total labor cost)	
Labor for cleaning up (2½% of total labor cost)	
Material Handling (10% of total material cost)	_____
	Grand total $

PLACEMENT OF COILS: A gravel fill of at least 6" thickness shall be placed over undisturbed earth or earth which has been rolled or tamped so as to preclude any possibility of settlement. In no instance shall cinders, slag, or other sulphur-bearing fill material be used.

Over the gravel fill shall be placed a moisture barrier of some suitable substance, lapped at the edges at least 6", and carried around to the edge of the slab. The overlapped sections and the surfaces of the foundation wall over which the moisture barrier is laid shall be mopped with pitch or asphaltum. A suitable type (Foamglas, Fiberglas, etc.) of insulation shall be placed over the moisture barrier and shall be thoroughly coated on all sides and edges with asphaltum or pitch before being applied. The coils shall be blocked up in position approximately 1" above the surface of the insulation. The concrete for the floor shall then be poured directly over the heating pipes in accordance with the drawings. In no instance shall the concrete above the top of the piping be less than 3".

Fill and drainage shall be adequate to maintain piping free from surface drainage and subsurface water.

Coils shall be placed in accordance with the plans and shall be laid level throughout.

The contractor shall be responsible for maintaining all heating pipes in their proper positions during the pouring of the concrete floors. This shall include all necessary instruction, supervision, and inspection.

BOILER: The boiler shall be a gas- or oil-fired, cast-iron, hot water boiler capable of delivering 50,000 Btu to the system.

The boiler shall be equipped with an insulating jacket, combination pressure gauge and thermometer, flow and return tappings, etc., to make a complete unit.

CIRCULATING PUMP: The water circulating pump shall be capable of pumping five gallons per minute against a 2.5' head. The pump is to be equipped with flanged connections, vibration dampers, and motor as supplied by the manufacturer.

EXPANSION TANK: The expansion tank shall be an all-steel, airtight tank of ten-gallon capacity. The tank shall have all the necessary tappings as indicated on the drawings.

BOILER ACCESSORIES: The boiler accessories shall include the following which shall be properly connected to the boiler and piping system:

The water filling line shall be provided with a pressure reducing valve.

The pressure relief valve shall be of the diaphragm type.

The boiler shall be provided with a drain valve in the conventional manner.

CONTROLS: The control system shall be the outdoor-indoor type with continuous circulation. (If one particular type is favored, the manufacturer of the control system should be consulted for specification, installation, and operation details.)

VALVES: All gate, globe, and square-head valves should be designed for 125 lb. working pressure.

All balancing valves shall be ____ Company #324 iron body brass mounted cock for $1\frac{1}{4}$" size and ____ V-53 for 1" size, or approved equals, and one shall be installed with each coil as shown on the drawings.

SCOPE OF THE WORK: The work to be done by the contractor consists of the complete installation of a radiant heating system within the building, with the furnishing of all labor and material to provide a complete and satisfactory system. The radiant heating system will use a floor panel with forced hot water as the heating medium. The system shall be installed in accordance with the plans and details shown on sheet ____ ____ of the drawings, and all work shall be done in a neat, thorough, and workmanlike manner.

CODES: All work in connection with the installation of this heating system must be performed in accordance with the applicable requirements of all pertinent building codes

and shall meet the requirements of the fire underwriters.
CUTTING, PATCHING, AND EXCAVATING: All cutting and patching necessary for the installation of the heating system shall be done by this contractor. No cutting of the work of other contractors shall be done without the consent of the architect, and no structural member shall be cut or weakened in any way unless proper arrangement is made to reinforce the portion so weakened. In such matters it will be necessary to receive the permission of the architect before proceeding.

PIPE AND FITTINGS: All pipe for the heating system mains, coils, risers, etc., shall be standard weight, black, wrought-iron pipe in accordance with ASTM Designation A72.

All screwed joint fittings shall be black, cast, or malleable iron, beaded or banded fittings of 125 lbs. rating.

HANGERS AND SUPPORTS: All pipe hangers for the heating system shall be of the split-ring pattern with hanger rods. Hangers and rods shall be of a size and weight to safely support the load.

JOINTS: All joints in the heating system piping shall be welded except those which are immediately accessible in the vicinity of the boiler. The welding shall be done in accordance with the standards of the American Society for Pipe Welds. All ends of pipe shall be properly reamed before welding.

All boiler connections shall be made with screwed joints only.

Joints on all valves and cocks shall be either flanged or screwed.

COIL AND GRID FABRICATION: All coils shall be fabricated in accordance with the shape shown on the plans and shall contain not less than the pipe quantity and pipe sizes shown.

Elbows and return bends may be either shop fabricated or bent on the job to the radius shown on the drawings, but shall be free of crushing or deformation.

All joints and connections in the coils are to be welded, and the welding shall be smooth and sound with particular attention given to maintaining a flush, even interior surface.

Any prefabricated coils shall be adequately braced (temporarily) during transportation to the job to prevent springing, misalignment, etc.

All field welds shall be done in a manner to protect other workmen on the job from glare and injury, and every precaution shall be taken to avoid the possibility of fire.

Finally, there is a schedule of heat losses for each room of the house. See Table III. This table is prepared by the architect or some other qualified, competent designer.

Materials Estimate. Since the boiler is the largest single expense in this estimate, it will be the first item estimated.

Boiler. The specifications and the schedule of heat losses are the basis for the estimate of the boiler.

The total heat loss is found to be

TABLE III
SCHEDULE OF HEAT LOSS FOR A RESIDENCE

Room	Room Size in Feet	Room Area in Square Feet	Heat Loss in Btu's
Living Room..............	14x13	182	9,600
Bedroom No. 1.............	10x10	100	3,500
Bedroom No. 2.............	11x12	132	5,800
Bedroom No. 3.............	10x12	120	5,700
Bath....................	5x10	50	2,100
Kitchen and Utility..........	10x16	160	5,700

32,400 *Btuh* (Btu's per hour) by adding the individual heat losses in Table III. To this is added 10% to compensate for heat loss to the ground under the slab, making a new total of 35,640 Btuh.

In average house heating systems, it is common practice to consider the *piping tax* (the estimated heat emission in Btuh of the piping connecting the radiators and other equipment to the boiler) as equal to approximately 25% of the net load (*radiation load*). However, since the supply and return mains in the example are, in a sense, a part of the panel system, a 10% figure is more reasonable. The piping tax, therefore, is approximately 3,564 Btuh.

The last item involved in selecting the boiler on the basis of its capacity is the allowance made for heating-up or picking-up. The usual allowance is about 20% of the *design load* (radiation load and the piping tax). This amounts to 7,841 Btuh.

The total capacity of the boiler, therefore, must be at least 47,045 Btuh in order to handle the heating load imposed upon it.

The specifications call for a gas- or oil-fired boiler. Since gas is mentioned first, the gas-fired boiler will be used for this estimate.

Flue Lining. The stack temperatures of a gas-fired boiler are lower than those where other fuels are used. Consequently, there is a greater amount of condensation present. The gases given off as a product of combustion combine with water vapor to form corrosive acids. It is therefore necessary to install a flue lining which will resist these acids. While a stainless-steel flue lining is usually used in new construction, vitreous, bellmouth sewer tile also gives a satisfactory installation. On old jobs, however, there is no substitute for the stainless-steel liner.

Assume that stainless steel is specified for the flue liner, and that the chimney height is 20'-0".

Drain Fitting and Tee. A means must be provided for the disposal of the condensation which will accumulate at the bottom of the chimney. A stainless-steel elbow is obtainable which is a combination cleanout and drain. This elbow is attached at the bottom of the stainless-steel tee. The smoke pipe or breeching enters the tee from the side. To the top of the tee is attached the flue lining. The condensation is led off from the stainless-steel drain fitting through a piece of plastic tubing to the nearest drain. A second tee is necessary for the purpose of connecting the domestic hot water heater to the flue.

Boiler Breeching. Assume the boiler to be conveniently located with respect to the chimney so that no problem is encountered in installing the smoke pipe or breeching.

Gas Piping. It is usually customary for the local gas company to bring the gas piping from the street into the meter free of charge. While the meter location in Fig. 19-30 is not shown, you can assume that not more than 15'-0" of 1" pipe is needed. The cost of the fittings is determined usually by picking them off the drawings, but in this case by adding 100% to the total pipe cost.

Circulator. There are a variety of circulating pumps available, all of which will handle the needs of this heating installation.

Relief and Reducing Valves. Quite a few manufacturers produce relief valves and reducing valves. In some cases, these valves are combined as one unit.

Modulating Valve. The modulating valve is a part of the entire automatic control system. As such, it is sold as a complete unit and includes the outdoor anticipating bulb and the room thermostat.

Expansion Tank. The size of the expansion tank can be determined by the total Btu capacity of the installation. The heat loss schedule indicates that approximately 50,000 Btu's are required. If an expansion tank has a capacity of one gallon for every 5,000 Btu's, a tank having a ten gallon capacity is needed for this job.

Valves. A radiant panel heating system requires a greater number of valves than do other types of heating systems. Fig. 19-30 shows that there are 20 gate valves of ¾" size, 8 balancing valves (¾"), 10 tees (¾"), and 11 elbows required for the entire system.

A simpler and quicker, though not as accurate, method of determining the cost of such fittings is to assume they will amount to 100% of the cost of piping.

Coil Piping. The number of linear feet of piping required for each room and the pattern or way in which it should be placed will have been determined by the architect or some other qualified designer. You can determine pipe requirements from these figures. However, if the quantity of coil piping has not been indicated, it is easily determined from the schedule of heat losses, Table III, and from Table IV. For example, the living room has a heat loss of 9,600 Btuh. From Table IV you can see that it will take 200 linear feet of piping to overcome the loss. This table is based on a 120° F average water temperature and a 50 Btuh emission rate per square foot.

The necessary piping is determined for the rest of the rooms in the residence in the same manner. The

TABLE IV
BTU REQUIREMENTS AND EQUIVALENT COIL SIZES

Btu* of Coil	Linear Feet of ¾" Pipe or Tubing Required	Btu* of Coil	Linear Feet of ¾ Pipe or Tubing Required
500	10'	5,500	110'
1,000	20'	6,000	120'
1,500	30'	6,500	130'
2,000	40'	7,000	140'
2,500	50'	7,500	150'
3,000	60'	8,000	160'
3,500	70'	8,500	170'
4,000	80'	9,000	180'
4,500	90'	9,500	190'
5,000	100'	*10,000	200'

* Maximum recommended output per coil is 10,000 Btu because of friction or resistance limitation. If more than 10,000 Btu is required, use two (2) or more coils. For example, if you have a room with a heat loss of 18,000 Btu, use two (2) 9,000 Btu coils, 180' in each.

lengths of the supply and return mains are scaled off the drawings. The total piping requirements are found to total approximately 900 linear feet — 680 linear feet for coil piping, 200 linear feet for supply and return mains.

Material for the gravel fill, moisture barrier, asphalt topping, and slab insulation are not figured as a cost in this particular estimate since they are not actually a part of the heating plant. The amount of acetylene, oxygen, and welding rod used is not sufficient to merit an accurate estimate.

Labor Estimate. It was pointed out that the experienced estimator rarely itemizes each factor of a job as a means of determining its cost in advance. The estimate will more likely be made on the basis of previous installations which were similar. One of the main reasons for this condition is the difficulty in making accurate estimates of the amount of labor required for a particular job. While some of the following material on determining labor costs may seem to the experienced estimator as needlessly detailed, the information is intended only as a temporary substitute for experience.

Boiler. Experienced estimators have found that to set and connect a boiler for a typical, medium-sized residence such as this will take two journeymen fitters one day—about 16 man-hours.

Flue Lining. No additional labor is necessary as a result of using stainless-steel flue lining. A flue lining of some sort should be installed in every chimney. All that is needed, therefore, is a substitution of materials.

Drain Fitting and Tee. No additional labor is figured here. The stainless-steel parts are merely furnished to the mason who proceeds to lay up the chimney around them.

Boiler Breeching. Depending on the location and position of the chimney and boiler, the boiler breeching (smokepipe) may be simple or complicated. An average time for this installation is one-half day for two journeymen fitters. This is about 8 man-hours.

Gas Piping. The labor involved in running the gas piping is determined to some extent by the distance from the meter to the boiler. As a general rule, it is safe to assume that the gas piping can be run and the boiler connected by two journeymen fitters working one day—16 man-hours.

Circulator. To connect the circulator requires the work of two fitters working one-half day—8 man-hours.

Relief and Reducing Valves. The installation of these valves is considered as a portion of the work done when connecting the circulator and setting the boiler. Likewise, the tempering valve, modulating valve, and expansion tank are installed as a part of the boiler labor and circulator labor.

Supply and Return Piping. It is customary to figure that two journeymen fitters require one day to run the supply and return piping—16 man-hours.

Coil Piping. Estimators allow two journeymen fitters four days in which to form, place, and weld the piping for the floor coils in a typical, average-sized residence such as this one. This is about 64 man-hours.

Finishing. Once the material and labor costs have been determined for a job, it is possible to complete the estimate by adding the customary percentages for overhead, profit, labor insurance and welfare, etc. These items are listed in their proper place on the estimating sheets.

See Appendix D, page 479, for Hourly Estimates for:
Insulation: Rigid Board Types; Non-Rigid Types; Reflective Types; Pouring Types; Semi-Rigid Types; Vapor Seals; Moisture Barriers.

Questions and Problems

1. A gas fired hot water system including boiler, thermostat, expansion tank, automatic fill valve, smoke pipe, copper tubing, fittings, and convector baseboards costs $740. Labor at $9.00 an hour runs as follows:

   16 hours for setting the boiler
   16 hours for gas piping
   8 hours for boiler breeching
   8 hours for connecting
      circulator
   16 hours for piping.

   Add 15% of the total cost for overhead. What is the total estimated cost?
   (Answer: $1513.40)

2. 1640 sq. ft. of reflective insulation is to be installed between studs. Allowing 100 sq. ft. per man hour, how much would the labor cost be at $9.00 per hour?
   (Answer: $147.60)

20

RECAPITULATION

~~~~~~

After all the time and effort you put in to compute the quantities of material and the amount of labor, and to procure the sub-contractor's estimates for the trades that must be sublet, comes the finale known as the firm bid. This is the recap or total sum of the cost of all the items that must be included in the building of the project. It is the price which the building contractor must get from the owner in order to pay for all the material, labor, and services which are necessary to complete the building or remodeling job for which you are bidding.

PREPARATION OF THE BID

If the owner is taking bids from numerous contractors and you are the lowest or successful bidder, then you must enter into a contract with the owner to build and complete the building or remodeling job for the amount which you have bid. This contract binds you to furnish all the labor, materials, and subcontracted trades to completely finish the project which you and the owner have agreed to. If you are not to do the job in its entirety, then you state in the contract which items are excluded. It also binds the owner to pay you for the satisfactory completion of the structure according to the terms and price which the contractor and owner agree on.

Therefore, you must be as accurate as possible with your figures because once the contract is signed you are obligated to do the job completely as specified for the sum of money shown in the contract. The most common of all errors in bidding is that of unintentionally leaving out (forgetting) one or more items which are necessary in order to complete the building. Once you have signed the contract with the owner to do the job for a given price, the owner is not interested in whether or not you make a profit or loss. In order to eliminate the chances of losing money on the job, you must have a form which also acts as a check list which you should use when preparing your bid. Fig. 20-1 is a bid form which is the best for this purpose. It also has

GENERAL CONTRACTOR'S RE-CAP SHEET

OWNER: _____ ADDRESS: _____ PHONE: _____

BUILDING TYPE: _____ LOCATION: _____

ARCHITECT OR ENGINEER: _____ ADDRESS: _____ PHONE: _____

PROJECT NO. : _____ STARTED: _____ COMPLETED: _____

	ESTIMATE	ACTUAL COST
Permits: Street Bond: Completion Bond:		
Wrecking:		
Excavating: Backfilling & Grading:		
Concrete Foundations: Concrete Work:		
Waterproofing:		
Masonry:		
Cut Stone: Terra Cotta or Cast Stone:		
Structural Steel: Ornamental Iron Work:		
Steel Sash: Fire Doors:		
Carpentry (Incl. Rough & Finish Hardware):		
Floor Sanding:		
Calking:		
Sheet Metal Work: Steel Medicine Case:		
Roofing:		
Lathing & Plastering or Drywall:		
Glass & Glazing:		
Painting& Decorating:		
Plumbing: Extra Sewer & Water from Main:		
Heating:		
Electric Wiring: Lighting Fixtures:		
Ceramic Tile Work:		
Resilient Floors:		
Metal Weatherstrips:		
Insulation:		
Screens: Storm Windows:		
Shades: Curtain Rods:		
Landscaping:		
Compensation Insurance:		
Liability Insurance:		
Fire Insurance:		
Cleaning Up Bldg., Yard & Washing Windows:		
Hauling Scaffold to and from Job:		
Sales Tax:		
Fuel:		
Interest:		
Taxes:		
Loan Commission:		
Guarantee Policy:		
Title Expense:		
Legal Expense:		
Sales Commission:		

Total Cost: _____
Overhead: _____

Total No. sq. ft.:   Cost per sq. ft.:   Profit: _____

Total No. cu. ft.:   Cost per cu. ft.:   Bid: _____

Profit: Loss:

Fig. 20-1. Checklist for preparing the total estimate.

a column in which you can enter the exact cost of each item after the job has been completed. This will enable you to compare the actual costs with the estimated costs which will guide you in your future estimating. Now, let us break down this estimate sheet item for item.

Permits. For every building construction job that is to be done a per-

CONCRETE WORK				PLASTERING			
Walls				Ext. on Metal Lath			
Floors				Int. on Gypsum Lath			
Walks				Int. on Metal Lath			
Drives				Arches			
Steps				Corner Beads			
Reinforcing				Labor			
Digging Trenches				Liability Insurance			
Labor				Miscellaneous			
Liability Insurance							
Miscellaneous					Total		
				DRYWALL			
	Total						
					Total		
MASONRY							
				ROOFING			
Common Brick							
Face Brick				Tar & Gravel			
Tile				Wood or Comp. Shingle			
Coping				Labor			
Flue Lining				Liability Insurance			
Fireplace				Miscellaneous			
Chimneys							
Setting Cut Stone							
Cleaning & Pointing					Total		
Labor							
Liability Insurance				PAINTING & DECORATING			
Miscellaneous							
				Ext. Openings			
	Total			Ext. Walls			
				Ext. Porches & Stairs			
				Int. Trim ( ) Coats			
ROUGH & FINISH CARPENTRY				Wood Floors--Soft			
				( ) Rms.--Ceilings Tinted			
				( ) Rms.--Walls Tinted			
Lumber				( ) Rms.--Ceilings Painted			
Trusses				( ) Rms.--Walls Painted			
Millwork--Cabinet Work				( ) Rms.--Walls Papered			
Stairs				( ) Rms.--Hard Oil or Sizing			
Rough Hardware				Labor			
Finish Hardware				Liability Insurance			
Labor				Miscellaneous			
Liability Insurance							
Miscellaneous							
	Total				Total		

SUB-CONTRACTOR'S DRAW FROM MORTGAGOR				
Date	No.	Issued to	For	Amount

Fig. 20-1 (cont.). Sub-estimate form for final bid.

mit must be taken out and paid for. This is obtained from your local Building Department.

Street Bond. This is a bond which must be taken out and paid for in order to guarantee repair of the street if it must be broken up for some utility work.

Completion Bond. This is a bond, sometimes demanded by the owner, which the contractor has to pay for, taken out to insure the completion of the building. In case, for some reason, the contractor fails to complete the building as per the contract, then the owner or the bonding company will get somebody else to complete the building.

Wrecking. On some building sites, there may be an old structure which has to be wrecked and removed. The cost of the wrecking must be included, and also the cost of hauling away the debris from the premises.

Excavating. This is the process of digging the hole in the ground for the basement of the building. You must also include the cost of hauling away the excavated material.

Backfilling. When a foundation or footing is built, the trench must be larger than the actual foundation in order for the men to have space to get in and install the forms and remove the forms. Therefore, after the cement is hardened and the forms are removed, this additional space must be filled with earth. This is known as back-filling.

Grading. This is the process of bringing down the high spots and filling in the low spots of earth around the structure.

Concrete Foundations and Concrete Work. Here you enter the cost of the cement labor and material if you do the work yourself or the cost which you get from a subcontractor in case you decide to sublet this work. If you are going to do this work yourself, you will find a sub-estimate form on page 445.

Waterproofing. This is the process of painting the outside of the foundation with a liquid asphalt. This is usually done by the roofing contractor.

Masonry. Here you enter the cost of all the mason work and brickwork required for the completion of the project. If you are going to sublet this work to a mason contractor, you enter his bid in the column, but if you are going to do the masonry work yourself, you will find a sub-estimate form for masonry on page 445.

Cut Stone, Terra Cotta, or Pre-Cast Work. Usually, this material is included in the masonry bid, but if you want to figure any of this material separately you can enter it on this line.

Structural Steel or Ornamental Iron Work. This trade is usually done by a specialist subcontractor who has the equipment for installing it. You get a bid from him and insert it in this line.

Steel Sash—Fire Doors. Here you enter the cost for the material and/or installation if either of these items is required for this building.

Carpentry. This trade is usually handled by the general contractor. On page 445 you will find a sub-estimate form and checklist for this trade.

Floor Sanding. All new hardwood floors must be sanded before finishing. This item is usually handled by the painting contractor. There are also specialists who do this work, and here you enter the cost.

Calking. On a small job, this is done by the carpenter. On a large job, it is done by a specialty contractor. Enter the cost on this line.

Sheet Metal Work. Enter the bid you get from the sheet metal contractor.

Steel Medicine Case. Usually, this is furnished by the owner.

Roofing. On a small job, this is done by the carpenter. On a large job, it is done by a roofing contractor. Enter the cost on this line.

Lathing and Plastering or Drywall. Enter the cost of the bid that you get from the plastering contractor. If the building is to be drywall instead of plaster, enter the cost of the drywall on this line. Sub-estimate forms are shown on page 445.

Glass and Glazing. Nowadays the windows and doors usually come from the lumber yard pre-glazed. However, if there is any additional glass to be used on the job, enter the cost of it on on this line.

Painting and Decorating. Here you enter the price you get from the painting contractor. If you do the painting yourself, you will find a sub-estimate form on page 445.

Plumbing. Here you enter the bid you receive from your plumbing contractor. Be sure he figures the job according to the local plumbing ordinance.

Heating. Here you enter the bid from the heating contractor for the type of system specified.

Electric Wiring. Enter the bid you get from a licensed electrical contractor for the job to be done according to your local building code. It should include the cost of installing the lighting fixtures which are usually furnished by the owner.

Ceramic Tile Work. Enter here the bid you receive from a tiling contractor for the tile floors and walls as specified or the cost if you do it yourself.

Resilient Floors. Usually, this is included in the tile bid. If it is purchased separately, insert the cost on this line.

Metal Weatherstrips. Usually, the windows come weather-stripped, but if additional stripping is required for the windows or doors enter the cost on this line.

Insulation. On a small job, this is done by the carpenter. On a large job, it is done by a specialty contractor. Enter the cost on this line.

Screens, Storm Windows, Shades, Curtain Rods. These items are usually taken care of by the owner. If you are to furnish any of these items, enter the cost on these lines.

Landscaping. Usually, this item is taken care of by the owner. If you are to furnish this, a limit on the amount to be spent must be agreed on with the owner.

Compensation Insurance. Here you enter the cost of the insurance in case of any injuries sustained by your workmen.

Liability Insurance. Enter here the cost of the insurance which protects you in case of any damage or injury to the life and property of others.

Fire Insurance. Include the cost of fire insurance for your material and equipment.

Cleaning up Building and Yard and Washing Windows. Enter here the cost of cleaning up after the job is completed.

Hauling. Enter here the cost of hauling your equipment to and from the job.

Sales Tax. Enter here the cost of any sales tax on materials or services for which you are liable.

Fuel. Enter here the cost of the fuel for temporary heating and for your mechanizing equipment.

Interest. Enter here any interest which you may have to pay providing you are forced to make a loan to help you finance the job until you are paid for it.

Taxes. Enter here the amount you will have to pay for the Social Security and Unemployment Taxes for your employees for the duration of this job.

Loan Commission. This is usually the owner's obligation, but if you are building for yourself, include the amount of commission you paid to the broker or the loan company.

Guarantee Policy and Title Expense. This is usually the owner's obligation, but if you are building this building for yourself include these costs for the total cost of your building.

Legal Expense. If you are to employ a lawyer to make out your contract, include the lawyer's fees.

Sales Commission. If you employ a salesman to sell this job for you, include his commission.

Total Number of Square Feet. Enter here the total of the floor measurements of the building.

Total Number of Cubic Feet. Enter here the total of the cubic feet of all floors of the building.

Cost Per Square Foot. Divide the amount of your bid by the total number of square feet in the building.

Cost Per Cubic Foot. Divide the sum of the total bid by the total number of cubic feet. This will give you a process by which you can make a fast, approximate estimate on future buildings by multiplying the total number of square feet or cubic feet in your proposed building by the cost per square foot or cubic foot.

Total Cost. Here you insert the total of all the costs above this line.

Overhead. This is an extremely important item. It includes compensation for the following:

Office expense

Temporary light and power

Watchmen

Winter protection

Supervisory personnel

Transportation

and other items which are not included in the above figures

In time, your auditor will figure out a regular percentage to add to the total cost for this item.

Profit. One must not forget this extremely important item. In order to stay in business, one must make a profit.

Bid. This is the sum total of all the figures above this line. This is the price with which you go to the owner to negotiate the sale of this job.

Profit—Loss. These spaces are filled in after the job is completed by subtracting the actual cost from the estimate or vice versa.

APPENDIX A

GRADING OF LUMBER AND SOFTWOOD LUMBER STANDARDS

GRADING OF LUMBER

The various defects and blemishes found in lumber necessitate the establishment of certain classification and grading rules. The American Lumber Standards for grading lumber were formulated by the National Bureau of Standards of the United States Department of Commerce. The purpose of setting up such standards was to insure uniform grading throughout the country. Lumber is normally classified in three different ways: by *use,* by method of *manufacture;* and by *size.*

Use Classifications. Use classification is broken down into three principal categories: *structural lumber, factory and shop lumber* and *yard lumber.*

Structural Lumber. Lumber, sometimes termed structural timber, is 5 inches or more in both thickness and width. It is graded according to its strength and to the use which is to be made of an entire piece. Such lumber is used principally for bridge or trestle timbers, for car and ship timbers, for ship decking, and for framing of buildings. (Much of the structural timber used today is formed of glued laminated members. Smaller lumber pieces are glued and laminated together to form larger beams and arches.)

Factory or Shop Lumber. Lumber intended for additional cutting in the process of further manufacturing is known to the trade as factory lumber or shop lumber. Such lumber is used principally in window sashes, doors and door frames, in different types of millwork, and in furniture factories. This lumber is graded on the basis of the percentage of area which will produce a limited number of clear cuttings of a given minimum, or specified, size and quality.

Yard Lumber. The lumber known as yard lumber is less than 5 inches in thickness and is intended for general building purposes.

(a) *Strips.* The yard lumber known to the trade as strips is less than 2 inches in thickness and less than 8 inches in width.

(b) *Boards.* Yard lumber, commonly called boards, is less than 2 inches thick and 8 inches or more in width.

(c) *Dimension Lumber.* Yard lumber, of any width and at least 2 inches but not more than 5 inches thick, when cut to specified sizes, is called *dimension lumber.*

Manufacturing Classifications. Manufacturing classifications are broken down into three categories: *rough lumber, dressed lumber* and *worked lumber.*

Rough Lumber. Lumber that has *not* been dressed (surfaced) but which has been sawed, edged, and trimmed at least to the extent of showing saw marks in the wood on the four longitudinal surfaces of each piece for its entire length.

Dressed (Surfaced) Lumber. Lumber that has been dressed by a planing machine, for purpose of attaining smoothness of surface and uniformity of sizes, on one side, two sides, one edge, two edges, or a combination of sides and edges.

Worked Lumber. Lumber which in

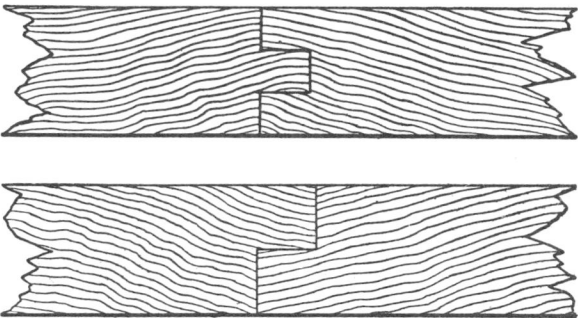

Fig. 1. Types of boards: tongued-and-grooved (top), shiplapped (bottom).

addition to being dressed has been matched, shiplapped, or patterned.

(a) *Matched Lumber*. Lumber that has been worked with a tongue on one edge of each piece and a groove on the opposite edge, to provide a close tongue-and-groove joint by fitting two pieces together; when end-matched, the tongue and groove are worked in the ends also, Fig. 1, *top*.

(b) *Shiplapped Lumber*. Lumber that has been worked or rabbeted on both edges of each piece to provide a close lapped joint by fitting two pieces together, Fig. 1, *bottom*.

The purpose of having the edges of boards shiplapped or matched together is to prevent an open joint between any two boards after they shrink.

(c) *Patterned Lumber*. Lumber that is shaped to a pattern or to a molded form, in addition to being dressed, matched, or shiplapped, or any combination of these workings.

There is a great variety in the shapes and sizes of patterned lumber. The exact dimensions of some of the most common stock lumber patterns are illustrated in Fig. 2. Lumberyard

stock of patterned lumber also includes moldings. See Fig. 3. Moldings and other trim members can be obtained in different shapes and sizes, also in a variety of stock designs. When building specifications call for designs not carried in the stock-lumber patterns, a special job of millwork is required to handle this order. Any specification which makes additional work necessarily increases the cost.

In ordering, the particular pattern and the sizes should be specified. In some cases, such as flooring, prefinished worked lumber is available. Consult manufacturer's catalogs for prefinished lumber specifications.

Size Classifications. Lumber is further classified as to size. There are two basic sizes: *nominal* or *rough size* and *actual* or *dressed size*. Nominal ("not real or actual") refers to the rough size of a board as compared to the finished, actual size. A board may be cut to a nominal size of $2'' \times 4''$ but after finishing and planing and, if green, after drying, the actual size would be much less; for example, $1\frac{1}{2}'' \times 3\frac{1}{2}''$, depending upon the end use

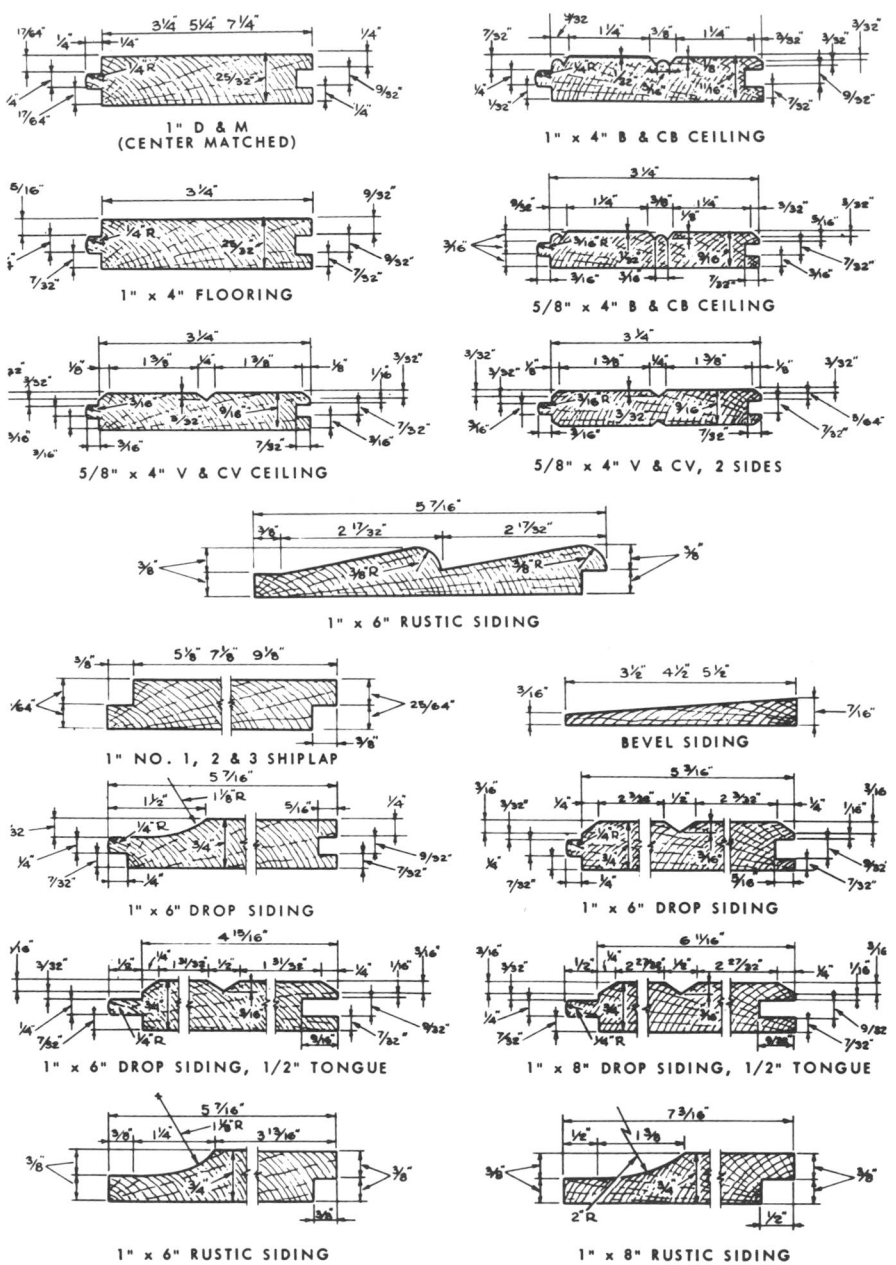

Fig. 2. Typical stock-lumber patterns of dressed and matched flooring, ceiling, and siding (softwood); and lumber patterns of shiplap and other types of siding (softwood). (West Coast Lumbermen's Assoc.)

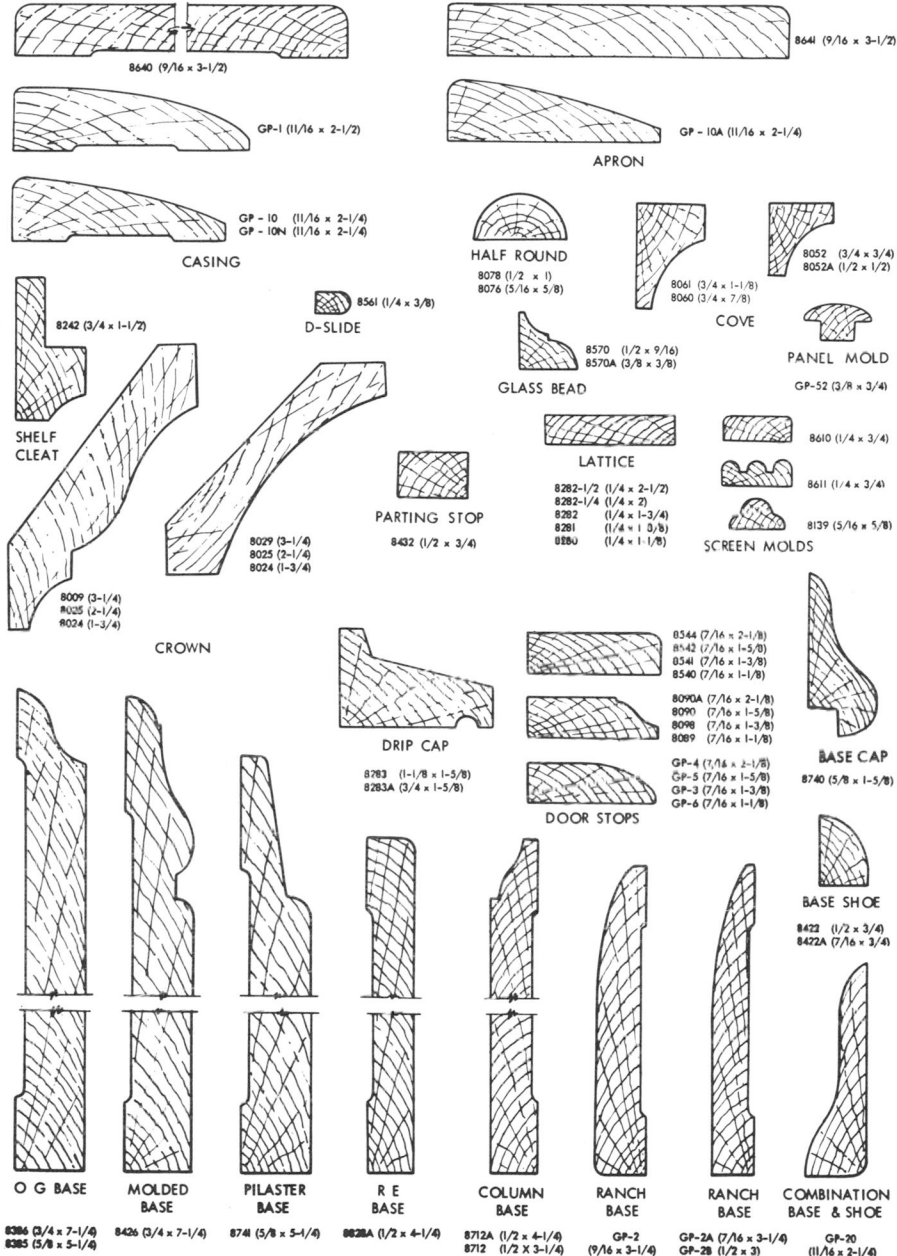

Fig. 3. A typical page from a manufacturer's catalog. Molding patterns are shown. The numbers are the manufacturer's stock numbers. (Georgia-Pacific Corp.)

and the lumber standard used. This does not mean, however, that a nominal size 2″ × 4″ board is originally cut by the mill at exactly 2″ × 4″ dimensions. The cut is slightly less. It is cut so that after finishing and drying the board will have specific predetermined dimensions. The actual size for the various kinds of lumber is established by commercial standards, which the lumber manufacturers agree to follow. Consult the applicable standards (softwood or hardwood) for the sizes of a particular type of wood.

Shingles. Western red cedar, white cedar, redwood, and cypress are woods commonly used for making shingles. The grading varies with the kind of wood used. The Western cedar is graded as: No. 1, No. 2, and No. 3. In cypress the grades include: No. 1, bests, prime, economies, and clippers. In white cedar the grades are: extra star A star, standard star A star, and sound butts. Redwood comes in two grades: No. 1 and No. 2. Shingles of the highest quality are all clear, all heartwood, and all edge grain.

Shingles come in three lengths—16, 18, and 24 inches—and in random widths; or *dimension* widths all cut to the same width. The thickness of shingles is indicated as 4/2, 5/2, 5/2½; that is, 4 shingles to 2 inches of butt thickness; 5 shingles to 2 inches of butt thickness; and 5 shingles to 2½ inches of butt thickness.

Shingles are usually ordered in *squares*. A square of shingles consists of four bundles of shingles: this is called a "square" because it will normally cover 100 square feet (10′

× 10′) of roof area. The amount covered will vary, however, depending on the shingle overlap. The shingle overlap is determined by the pitch of the roof and the length of the shingle.

SOFTWOOD LUMBER STANDARDS

Lumber standards both softwood and hardwood, are sponsored by the U.S. Department of Commerce, National Bureau of Standards, and voted on by the affected lumber manufacturers, lumber mills, home builders, etc. When agreement is reached as to the acceptability of a proposed standard, it is adopted and adhered to by the industry. More than a majority acceptance, however, is needed before a lumber standard is adopted. The Department of Commerce requires that a *consensus* within the industry must be reached.

The Department of Commerce, in proposing new lumber standards, follows the recommendation of the American Lumber Standards Committee. This committee is composed of representatives from the lumber industry.

Recently a new standard for softwood lumber was adopted, (PS20-70). The major effect of the new softwood lumber standard is to reduce the *actual* sizes of lumber. For example, a nominal 2-inch dry dimension would have an actual size of 1½ inch in the new standard, as opposed to 1⅝ inch in the old standard. The new softwood standard also establishes the moisture content of *dry lumber* at a maximum of 19 percent.

The following tables are based on the new American Softwood Lumber Standard, PS20-70.

TABLE I. NOMINAL AND MINIMUM DRESSED DRY SIZES OF FINISH, FLOORING, CEILING, PARTITION, AND STEPPING AT 19 PERCENT MAXIMUM MOISTURE CONTENT.
(The thicknesses apply to all widths and all widths to all thicknesses except as modified.)

Item	Thicknesses		Face Widths	
	Nominal [1] Inches	Minimum Dressed Inches	Nominal Inches	Minimum Dressed Inches
Finish	3/8	5/16	2	1-1/2
	1/2	7/16	3	2-1/2
	5/8	9/16	4	3-1/2
	3/4	5/8	5	4-1/2
	1	3/4	6	5-1/2
	1-1/4	1	7	6-1/2
	1-1/2	1-1/4	8	7-1/4
	1-3/4	1-3/8	9	8-1/4
	2	1-1/2	10	9-1/4
	2-1/2	2	11	10-1/4
	3	2-1/2	12	11-1/4
	3-1/2	3	14	13-1/4
	4	3-1/2	16	15-1/4
Flooring [2]	3/8	5/16	2	1-1/8
	1/2	7/16	3	2-1/8
	5/8	9/16	4	3-1/8
	1	3/4	5	4-1/8
	1-1/4	1	6	5-1/8
	1-1/2	1-1/4		
Ceiling [2]	3/8	5/16	3	2-1/8
	1/2	7/16	4	3-1/8
	5/8	9/16	5	4-1/8
	3/4	11/16	6	5-1/8
Partition [2]	1	23/32	3	2-1/8
			4	3-1/8
			5	4-1/8
			6	5-1/8
Stepping [2]	1	3/4	8	7-1/4
	1-1/4	1	10	9-1/4
	1-1/2	1-1/4	12	11-1/4
	2	1-1/2		

[1] For nominal thicknesses under 1 inch, the board measure count is based on the nominal surface dimensions (width by length). With the exception of nominal thicknesses under 1 inch, the nominal thicknesses and widths in this table are the same as the board measure or count sizes.

[2] In tongued-and-grooved flooring and in tongued-and-grooved and shiplapped ceiling of 5/16 inch, 7/16 inch, and 9/16 inch dressed thicknesses, the tongue or lap shall be 3/16 inch wide, with the overall widths 3/16 inch wider than the face widths shown in the table above. In all other worked lumber of dressed thicknesses of 5/8 inch to 1-1/4 inch, the tongue shall be 1/4 inch wide or wider in tongued-and-grooved lumber, and the lap 3/8 inch wide or wider in shiplapped lumber, and the overall widths shall be not less than the dressed face widths shown in the above table plus the width of the tongue or lap.

TABLE II. NOMINAL AND MINIMUM DRESSED DRY SIZES OF SIDING AT 19 PERCENT MAXIMUM MOISTURE CONTENT.

(The thicknesses apply to all widths and all widths to all thicknesses.)

Item	Thicknesses		Face Widths	
	Nominal [1] Inches	Minimum Dressed Inches	Nominal Inches	Minimum Dressed Inches
Bevel Siding	1/2 9/16 5/8 3/4 1	7/16 butt, 3/16 tip 15/32 butt, 3/16 tip 9/16 butt, 3/16 tip 11/16 butt, 3/16 tip 3/4 butt, 3/16 tip	4 5 6 8 10 12	3-1/2 4-1/2 5-1/2 7-1/4 9-1/4 11-1/4
Bungalow Siding	3/4	11/16 butt, 3/16 tip	8 10 12	7-1/4 9-1/4 11-1/4
Rustic and Drop Siding (shiplapped, 3/8-in. lap)	5/8 1	9/16 23/32	4 5 6	3 4 5
Rustic and Drop Siding (shiplapped, 1/2-in. lap)	5/8 1	9/16 23/32	4 5 6 8 10 12	2-7/8 3-7/8 4-7/8 6-5/8 8-5/8 10-5/8
Rustic and Drop Siding (dressed and matched)	5/8 1	9/16 23/32	4 5 6 8 10	3-1/8 4-1/8 5-1/8 6-7/8 8-7/8

[1] For nominal thicknesses under 1 inch, the board measure count is based on the nominal surface dimensions (width by length). With the exception of nominal thicknesses under 1 inch, the nominal thicknesses and widths in this table are the same as the board measure or count sizes.

TABLE III.   NOMINAL AND MINIMUM DRESSED SIZES OF BOARDS, DIMENSION, AND TIMBERS.

(The thicknesses apply to all widths and all widths to all thicknesses.)

ITEM	Thicknesses			Face Widths		
	Nominal	Minimum Dressed		Nominal	Minimum Dressed	
		Dry[1]	Green[1]		Dry[1]	Green[1]
Boards[2]	1	3/4	25/32	2	1-1/2	1-9/16
				3	2-1/2	2-9/16
				4	3-1/2	3-9/16
				5	4-1/2	4-5/8
	1-1/4	1	1-1/32	6	5-1/2	5-5/8
				7	6-1/2	6-5/8
				8	7-1/4	7-1/2
	1-1/2	1-1/4	1-9/32	9	8-1/4	8-1/2
				10	9-1/4	9-1/2
				11	10-1/4	10-1/2
				12	11-1/4	11-1/2
				14	13-1/4	13-1/2
				16	15-1/4	15-1/2
Dimension	2	1-1/2	1-9/16	2	1-1/2	1-9/16
	2-1/2	2	2-1/16	3	2-1/2	2-9/16
	3	2-1/2	2-9/16	4	3-1/2	3-9/16
	3-1/2	3	3-1/16	5	4-1/2	4-5/8
				6	5-1/2	5-5/8
				8	7-1/4	7-1/2
				10	9-1/4	9-1/2
				12	11-1/4	11-1/2
				14	13-1/4	13-1/2
				16	15-1/4	15-1/2
Dimension	4	3-1/2	3-9/16	2	1-1/2	1-9/16
	4-1/2	4	4-1/16	3	2-1/2	2-9/16
				4	3-1/2	3-9/16
				5	4-1/2	4-5/8
				6	5-1/2	5-5/8
				8	7-1/4	7-1/2
				10	9-1/4	9-1/2
				12	11-1/4	11-1/2
				14		13-1/2
				16		15-1/2
Timbers	5 and thicker	1/2 off		5 and thicker	1/2 off	

1/
  Dry Lumber. For the purposes of this standard, dry lumber is defined as lumber which has been
  seasoned or dried to a moisture content of 19 percent or less.

  Green Lumber. For the purpose of this standard, green lumber is defined as lumber having a
  moisture content in excess of 19 percent.

2/
  Boards less than the minimum thickness for 1 inch nominal but 5/8 inch or greater thickness dry (11/16 inch green) may be regarded as American Standard Lumber, but such
  boards shall be marked to show the size and condition of seasoning at the time of dressing.
  They shall also be distinguished from 1 inch boards on invoices and certificates.

TABLE IV. NOMINAL AND MINIMUM DRESSED SIZES OF (2-inch and under) SHIPLAP, CENTERMATCH AND D & M.

(The thicknesses apply to all widths and all widths to all thicknesses)

Item	Thicknesses			Face Widths		
	Nominal Inches	Minimum Dressed		Nominal Inches	Minimum Dressed	
		Dry[1] Inches	Green[1] Inches		Dry[1] Inches	Green[1] Inches
Shiplap . . . . . 3/8-inch lap	1	3/4	25/32	4	3-1/8	3-3/16
				6	5-1/8	5-1/4
				8	6-7/8	7-1/8
				10	8-7/8	9-1/8
				12	10-7/8	11-1/8
				14	12-7/8	13-1/8
				16	14-7/8	15-1/8
Shiplap . . . . . 1/2-inch lap	1	3/4	25/32	4	3	3-1/16
				6	5	5-1/8
				8	6-3/4	7
				10	8-3/4	9
				12	10-3/4	11
				14	12-3/4	13
				16	14-3/4	15
Centermatch . . . 1/4 -inch tongue	1 1-1/4 1-1/2	3/4 1 1-1/4	25/32 1-1/32 1-9/32	4	3-1/8	3-3/16
				5	4-1/8	4-1/4
				6	5-1/8	5-1/4
				8	6-7/8	7-1/8
				10	8-7/8	9-1/8
				12	10-7/8	11-1/8
2" D & M 3/8-inch tongue	2	1-1/2	1-9/16	4	3	3-1/16
				6	5	5-1/8
				8	6-3/4	7
				10	8-3/4	9
				12	10-3/4	11
2" Shiplap 1/2-inch lap	2	1-1/2	1-9/16	4	3	3-1/16
				6	5	5-1/8
				8	6-3/4	7
				10	8-3/4	9
				12	10-3/4	11

[1]/

Dry Lumber. For the purposes of this standard, dry lumber is defined as lumber which has been seasoned or dried to a moisture content of 19 percent or less.

Green Lumber. For the purpose of this standard, green lumber is defined as lumber having a moisture content in excess of 19 percent.

TABLE V. WORKED LUMBER, SUCH AS FACTORY FLOORING, HEAVY ROOFING DECKING, AND SHEET PILING.

(The thicknesses apply to all widths and all widths to all thicknesses.)
See "NOTE"

THICKNESSES[1]			FACE WIDTHS		
NOMINAL	MINIMUM DRESSED		NOMINAL	MINIMUM DRESSED	
	Inches			Inches	
TONGUE AND GROOVED					
	Dry	Green		Dry	Green
2-1/2	2	2-1/16	4	3	3-1/16
3	2-1/2	2-9/16	6	5	5-1/8
3-1/2	3	3-1/16	8	6-3/4	7
4	3-1/2	3-9/16	10	8-3/4	9
4-1/2	4	4-1/16	12	10-3/4	11
SHIPLAP					
	Dry	Green		Dry	Green
2-1/2	2	2-1/16	4	3	3-1/16
3	2-1/2	2-9/16	6	5	5-1/8
3-1/2	3	3-1/16	8	6-3/4	7
4	3-1/2	3-9/16	10	8-3/4	9
4-1/2	4	4-1/16	12	10-3/4	11
GROOVED-FOR-SPLINES					
	Dry	Green		Dry	Green
2-1/2	2	2-1/16	4	3-1/2	3-9/16
3	2-1/2	2-9/16	6	5-1/2	5-5/8
3-1/2	3	3-1/16	8	7-1/4	7-1/2
4	3-1/2	3-9/16	10	9-1/4	9-1/2
4-1/2	4	4-1/16	12	11-1/4	11-1/2

NOTE: In worked lumber of nominal thicknesses of 2 inch and over, the tongue shall be 3/8 inch wide in tongued-and-grooved lumber and the lap 1/2 inch wide in shiplapped lumber, with the overall widths 3/8 inch and 1/2 inch wider, respectively, than the face widths shown in the above table. Double tongued-and-grooved decking may be manufactured with a 3/16-inch tongue.

[1] See Table 3 for information on 2 inch dimension.

APPENDIX B

METRIC-TO-ENGLISH AND ENGLISH-TO-METRIC CONVERSION

METRIC-TO-ENGLISH AND ENGLISH-TO-METRIC CONVERSIONS

Rapid expansion of trade and industry on an international basis in the past two decades has increased the need for understanding of both the *metric or* CGS (Centimeter-Gram-Second) system used by nearly all countries of the world and the *English* or FPS (Foot-Pound-Second) system used by the United States and some other English-speaking countries.

If the co-existence of two systems seems inconvenient, as it is, remember that in respect to worldwide agreement we are the exception. In view of the increasing need for a universal system to measure lengths, areas, volumes, weights, temperatures, etc., it now seems likely that the CGS system will ultimately replace the FPS system despite immense costs and problems that will be involved in making the change-over.

Table I lists factors for converting units from metric to English, while Table II lists factors for converting from English to metric units.

To convert a quantity from *metric* to *English* units:

1. Multiply by the factor shown in Table I.
2. Use the resulting quantity "rounded off" to the number of decimal digits needed for practical application.
3. Wherever practical in semi-precision measurements, convert the decimal part of the number to the nearest common fraction.

TABLE I. CONVERSION OF METRIC TO ENGLISH UNITS

LENGTHS:		WEIGHTS:	
1 MILLIMETER (mm)	= 0.03937 IN.	1 GRAM (g)	= 0.03527 OZ (AVDP)
1 CENTIMETER (cm)	= 0.3937 IN.	1 KILOGRAM (kg)	= 2.205 LBS
1 METER (m)	= 3.281 FT OR 1.0937 YDS	1 METRIC TON	= 2205 LBS
1 KILOMETER (km)	= 0.6214 MILES	LIQUID MEASUREMENTS:	
AREAS:		1 CU CENTIMETER (cc)	= 0.06102 CU IN.
1 SQ MILLIMETER	= 0.00155 SQ IN.	1 LITER (= 1000 cc)	= 1.057 QUARTS OR 2.113 PINTS OR 61.02 CU INS.
1 SQ CENTIMETER	= 0.155 SQ IN.	POWER MEASUREMENTS:	
1 SQ METER	= 10.76 SQ FT OR 1.196 SQ YDS	1 KILOWATT (kw)	= 1.341 HORSEPOWER
VOLUMES:		TEMPERATURE MEASUREMENTS:	
1 CU CENTIMETER	= 0.06102 CU IN.	TO CONVERT DEGREES CENTIGRADE TO DEGREES FAHRENHEIT, USE THE FOLLOWING FORMULA: DEG F = (DEG C X 9/5) + 32	
1 CU METER	= 35.31 CU FT OR 1.308 CU YDS		

SOME IMPORTANT FEATURES OF THE CGS SYSTEM ARE:

1 cc OF PURE WATER = 1 GRAM. PURE WATER FREEZES AT 0 DEGREES C AND BOILS AT 100 DEGREES C.

TABLE II. CONVERSION OF ENGLISH TO METRIC UNITS

LENGTHS:		WEIGHTS:	
1 INCH	= 2.540 CENTIMETERS	1 OUNCE (AVDP)	= 28.35 GRAMS
1 FOOT	= 30.48 CENTIMETERS	1 POUND	= 453.6 GRAMS OR 0.4536 KILOGRAM
1 YARD	= 91.44 CENTIMETERS OR 0.9144 METERS	1 (SHORT) TON	= 907.2 KILOGRAMS
1 MILE	= 1.609 KILOMETERS	LIQUID MEASUREMENTS:	
AREAS:		1 (FLUID) OUNCE	= 0.02957 LITER OR 28.35 GRAMS
1 SQ IN.	= 6.452 SQ CENTIMETERS	1 PINT	= 473.2 CU CENTIMETERS
1 SQ FT	= 929.0 SQ CENTIMETERS OR 0.0929 SQ METER	1 QUART	= 0.9463 LITER
1 SQ YD	= 0.8361 SQ METER	1 (US) GALLON	= 3785 CU CENTIMETERS OR 3.785 LITERS
VOLUMES:		POWER MEASUREMENTS:	
1 CU IN.	= 16.39 CU CENTIMETERS	1 HORSEPOWER	= 0.7457 KILOWATT
1 CU FT	= 0.02832 CU METER	TEMPERATURE MEASUREMENTS:	
1 CU YD	= 0.7646 CU METER	TO CONVERT DEGREES FAHRENHEIT TO DEGREES CENTIGRADE USE THE FOLLOWING FORMULA: DEG C = 5/9 (DEG F − 32)	

To convert a quantity from *English* to *metric* units:

1. If the English measurement is expressed in fractional form, change this to an equivalent decimal form.
2. Multiply this quantity by the factor shown in Table II.
3. Round off the result to the precision required.

Relatively small measurements, such as 17.3 cm, are generallly expressed in equivalent millimeter form. In this example the measurement would be read as 173 mm.

APPENDIX C

FUNCTIONS OF NUMBERS

Functions of Numbers

| 1 to 59 | | | | | 60 to 119 | | | | |

No.	Square	Cube	Square Root	Cubic Root	No.	Square	Cube	Square Root	Cubic Root
1	1	1	1.0000	1.0000	60	3600	216000	7.7460	3.9149
2	4	8	1.4142	1.2599	61	3721	226981	7.8102	3.9365
3	9	27	1.7321	1.4422	62	3844	238328	7.8740	3.9579
4	16	64	2.0000	1.5874	63	3969	250047	7.9373	3.9791
5	25	125	2.2361	1.7100	64	4096	262144	8.0000	4.0000
6	36	216	2.4495	1.8171	65	4225	274625	8.0623	4.0207
7	49	343	2.6458	1.9129	66	4356	287496	8.1240	4.0412
8	64	512	2.8284	2.0000	67	4489	300763	8.1854	4.0615
9	81	729	3.0000	2.0801	68	4624	314432	8.2462	4.0817
10	100	1000	3.1623	2.1544	69	4761	328509	8.3066	4.1016
11	121	1331	3.3166	2.2240	70	4900	343000	8.3666	4.1213
12	144	1728	3.4641	2.2894	71	5041	357911	8.4261	4.1408
13	169	2197	3.6056	2.3513	72	5184	373248	8.4853	4.1602
14	196	2744	3.7417	2.4101	73	5329	389017	8.5440	4.1793
15	225	3375	3.8730	2.4662	74	5476	405224	8.6023	4.1983
16	256	4096	4.0000	2.5198	75	5625	421875	8.6603	4.2172
17	289	4913	4.1231	2.5713	76	5776	438976	8.7178	4.2358
18	324	5832	4.2426	2.6207	77	5929	456533	8.7750	4.2543
19	361	6859	4.3589	2.6684	78	6084	474552	8.8318	4.2727
20	400	8000	4.4721	2.7144	79	6241	493039	8.8882	4.2908
21	441	9261	4.5826	2.7589	80	6400	512000	8.9443	4.3089
22	484	10648	4.6904	2.8020	81	6561	531441	9.0000	4.3267
23	529	12167	4.7958	2.8439	82	6724	551368	9.0554	4.3445
24	576	13824	4.8990	2.8845	83	6889	571787	9.1104	4.3621
25	625	15625	5.0000	2.9240	84	7056	592704	9.1652	4.3795
26	676	17576	5.0990	2.9625	85	7225	614125	9.2195	4.3968
27	729	19683	5.1962	3.0000	86	7396	636056	9.2736	4.4140
28	784	21952	5.2915	3.0366	87	7569	658503	9.3274	4.4310
29	841	24389	5.3852	3.0723	88	7744	681472	9.3808	4.4480
30	900	27000	5.4772	3.1072	89	7921	704969	9.4340	4.4647
31	961	29791	5.5678	3.1414	90	8100	729000	9.4868	4.4814
32	1024	32768	5.6569	3.1748	91	8281	753571	9.5394	4.4979
33	1089	35937	5.7446	3.2075	92	8464	778688	9.5917	4.5144
34	1156	39304	5.8310	3.2396	93	8649	804357	9.6437	4.5307
35	1225	42875	5.9161	3.2711	94	8836	830584	9.6954	4.5468
36	1296	46656	6.0000	3.3019	95	9025	857375	9.7468	4.5629
37	1369	50653	6.0828	3.3322	96	9216	884736	9.7980	4.5789
38	1444	54872	6.1644	3.3620	97	9409	912673	9.8489	4.5947
39	1521	59319	6.2450	3.3912	98	9604	941192	9.8995	4.6104
40	1600	64000	6.3246	3.4200	99	9801	970299	9.9499	4.6261
41	1681	68921	6.4031	3.4482	100	10000	1000000	10.0000	4.6416
42	1764	74088	6.4807	3.4760	101	10201	1030301	10.0499	4.6570
43	1849	79507	6.5574	3.5034	102	10404	1061208	10.0995	4.6723
44	1936	85184	6.6332	3.5303	103	10609	1092727	10.1489	4.6875
45	2025	91125	6.7082	3.5569	104	10816	1124864	10.1980	4.7027
46	2116	97336	6.7823	3.5830	105	11025	1157625	10.2470	4.7177
47	2209	103823	6.8557	3.6088	106	11236	1191016	10.2956	4.7326
48	2304	110592	6.9282	3.6342	107	11449	1225043	10.3441	4.7475
49	2401	117649	7.0000	3.6593	108	11664	1259712	10.3923	4.7622
50	2500	125000	7.0711	3.6840	109	11881	1295029	10.4403	4.7769
51	2601	132651	7.1414	3.7084	110	12100	1331000	10.4881	4.7914
52	2704	140608	7.2111	3.7325	111	12321	1367631	10.5357	4.8059
53	2809	148877	7.2801	3.7563	112	12544	1404928	10.5830	4.8203
54	2916	157464	7.3485	3.7798	113	12769	1442897	10.6301	4.8346
55	3025	166375	7.4162	3.8030	114	12996	1481544	10.6771	4.8488
56	3136	175616	7.4833	3.8259	115	13225	1520875	10.7238	4.8629
57	3249	185193	7.5498	3.8485	116	13456	1560896	10.7703	4.8770
58	3364	195112	7.6158	3.8709	117	13689	1601613	10.8167	4.8910
59	3481	205379	7.6811	3.8930	118	13924	1643032	10.8628	4.9049
					119	14161	1685159	10.9087	4.9187

Table from *How to Estimate for the Bldg. Trades*

Functions of Numbers (continued)

120 to 179

No.	Square	Cube	Square Root	Cubic Root
120	14400	1728000	10.9545	4.9324
121	14641	1771561	11.0000	4.9461
122	14884	1815848	11.0454	4.9597
123	15129	1860867	11.0905	4.9732
124	15376	1906624	11.1355	4.9866
125	15625	1953125	11.1803	5.0000
126	15876	2000376	11.2250	5.0133
127	16129	2048383	11.2694	5.0265
128	16384	2097152	11.3137	5.0397
129	16641	2146689	11.3578	5.0528
130	16900	2197000	11.4018	5.0658
131	17161	2248091	11.4455	5.0788
132	17424	2299968	11.4891	5.0916
133	17689	2352637	11.5326	5.1045
134	17956	2406104	11.5758	5.1172
135	18225	2460375	11.6190	5.1299
136	18496	2515456	11.6619	5.1426
137	18769	2571353	11.7047	5.1551
138	19044	2628072	11.7473	5.1676
139	19321	2685619	11.7898	5.1801
140	19600	2744000	11.8322	5.1925
141	19881	2803221	11.8743	5.2048
142	20164	2863288	11.9164	5.2171
143	20449	2924207	11.9583	5.2293
144	20736	2985984	12.0000	5.2415
145	21025	3048625	12.0416	5.2536
146	21316	3112136	12.0830	5.2656
147	21609	3176523	12.1244	5.2776
148	21904	3241792	12.1655	5.2896
149	22201	3307949	12.2066	5.3015
150	22500	3375000	12.2474	5.3133
151	22801	3442951	12.2882	5.3251
152	23104	3511808	12.3288	5.3368
153	23409	3581577	12.3693	5.3485
154	23716	3652264	12.4097	5.3601
155	24025	3723875	12.4499	5.3717
156	24336	3796416	12.4900	5.3832
157	24649	3869893	12.5300	5.3947
158	24964	3944312	12.5698	5.4061
159	25281	4019679	12.6095	5.4175
160	25600	4096000	12.6491	5.4288
161	25921	4173281	12.6886	5.4401
162	26244	4251528	12.7279	5.4514
163	26569	4330747	12.7671	5.4626
164	26896	4410944	12.8062	5.4737
165	27225	4492125	12.8452	5.4848
166	27556	4574296	12.8841	5.4959
167	27889	4657463	12.9228	5.5069
168	28224	4741632	12.9615	5.5178
169	28561	4826809	13.0000	5.5288
170	28900	4913000	13.0384	5.5397
171	29241	5000211	13.0767	5.5505
172	29584	5088448	13.1149	5.5613
173	29929	5177717	13.1529	5.5721
174	30276	5268024	13.1909	5.5828
175	30625	5359375	13.2288	5.5934
176	30976	5451776	13.2665	5.6041
177	31329	5545233	13.3041	5.6147
178	31684	5639752	13.3417	5.6252
179	32041	5735339	13.3791	5.6357

180 to 239

No.	Square	Cube	Square Root	Cubic Root
180	32400	5832000	13.4164	5.6462
181	32761	5929741	13.4536	5.6567
182	33124	6028568	13.4907	5.6671
183	33489	6128487	13.5277	5.6774
184	33856	6229504	13.5647	5.6877
185	34225	6331625	13.6015	5.6980
186	34596	6434856	13.6382	5.7083
187	34969	6539203	13.6748	5.7185
188	35344	6644672	13.7113	5.7287
189	35721	6751269	13.7477	5.7388
190	36100	6859000	13.7840	5.7489
191	36481	6967871	13.8203	5.7590
192	36864	7077888	13.8564	5.7690
193	37249	7189057	13.8924	5.7790
194	37636	7301384	13.9284	5.7890
195	38025	7414875	13.9642	5.7989
196	38416	7529536	14.0000	5.8088
197	38809	7645373	14.0357	5.8186
198	39204	7762392	14.0712	5.8285
199	39601	7880599	14.1067	5.8383
200	40000	8000000	14.1421	5.8480
201	40401	8120601	14.1774	5.8578
202	40804	8242408	14.2127	5.8675
203	41209	8365427	14.2478	5.8771
204	41616	8489664	14.2829	5.8868
205	42025	8615125	14.3178	5.8964
206	42436	8741816	14.3527	5.9059
207	42849	8869743	14.3875	5.9155
208	43264	8998912	14.4222	5.9250
209	43681	9129329	14.4568	5.9345
210	44100	9261000	14.4914	5.9439
211	44521	9393931	14.5258	5.9533
212	44944	9528128	14.5602	5.9627
213	45369	9663597	14.5945	5.9721
214	45796	9800344	14.6287	5.9814
215	46225	9938375	14.6629	5.9907
216	46656	10077696	14.6969	6.0000
217	47089	10218313	14.7309	6.0092
218	47524	10360232	14.7648	6.0185
219	47961	10503459	14.7986	6.0277
220	48400	10648000	14.8324	6.0368
221	48841	10793861	14.8661	6.0459
222	49284	10941048	14.8997	6.0550
223	49729	11089567	14.9332	6.0641
224	50176	11239424	14.9666	6.0732
225	50625	11390625	15.0000	6.0822
226	51076	11543176	15.0333	6.0912
227	51529	11697083	15.0665	6.1002
228	51984	11852352	15.0997	6.1091
229	52441	12008989	15.1327	6.1180
230	52900	12167000	15.1658	6.1269
231	53361	12326391	15.1987	6.1358
232	53824	12487168	15.2315	6.1446
233	54289	12649337	15.2643	6.1534
234	54756	12812904	15.2971	6.1622
235	55225	12977875	15.3297	6.1710
236	55696	13144256	15.3623	6.1797
237	56169	13312053	15.3948	6.1885
238	56644	13481272	15.4272	6.1972
239	57121	13651919	15.4596	6.2058

FUNCTIONS OF NUMBERS

Functions of Numbers (continued)

240 to 299   **300 to 359**

No.	Square	Cube	Square Root	Cubic Root	No.	Square	Cube	Square Root	Cubic Root
240	57600	13824000	15.4919	6.2145	300	90000	27000000	17.3205	6.6943
241	58081	13997521	15.5242	6.2231	301	90601	27270901	17.3494	6.7018
242	58564	14172488	15.5563	6.2317	302	91204	27543608	17.3781	6.7092
243	59049	14348907	15.5885	6.2402	303	91809	27818127	17.4069	6.7166
244	59536	14526784	15.6205	6.2488	304	92416	28094464	17.4356	6.7240
245	60025	14706125	15.6525	6.2573	305	93025	28372625	17.4642	6.7313
246	60516	14886936	15.6844	6.2658	306	93636	28652616	17.4929	6.7387
247	61009	15069223	15.7162	6.2743	307	94249	28934443	17.5214	6.7460
248	61504	15252992	15.7480	6.2828	308	94864	29218112	17.5499	6.7533
249	62001	15438249	15.7797	6.2912	309	95481	29503629	17.5784	6.7606
250	62500	15625000	15.8114	6.2996	310	96100	29791000	17.6068	6.7679
251	63001	15813251	15.8430	6.3080	311	96721	30080231	17.6352	6.7752
252	36504	16003008	15.8745	6.3164	312	97344	30371328	17.6635	6.7824
253	64009	16194277	15.9060	6.3247	313	97969	30664297	17.6918	6.7897
254	64516	16387064	15.9374	6.3330	314	98596	30959144	17.7200	6.7969
255	65025	16581375	15.9687	6.3413	315	99225	31255875	17.7482	6.8041
256	65536	16777216	16.0000	6.3496	316	99856	31554496	17.7764	6.8113
257	66049	16974593	16.0312	6.3579	317	100489	31855013	17.8045	6.8185
258	66564	17173512	16.0624	6.3661	318	101124	32157432	17.8326	6.8256
259	67081	17373979	16.0935	6.3743	319	101761	32461759	17.8606	6.8328
260	67600	17576000	16.1245	6.3825	320	102400	32768000	17.8885	6.8399
261	68121	17779581	16.1555	6.3907	321	103041	33076161	17.9165	6.8470
262	68644	17984728	16.1864	6.3988	322	103684	33386248	17.9444	6.8541
263	69169	18191447	16.2173	6.4070	323	104329	33698267	17.9722	6.8612
264	69696	18399744	16.2481	6.4151	324	104976	34012224	18.0000	6.8683
265	70225	18609625	16.2788	6.4232	325	105625	34328125	18.0278	6.8753
266	70756	18821096	16.3095	6.4312	326	106276	34645976	18.0555	6.8824
267	71289	19034163	16.3401	6.4393	327	106929	34965783	18.0831	6.8894
268	71824	19248832	16.3707	6.4473	328	107584	35287552	18.1108	6.8964
269	72361	19465109	16.4012	6.4553	329	108241	35611289	18.1384	6.9034
270	72900	19683000	16.4317	6.4633	330	108900	35937000	18.1659	6.9104
271	73441	19902511	16.4621	6.4713	331	109561	36264691	18.1934	6.9174
272	73984	20123648	16.4924	6.4792	332	110224	36594368	18.2209	6.9244
273	74529	20346417	16.5227	6.4872	333	110889	36926037	18.2483	6.9313
274	75076	20570824	16.5529	6.4951	334	111556	37259704	18.2757	6.9382
275	75625	20796875	16.5831	6.5030	335	112225	37595375	18.3030	6.9451
276	76176	21024576	16.6132	6.5108	336	112896	37933056	18.3303	6.9521
277	76729	21253933	16.6433	6.5187	337	113569	38272753	18.3576	6.9589
278	77284	21484952	16.6733	6.5265	338	114244	38614472	18.3848	6.9658
279	77841	21717639	16.7033	6.5343	339	114921	38958219	18.4120	6.9727
280	78400	21952000	16.7332	6.5421	340	115600	39304000	18.4391	6.9795
281	78961	22188041	16.7631	6.5499	341	116281	39651821	18.4662	6.9864
282	79524	22425768	16.7929	6.5577	342	116964	40001688	18.4932	6.9932
283	80089	22665187	16.8226	6.5654	343	117649	40353607	18.5203	7.0000
284	80656	22906304	16.8523	6.5731	344	118336	40707584	18.5472	7.0068
285	81225	23149125	16.8819	6.5808	345	119025	41063625	18.5742	7.0136
286	81796	23393656	16.9115	6.5885	346	119716	41421736	18.6011	7.0203
287	82369	23639903	16.9411	6.5962	347	120409	41781923	18.6279	7.0271
288	82944	23887872	16.9706	6.6039	348	121104	42144192	18.6548	7.0338
289	83521	24137569	17.0000	6.6115	349	121801	42508549	18.6815	7.0406
290	84100	24389000	17.0294	6.6191	350	122500	42875000	18.7083	7.0473
291	84681	24642171	17.0587	6.6267	351	123201	43243551	18.7350	7.0540
292	85264	24897088	17.0880	6.6343	352	123904	43614208	18.7617	7.0607
293	85849	25153757	17.1172	6.6419	353	124609	43986977	18.7883	7.0674
294	86436	25412184	17.1464	6.6494	354	125316	44361864	18.8149	7.0740
295	87025	25672375	17.1756	6.6569	355	126025	44738875	18.8414	7.0807
296	87616	25934336	17.2047	6.6644	356	126736	45118016	18.8680	7.0873
297	88209	26198073	17.2337	6.6719	357	127449	45499293	18.8944	7.0940
298	88804	26463592	17.2627	6.6794	358	128164	45882712	18.9209	7.1006
299	89401	26730899	17.2916	6.6869	359	128881	46268279	18.9473	7.1072

Functions of Numbers (continued)

360 to 419   420 to 479

No.	Square	Cube	Square Root	Cubic Root	No.	Square	Cube	Square Root	Cubic Root
360	129600	46656000	18.9737	7.1138	420	176400	74088000	20.4939	7.4889
361	130321	47045881	19.0000	7.1204	421	177241	74618461	20.5183	7.4948
362	131044	47437928	19.0263	7.1269	422	178084	75151448	20.5426	7.5007
363	131769	47832147	19.0526	7.1335	423	178929	75686967	20.5670	7.5067
364	132496	48228544	19.0788	7.1400	424	179776	76225024	20.5913	7.5126
365	133225	48627125	19.1050	7.1466	425	180625	76765625	20.6155	7.5185
366	133956	49027896	19.1311	7.1531	426	181476	77308776	20.6398	7.5244
367	134689	49430863	19.1572	7.1596	427	182329	77854483	20.6640	7.5302
368	135424	49836032	19.1833	7.1661	428	183184	78402752	20.6882	7.5361
369	136161	50243409	19.2094	7.1726	429	184041	78953589	20.7123	7.5420
370	136900	50653000	19.2354	7.1791	430	184900	79507000	20.7364	7.5478
371	137641	51064811	19.2614	7.1855	431	185761	80062991	20.7605	7.5537
372	138384	51478848	19.2873	7.1920	432	186624	80621568	20.7846	7.5595
373	139129	51895117	19.3132	7.1984	433	187489	81182737	20.8087	7.5654
374	139876	52313624	19.3391	7.2048	434	188356	81740504	20.8327	7.5712
375	140625	52734375	19.3649	7.2112	435	189225	82312875	20.8567	7.5770
376	141376	53157376	19.3907	7.2177	436	190096	82881856	20.8806	7.5828
377	142129	53582633	19.4165	7.2240	437	190969	83453453	20.9045	7.5886
378	142884	54010152	19.4422	7.2304	438	191844	84027672	20.9284	7.5944
379	143641	54439939	19.4679	7.2368	439	192721	84604519	20.0523	7.6001
380	144400	54872000	19.4936	7.2432	440	193600	85184000	20.9762	7.6059
381	145161	55306341	19.5192	7.2495	441	194481	85766121	21.0000	7.6117
382	145924	55742968	19.5448	7.2558	442	195364	86350888	21.0238	7.6174
383	146689	56181887	19.5704	7.2622	443	196249	86938307	21.0476	7.6232
384	147456	56623104	19.5959	7.2685	444	197136	87528384	21.0713	7.6289
385	148225	57066625	19.6214	7.2748	445	198025	88121125	21.0950	7.6346
386	148996	57512456	19.6469	7.2811	446	198916	88716536	21.1187	7.6403
387	149769	57960603	19.6723	7.2874	447	199809	89314623	21.1424	7.6460
388	150544	58411072	19.6977	7.2936	448	200704	89915392	21.1660	7.6517
389	151321	58863869	19.7231	7.2999	449	201601	90518849	21.1896	7.6574
390	152100	59319000	19.7484	7.3061	450	202500	91125000	21.2132	7.6631
391	152881	59776471	19.7737	7.3124	451	203401	91733851	21.2368	7.6688
392	153664	60236288	19.7990	7.3186	452	204304	92345408	21.2603	7.6744
393	154449	60698457	19.8242	7.3248	453	205209	92959677	21.2838	7.6801
394	155236	61162984	19.8494	7.3310	454	206116	93576664	21.3073	7.6857
395	156025	61629875	19.8746	7.3372	455	207025	94196375	21.3307	7.6914
396	156816	62099136	19.8997	7.3434	456	207936	94818816	21.3542	7.6970
397	157609	62570773	19.9249	7.3496	457	208849	95443093	21.3776	7.7026
398	158404	63044792	19.9499	7.3558	458	209764	96071912	21.4009	7.7082
399	159201	63521199	19.9750	7.3610	459	210681	96702579	21.4243	7.7138
400	160000	64000000	20.0000	7.3681	460	211600	97336000	21.4476	7.7194
401	160801	64481201	20.0250	7.3742	461	212521	97972181	21.4709	7.7250
402	161604	64964808	20.0499	7.3803	462	213444	98611128	21.4942	7.7306
403	162409	65450827	20.0749	7.3864	463	214369	99252847	21.5174	7.7362
404	163216	65939264	20.0998	7.3925	464	215296	99897344	21.5407	7.7418
405	164025	66430125	20.1246	7.3986	465	216225	100544625	21.5639	7.7473
406	164836	66923416	20.1494	7.4047	466	217156	101194696	21.5870	7.7529
407	165649	67419143	20.1742	7.4108	467	218089	101847563	21.6102	7.7584
408	166464	67917312	20.1990	7.4169	468	219024	102503232	21.6333	7.7639
409	167281	68417929	20.2237	7.4229	469	219961	103161709	21.6564	7.7695
410	168100	68921000	20.2485	7.4290	470	220900	103823000	21.6795	7.7750
411	168921	69426531	20.2731	7.4350	471	221841	104487111	21.7025	7.7805
412	169744	69934528	20.2978	7.4410	472	222784	105154048	21.7256	7.7860
413	170569	70444997	20.3224	7.4470	473	223729	105823817	21.7486	7.7915
414	171396	70957944	20.3470	7.4530	474	224676	106496424	21.7715	7.7970
415	172225	71473375	20.3715	7.4590	475	225625	107171875	21.7945	7.8025
416	173056	71991296	20.3961	7.4650	476	226576	107850176	21.8174	7.8079
417	173889	72511713	20.4206	7.4710	477	227529	108531333	21.8403	7.8134
418	174724	73034632	20.4450	7.4770	478	228484	109215352	21.8632	7.8188
419	175561	73560059	20.4695	7.4829	479	229441	109902239	21.8861	7.8243

Functions of Numbers (continued)

480 to 539   540 to 599

No.	Square	Cube	Square Root	Cubic Root	No.	Square	Cube	Square Root	Cubic Root
480	230400	110592000	21.9089	7.8297	540	291600	157464000	23.2379	8.1433
481	231361	111284641	21.9317	7.8352	541	292681	158340421	23.2594	8.1483
482	232324	111980168	21.9545	7.8406	542	293764	159220088	23.2809	8.1533
483	233289	112678587	21.9773	7.8460	543	294849	160103007	23.3024	8.1583
484	234256	113379904	22.0000	7.8514	544	295936	160989184	23.3238	8.1633
485	235225	114084125	22.0227	7.8568	545	297025	161878625	23.3452	8.1683
486	236196	114791256	22.0454	7.8622	546	298116	162771336	23.3666	8.1733
487	237169	115501303	22.0681	7.8676	547	299209	163667323	23.3880	8.1783
488	238144	116214272	22.0907	7.8730	548	300304	164566592	23.4094	8.1833
489	239121	116930169	22.1133	7.8784	549	301401	165469149	23.4307	8.1882
490	240100	117649000	22.1359	7.8837	550	302500	166375000	23.4521	8.1932
491	241081	118370771	22.1585	7.8891	551	303601	167284151	23.4734	8.1982
492	242064	119095488	22.1811	7.8944	552	304704	168196608	23.4947	8.2031
493	243049	119823157	22.2036	7.8998	553	305809	169112377	23.5160	8.2081
494	244036	120553784	22.2261	7.9051	554	306916	170031464	23.5372	8.2130
495	245025	121287375	22.2486	7.9105	555	308025	170953875	23.5584	8.2180
496	246016	122023936	22.2711	7.9158	556	309136	171879616	23.5797	8.2229
497	247009	122763473	22.2935	7.9211	557	310249	172808693	23.6008	8.2278
498	248004	123505992	22.3159	7.9264	558	311364	173741112	23.6220	8.2327
499	249001	124251499	22.3383	7.9317	559	312481	174676879	23.6432	8.2377
500	250000	125000000	22.3607	7.9370	560	313600	175616000	23.6643	8.2426
501	251001	125751501	22.3830	7.9423	561	314721	176558481	23.6854	8.2475
502	252004	126506008	22.4054	7.9476	562	315844	177504328	23.7065	8.2524
503	253009	127263527	22.4277	7.9528	563	316969	178453547	23.7276	8.2573
504	254016	128024064	22.4499	7.9581	564	318096	179406144	23.7487	8.2621
505	255025	128787625	22.4722	7.9634	565	319225	180362125	23.7697	8.2670
506	256036	129554216	22.4944	7.9686	566	320356	181321496	23.7908	8.2719
507	257049	130323843	22.5167	7.9739	567	321489	182284263	23.8118	8.2768
508	258064	131096512	22.5389	7.9791	568	322624	183250432	23.8328	8.2816
509	259081	131872229	22.5610	7.9843	569	323761	184220009	23.8537	8.2865
510	260100	132651000	22.5832	7.9896	570	324900	185193000	23.8747	8.2913
511	261121	133432831	22.6053	7.9948	571	326041	186169411	23.8956	8.2962
512	262144	134217728	22.6274	8.0000	572	327184	187149248	23.9165	8.3010
513	263169	135005697	22.6495	8.0052	573	328329	188132517	23.9374	8.3059
514	264196	135796744	22.6716	8.0104	574	329476	189119224	23.9583	8.3107
515	265225	136590875	22.6936	8.0156	575	330625	190109375	23.9792	8.3155
516	266256	137388096	22.7156	8.0208	576	331776	191102976	24.0000	8.3203
517	267289	138188413	22.7376	8.0260	577	332929	192100033	24.0208	8.3251
518	268324	138991832	22.7596	8.0311	578	334084	193100552	24.0416	8.3300
519	269361	139798359	22.7816	8.0363	579	335241	194104539	24.0624	8.3348
520	270400	140608000	22.8035	8.0415	580	336400	195112000	24.0832	8.3396
521	271441	141420761	22.8254	8.0466	581	337561	196122941	24.1039	8.3443
522	272484	142236648	22.8473	8.0517	582	338724	197137368	24.1247	8.3491
523	273529	143055667	22.8692	8.0569	583	339889	198155287	24.1454	8.3539
524	274576	143877824	22.8910	8.0620	584	341056	199176704	24.1661	8.3587
525	275625	144703125	22.9129	8.0671	585	342225	200201625	24.1868	8.3634
526	276676	145531576	22.9347	8.0723	586	343396	201230056	24.2074	8.3682
527	277729	146363183	22.9565	8.0774	587	344569	202262003	24.2281	8.3730
528	278784	147197952	22.9783	8.0825	588	345744	203297472	24.2487	8.3777
529	279841	148035889	23.0000	8.0876	589	346921	204336469	24.2693	8.3825
530	280900	148877000	23.0217	8.0927	590	348100	205379000	24.2899	8.3872
531	281961	149721291	23.0434	8.0978	591	349281	206425071	24.3105	8.3919
532	283024	150568768	23.0651	8.1028	592	350464	207474688	24.3311	8.3967
533	284089	151419437	23.0868	8.1079	593	351649	208527857	24.3516	8.4014
534	285156	152273304	23.1084	8.1130	594	352836	209584584	24.3721	8.4061
535	286225	153130375	23.1301	8.1180	595	354025	210644875	24.3926	8.4108
536	287296	153990656	23.1517	8.1231	596	355216	211708736	24.4131	8.4155
537	288369	154854153	23.1733	8.1281	597	356409	212776173	24.4336	8.4202
538	289444	155720872	23.1948	8.1332	598	357604	213847192	24.4540	8.4249
539	290521	156590819	23.2164	8.1382	599	358801	214921799	24.4745	8.4296

APPENDIX D

*The following pages, 469 through 484, are direct reproductions from **Professional Builder.**

HOURLY ESTIMATES OF LABOR

INDEX TO LABOR ESTIMATING SECTION

Labor estimating*

For residential and light commercial construction; modernization

The following labor estimating tables provide a quick method for determining the amount of time required to complete various types of construction.

Time requirements shown are based on average production per man hour, for quality workmanship, under average conditions. Time requirements do not include supervision.

How to use labor estimating tables

Locate in the tables the type of work to be done. Tabulate skilled and unskilled hours. Add these figures to get the number of hours required per estimating unit, (100 sq. ft., 100 lin. ft., etc.) Multiply this total by the wage rate per hour prevailing in your area. This is the labor cost per estimating unit. Multiply the unit cost by the number of units involved. Repeat this process for all phases of the job to get total labor cost.

If you desire you can also develop a "price per sq. ft." for labor on each type of construction. Or, by combining the various totals of a given job you may develop a "per cu. ft." price for the labor required. Once these figures are prepared they may be used for future projects of the same kind of construction.

EXAMPLE: Extra room, size 10 ft. x 16 ft., 160 square feet area.

Construction, Wood floor with 2 in. x 8 in. joist 12 in. on center with braces. Plywood subfloor with paper, ⅜ in. or ½ in. prefinished oak flor.

Labor rates, skilled $8.00, unskilled $4.80 per hr.

From the tables use the following figures:

	Skilled	Unskilled
Joist labor	5 hrs. (one unit)	2 hrs. (one unit)
Subfloor labor	1 hr. "	½ hr. "
Finished floor labor	4 hrs. "	½ hr. "
(160 sq. ft. =	10 hrs.	3 hrs.
1.6 estimating units)	x1.60 S.F.	x1.60 S.F.
	16 hrs.	4.8 hrs.

16 hrs. skilled @ $8.00 $128.00
4.8 hrs. unskilled @ 4.80   23.04
Total $151.04, labor cost for 160 sq. ft.

Reduced to a square ft. price, the labor cost for this type floor construction is $0.944.

EXCAVATIONS, BACKFILLS, CONCRETE WALLS AND SLABS

	HOURS PER UNIT	
	Machine	Unskilled
EXCAVATION, 100 cubic yards. Rates based on average type dry, solid soil.		
Handwork		130
Machine work (3/4 yard dipper continuous operation)		
Power shovel	1	
Backhoe	1½	
Bulldozer (medium size tractor and blade)	3	
BACKFILLING, 100 cubic yards. Loose soil Machine work (see above)		
Bulldozer or backhoe	1	
Power tamping	2	
Handwork		65
Hand tamping		80
DITCHING, 100 lineal feet. Based on trench size of 12" x 24" or 2 cubic feet per lineal foot.		
Machine work (trencher)	2/3	
Handwork		10
SEWERS AND DRAINS, 100 lineal feet. Laying in 2' ditch and covered		
3 & 4" vitrified, 2' lengths		6
6" vitrified, 2' lengths		7
8" vitrified, 2' lengths		8
10" vitrified, 2' lengths		9

12" vitrified, 2' lengths		10
(Deduct 25% if 4 ft. lengths. Add 10% if asphalt, rubber or cement joints.)		
3-4-6" plastic, fibre 10' joints		43/4
3 & 4" drain tile, 1' lengths		5
6" drain tile, 1' lengths		5½
8" drain tile, 1' lengths		6
FOOTERS (Excavating), 100 lineal feet. Based on 8" x 16" footer.		
Machine work (trencher)	1/3	
Handwork		5
(Add or deduct for other dimensions of ditches or footers.)		

	Skilled	Unskilled
FOOTERS (Placing), 100 lineal feet. Based on 8" x 16" footer.		
Setting forms to level grade. Wood	2	2
Steel	1½	1½
Placing reinforcing rods		1/4
Placing key forms	1	
Placing ready mixed concrete (Average conditions and wheeling distance to forms)	1/2	4
Laying drain tile		1
Placing 12" porous fill over tile		3
FOUNDATIONS (Concrete), 100 square feet. Average type construction to 8' heights. 8" to 12" thick walls. Normal openings included.		
Setting 2' x 8' sectional forms. Wood	1¼	3/4
Steel	1	3/4

---

*The *Labor Estimating* Tables courtesy of **Professional Builder.**

EXCAVATIONS, BACKFILLS, CONCRETE WALLS AND SLABS — Continued

FOUNDATIONS (Concrete) — Cont.	Skilled	Unskilled
HOURS PER UNIT		
Building forms		
(plywood construction)	2	3/4
(1" x 8" sheathing and 2" x 4" or 2" x 6")	4	1½
Placing reinforcing steel		
(½" rods spaced 12" x 12" or wire mesh reinforcing)		1½
Corbeling, chamfering or setbacks		
(up to 4" x 6")	1/2	
Placing concrete (Ready mixed concrete under average conditions)		
8" walls	1	3
12" walls	1	4
Removing forms, ties, etc.		
Sectional forms	1/2	1/2
Built in place forms	1	1/2
Hand rubbing walls, (minor blemishes)		1/2
Cleaning, oiling sectional forms		1
(Adjust rates for extra wheeling distance and handling or foundation heights over 8')		

FOUNDATIONS (Masonry), 100 square feet.
(Average conditions, struck joints, common bond, openings included)

	Skilled	Unskilled
8 x 8 x 16 concrete masonry units	6	6
10 x 8 x 16 concrete masonry units	6½	6½
12 x 8 x 16 concrete masonry units	7	7
(Rate based on heavy units. Deduct 10% for medium weight and 20% for lightweight units)		
Placing masonry reinforcing		1/2
8" solid brick walls	11	11
12" solid brick walls	16	16
Loadbearing structural tile		
5 x 8 x 12 laid flat	5½	5½
8 x 12 x 12	7	7
10 x 12 x 12	7½	7½
12 x 12 x 12	8	8
Waterproofing		
Cement plaster, 1 coat	2	
Membrane (felt, polyethylene cemented to wall)		3
Tar or asphalt, 1 coat Brush coat		1¼
Trowel coat		2½

CONCRETE (Walls), 1 cubic yard.
Based on sections 12" to 36" thick. 20 to 100 cubic yard projects. Average type construction. Ready mixed concrete dumped in forms to 8' heights.

	Skilled	Unskilled
Setting 2' x 8' sectional forms wood	3/4	1/4
steel	1/2	1/4
Building forms		
Plywood construction	1	1/2
1" x 8" sheathing and 2" x 4" or 2" x 6"	2	1

CONCRETE (Walls) — Cont.	Skilled	Unskilled
HOURS PER UNIT		
Placing reinforcing steel		
12" x 12" spacing 3/4" rods wired		1/2
Placing concrete	1/4	1½

CONCRETE SLAB CONSTRUCTION, 100 sq.ft.

Base Preparation	Skilled	Unskilled
Handwork after machine grading		1/2
(Not required if hand excavated)		
Placing, grading, tamping base material. Stone, Slag, Sand, Gravel, Cinders, etc.		1¼
Placing reinforcing wire mesh or rods		1/2
Vapor Barrier—Polyethylene, sisal reinforced paper felt or other non-rigid material		1/2
Perimeter Insulation		
1" x 12" to 2" x 24" rigid in 8' strips or sheets		1/2
Slab Insulation		
Up to 2" rigid or semi rigid insulation laid on base under entire floor area	1/2	1/4

INSULATING CONCRETE, 100 square feet.
Lightweight vermiculite, perlite, pumice (Not finish troweled)

	Skilled	Unskilled
4" thick	3/4	3/4
6" thick	1	1
8" thick	1½	1½
(Finish troweled floor use figures below for concrete slab installation)		

CONCRETE SLAB POURING, 100 square feet.

	Skilled	Unskilled
3" thick	1½	1½
4" thick	1½	1½
5" thick	1½	1½
6" thick	2	2
8" thick	2½	2½

(Time based on the use of ready mixed concrete delivered to the site, no forming, finished or troweled surface on prepared base. Includes placing of ¼" or ½" asphalt, rubber or fibre expansion joint and normal blocking of 4' squares. Add for wheeling concrete if required. Adjust for machine finishing.)

CONCRETE FLATWORK, STEPS

WALKS, DRIVEWAYS, PATIOS, 100 square feet.
Based on hand work under average conditions. Placed on prepared grade.

Walks (4" concrete) Average 4' width	Skilled	Unskilled
Grading, leveling 4" to 6" base materials		1¼
Forming: metal or wood and placing joints	3/4	1/2
Pouring and finishing ready mixed concrete	1½	1½
Removing forms		1/4
For other walk surfaces see brick and patio block under heading of Floors.		

Driveways 6" concrete, to 16' widths on prepared grade	Skilled	Unskilled
Grading, leveling 4" to 8" base materials		1½
Forming: metal or wood & placing asphalt, rubber or fibre expansion joints.	1	1/2
Placing reinforcing mesh	1/2	
Pouring and finishing using ready mixed concrete	2	2

CONCRETE FLATWORK, STEPS – Continued

	HOURS PER UNIT Skilled	Unskilled
WALKS, DRIVEWAYS, PATIOS – Cont.		
Removing forms		1/2
Adjust for machine work & special curing methods, weather protection, etc.		
Porches and Patios		
Concrete slab construction on prepared base, shored wood formed or metal deck.		
Placing reinforcing mesh		1/2
Placing reinforcing rods 8" x 8" centers 3/8" rods		1
Pouring ready mixed concrete troweled finish		
4" thick	1½	1½
5" thick	1¾	1¾
6" thick	2	2
Placing corrugated metal deck over steel or concrete joists	1/2	1/2
Complete forming and shoring to 8' heights	8	4
Removing forms	1/2	1
STEPS, 10 sq. ft. tread area. (On roughed-in concrete base) 12" treads, 8" risers.		
Brick treads and risers, tooled joints		
Treads and risers	5	2½
Treads only	3	1½
Concrete masonry treads and risers (4 x 8 x 12 solids or equivalent)	2	1
Stone or precast concrete treads on concrete base.		
one piece to 4" thick cut to size	1	1
Rough slate or stone treads to 4" thick	2	2
Concrete step treads to 12" wide, 4" to 6" thick		
Forming	1	
Pouring ready mixed concrete	1/2	1/2
Removing forms and finishing	1/2	1/2
Concrete steps, risers & treads (see labor required for cheek walls under Foundations, concrete or masonry)		
Forming steps 8" rise 12" treads	2	1
Pouring steps and risers	1	1/2
Removing forms and finishing	1	1/2
Reinforced concrete steps 8" rise to 12" treads to 20" length, 4 ft. wide to 8 ft. height		
Complete forming and placing steel	6	1½
Pouring ready mixed concrete	2	1
Removing forms and finishing	1	1½
Wood steps and stairs (see frame construction)		

MASONRY

	HOURS PER UNIT Skilled	Unskilled
MASONRY, 100 square feet. (All rates under masonry cover standard or modular units)		
Face brick – based on 3/8" flush joints. Normal openings, sills, headers. To 16" heights.		
4" brick veneer		
Stretcher bond	12	8
Stacked bond	15	10
Soldier course	15	10
Header course	18	12
Roman size (12") Stretcher bond	12	18
Roman size (12") Stacked bond	15	10
Norman size (12") Stretcher bond	10	7
Norman size (12") Stacked bond	13	9
SCR (½" joints) Stretcher bond	9	6
Double brick Stretcher bond	8	6
Adjust for special bonds, tooled, raked, struck or special joints, special color patterns, curved or fancy walls or brick sizes. Add 10% for glazed brick.		
Glass block – ¼" joints. 3-7/8" thick block		
5¾" x 5¾"	20	10
7¾" x 7¾"	15	8
11¾" x 11¾"	11	6
4¾" x 11¾"	20	10
Cleaning masonry – brick, tile, concrete, stone etc. using muriatic acid and water, (no scaffolding included).		
Rough surfaced walls	1½	3/4
Smooth surfaced walls	1	1/2
Concrete masonry (other than foundation work)–average weight regular or patterned units, stretcher bond hand tooled joints. Reinforcing mesh. To 16' heights. Normal openings, sills, headers, control joints included.		
4 x 8 x 12	5	5
2 x 8 x 16	3	3
3 x 8 x 16	3½	3½
4 x 8 x 16	4	4
6 x 8 x 16	4½	4½
8 x 8 x 16	5	5
Add 25% for stacked, ashler or other bonds.		
Masonry fill – Vermiculite insulation		
8" Walls		1/2
12" Walls		3/4
Figure cavity walls based on space to be filled at the rate of 5 minutes per 4 cubic feet for openings over 2" wide.		
Stone work–coursed ashler 4"–5" widths. Random lengths. End cuts made on job.		
2¼" thick	16	16
5" thick	12	12
Random widths 2¼" to 10½" thick	14	14
Random widths 5" to 10½" thick	10	10
Random ashler cut to size		
2¼" to 5" thick	10	10
5" to 10½"	7	7

MASONRY – Continued

		HOURS PER UNIT	
		Skilled	Unskilled
Rustic rubble stone 6" to 18" size cut and fit on job		16	16
Back plastering masonry work (all types)			
1 coat work		1½	1
Backcoating masonry work (all types)			
1 coat work asphalt type	trowell		2½
	brush		1¼
Liquid (transparent waterproofing) – brick, concrete, masonry, stone			
1 coat application	brush coat		1
	spray coat		1/2
Painting concrete masonry, brick, tile, stone, concrete.			
Cement base paints 1 coat	brush work		1½
	spray work		3/4
Back-up walls – normal openings.			
4" Common brick		8	6
8" Common brick		10	8
Concrete masonry Average weight units Reinforcing mesh included			
4 x 8 x 12 " " "		5	5
4 x 8 x 16 " " "		4	4
6 x 8 " " "		4½	4½
8 x 8 " " "		5	5
10 x 8 " " "		6	6
12 x 8 " " "		7	7
Structural tile			
4 x 5 x 12		5½	5½
5 x 8 x 12		4½	4½
4 x 12 x 12		4½	4½
6 x 12 x 12		5	5
8 x 12 x 12		6	6
10 x 12 x 12		6½	6½
12 x 12 x 12		7	7
Add 10% to all types for cavity walls, adjust for special sizes, shapes, unusual wall designs.			
Gypsum partition tile (partition walls) 3/8" joints 12 x 30" units, unusual openings			
2" thickness		3	2½
3" thickness		3½	3
4" thickness		4	3½
6" thickness		5	4
Screen or decorative walls, flush joints.			
8 x 8 to 8 x 12 Face 4" thick concrete		5	5
8" thick concrete		6	6
8 x 16 to 16 x 16 Face 4" thick concrete		6	6
8" thick concrete		7	7
20 x 20 to 24 x 24 Face 8" thick concrete		6	6
8 x 8 to 8 x 12 Face 4" thick clay		5	5
Adjust for unusual wall designs, joints, patterns or unit types.			

GLAZED MASONRY WORK, 100 pieces.

Interior or exterior. Tooled joints. Up to 8' heights.

	HOURS PER UNIT	
	Skilled	Unskilled
5-1/16" x 7-3/4" Face		
1-3/4" Soap Stretcher	4½	4½
3-3/4" Stretcher	5	5
5-3/4" Stretcher	5½	5½
7-3/4" Stretcher	6	6
5-1/16" x 7-3/4" Face		
1-3/4" Soap Stretcher	6½	6½
3-3/4" Stretcher	7	7
5-3/4" Stretcher	7½	7½
7-3/4" Stretcher	8	8
7-3/4" x 15-3/4" Face		
1-3/4" Soap Stretcher	10	10
3-3/4" Stretcher	11	11
Glazed concrete masonry		
2 x 8 x 16	10½	10½
4 x 8 x 16	11½	11½
6 x 8 x 16	12½	12½
8 x 8 x 16	14	14
12 x 8 x 16	17	17
Both types add 25% if glazed two sides. Double rates for all shapes.		
Cutting masonry units (Average power saw) Clay tile, brick, concrete masonry.		
To 32" perimeter units	1	
32" to 48" perimeter units	1½	

COPING, LINTELS, BEAMS, COLUMNS

WALL COPING, 100 lineal feet.

	Skilled	Unskilled
Vitrified		
9" – 13"	6	6
18"	7	7
Includes placing corners and ends.		
Precast concrete or cut stone		
4" thick to 16" widths; 4–5' lengths	6	5

SILLS–LINTELS, 100 lineal feet.

	Skilled	Unskilled
Door – window. Precast concrete or stone cut to size to 5' lengths.		
To 4" x 8"	5	2
6" x 8" – 8" x 8"	5½	2
8" x 10" – 8" x 12"	6	3

CONCRETE BEAMS, LINTELS, 100 lineal ft.

	Skilled	Unskilled
Poured in place to 8' heights		
8' x 8" setting forms and placing steel rods	7½	3½
8" x 12" setting forms and placing steel rods	8	4
Pouring ready mixed concrete		
8" x 8"	1	3
8" x 12"	1	4
Removing forms	1	1/2
Finishing concrete		2

COPING, LINTELS, BEAMS, COLUMNS – Continued	HOURS PER UNIT	
	Skilled	Unskilled
CONCRETE MASONRY PILASTERS, 100 lineal feet.		
Laying units. Placing reinforcing rods. Filling with ready mixed concrete to 16' heights. Hand work.		
8'' x 8'' Core types	9½	17
12'' x 12'' Core types	11½	23½
Adjust for additional heights and mechanical filling methods.		
CONCRETE COLUMNS, 100 lineal feet.		
Poured in place to 8'' heights, setting forms and placing steel.		
8'' x 8'' – 8'' x 12'' – 12'' x 12''	6	3
Pouring ready mixed concrete		
8'' x 8''	1	3
8'' x 12''	1	4
12'' x 12''	1	5
Removing forms	3/4	1/2
Finishing concrete		2
Adjust for additional heights		
FOUNDATION BLOCK, (For slab construction) 100 lineal feet.		
6'' x 12'' x 16'' U and J types or		
8'' x 8'' x 16'' header block	2	3
CONCRETE MASONRY BEAMS and LINTELS, 100 lineal feet.		
Laying units. Placing reinforcing rods. Filling with ready mixed concrete. Hand work to 8' heights.		
8'' x 8'' x 16'' units continuous beam	5	7
8'' x 8'' x 16'' units over openings. Shoring included.	6	8
8'' x 16'' x 12'' units continuous beam	12	16
8'' x 16'' x 12'' units over openings. Shoring included.	13	16½
Adjust for additional heights and mechanical filling methods.		

CHIMNEYS, FIREPLACES

CHIMNEYS, per foot of height with flue liner, to 30' heights. Inside chimneys with normal face brick topping. No scaffolding provided. Footer or Base (use rates under Footers).

Brick		
1 8'' x 8'' flue	3/4	3/4
2 8'' x 8'' flues	1¼	1¼
1 8'' x 12'' flue	3/4	3/4
2 8'' x 12'' flues	1½	1½
1 12'' x 12'' flue	1	1
2 12'' x 12'' flues	1¾	1¾

Adjust for other sizes and combinations. Add 25% for average face brick outside chimneys. Flue liner sizes may vary. Use nearest size.

	HOURS PER UNIT	
	Skilled	Unskilled
Concrete masonry chimney units		
8'' x 8'' flue size	1/4	1/4
8'' x 12'' flue size	1/3	1/3
Deduct 10% for units without flue liner or for unlined round types. Add 10% for solid masonry or 2 unit types.		
FIREPLACES, per fireplace. Based on average design and type 6' wide and 5' high, 36'' x 30'' opening.		
Hearth construction		
Concrete base	1/2	1
Brick hearth on concrete base	2	2
Tile hearth on concrete base 2'' x 2'' to 6'' x 6''	2	2
Brick work	10	10
Firebrick lining and damper	4	2
Setting steel circulator	1	1
Double above time for double faced or 3 way installation. Use 1½ times above for corner projecting fireplace. Adjust for unusual brick patterns or other facing materials.		
Tile or glass facing 4'' x 4'' to 8'' x 8'' sizes	4	4
Setting factory mantel (to prepared wall)	1	1/2

FRAMING, SHEATHING, DECKING

FLOOR JOISTS, WOOD, 100 square feet.

Joist Size	Spacing	Approx. Span*	Skilled	Unskilled
2'' x 6''	12''	8'	4½	1½
	16''	7'	4	1½
	18''	6'–6''	4	1¼
	24''	6'	3½	1
2'' x 8''	12''	11'	6	2
	16''	10'	4½	1¾
	18''	9'	4	1½
	24''	8'	3½	1
2'' x 10''	12''	14'	6½	2½
	16''	12'	6	2¼
	18''	11'	5½	2
	24''	10'	5	1¾
2'' x 12''	12''	17'	7½	2¾
	16''	15'	7	2½
	18''	14'	6½	2¼
	24''	12'	6	2

Labor rates cover both platform or balloon type construction, also sill, plate, edge joist and bridging. Add 25% for spans over 12', wood girder if required also if above 1st floor level.
*Spanning based on 100 pounds per square foot total load using Douglas Fir or equal.

STEEL BAR JOIST, 100 square feet. Based on 36'' spacing to 20' spans. Types 2 & 3 lightweight. To 2nd floor levels.

8''	1	1
10''	1¼	1
12''	1½	1¼

FRAMING, SHEATHING, DECKING — Continued

	HOURS PER UNIT	
	Skilled	Unskilled
STEEL BEAMS, 100 square feet. Light weight type to 20' spans; to 2nd floor levels.		
6" – 7" widths	1	1
8" – 10" widths	1¼	1
12"	1½	1¼
CONCRETE JOISTS, 100 square feet. Based on 24" spacing to 12' lengths. at ground level.		
3" x 8"	1	2
3" x 10"	1¼	2¼
4" x 12"	1½	3
Add 25% for 14 to 20' lengths. Add 50% for 2nd floor construction.		
PRECAST CONCRETE FLOOR SLABS, 100 square feet. On concrete joist or steel beams.		
1" to 2" thick 24" x 30"	1¼	1¼
1" to 2" thick 24" x 60"	3/4	1
Adjust for various sizes and types.		
PRECAST AND PRESTRESSED CONCRETE FLOOR BEAMS, 100 square feet. Over steel or concrete beams. To 20' lengths. Erected to second floor level. Includes top grouting. Hollow beams.		
6" x 12" Beams	1/2	1
6"–8" x 16" Beams	1/3	2/3
6"–8" x 48" Beams	1/4	1/2
Adjust for other sizes. Add 25% for spans 20' to 30'. Add machine or crane time.		
COMMERCIAL STORE FRONTS AND WINDOW CONSTRUCTION. Based on total front or window area. Average type construction. Excluding glass.		
Rough framing wood	8	2
Metal work over rough framing using rolled light metal stock	20	5
Metal work over rough framing or free standing using medium or heavy extruded stock	30	10
Adjust for unusual types, patterns and designs or extra metal cutting and fitting.		
STUDDING, 100 square feet. Includes normal openings, average type outside walls, frame or veneer construction. Plates, headers, fillers, bracing, firestops, girts included.		
2" x 4" 12" centers 8' to 12' heights	3	1/2
12" centers 12' to 20' heights	4	3/4
16" centers 8' to 12' heights	2½	1/2
16" centers 12' to 20' heights	3½	3/4
24" centers 8' to 12' heights	2	1/2
24" centers 12' to 20' heights	3	3/4
2" x 6" 12" centers 8' to 12' heights	4½	3/4
12" centers 12' to 20' heights	5½	1
16" centers 8' to 12' heights	3½	3/4
16" centers 12' to 20' heights	4½	1
24" centers 8' to 12' heights	2½	3/4
24" centers 12' to 20' heights	3½	1

	HOURS PER UNIT	
	Skilled	Unskilled
Add 25% for irregular or cut up walls and shed or gable dormers. Deduct 25% for interior stud partition walls.		
SCR construction (Furring)		
2" x 2" 12" centers 8' to 12' heights	2½	1/2
2" x 2" 16" centers 8' to 12' heights	2	1/2
CEILING JOISTS, 100 square feet. Normal type frame construction. 1st and 2nd floor levels. Includes bridging, trimmers to 16' spans. Normal openings.		
2" x 6" 12" centers	5½	2
16" centers	5	2
18" centers	4½	1½
24" centers	3	1
2" x 8" 12" centers	7	2
16" centers	6	2
18" centers	5½	1½
24" centers	4	1
2" x 10" 12" centers	8	2½
16" centers	7	2¼
18" centers	6½	2
24" centers	5	1½
2" x 12" 12" centers	9	3
16" centers	8	3
18" centers	7½	2½
24" centers	6	2
For open beam ceilings		
3" x 6"–8" 16" centers	7	2½
18" centers	6½	2
24" centers	5	1½
4" x 6"–8" 16" centers	7½	2½
18" centers	7	2
24" centers	5	1½
All of above add 25% for spans over 16' and heights above normal 2nd floor ceiling level.		
RAFTERS, 100 square feet. Average type construction, normal pitch and flat, gable ends, to 22' lengths.		
2" x 4" 12" centers	3	3/4
16" centers	2½	3/4
2" x 6" 12" centers	3½	1
16" centers	3	1
18" centers	2½	1
24" centers	2	3/4
2" x 8" 12" centers	4	1
16" centers	3½	1
18" centers	3	1
24" centers	2½	3/4
Open beam rafters		
3" x 6"–8" 16" centers Flat construction	7	3½
24" centers " "	6	3
4" x 6"–8" 16" centers " "	8	4
24" centers " "	7	3½
(Add 25% for pitched roofs over 2" pitch)		
All of above add 50% for cut up and hip roof types.		

FRAMING, SHEATHING, DECKING — Continued

	HOURS PER UNIT	
	Skilled	Unskilled
Add 25% for rafters on shed, gable or hip type roofs on dormers and for rafters over 22' long. Deduct 10% for rafters with short or plain overhang.		
SHEATHING, 100 square feet. Average type construction, frame or veneer, normal openings.		
1" x 6"–8" Wood sheathing flat roofs	1	1/2
pitched roofs	1½	3/4
steep, cut-up or hip roofs	3	1½
sidewalls	1½	3/4
(Add 25% for diagonal sheathing on sidewalls)		
2" x 6"–8"/3" x 6" flat roofs	3	1½
pitched roofs	4	2½
(Add 25% if exposed underside on beamed ceiling construction)		
Plywood, rigid insulating sheathing to 25/32" thick		
4' x 8' sheets flat roofs	1/2	1/4
pitched roofs	3/4	1/2
steep, cut-up or hip roofs	1¼	1
sidewalls	1	1/2
Gypsum sheathing		
2' x 8' panels sidewalls	1¼	1/2
Strip sheathing		
1" x 2"–3"–4" pitched roofs	1	1/2
Insulating sheathing		
1" to 3" thick flat roofs	3/4	1/2
pitched roofs	1¼	1
(Add 25% if finished open beam ceiling construction)		
Asbestos covered insulating sheathing 19/16" to 2" thick.		
4 x 8 panels. Frame const. flat roofs	1½	1
pitched roofs	2	1½
BAR RIB LATH, 100 square feet. For poured concrete or gypsum over concrete, bar joist or steel beams.		
3/4" rib lath.*	1/2	1/2
STEEL DECKING, 100 square feet. Sheets 2' x 8' for floor or roof construction.		
18 or 20 gauge. *	1/2	3/4
PAPER BACKED WIRE MESH, 100 square feet. For floor construction.		
3" x 4" mesh, 12 gauge, rolls 4' x 125'	1/4	1/4
*(Add 25% for 2nd or 3rd floor construction)		
STAIRWAYS, (Frame), per stairway Average type straight flight to 4' wide and 12' long.		
Rough cutting, framing, placing	5	2
(Add 50% for 2 flight type)		

	HOURS PER UNIT	
	Skilled	Unskilled
ROOF TRUSSES, (wood) per truss. Average type 1 floor construction, using precut lumber.		
24" spacing 14' to 20' length	1/2	1/2
22' to 32' length	3/4	3/4

SUBFLOORS, FINISHED FLOORS

	HOURS PER UNIT	
	Skilled	Unskilled
SUBFLOORS ON WOOD JOISTS, 100 sq.ft.		
1" x 8" Lumber	1½	1/2
Plywood 3/8" to 3/4"	1	1/2
INSULATION, 100 sq.ft.		
Blanket type placed between joists	3/4	
UNDERLAYMENT, 100 sq.ft.		
Board-type (Hardboard, Fiber Board, Plywood, etc.) ¼" to ½" thick	1	
WOOD FINISH FLOORING, 100 sq.ft.		
Softwood strip		
2¼" face	3	1/2
3½" face	2	1/2
Hardwood strip 7/8"		
1¼" face	4½	3/4
2½" face	4	1
Hardwood strip ½"–3/8"		
1½" face	4	1/2
2¼" face	3½	1/2
(Includes placing paper or felt under floor.)		
Sanding, machine	1	
Finishing		
Three liquid applications to all unfinished wood floors	1½	
Prefinished hardwood (Strip, plank, block)		
½" – 3/8" x 1½" face	4	1/2
Plank 6" wide	2½	3/4
Block 9" x 9"	2	3/4
(Either nailed or cemented on prepared subfloor)		
RESILIENT TYPE FLOORING, 100 sq.ft.		
Roll linoleum (plain)	3	1/2
Patterns, borders	4	1
Tile, (asphalt, vinyl, rubber, linoleum, cork, etc.)		
4" x 4" size	5	1
6" x 6"	4	1
9" x 9"	3	1
6" x 12"	3	1
12" x 12"	2½	1
9" x 18"	2½	1
Strip (to 8")	3	1
(Includes laying felt or underlayment)		

SUBFLOORS, FINISHED FLOORS – Continued

	HOURS PER UNIT	
	Skilled	Unskilled
CERAMIC TILE ON CONCRETE, 100 sq.ft.		
½" to 2", paper backing	4	4
2" x 2"	5	5
4" x 4"	4	4
6" x 6"	3	3
(Includes laying the necessary underlayment on any base.)		
SLATE ON CONCRETE, 100 sq.ft.		
Up to 1" thick.		
Random cut sizes to 12" x 12"	2	2
Rough uncut slabs	1½	1½
MARBLE ON CONCRETE, 100 sq.ft.		
4 x 6 to 12 x 12 tile		
Cement bed preparation	2	4
Setting tile	8	8
Machine finishing	8	
BRICK FLOOR ON CONCRETE, 100 sq.ft.		
(Standard size on edge)		
Laid in mortar, mortared joints, basket weave or common bond.	12	6
Herringbone or fancy patterns	16	8
(Standard size laid flat)		
Basket weave or common bond	8	4
Herringbone or fancy patterns	12	6
(Use 1/3 of above rates if laid in sand without mortar.)		
Adjust rates for other brick sizes.		
PATIO BLOCKS ON CONCRETE, 100 sq.ft.		
Laid in mortar, mortared joints up to 4" thick. Square, rectangular, octagonal or odd shapes.		
6 x 12 to 8 x 16	4	2½
12 x 12 to 12 x 18	3½	2
12 x 24	2½	2
24 x 24	2	2
Use 1/3 of above rates if laid in sand base without mortar.		

ROOFING

	HOURS PER UNIT	
	Skilled	Unskilled
ASPHALT ROOFING, 100 sq.ft. (Square)		
Roll, Plain Surface Pitched roofs	3/4	1/4
Mineral Surface Pitched roofs	1	1/4
Felt Underlayment		1/4
Shingles, Individual type	3	1
Individual type Dutch lap	1¾	1
Interlocking	2	1
Strip 10" x 36"	2	1
Strip 12" x 36"	2¼	
ASBESTOS SHINGLES, 100 sq.ft. (Square)		
American method 8" x 16"	4½	2½
Dutch lap hexagonal 16" x 16"	2½	1½
Colonial method 10" x 24" to 12" x 30"	2¾	1

	HOURS PER UNIT	
	Skilled	Unskilled
WOOD SHINGLES, 100 sq.ft. (Square)		
16" Approx. 5" exposure	4	1½
18" Approx. 6" exposure	3½	1¼
24" Approx. 8" exposure	3	1
(Add 35% for staggered or thatched butts or double coursing.)		
SLATE, 100 sq.ft. (Square)		
16" x 8"	5	2½
18" x 9"	4	2
20" x 10"	3½	2
22" x 12"	3	1½
Random widths to 3/8" thick	7	3½
Graduated slate to 3/4" thick	8	4
CLAY TILE, 100 sq.ft. (Square)		
Spanish, Mission or Shingle type	5	2½
Interlocking tile type	6	3
(Add for stripping and underlayment)		
METAL SHEETS, 100 sq.ft. (Square)		
Pitched roofs.		
Corrugated aluminum or galvanized steel in sheets. Over wood framing.	1¼	3/4
Crimped types, aluminum or steel in sheets over wood framing.	1½	3/4
Rolls over flat areas. Copper or tin 14" widths. Flat seams	2½	2½
CORRUGATED ASBESTOS SHEETS, 100 sq.ft. (Square)		
¼ x 42–8' length over open rafters or steel	2	2
BUILT-UP, 100 sq.ft. (Square)		
(Includes topping of slag, stone, gravel or chips)		
3 Ply over wood or insulating deck	1½	3/4
3 Ply over gypsum or concrete deck	1¾	1
4 Ply over wood or insulating deck	1¾	1
4 Ply over gypsum or concrete deck	2	1¼
CONCRETE SLABS, LIGHTWEIGHT, 100 sq.ft. (Square)		
On steel beams or bar joist. 2nd floor levels. Add for mechanical hoisting equipment.		
Flat type 1" to 2" thick 24" x 5'	1	2
Channel type.		
2¾" thick 24" x 8'	1¼	2½
3¾" thick 18" x 8'	1½	3
3¾" thick 24" x 8'	1¼	2½

NOTE: Add 25% to all roofing estimates for steep pitched or cut-up roofs and unusual conditions. Above rates include time for average type ridges, hips and valleys. Underlayment or scaffolding not included. Add for preparation of surface, eaves, etc. when estimating reroofing jobs.

ROOFING – Continued

		HOURS PER UNIT	
		Skilled	Unskilled
METAL WORK, 100 lineal feet			
Eaves and gutters, metal, standard sizes.		5	5
(Deduct 1/3 for interlocking prefit types)			
Downspouts	2" to 6" round or square	4	2
Valleys	20" width metal	4	
Flashing	Parapet walls, chimneys and dormer sides, etc.	4	
Raggles for flashing (cutting only in set up masonry walls.)		1	4
WOOD GUTTERS, 100 lineal feet			
3" x 5" and 4" x 6"		5½	4
5" x 7"		6½	5
CANT STRIPS, 100 lineal feet			
3" to 6"		1½	1/4

SIDING

		HOURS PER UNIT	
		Skilled	Unskilled
WOOD SIDING (horizontal), 100 sq.ft.			
Shiplap, patterns, rustic types			
1" x 3½–4"		2½	3/4
1" x 4½–5"		2¼	3/4
1" x 6–8"		2	1
Lap, bevel or bungalow types			
½" x 8"		2½	3/4
¾" x 10"		2¼	3/4
¾" x 12"		2	3/4
Add 25% for cut to fit ends and mitred corners. Includes metal corners if required.			
Vertical patterned types			
1" x 6–8"		3	1½
1" x 10–12"		2¾	1¼
Board and batten			
1" x 6–8"		3½	1½
1" x 10–12"		3	1¼
(Includes horizontal stripping over studs)			
Wood shingles			
16" approximately 5" exposure		5	1½
18" approximately 6" exposure		4½	1½
24" approximately 8" exposure		3½	1½
Add 25% for special patterns or double coursing.			
ASPHALT SIDING*, 100 sq.ft.			
Brick, Stone, Shingle patterns.			
Roll types (15" widths)		2	1/2
Panel types (insulating)			
10-7/8" x 43"		3	1
14" x 43"		2½	1

	HOURS PER UNIT	
	Skilled	Unskilled
COMPOSITION SIDING*, 100 sq.ft. Hardboard, wood fibre etc., ¼"–5/16", plain or factory painted.		
10"–12" widths. Horizontal. Self venting and spacing metal furring.	1¾	1/2
10"–12" widths. Horizontal	1¼	1/2
16" " "	1	1/2
10"–12" widths. Horiz. Shadow edge furring	2	1/2
16" " " " " "	1¾	1/2
10"–12" widths. Vertical Board & Batten	2	1
16" " " " "	1¾	1
ASBESTOS SIDING*, 100 sq.ft.		
8" x 32" to 12" x 24"	2	1
Plain Sheet Types*		
3/16" and ¼" thick, 48" wide 8' long	1½	1½
Corrugated Sheets*		
¼" x 42"–8' lengths	1¼	1¼
* Includes corners, trim, etc.		
SHINGLE BACKER, 100 sq.ft.		
8¾" to 11¾" 48" lengths	1¼	1/2
13½" to 15½" 48" lengths	1	1/2
Lattice strip furring	1/4	1/4
¾" x 2" strip furring	1/2	1/4
PAPER, FELT, ETC., 100 sq.ft.	1/4	1/4
ALUMINUM SIDING, 100 sq.ft.		
Clapboard types 8"		
Horizontal plain	1½	1½
insulated	2	2
Vertical plain	2	2
insulated	2½	2½
(Includes felt, paper, foil, moisture barrier, corners, fittings and trim)		
Corrugated sheets, 8' lengths		
32" coverage sheets, over wood frame	1	1
over steel frame	2	1½
(Deduct 10% for 45" coverage sheets)		
GALVANIZED STEEL SIDING, Corrugated Sheets, 100 sq.ft.		
8' sheets, 26" coverage, over wood frame	1¼	1
over steel frame	2	2
Adjust for unusual cutting and fitting and heights over 20'.		
To all siding jobs add 25% for unusual conditions, cut up walls, bays and gables.		
Adjust for wall preparation on all types of remodeling projects.		
STUCCO, 100 square yards. Cement stucco 3 coats. Float finish over tile, concrete, brick, metal lath or stucco base. Average openings included to 16' heights. Scaffolding provided.		
Gray, White portland cement or colored prepared stucco	25	18
Add 10% skilled time for textured finish. Add 15% skilled time for troweled finish, unusual finished surfaces or patterns.		

INSULATION

	HOURS PER UNIT	
Based on average conditions, normal openings. New frame construction.	**Skilled**	**Unskilled**
RIGID BOARD TYPES, 100 sq.ft.		
4' x 8' to 1" thick		
Flat roofs	1/2	1/4
Pitched roofs	3/4	1/2
Steep pitched or cut-up roofs	1	3/4
(Add 10% for 2" & 3" thickness)		
Sidewalls	1	1/2
NON-RIGID TYPES, 100 sq.ft.		
Batt type 2 to 4" thick between studs		
15—19—23" x 24" batts	3/4	1/2
15—19—23" x 48" batts	1/2	1/2
Strip types 2 to 4" thick, between studs		
15" widths	1/2	1/2
Blanket types 1 to 4" thick, between studs		
16—20" widths, 4 to 8' lengths	3/4	1/2
24—33" widths, 4 to 8' lengths	1/2	1/2
Wide stock over studs or sheathing	1/2	1/2
(Add 25% for ceiling installations)		
REFLECTIVE TYPES, 100 sq.ft.		
1 sheet stripped in place between studs	1/2	1/4
Multiple, accordion sheets	3/4	1/4
POURING TYPES, 100 sq.ft.		
Between 4" studs when accessible	1	1
Over ceilings 4 to 6" thick	1	1/2
Poured in cavities, cryptite masonry or cavity type walls to 4" space	1	1/2
SEMI-RIGID TYPES, 100 sq.ft.		
Applied over frame or masonry walls using adhesives or asphaltic cements	2	2
(Add 25% for each additional 2" in thickness)		
(Add 25% for ceiling or unusual wall installations all thicknesses)		
VAPOR SEALS, MOISTURE BARRIERS, 100 sq.ft.		
Roll or sheet types, aluminum, polyethylene, papers		
Tacked in place	1/2	1/4
Cemented on frame or masonry surfaces	3/4	1/2
Adjust all rates for extra time and conditions on unusual projects and remodeling jobs.		

WALLS AND CEILINGS

Average type construction, normal openings. Walls and ceilings combined. Necessary working level planking or scaffolding included.		
PLASTER BASES, 100 square yards.		
Gypsum lath 3/8"—1/2" 16" x 48"	6	1½
3/8"—1/2" 24" to 12'	4	1½
Insulating lath 1/2" 18" x 48"	7	2
Metal lath, average size sheets, over studs	4	1½
Metal lath, average size sheets, over masonry	5	1½
Paper backed wire or expanded mesh	3½	1½
Foamed plastic, 12"x9' planks, 1—2" thick cemented	8	1½

	HOURS PER UNIT	
PLASTER, average 1/2"—5/8" grounds, 100 square yards.	**Skilled**	**Unskilled**
Scratch coat over gypsum or insulating lath	2	2
masonry or gypsum tile	2½	2
metal lath or mesh	3	2
Brown coat over average scratch coats	3	2
Sand finish coat*	4	2
White coat lime finish*	6	2
*Add 10% if color mixed on job.		
Prepared color finish	5	2
Keenes cement finish, smooth	4½	2
Keenes cement finish, 4"x4" tile pattern	5½	3
2 coat work over gypsum or insulating lath	10	5
3 coat work over gypsum or insulating lath	13	6
2 coat work over metal lath, mesh	11	6
3 coat work over metal lath, mesh	14	6
2 coat work over masonry, gypsum tile concrete	10½	5
3 coat work over masonry, gypsum tile concrete	13½	6
2 coat work over foamed plastic base	10	5
3 coat work over foamed plastic base	13	6
Deduct 25% unskilled time for factory mixed plasters. Add 10% for heavily wood fibred plasters or heavy plasters.		
Adjust for special finishes of all types, machine application and for work over radiant heating cables or tubing, etc.		
METAL LATH AND PLASTER PARTITIONS, 100 square yards.		
Hollow 4" thick installed in masonry or concrete work		
2 coat work plastered 1 side only	41	10
2 coat work plastered both sides	52	16
3 coat work plastered 1 side only	44	10
2 coat work plastered both sides	55	18
(Add for various partition widths)		
2 coat work both sides	46	12
3 coat work both sides	52	18
GYPSUM LATH AND PLASTER PARTITIONS, 100 square yards.		
Solid 2" thick, includes floor and ceiling runners, concrete or masonry construction		
2 coat work, 2 sides	50	16
3 coat work, 2 sides	53	18
SUSPENDED CEILINGS, 100 square yards.		
Hung from concrete or steel, 1½" main runners with 3/4" channels on 12" centers, metal lath wired or clipped on.		
2 coat work complete ceiling	45	15
3 coat work complete ceiling	48	17
Less 10% if hung from wood joist or rafters.		
PLASTER BONDING, 100 square yards.		
Over concrete and masonry surfaces		
Asphaltic type brushed on		5
Cement base types brushed on		6
Cement base types troweled on 1/8" to 1/2"	4	4
Adjust and add for chipping, roughing up or other wall preparations if required.		

WALLS AND CEILINGS — Continued

	HOURS PER UNIT	
	Skilled	Unskilled
GYPSUM BOARD, WALLS, 100 square feet.		
3/8"–½"–5/8" thick 4' x 8' to 10' panels	1	1
Finishing joints	1	
Finishing joints 2 ply application 2nd layer cemented in place finished	3½	1½
GYPSUM BOARD, CEILINGS, 100 square feet.		
Plain 4'x6' to 12' lengths	1½	1½
Wood veneer or other finishes	2	1½
Plank type 16" widths	2½	1½
PLYWOOD PANELS to 4'x12' panels, 100 square feet.		
Plain joints or tongue and grooved	1	1
Covered joints with moulding	3	1
Fitted joints cemented	4	1
Plank type paneling tongue and grooved 6" to 16" widths	3	1
PATTERNED PANELING, 100 square feet.		
¾" x 6" to 12" widths, horiz. or vert.	3½	1
BEADED WOOD, 100 square feet.		
½" x 3½"	3	1
INSULATING BOARD, 100 square feet.		
Rigid panels 4' x 8' to 12'	1	1
Plank type	1½	1
INSULATING OR ACOUSTICAL TILE, 100 sq. ft.		
Plain or decorated		
12" x 12" nailed or tacked	3	1
12" x 24" nailed or tacked	2	1
24" x 24" nailed or tacked	1½	1
16" x 32" nailed or tacked	1	1
(Furring not included)		
Add 25% for patterns and adhesive application.		
WALL TILE, 100 square feet.		
Applied with adhesives to prepared walls, plain patterns. Includes base and cap. Metal trim.		
Plastic 4" x 4"	5	1
9" x 9"	3	1
Metal 4" x 4"	6	1
Ceramic 4" x 4"	8	2
(Add 50% if applied in mortar bed)		
Adjust all work for unusual conditions, materials, designs, etc.		
HARDBOARD, 3/16"–¼"–5/16", 100 sq.ft.		
Plain 4' x 8', 8' to 12' lengths. Nailed	1¼	1
4' x 8', 8' to 12' lengths. Nailed with metal fittings	2½	1
METAL SIDEWALLS AND CEILINGS, 100 square feet.		
Patterned types	5	2
(Add for furring and intricate designs)		

	HOURS PER UNIT	
	Skilled	Unskilled
CURTAIN WALL PANELS, 100 square feet.		
1-9/16" and 2" thickness		
4' x 8' –9'–10'–12' panels	4½	4
(Includes 2" x 4" studs and plates floor and ceiling)		
FURRING, PLASTER GROUNDS. Up to ¾" x 4" wood strips, 100 square feet.		
Frame construction 12" centers	1	
16" centers	3/4	
24" centers	1/2	
Nailable masonry construction 12" centers	1¼	
16" centers	1	
24" centers	3/4	
Adjust for installation on concrete walls. Add 25% all types of work if figuring ceilings only.		
MOLDINGS, TRIM, AND ACCESSORIES, 100 lineal feet.		
Plastering and corner beads, corner lath, metal furring, screeds, picture mold, bullnose beads, etc.	2/3	1/3
Door and window casing beads (metal) metal trim, corners, mitred and fitted before plastering or used with dry wall construction	2½	1/2

WINDOWS AND DOORS

Window Sizes:
A To 3' x 3'. Single unit
B Over 3' to 3' x 5'–6" Single unit
C Over 3' to 6' x 5'–6" Double unit
D Over 3' to 6' to 8' x 6' Triple unit and picture windows

	Skilled	Unskilled
WOOD WINDOWS, per each unit		
Window frames, assembled from stock sections		
Size A	1/2	1/4
B	3/4	1/2
C	1¼	1/2
D	1½	1/2
Setting window frames, includes handling and bracing when required.		
Size A	1/2	1/4
B	3/4	1/2
C	1	1/2
D	1½	1/2
Fitting and hanging double hung sash.		
Size A	1/2	1/4
B	3/4	1/2
C	1½	1/2
D	2½	3/4
Fitting and hanging casement sash. Per opening.		
Size A 1 sash	1/2	1/4
B 2 sash	1	1/4
C 3 sash	1½	1/2
D 4 sash	2	1/2
Setting complete window units. All types glazed. To prepared openings.		
Size A	3/4	1/4
B	1	1/2
C	1½	1/2
D	1¾	3/4

WINDOWS AND DOORS — Continued

	HOURS PER UNIT	
	Skilled	Unskilled

METAL WINDOWS, per each unit.
All residential types, fins or wood surround attached. Setting sash units. To prepared openings.

	Skilled	Unskilled
Size A	3/4	1/2
B	1¼	1/2
C	1¾	3/4
D	2	1

(Adjust for unusual installations and glazed units.)

BASEMENT SASH, wood or metal, per each unit 1/4

Setting poured in place basement frames & sash 1/2

Setting utility metal sash 4 light open	1/2
6 light open	3/4

COMMERCIAL PROJECTED.
Pivoted or security type metal sash. Bracing and handling to second floor heights. Fixed or vented.

	Skilled	Unskilled
Small units to 2' x 4'	3/4	1/4
2' x 4' to 4' x 8'	1½	1/2
4' x 8' to 6' x 8'	2	3/4
6' x 8' to 8' x 8'	2½	1

(Add 25% for architectural projected or heavy awning types.)
Add 25% to all work if second floor installation only.
Adjust for plate glass or twin glass glazing where work is figured glazed.

TRIM, INTERIOR, per each unit.
Single member type, casings, stool, apron, stops. Soft wood.

	Skilled	
Size A	1	
B	1¼	
C	1½	
D	2	

(Add 25% for wood jambs over plaster returns)

Plain wood Stool and Apron. All sizes	1/2
Plain Wood Stool only. All sizes	1/4

Adjust for various types trim.

Metal Stools to 4'	3/4
4' to 8'	1
Glass, Marble, 1 piece precut stools to 4'	1
Ceramic tile stools to 4' plain	1½

TRIM, EXTERIOR, per each window

	Skilled	Unskilled
Size A	1/2	
B	3/4	1/4
C	1	1/4
D	1½	1/4

(Includes flashing at head if required)

SHUTTERS, per pair
Wood, Metal

	Skilled
Small	1/2
Medium	3/4
Large	1

(Add 50% if over masonry construction)

STORM SASH & SCREENS, (Prefit wood comb.) per unit

	Skilled	Unskilled
Size A	1/4	1/4
B	1/2	1/4

	HOURS PER UNIT	
	Skilled	Unskilled
C	3/4	1/4
D	1	1/4

(Add 25% for metal or plastic)

DOORS (Residence types)

	Skilled	Unskilled
Setting exterior door frames. Wood residential types. Standard sizes	1	1/4
Setting front entrance door frames with patterned side panels to 5' widths	2	1/4
Setting front entrance door frames with side light panels to 5' widths	2½	1/4
Cutting and setting 2" x 6"–8" frames from stock material 3' x 7'	1	1/4
same as above 8' x 8' to 10' x 8'	1½	1/2
10' x 10' to 16' x 10'	2	3/4
Interior door jambs and heads. Assembling from stock sections. Standard sizes	3/4	1/4
Setting interior door frames. Standard sizes	1	1/4
same as above small closet	1/2	
large closet	3/4	
Setting sliding door pockets in stud walls. Standard sized doors	3/4	1/4

Building sliding door pockets & setting track

	Skilled
1 pocket, single door	2½
2 pockets, double door	4½

GLASS SLIDING DOORS, setting complete units.

	Skilled
To 4' widths	2
4' to 6' widths	2½
6' to 8' widths	3

METAL DOOR FRAMES, setting each.

	Skilled	Unskilled
Standard size	3/4	1/4
Double size	1½	1/2

OUTSIDE DOOR TRIM, 1 piece, setting each.

	Skilled	Unskilled
Standard size	3/4	1/2

INSIDE DOOR TRIM, 1 piece, setting each.

	Skilled
Standard size doors	1¼
Small closet doors	1/2
Large closet doors	3/4

THRESHOLDS, setting each.

	Skilled
Wood	1/4
Metal	1/4
Ceramic, Marble	3/4

HANGING DOORS, per each.
Exterior, wood, standard sizes. All types 3 butts.

	Skilled	Unskilled
1¾"	1	1/2
metal, prefit	3/4	1/4
Interior metal, prefit 1-3/8"	1/2	1/4
wood 1-3/8"	3/4	1/2
wood sliding 1-3/8"	1/2	1/2
metal sliding 1-3/8"	1/2	1/4
metal bifolding, pair	3/4	1/4
wood bifolding, pair	3/4	1/4
wood french, pair	1	1/4
closet small, 1-1/8"–1-3/8" thick	1/2	
closet large, 1-1/8"–1-3/8" thick	3/4	
folding fabric or slatted types	3/4	

WINDOWS AND DOORS — Continued

	HOURS PER UNIT	
	Skilled	Unskilled
Exterior combination storm/screen wood complete	1	1/4
Complete prefit metal or plastic	3/4	1/4
(Adjust all door rates for special designs and types)		
Garage and heavy doors.		
1¾'' hinged 4' x 8'	2	1/2
sliding 4' x 8'	2½	1/2
sliding 8' x 8'	3	1/2
Setting overhead doors, wood 1-3/8'' (complete)		
8' x 6'–6'' to 8' x 8'	3	1
to 10' x 7' & 8'	3½	1
12' x 7' & 8'	4	1
12' x 10' & 12'	4½	1½
16' x 6'–6'' to 8'	5½	2
(Adjust for metal doors, add 25% for 1¾'' doors)		

WEATHERSTRIPPING, per opening.

	Skilled	Unskilled
Standard type metal for average size door.	3/4	

LOUVERS, VENTS, per each.

Metal, screened or regular.

	Skilled	Unskilled
Small medium sizes	1/4	
Large sizes	1/2	
Half circle types	1	

ACCESS DOORS, prefit metal.

	Skilled	Unskilled
To 24'' x 30''	1/2	
Over to 36'' x 48''	3/4	

METAL-CLAD DOORS, complete masonry or steel construction. (Underwriters)

		Skilled	Unskilled
Swinging	Single	4	2
	Double	6	2
Sliding	Single	6	2
	Double	8	2

Adjust all window and door rates for transoms if required. Add 25% for hardwood trim. Adjust for masonry construction unless indicated and for unusual conditions. Rates do not include hardware unless shown as complete.

MILLWORK AND TRIM

Normal openings. Frame construction. Soft wood. First class work. Rooms 10' x 10' or over. Add 25% for hardwood. Adjust if drilling is required.

WOOD INTERIOR TRIM, 100 lineal feet.

	Skilled	Unskilled
Base — 1 piece	2½	1
Shoe mold	1½	
Chair rail	2½	1
Picture mold	3	1
Cornices, crown, bed, cornice mold		
single member to 3½'' widths	6	2
single member over 3½'' widths	7	2
2 to 4 member	12	2
Ceiling beams, built in place 3'' x 6'' to 6'' x 12''	18	2
Closet shelving 3/4'' x 12'' stock includes cleats	3	1

METAL INTERIOR TRIM, 100 lineal feet.
Applied either before or after plastering. Applied with screws or clipped-on.

	Skilled	Unskilled
1 piece base	3½	1
2 piece base	4½	1
Cove mold	2	1
Chair rail	2½	1
Picture mold	3	1
Cornice to 8'' width with preformed mitred corners	6	2
Closet shelving. Adjustable lengths 12'' and 16'' widths	2	1
Adjust time for all trim if applied over masonry walls or over plaster over masonry.		

WOOD EXTERIOR TRIM, 100 lineal feet.

	Skilled	Unskilled
Corner board, verge boards, fascia, frieze to 3/4'' x 8''	3	2
Shingle mold, bed mold	1½	1
Soffits, 3/4'' x 6''–8'' to 24'' widths	7	1
Plywood, hardboard, asbestos pegboard ¼'' x 12'' x 24''	4	1
(Add for screens if used)		
Metal with screens or vents	3	1
Cornices, 2 member to 12'' widths	6	1

PORCH RAIL, 100 lineal feet.

	Skilled	Unskilled
Top, bottom, balusters	3	1½

STAIRS, Exterior Wood, per flight.

Plain open stairs. 4' width to 12' lengths,

	Skilled	Unskilled
single flight	4	2
double flights	6	2

PORCH COLUMNS to 10' heights with caps and bases where required.

	Skilled	Unskilled
4'' x 4''–6'' x 6'' solid	1/2	1/2
built up square to 8'' x 8''	1½	1
round hollow to 12''	1	1
round turned solid	3/4	1/2

MANTELS, per unit.

	Skilled	Unskilled
Setting average type factory built mantel units. To prepared walls	2½	1½

CABINETS, CUPBOARDS, 100 sq. ft. face area
Average type work. Includes base and mold as required.

	Skilled	Unskilled
Setting factory built base cabinets, cases, range, oven sections. Also broom and utility cabinets, vanities	4	2
Hanging top cabinets	3	2

COUNTER TOPS, 10 sq. ft. surface area

Placing factory built tops over cabinet bases without sinks

	Skilled	Unskilled
3/4'' Plywood, lumber	1/6	1/6
1'' to 2'' Maple	1/2	1/4
Stainless Steel	1/5	1/5
Covering sink, base, vanity tops, plain type using metal trim. Cutting, fitting, cementing		
Laminated plastic	3/4	

MILLWORK AND TRIM — Continued

	HOURS PER UNIT	
	Skilled	Unskilled
Linoleum	1/2	
Ceramic tile 4" x 4"	1½	
(Add 50% if sinks or lavatories included)		

MISC. FACTORY—BUILT CABINETS, per each unit.

	Skilled	Unskilled
Placing or setting to rough openings		
Package receivers, through-wall types. Average size	1	
Medicine and wall cabinets — Small	1/2	
Large	3/4	
Ironing board, broom, shoe cabinets, etc.	1/2	
Clothes chutes, complete, (except basement bin)	1½	
Bath and Kitchen Accessories. (Soap dish, towel racks, paper holders, hand bars, etc.	1/4	
Setting prefit lightweight shower doors — plastic glazing	1	
heavy weight plate glass	2	

GLAZING

Average type work, to second floor heights. Single strength, double strength.

GLAZING WOOD SASH AND DOORS.
Using putty, per 10 lights.

Approximate glass size	Skilled	Unskilled
8" x 10" to 12" x 14"	1	
16" x 20" to 20" x 28"	2	
30" x 36" to 36" x 40"	2½	
40" x 48" to 48" x 60"	4½	

Deduct 1/3 if glazed with wood stops.

GLAZING METAL SASH.
Using putty or plastic glazing compound, per 10 lights.

Approximate glass size. Square inches per light	Skilled	Unskilled
Up to 300 square inches	1¼	
300 to 600 square inches	2½	
600 to 900 square inches	3¼	
900 to 1200 square inches	4½	
1200 to 1800 square inches	5½	
1800 to 2400 square inches	6½	

Deduct 1/3 if glazed with metal or plastic stops or strips.
Adjust all rates for twin-pane lights, plate, wired, ribbed or special types of glass and other unusual conditions.

STORE OR COMMERCIAL BUILDING FRONTS, 100 square feet.

	Skilled	Unskilled
Setting plate glass over 4' x 5' to 8' x 10'	10	
Over 8' x 10' to 10' x 15'	15	

PAINTING

	Skilled
Based on 1st coat over new work unless indicated otherwise. Surface areas.

OUTSIDE WALLS, 100 square feet.
Oil, stain or rubberized types.

	Skilled	Unskilled
Wood sidings	1	
Wood shingle (stain)	1¼	
Burned smooth brick	1¼	
Burned rough brick	1¾	
Concrete masonry, concrete, stucco	1	
(Add 10% if masonry paints)		

INSIDE WALLS AND CEILINGS, 100 square feet.
Flat, casein, or rubberized types.

	Skilled	Unskilled
Sizings	1/4	
Plaster, white coat, dry wall	1/2	
Insulating plank, panels, tile, rough raw finish	1	
Plywood, lumber, plasterboard, composition board, smooth finishes	3/4	
	3/4	

(Deduct 10% if calcimine or similar types. Add 25% all work if ceilings only.)

CABINETS, CUPBOARDS.
Vanities, cabinet or closet doors, bookcases.

	Skilled	Unskilled
1 side	1½	

WALL PAPER, 100 square feet.
Ordinary type walls.

	Skilled	Unskilled
Butt joint work. Average type paper	2	
Special papers and coated fabrics, etc.	3	

FLOORS, 100 square feet.

	Skilled	Unskilled
Filling, wiping	2/3	
Shellac, varnish, stain	1/2	
Painting — Wood types	1/2	
Concrete	1/3	

ROOFS, 100 square feet.
Average conditions, smooth surfaces.

	Skilled	Unskilled
Asphalt, aluminum and other free flowing types	1/2	
Fibred semi-plastic types	1	
Wood shingles — stained	3/4	
painted	1	

TRIM, 1 coat per opening.
Inside and Outside

	Skilled	Unskilled
Windows and doors, average size	1/4	
Picture mold, chair rail, base, cornice, etc. Figure each item for each average size room as 1 opening	1/4	

DOORS, 1 coat per opening.

	Skilled	Unskilled
All types, inside, outside 2 sides average size	1/2	
Combination storm doors	3/4	
Inside, medium size closet & cupboard	1/3	
Inside, small size closet & cupboard	1/4	

WINDOWS — Based on wood double hung, casements, sliding, projected, awning, window-wall types with large lights. 1 side.

	Skilled	Unskilled
Small sizes	1/4	
Medium sizes	1/3	
Large sizes	1/2	

Metal. Residential types average sized lights. 1 side. All types.

	Skilled	Unskilled
Small size units	1/5	

PAINTING — Continued

HOURS PER UNIT
Skilled

	Skilled
Medium size units	1/4
Large size units	1/3
Adjust both wood and metal residential rates for small lights of glass where found.	

Metal. Commercial types, average sized lights. 1 side.

	Skilled
Small openings	1/4
Medium openings	1/3
Large openings	1/2
All window units figure each unit and combine time when two or more units are combined by mullions.	

	Skilled
SHUTTERS. All types, per 2 sides	1/4
STAIRWAYS. Complete open types, per stairway	4
WOOD MANTELS, per unit	1/2

CAULKING, 100 lineal feet.

	Skilled
Average type work, using gun, around windows, door trim, etc.	1¼

Adjust all painting time for enamels, special type paints, finishes and unusual conditions, old work, and spray painting. Deduct 10% for successive coats on large areas of inside walls, ceilings and outside walls.

HARDWARE

HOURS PER UNIT
Skilled

LOCK SETS, per each unit.

	Skilled
Outside doors, plain type sets	1/2
Outside doors, front, fancy type sets	3/4
Outside storm doors	1/2
Outside door closers	1/2
Inside doors softwood	1/5
Inside doors hardwood	1/4
Inside closet, cupboard, cabinet, etc.	1/10
Inside sliding doors, swinging, single	1/5
Inside sliding doors, swinging, double	1/4

DOOR ACCESSORIES, per 10 units.

	Skilled
Door bumpers, stops, surface bolts, night latches, closet hooks, handles, pulls, catches, and misc. small items. Simple installation	1/2

SASH HARDWARE, per each unit.

	Skilled
Double hung, locks and lifts per window	1/10
Casement sash, locks and operators per sash	1/4
CLOSET RODS, per rod	1/6

	Skilled
GARAGE or heavy swinging doors, latches and lock sets, holders and top or bottom bolts, per door	1/2

INDEX